ORGANIC REACTION MECHANISMS · 1965

ORGANIC REACTION MECHANISMS · 1965

An annual survey covering the literature
dated December 1964 through November 1965

B. CAPON University of Leicester

M. J. PERKINS King's College, University of London

C. W. REES University of Leicester

INTERSCIENCE PUBLISHERS a division of

John Wiley & Sons   London · New York · Sydney

First printed July 1966
Reprinted August 1967

Made and printed in Great Britain by
William Clowes and Sons Limited, London and Beccles

Preface

This book is a survey of the work on organic reaction mechanisms published in 1965.* For convenience, the literature dated from December 1964 to November 1965, inclusive, was actually covered. The principal aim has been to scan all the chemical literature and to summarize the progress of work on organic reaction mechanism generally and fairly uniformly, and not just on selected topics. Therefore, certain of the sections are somewhat fragmentary and all are concise. Of the 2000 or so papers which have been reported, those which seemed at the time to be the more significant are normally described and discussed, and the remainder are listed.

Our other major aim, second only to comprehensive coverage, has been early publication since we felt that the immediate value of such a survey as this, that of "current awareness", would diminish rapidly with time. In this we have been fortunate to have the expert cooperation of the London office of John Wiley and Sons.

If this book proves to be generally useful, we will continue these annual surveys, and then hope that the series will have some lasting value; some form of cumulative reporting or indexing may even be desirable.

It is not easy to deal rigidly and comprehensively with so ubiquitous and fundamental a subject as reaction mechanism. Any subdivision is a necessary encumbrance and our system, exemplified by the chapter headings, has been supplemented by cross-references and by the form of the subject index. We should welcome suggestions for improvements in future volumes.

February 1966

B.C.
M.J.P.
C.W.R.

* In Chapter 1, only, an account of earlier work is given in some detail, since an introduction to the current controversy on classical and non-classical carbonium ions seemed particularly timely.

Contents

Classical and Non-classical Carbonium Ions

Introduction

The questions which at present stimulate most interest in this field are whether or not certain carbonium ions have bridged or non-classical structures and whether this bridging is developed in the transition states of reactions in which these ions are thought to be intermediates.[1] One of the earliest proposals that a carbonium ion does not have a classical Kekulé structure was that of Nevell, de Salas, and Wilson[2] for the ion from camphene hydrochloride (1). In aprotic solvents this compound undergoes a stereospecific rearrangement to yield isobornyl chloride (2) and not the more stable bornyl chloride (3). It was therefore suggested that, rather than there being two

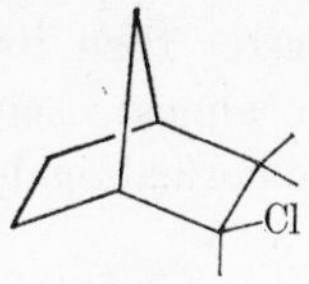

intermediate ions (4) and (5) with classical structures, there was a single ion, mesomeric between them. This ion can then be written as (6)[3] and it has

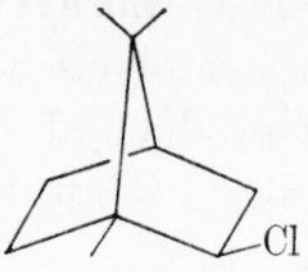

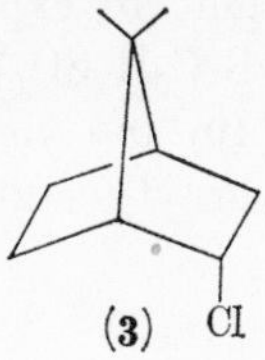

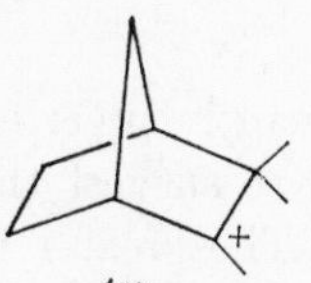

become customary to ascribe to the split bond, or bridge, the property of preventing attack from the *endo*-direction. This idea was developed by Hughes, Ingold, and their colleagues, who suggested that the 600-fold greater

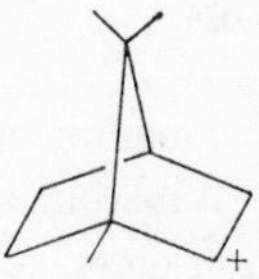

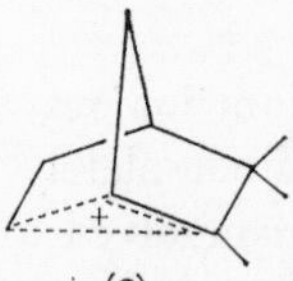

[1] For reviews on the subject of carbonium ions and their reactions see D. Bethell and V. Gold, *Quart. Rev.* (*London*), **12**, 173 (1958); N. C. Deno, *Progr. Phys. Org. Chem.*, **2**, 129 (1964); *Chem. Eng. News*, 5th Oct., **42**, No. 40, p. 88 (1964).

[2] T. P. Nevell, E. de Salas, and C. L. Wilson, *J. Chem. Soc.*, **1939**, 1188.

[3] Suggestion of C. K. Ingold quoted by H. B. Watson, *Ann. Rep. Progr. Chem.* (*Chem. Soc. London*), **36**, 197 (1939).

rate of solvolysis of camphene hydrochloride compared with *t*-butyl chloride and the 10^5-fold greater rate of solvolysis of isobornyl chloride compared with bornyl chloride and pinacolyl chloride could be explained if this bridging were developed in the transition states (7) and (8) of these reactions.[4] They

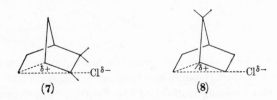

referred to these bridged ions as "synartetic" ions and to the rate enhancements thought to be caused by this bridging in a transition state as "synartetic acceleration". The synartetic ion (6) was discussed as a resonance hybrid between the two classical ions (4) and (5), despite there being substantial differences in the positions of some of the nuclei in the two structures.

The concept of bridged or non-classical ions was also taken up by Winstein and Trifan[5] to explain why *exo*-norbornyl *p*-bromobenzenesulphonate (9; $Bs = p\text{-}BrC_6H_4SO_2$) undergoes acetolysis 350 times faster than its *endo*-isomer (10) and yields, as the product of substitution, almost exclusively *exo*-norbornyl acetate.[6] This striking result, in which the normal tendency of

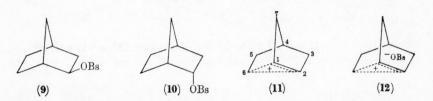

carbonium ion reactions to yield racemic or predominantly inverted products is completely reversed, led Winstein and Trifan to suggest that the reaction involved a bridged or non-classical ion (11). The greater rate of solvolysis of the *exo*-isomer was explained by supposing that this bridging was developed in the transition state: i.e. that the reaction involved partici-

---

[4] F. Brown, E. D. Hughes, C. K. Ingold, and J. F. Smith, *Nature*, **168**, 65 (1951); C. K. Ingold, "Structure and Mechanism in Organic Chemistry," Cornell University Press, Ithaca, N.Y., 1953, pp. 514—523.

[5] S. Winstein and D. S. Trifan, *J. Am. Chem. Soc.*, **71**, 2953 (1949); **74**, 1147, 1154 (1953).

[6] The most recent figure is that acetolysis of *exo*-norbornyl toluene-*p*-sulphonate yields $0.05 \pm 0.02\%$ of *endo*-acetate at 100° and ca. 0.01% at 30° [H. L. Goering and C. B. Schewene, *J. Am. Chem. Soc.*, **87**, 3516 (1965)]. It has also been reported that solvolysis of the *exo-p*-bromobenzenesulphonate in 75% aqueous acetone yields less than 0.02% of *endo*-norbornanol [S. Winstein, E. Clippinger, R. Howe, and E. Vogelfanger, *J. Am. Chem. Soc.*, **87**, 376 (1965)].

pation by the $C_{(1)}$–$C_{(6)}$ bonding electrons. Winstein refers to the rate enhancement brought about in this way as "anchimeric assistance".[7]

Supporting evidence for this proposal was provided by the observation that the *exo*-norbornyl acetate obtained from optically active *exo*-norbornyl *p*-bromobenzenesulphonate was, within experimental error, completely racemic.[5,8] This is, of course, to be expected if ion (11) is an intermediate since it has a plane of symmetry. It was also observed that the rate of racemization was about 3.5 times greater than the rate of formation of titratable acid. This means that the *exo*-norbornyl *p*-bromobenzenesulphonate is itself undergoing racemization and it was suggested that this involves dissociation to, and recombination of, an ion pair (12), the whole process being referred to as "ion-pair return". Thus only 29% of the acetate in the product comes directly from optically active starting material, but this also was shown to be racemic within experimental error.

That the processes which occur are even more complex than this was shown by Roberts, Lee, and Saunders[9] who studied the acetolysis of *exo*-norbornyl *p*-bromobenzenesulphonate labelled specifically in the 2- and the

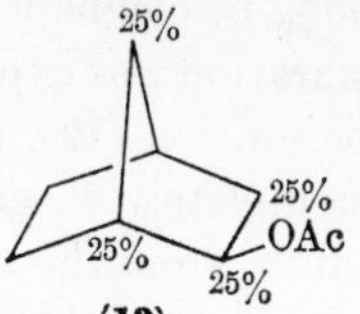

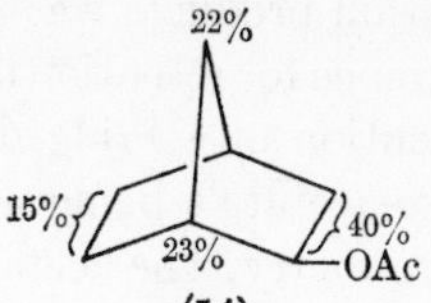

3-position with carbon-14. Instead of obtaining *exo*-acetate labelled only in the 1-, 2-, 3-, and 7-positions (as 13) which would be expected if there was just one intermediate bridged ion they found that the label was in positions 1, 2, 3, 5, 6, and 7 as (14). This additional scrambling of the label must result from a $C_{(6)} \rightarrow C_{(2)}$ hydride ion shift to give a second ion (15). The postulated

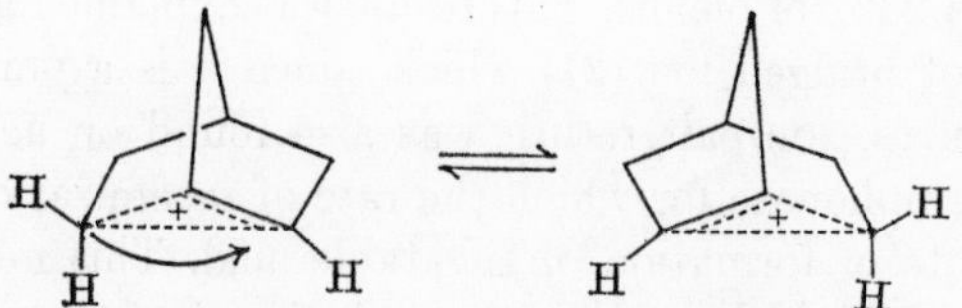

[7] S. Winstein, C. R. Lindegren, H. Marshall, and L. L. Ingraham, *J. Am. Chem. Soc.*, **75**, 147 (1953).

[8] The latest value is that the survival of optical activity on solvolysis of optically active *exo-p*-bromobenzenesulphonate (9) in acetic acid, 80% dioxan, or 75% acetone at 25°, and 90% dioxan at 50°, is less than 0.05% [S. Winstein, E. Clippinger, R. Howe, and E. Vogelfanger, *J. Am. Chem. Soc.*, **87**, 376 (1965)].

[9] J. D. Roberts and C. C. Lee, *J. Am. Chem. Soc.*, **73**, 5009 (1951); J. D. Roberts, C. C. Lee, and W. H. Saunders, *ibid.*, **76**, 4501 (1954).

non-classical ion therefore appears to have the, at first sight, unusual property of being attacked by external nucleophiles only in the *exo*-direction, but by the migrating hydride ion in the *endo*-direction. The migration was, however, rationalized as involving an intermediate face-protonated, (16),[9] or edge-protonated, (17),[5] cyclopropane, the former being also referred to as a nortricyclonium ion.

(16)   (17)

At about the same time another striking stereochemical result was obtained by Cram,[10] who showed that the product of substitution from the acetolysis of optically active L-*threo-αβ*-dimethylphenethyl toluene-*p*-sulphonate (18) contained 95% of racemic *threo*-acetate, and 4% of *erythro*-acetate. The total yield of substitution products was 53%; 35% of olefins were also obtained. The strong preference for retention of configuration was explained as resulting from the intervention of a bridged phenonium ion (19), and this accounts also for the *threo*-acetate's being almost completely racemic since this ion has a plane of symmetry. The acetates obtained from the D-*erythro*-ester (20) consisted of 94% D-*erythro*- and 5% D-*threo*-acetate, obtained in total yield of 68% along with 23% of olefins. This result was explained as being due to the intervention of bridged ion (21) which, since it is asymmetric, should yield active products. Ion-pair return was also found on acetolysis of the L-*threo*-toluene-*p*-sulphonate for which the rate of racemization is five times as great as the rate of formation of titratable acid. This means that only 20% of the product could have been formed directly from optically active toluene-*p*-sulphonate. It is not possible to study ion-pair return in the acetolysis of the *erythro*-toluene-*p*-sulphonate by polarimetric measurements because this does not result in racemization; but ion-pair return has recently been detected and its rate measured by a carbon-14 labelling technique.[11]

Cram has obtained similar results for the formolyses of these compounds, and in these reactions the stereospecificity is even greater.[10]

The anchimeric assistance that can be associated with participation by the phenyl groups in these systems is small. Thus the rates of the acetolyses

[10] D. J. Cram, *J. Am. Chem. Soc.*, **71**, 3863 (1949); **74**, 2129, 2137 (1952); for a recent summary of this work see D. J. Cram, *J. Am. Chem. Soc.*, **86**, 3767 (1964).
[11] W. B. Smith and M. Showalter, *J. Am. Chem. Soc.*, **86**, 4136 (1964).

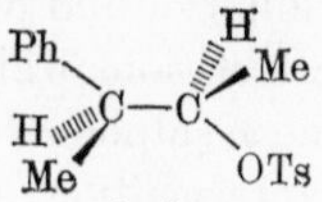

$$(18) \quad (19)$$

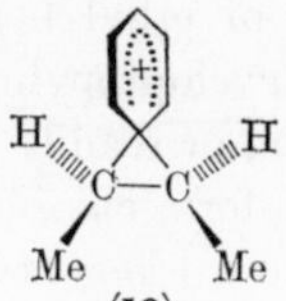

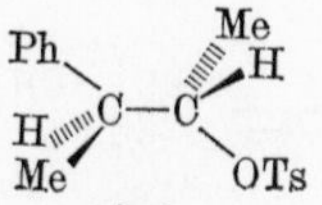

$$(20) \quad (21)$$

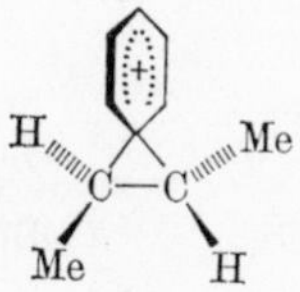

of both the $\alpha\beta$-dimethylphenethyl toluene-p-sulphonates are less than that of s-butyl toluene-p-sulphonate;[12] but when allowance is made for the electron-withdrawing inductive effect of the phenyl group, and when the rate of ionization (as determined polarimetrically) of the *threo*-toluene-p-sulphonate rather than its rate of acetolysis is used, the rate enhancement due to phenyl participation is computed to be 24-fold.[10] The validity of this procedure is uncertain since there is the possibility that ionization to, and recombination from, an ion pair occurs without racemization, and since also the extent of ion-pair return occurring in the acetolysis of s-butyl toluene-p-sulphonate is unknown.

Another class of compound in whose solvolyses the intervention of non-classical ions has been postulated are derivatives of cyclopropylmethyl alcohol. The evidence for this is that the solvolyses of these compounds frequently yield cyclobutyl and 3-butenyl derivatives. Thus the acetolysis of cyclopropylmethyl chloride yields cyclopropylmethyl and cyclobutyl acetate in the ratio 2.6:1, a small amount of 3-butenyl acetate, and a 1.7:1 mixture of cyclobutyl and 3-butenyl chloride.[13] Acetolysis of cyclobutyl toluene-p-sulphonate[14] and formolysis of 3-butenyl toluene-p-sulphonate[15] yield similar mixtures. An experiment with specifically deuterated cyclopropylmethyl chloride indicated considerable skeletal rearrangement in the cyclopropylmethyl chloride isolated from a partly solvolysed reaction mixture.[16] The solvolyses of the cyclopropylmethyl compounds also proceed at enhanced rates,[13,17] ethanolysis of the benzenesulphonate, for instance, being 500 times

[12] S. Winstein, B. K. Morse, E. Grunwald, K. C. Schreiber, and J. Corse, *J. Am. Chem. Soc.*, **74**, 1113 (1952).

[13] J. D. Roberts and R. H. Mazur, *J. Am. Chem. Soc.*, **73**, 2509 (1951).

[14] J. D. Roberts and V. C. Chambers, *J. Am. Chem. Soc.*, **73**, 5034 (1951).

[15] K. L. Servis and J. D. Roberts, *J. Am. Chem. Soc.*, **86**, 3773 (1964).

[16] M. C. Caserio, W. H. Graham, and J. D. Roberts, *Tetrahedron*, **11**, 171 (1960).

[17] C. G. Bergstrom and S. Siegel, *J. Am. Chem. Soc.*, **74**, 145 (1952).

faster than that of ethyl benzenesulphonate. These results suggest that ionization of the cyclopropylmethyl, cyclobutyl, and 3-butenyl compounds yield the same ion, or readily interconvertible ions which may either react with solvent or re-form chloride or toluene-*p*-sulphonate by ion-pair return. Roberts and his co-workers favoured the intervention of an equilibrating set of bicyclobutonium ions (**22**) to (**24**) rather than a single tricyclobutonium

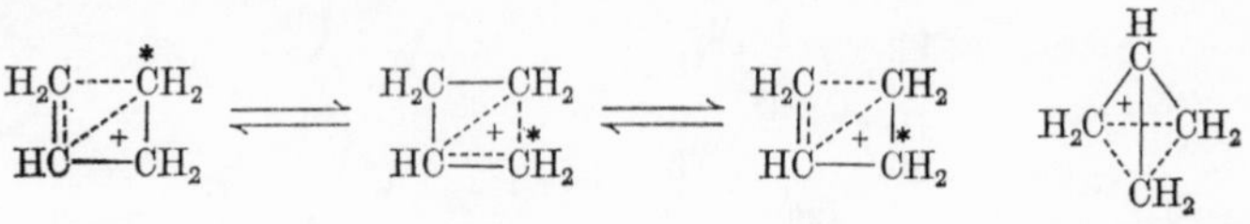

(**22**)   (**23**)   (**24**)   (**25**)

ion (**25**) because in certain non-solvolytic reactions there was incomplete equilibration of the methylene groups.

Non-classical ions have also been postulated as intervening in reactions in which there is participation by double bonds.[18] Thus *exo*-norbornenyl *p*-bromobenzenesulphonate (**26**) undergoes acetolysis 8000 times more rapidly than its *endo*-isomer[19] and yields about 80% of 3-acetoxynortricyclene (**29**) and some *exo*-norbornenyl acetate (**30**).[20,21] The reaction may therefore

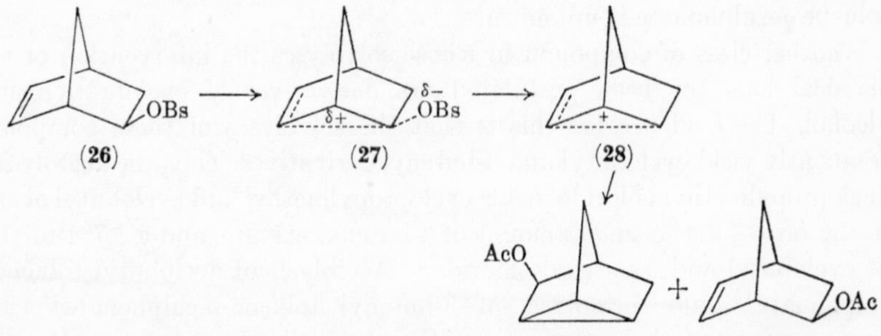

(**26**)   (**27**)   (**28**)

(**29**)   (**30**)

be formulated as involving a "non-classical transition" state (**27**), leading to a non-classical ion (**28**). Acetolysis of *exo*-norbornenyl *p*-bromobenzene-sulphonate labelled at $C_{(2)}$ and $C_{(3)}$ with carbon-13 yields *exo*-acetate which

[18] For reviews see B. Capon, *Quart Rev. (London)*, **18**, 97 (1964); B. Capon and C. W. Rees, *Ann. Rept. Progr. Chem. (Chem. Soc. London)*, **61**, 231 (1964).

[19] Unpublished work of H. J. Schmid and K. C. Schreiber, reported by S. Winstein and M. Shatavsky, *J. Am. Chem. Soc.*, **78**, 595 (1956).

[20] S. Winstein, H. M. Walborsky, and K. Schreiber, *J. Am. Chem. Soc.*, **72**, 5795 (1950).

[21] J. D. Roberts, W. Bennett, and R. Armstrong, *J. Am. Chem. Soc.*, **72**, 3329 (1950).

has lost about one-third of its label at these positions.[22] It thus became necessary to propose that ion (28) underwent rearrangement to its enantiomorph (31) or to a symmetrical ion (32).[22]

(31)   (32)

A much greater rate enhancement is observed in the acetolysis of *anti*-7-norbornenyl toluene-*p*-sulphonate (yielding *anti*-7-norbornenyl acetate) which proceeds 10^{11} times faster than that of the analogous saturated compound, 7-norbornyl toluene-*p*-sulphonate.[23] The π-electron cloud of the double bond of this compound is particularly well placed to interact with the

Scheme 1

developing carbonium ion at position 7 and the solvolysis has therefore been formulated as shown in Scheme 1. 7-Norbornadienyl derivatives undergo solvolysis even more readily; thus the chloride (33) reacts in aqueous acetone at a rate 750 times greater than that of *anti*-7-norbornenyl chloride and it was suggested that the reaction involved an ion of structure (34) or (35) as an intermediate.[24]

(33)   (34)   (35)

[22] J. D. Roberts, C. C. Lee, and W. H. Saunders, *J. Am. Chem. Soc.*, **77**, 3034 (1955).

[23] S. Winstein, M. Shatavsky, C. Norton, and R. B. Woodward, *J. Am. Chem. Soc.*, **77**, 4183 (1955); S. Winstein and M. Shatavsky, *ibid.*, **78**, 592 (1956).

[24] S. Winstein and C. Ordronneau, *J. Am. Chem. Soc.*, **82**, 2084 (1960).

This then summarizes the salient evidence available at the end of the 1950's for the existence of some of the more important non-classical ions and for their being formed and reacting by way of non-classical transition states. The concept had received almost universal acceptance among organic chemists, so much so that in the last years of that decade and in the early 1960's a large number of non-classical ions were postulated as reaction intermediates, frequently with little supporting evidence. Since about 1960, however, the view has been taken by H. C. Brown that, in many and possibly all of these reactions, it is unnecessary to postulate these non-classical or bridged ions and that it is possible to explain the stereochemical and sometimes the kinetic results without involving non-classical ions or transition states.[25] He has paid particular attention to *exo*-norbornyl systems and made

a detailed study of the effects of substituents in the 1- and 2-positions on the rates and products.[26-28] On ionization a 1- or 2-substituted *exo*-norbornyl derivative would give the same non-classical ion, as illustrated. Hence, if we assume that there is little difference in the free energies of the initial state, which is reasonable, then in the extreme case in which delocalization of the 1,6-bond is complete in the transition state the effect of a 1- and a 2-substituent on the rate should be almost the same, and the greater the difference between the effects of substituents in these two positions the less this delocalization must be. The effect on the rate of introducing phenyl and methyl substituents was therefore investigated and it was found that, whereas 2-phenyl-*exo*-norbornyl chloride underwent ethanolysis at a rate estimated to be 3.9×10^7 times greater than that for *exo*-norbornyl chloride, with the 1-phenyl isomer the corresponding rate enhancement was only 3.9.[26] Provided then that there is no steric inhibition of resonance between the 1-phenyl substituent and the developing cationic centre this result means either that there is little delocalization of the 1,6-bond in the transition state which is akin to saying that it is "classical", or that the stabilizing effect of a 2-phenyl substituent on the classical transition state (**36**) is so much greater than on the non-classical one (**37**) that reaction proceeds exclusively through the former. Nevertheless, it was also shown that the rate

[25] Cf. H. C. Brown, "The Transition State", *Chem. Soc. Special Publ.*, No. 16, p. 140 (1962).
[26] H. C. Brown, F. J. Chloupek, and Min-Hon Rei, *J. Am. Chem. Soc.*, **86**, 1246, 1247, 1248 (1964).

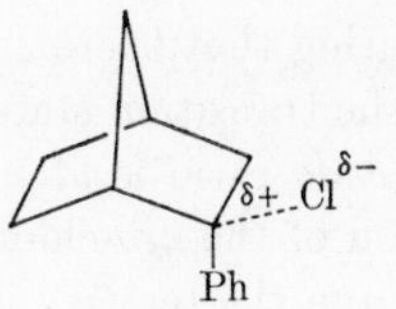

(36) 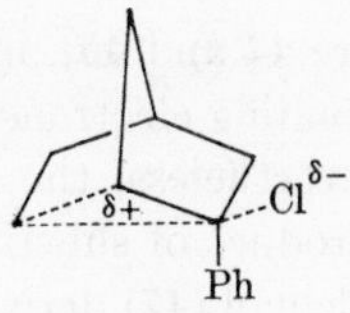 (37)

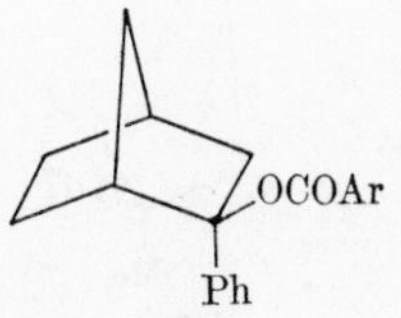

(38) 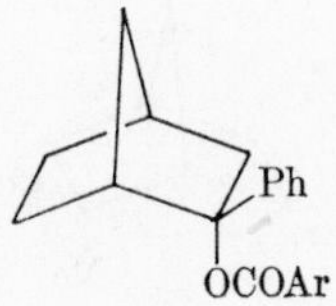 (39)

$Ar = p\text{-}NO_2C_6H_4$

of solvolysis of 2-phenyl-*exo*-norbornyl *p*-nitrobenzoate (38) in 60% ethanol is 140 times greater than that of the *endo*-isomer (39).

A similar situation is found when considering 2-methylnorbornyl derivatives.[27] A 2-methyl substituent increases the rate of ethanolysis of *exo*-norbornyl chloride by a factor of about 6×10^4 (compare 40 and 41), but

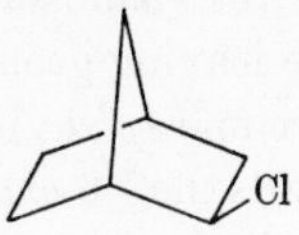

(40) (41)

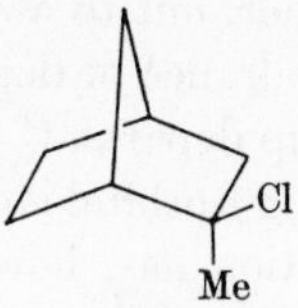

$10^6 k \, (\text{sec}^{-1})$
at 25°

$4\cdot72 \times 10^{-4}$ 30·0

an additional 1-methyl substituent has only a 4.5-fold effect on the rate of hydrolysis of 2-methyl-*exo*-norbornyl *p*-nitrobenzoate in aqueous dioxan (compare 42 and 43).[27] This is in fact slightly less than that observed in the

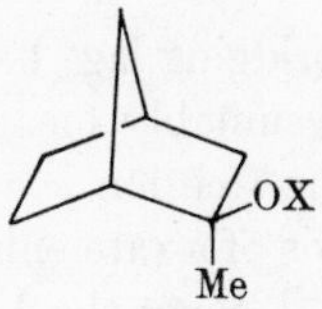

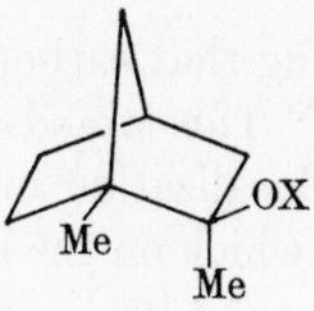

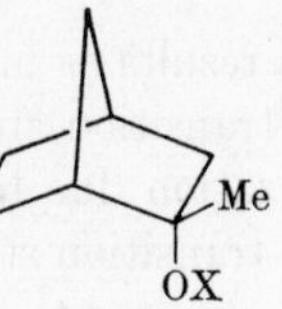

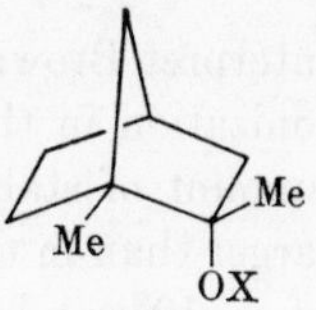

(42) (43) (44) (45)

$10^6 k \, (\text{sec}^{-1})$
at 50°

2·2 9·4 0·012 0·057

$X = p\text{-}NO_2C_6H_4CO$

[27] H. C. Brown and Min-Hon Rei, *J. Am. Chem. Soc.*, **86**, 5004 (1964).

1*

endo-series (compare **44** and **45**), indicating that there can be little delocalization of the 1,6-bonding electrons in the transition state for the reaction of the *exo*-isomer. Nevertheless, the *exo*:*endo* ratio is about 180 (compare **42** and **44**), and the product of substitution of the *exo*-chloride (**46**) in aqueous dioxan is the *exo*-alcohol (**47**) derived from the tertiary ion exclusively.[28]

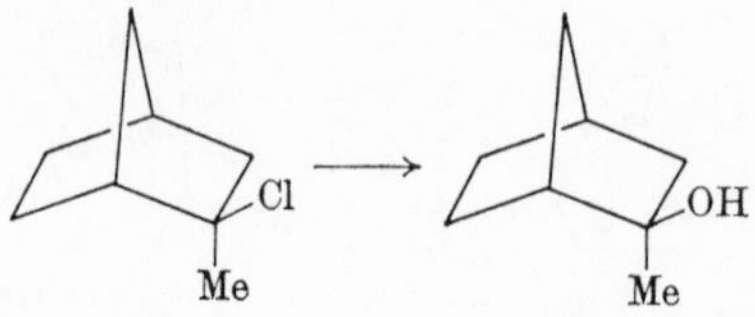

Brown has interpreted this large difference between the effects of substituents in positions 1 and 2 as meaning that the tertiary *exo*-norbornyl derivatives react via classical transition states and an equilibrating pair of classical ions. These reactions, however, show high *exo*:*endo* rate- and product-ratios and hence, if Brown's view is correct, these criteria lose the validity as tests for non-classical ions and transition states.

Brown explains the high *exo*:*endo* rate-ratio as due, not to a large rate for the *exo*-isomer, but to a small rate for the *endo*-isomer.[26] This is attributed to steric hindrance of departure of the leaving group from the *endo*-isomer. As this group departs, $C_{(2)}$ starts to become sp^2-hybridized and planar, and its developing p orbital is directed towards the 5,6-*endo*-hydrogens (see **48**). This position has, however, been criticized by Winstein who prefers to

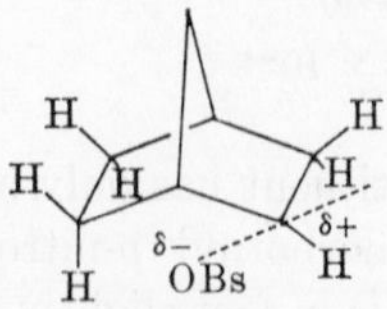

interpret Brown's results as meaning that carbon bridging lags behind C–X ionization in the transition state.[29] This would presumably mean that the amount of stabilization due to delocalization in the free ion is very much larger than in the transition state, which on the basis of a rate enhancement of ca. 10^3 can be computed to be about 4 kcal. mole^{-1}. Since the 2-norbornyl cation can be observed directly by NMR spectroscopy (see p. 23) it is to be hoped that some measure of its stability will be forthcoming.

[28] H. C. Brown and H. M. Bell, *J. Am. Chem. Soc.*, **86**, 5006 (1964).
[29] S. Winstein, *J. Am. Chem. Soc.*, **87**, 381 (1965).

It is interesting that 1-methylnorbornyl toluene-*p*-sulphonate undergoes acetolysis about 50 times faster than *exo*-norbornyl toluene-*p*-sulphonate.[30] This must mean that there is participation by the 1,6-bonding electrons in this reaction (i.e. a non-classical transition state), but the driving force for this could be the rearrangement to the more stable tertiary 1-methyl-norbornyl cation which could be classical. It is perhaps significant that, as outlined above, there was no similar rate enhancement in the solvolysis of the symmetrical 1,2-dimethyl derivatives, for here this driving force would be absent.

Brown's view that the solvolysis of *endo*-norbornyl derivatives is unusually slow is in disagreement with Schleyer's correlation.[31] This is a relationship between the rates of solvolyses and the carbonyl stretching frequencies of the corresponding ketones, and it includes correction terms for inductive effects and for torsional and non-bonded interaction strain. It predicts correctly the rates of solvolysis of a large number of compounds, but compounds whose reactions are considered to be anchimerically assisted show higher reactivities than are calculated. The observed reactivity of *endo*-norbornyl toluene-*p*-sulphonate is close to the calculated value but that of the *exo*-isomer is larger, suggesting that its reaction is anchimerically assisted. There is, however, always the possibility that some important steric factor unique to the norbornyl system (e.g. steric hindrance to ionization) has been overlooked.

It has been urged that these kinetic studies tell us nothing about the structures of the carbonium ions but only about the structure of the transition states,[32] and that it is the stereochemistry of the reaction products which provide information about the former. On this ground, therefore, Berson considers that the best evidence for the incursion of a non-classical ion in the norbornane system is that replacement occurs exclusively from the

---

[30] P. von R. Schleyer and D. C. Kleinfelter, 138th Meeting A.C.S., New York, September, 1960, Abstracts, p. 43P; see also J. A. Berson in "Molecular Rearrangements", P. de Mayo, ed., Interscience, New York, 1963, Part 1, p. 182.

[31] P. von R. Schleyer, *J. Am. Chem. Soc.*, **86**, 1854, 1856 (1964).

[32] An interesting dichotomy of viewpoint on the relevance of rate differences to the structure of the ion, as distinct from the transition state for its formation, has grown up. The original workers were frequently unconcerned with this distinction and assumed without comment that the rate enhancements could be discussed in terms of the structures of the ions themselves (cf. ref. 5). More recently the views have been taken (see ref. 81) that "Rate comparisons reflect only differences in activation free energies and therefore provide no information about structure after the rate-determining transition state is reached", and that rate comparisons are the "least cogent" evidence for non-classical ions (see ref. 38*b*). Alternatively it has been written: "Of the three unusual properties associated with non-classical ions, only one, enhanced reaction rate, can serve to distinguish between these structural alternatives" (see ref. 31), and "Some of the recent discussions of anchimeric assistance to ionization have sounded as if there were no connection between the structure of a carbonium ion and that of the transition state leading to it" (see ref. 100).

exo-direction even when $C_{(7)}$ carries a *gem*-dimethyl group.[33] This contrasts with the behaviour of the corresponding ketones towards hydride reduction, in which a tendency towards *exo*-attack changes on the introduction of a 7,7-dimethyl group to a tendency towards *endo*-attack. If it is valid to draw an analogy between this reaction and that of a classical carbonium ion with a nucleophile one would expect a 7,7-dimethylnorbornyl cation to react with substantial *endo*-attack if it were classical, but this has never been observed. This is a telling piece of evidence for the existence of non-classical carbonium ions, but at the moment it is rather narrowly based and it is to be hoped that more experimental evidence will be forthcoming to indicate the relative ease of *exo*- and *endo*-attack in 7,7-dimethylnorbornyl systems (see also p. 256).

Brown has also questioned Hughes and Ingold's assignment of the high rate of solvolysis of camphene hydrochloride (**49a**) compared with that of *t*-butyl chloride to synartetic acceleration (see p. 2).[34] In fact, the rate of solvolysis of camphene hydrochloride is not appreciably larger than that of the almost analogously substituted cyclopentyl chloride (**49b**). The high

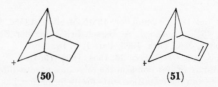

(**49a**)   (**49b**)

reactivity is therefore most probably caused by steric acceleration and not by participation of σ-bonding electrons.

Although accepting that the solvolyses of *anti*-7-norbornenyl and 7-norbornadienyl derivatives involve participation by the double bond (i.e.

(**50**)   (**51**)

involve non-classical transition states), Brown has questioned whether the ions could not better be represented as classical structures (**50**) and (**51**).[35] Some support for this formulation of the 7-norbornadienyl cation has come

[33] J. A. Berson in "Molecular Rearrangements", P. de Mayo, ed., Interscience, New York, 1963, Part 1, p. 130.

[34] H. C. Brown and F. J. Chloupek, *J. Am. Chem. Soc.*, **85**, 2322 (1963).

[35] H. C. Brown and H. M. Bell, *J. Am. Chem. Soc.*, **85**, 2324 (1963); see also S. Winstein, A. H. Lewin, and K. C. Pande, *ibid.*, **85**, 2324 (1963).

from the NMR spectrum of its hexafluoroantimonate.[36] The signal from the proton at position 7 occurs at -3.5 ppm and this is at rather a high field for a proton attached to carbon carrying a positive charge (usually -9.5 to -13.5 ppm) even when allowance is made for some drawing off of this charge by the double bond.[37] It is, however, at approximately the correct position for the β-proton of a cyclopropylmethyl cation.

The status of non-classical carbonium ions is thus a matter of considerable controversy, but there have been significant developments in this during the last year and these and other aspects of carbonium ion chemistry will now be reviewed.[38]

Bicyclic Systems[39]

This year interest has continued in the structure of the norbornyl cation and the transition states of its reactions. In a highly illuminating investigation Goering and Schewene[40] have measured the rate of the perchloric acid-catalysed loss of optical activity of *exo*-norbornyl acetate (52) and of its isomerization to the *endo*-isomer (53). These reactions are thought to involve reversible protonation of the substrate, followed by heterolysis of the conjugate acid of the acetate:

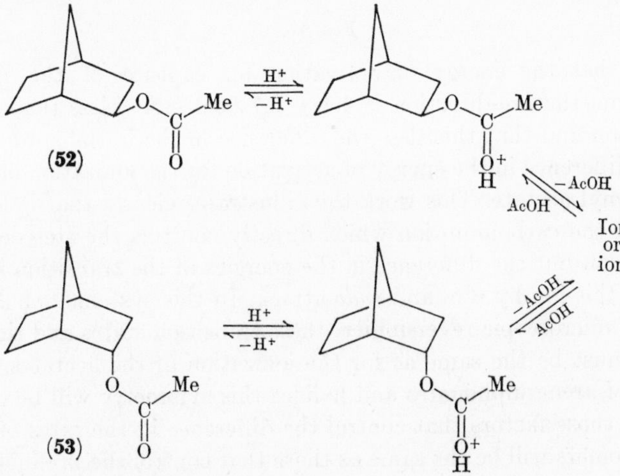

[36] P. R. Story, L. C. Snyder, D. C. Douglass, E. W. Anderson, and R. L. Kornegay, *J. Am. Chem. Soc.*, 85, 3630 (1963).

[37] N. C. Deno, *Progr. Phys. Org. Chem.*, 2, 248 (1964).

[38] For other recent reviews see: (*a*) W. Hückel, *J. Prakt. Chem.*, 28, 27 (1965); (*b*) M. J. S. Dewar and A. P. Marchand, *Ann. Rev. Phys. Chem.*, 16, 321 (1965).

[39] For a recent review see J. A. Berson in "Molecular Rearrangements", P. de Mayo, ed., Interscience, New York, N.Y., 1963, Part 1, p. 111.

[40] H. L. Goering and C. B. Schewene, *J. Am. Chem. Soc.*, 87, 3516 (1965).

The former rate is then the rate of ionization of the *exo*-acetate (**52**) and the latter the rate of capture of the ion (or ions) by *endo*-attack. It was thus estimated that *exo*-attack predominates to the extent of 99.99% and 99.95% at 25° and 100°, respectively. From these results and the equilibrium constant for *exo–endo* conversion a potential diagram (Figure 1) was constructed.

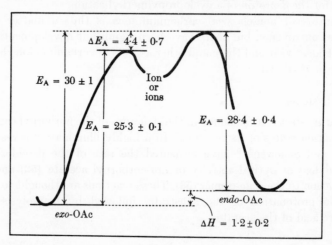

FIG. 1

It is seen that the energy of activation for capture of the norbornyl cation(s) from the *exo*-direction is 4.4 ± 0.7 kcal mol^{-1} less than from the *endo*-direction and that this (less the difference in the initial-state energies) is also the difference in the energy of activation for the ionization of *exo*- and *endo*-norbornyl acetate. This work thus illustrates clearly that it is not the structure of the carbonium ion which directly controls the stereochemistry of substitution but the difference in the energies of the transition states for reaction of this ion by *exo*- and *endo*-attack. In this system though, by the principle of microscopic reversibility, these transition states and this energy difference must be the same as for the ionization of the acetates.[41] In the solvolyses of arenesulphonates and halides this symmetry will be distorted, but clearly those factors that control the difference in the rates of the *exo*- and *endo*-isomers will be the same as those that control the product ratio. If therefore one subscribes to Berson's view[42] that in such reactions "kinetics alone give no information on the structure of the cationic intermediate (or intermediates)", one is forced to the conclusion that neither does the *exo*:*endo* product ratio.

In principle, however, if classical ions are reaction intermediates it should

[41] H. C. Brown and G. L. Tritle, *J. Am. Chem. Soc.*, in the press.
[42] See ref. 39, p. 118.

prove possible to trap them before the system becomes symmetrical; one reaction in which a substituted classical norbornyl cation derived from a *exo*-toluene-*p*-sulphonate has been trapped is reported by Takeuchi, Oshika, and Koga.[43]

These workers investigated the methanolysis of the four 5,6-trimethylene-2-norbornyl toluene-*p*-sulphonates (**54—57**). Compound (**56**), with the

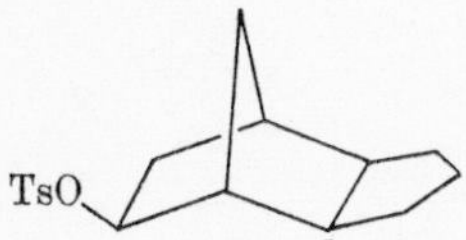

TsO (**54**)   TsO (**55**)   TsO (**56**)   OTs (**57**)

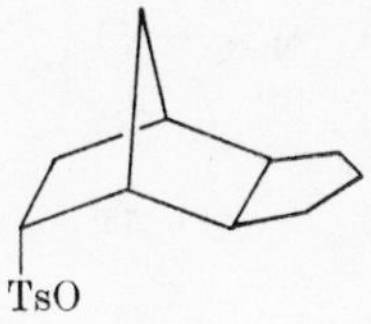

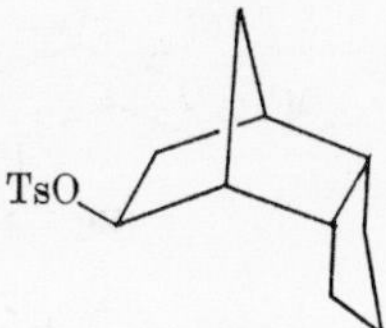

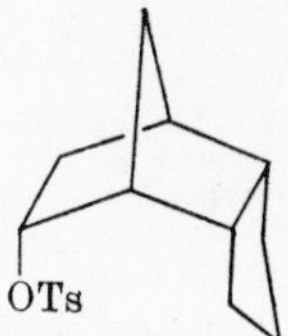

toluene-*p*-sulphonoxy group *exo* and the trimethylene bridge *endo*, is converted into compound (**54**) by ion pair return at a rate 2.2 times that of its solvolysis. Both compounds yield only the *exo*-methyl ethers (**58**) and (**59**) and their ratios at 100% reaction are, within experimental error, the same from the two toluene-*p*-sulphonates. Most of the product at 100% reaction has however always come from (**54**), since (**56**) is converted into (**54**); and to determine the direct solvolysis product of (**56**) it was necessary to examine the product early in the solvolysis. When this was done it became clear that compound (**56**) was yielding more of the corresponding methyl ether (**59**) than of its isomer (see Table 1). There cannot therefore be a single product-forming intermediate (i.e. a non-classical ion), and the reactions probably

Table 1. Rates and products for the methanolysis of the 5,6-trimethylene-2-norbornyl toluene-*p*-sulphonates.

Compound	54	55	56	57
Relative rate of methanolysis at 74.8°	1.00	0.0965	3.84	0.0246
% of (59) in product	3.2	1.6	7.3	23.4

involve two classical ions (**60**) and (**61**). Nevertheless, compounds (**54**) and (**56**) react faster than their epimers (**55**) and (**57**) which contain an *endo*-tosyloxy group, the rate ratios being 10.4 and 156, respectively (Table 1). Since the reactions of (**54**) and (**56**) involve classical ions, the transition states are presumably also classical; hence this rate difference cannot be ascribed to anchimeric assistance and, as it is much greater when the trimethylene bridge is in the *endo*-position, Brown's suggestion of steric

[43] K. Takeuchi, T. Oshika, and Y. Koga, *Bull. Chem. Soc. Japan*, **38**, 1318 (1965).

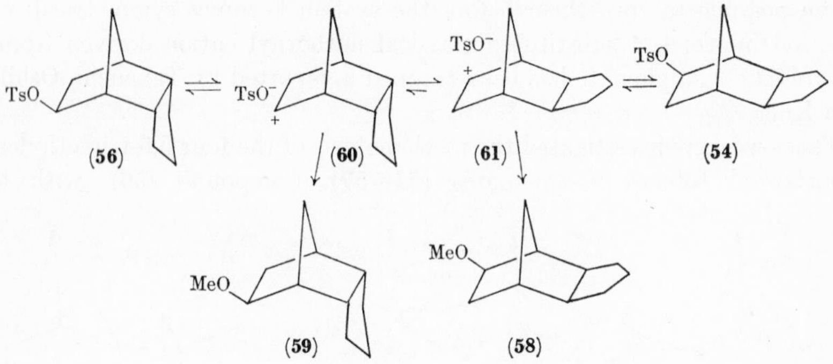

hindrance to ionization (see p. 10) of the *endo*-isomers (55) and (57) seems to offer a reasonable explanation. It should, however, be noted that the *exo*:*endo* rate ratio is rather small, so it is possible that the property of reacting via classical ions is unique to this system. It is to be hoped that other systems will be studied in a similar way.

The solvolyses of 2-norbornyl *p*-bromobenzene-sulphonates with a one-carbon bridge between positions 5 and 6 (i.e. the tricyclo[3.2.1.0^{2,4}]octane system) have also been investigated.[44,45] Compound (63) is converted into

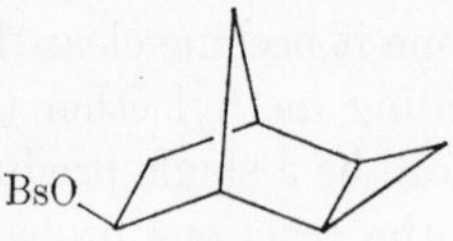

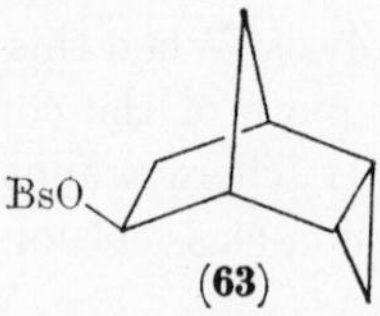

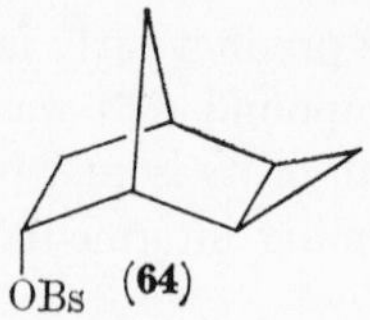

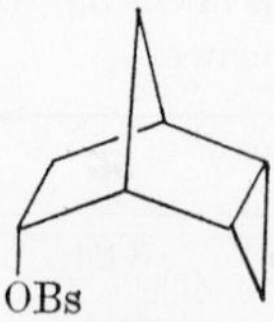

compound (62) by ion-pair return at a rate ($k = 2.57 \times 10^{-5} \sec^{-1}$ at 25°) about four times faster than its rate of acetolysis, and all three compounds (62), (63), and (64) are partially converted into 3-nortricyclylmethyl *p*-bromobenzenesulphonate (66) which undergoes acetolysis much more slowly than any of them. The products isolated at complete reaction from (62), (63), and (64) (Table 2) are, within experimental error, identical; clearly not too much emphasis should be placed on this since they all probably result, in the

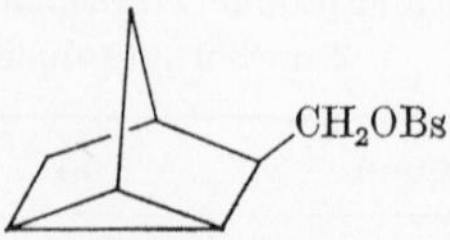

[44] K. B. Wiberg and G. R. Wenzinger, *J. Org. Chem.*, **30**, 2278 (1965).
[45] A. K. Colter and R. C. Musso, *J. Org. Chem.*, **30**, 2462 (1965).

Table 2. The rates and products of acetolysis of the 6-tricyclo[3.2.1.0²,⁴]octanyl *p*-bromobenzenesulphonates.

p-Bromobenzenesulphonate		$10^5 k_{25}$ (sec⁻¹)	$10^5 k_{85}$ (sec⁻¹)	Products (%)	
				67	Minor acetates mainly **68**
exo,exo-	**(62)**	1.36		67.5 ± 0.8	32.5
endo,exo-	**(63)**	0.68		68.5 ± 0.1	31.5
exo,endo-	**(64)**		3.35	67.5 ± 0.1	32.5
endo,endo-	**(65)**		3.36		
3-Nortricyclylmethyl	**(66)**		0.653	58.2 ± 0.5	41.8

main, from **(62)** which is formed by ion-pair return from **(68)** and **(64)**. It would be of considerable interest to know if the product ratio from **(63)** as determined early in the reaction, before there had been appreciable isomerization, is the same as that of the product isolated at complete reaction. The products from the *endo,endo*-isomer **(65)** were not reported but are being investigated.

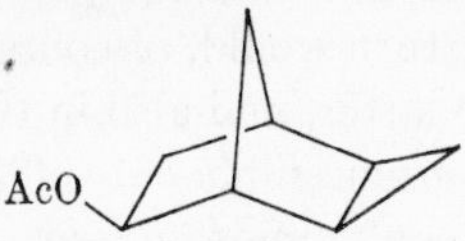

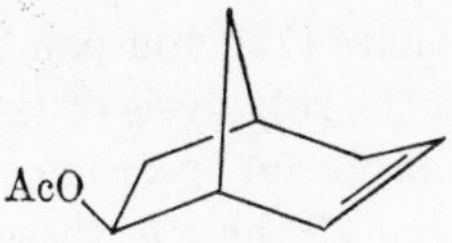

Schleyer, Donaldson, and Watts have shown that charge delocalization to position 6 in the transition state for the acetolysis of *exo*-norbornyl toluene-*p*-sulphonate [i.e. as indicated by resonance between structures **(69)** and **(70)** for the ion] must be of minor importance since the introduction of a *gem*-

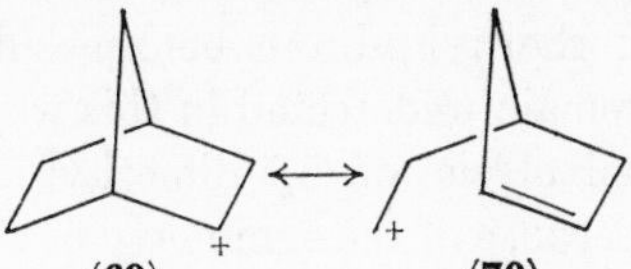

Table 3. The rates of acetolysis of norbornyl toluene-*p*-sulphonates at 74.84°.

	2-*endo*-Norbornyl	6,6-Dimethyl-2-*endo*-norbornyl	2-*exo*-Norbornyl	6,6-Dimethyl-2-*exo*-norbornyl
$10^5 k$ (sec⁻¹)	5.09	0.514	519	36.5

dimethyl grouping at this position caused a rate retardation rather than a rate enhancement (Table 3).[46] This retardation was attributed to an unfavourable steric interaction between the methyl groups and $C_{(1)}$ and $C_{(2)}$ in a non-classical transition state. A similar rate depression was observed in the *endo*-series and this was reported as the first case of significant steric deceleration in a unimolecular solvolysis.

[46] P. von R. Schleyer, M. M. Donaldson, and W. E. Watts, *J. Am. Chem. Soc.*, **87**, 375 (1965).

The effects of *gem*-dimethyl groupings at positions 3, 5, and 7 have been investigated by Winstein and his co-workers.[47] Ion-pair return was detected in the solvolysis of 7,7-dimethyl-*exo*-norbornyl *p*-bromobenzenesulphonate (71) (*a*) because the kinetics were not of the first order and (*b*) by isolation

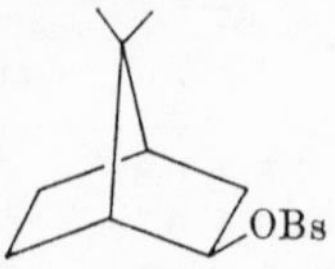

(71) 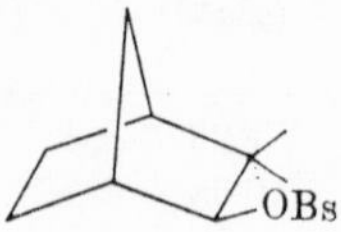   (72) 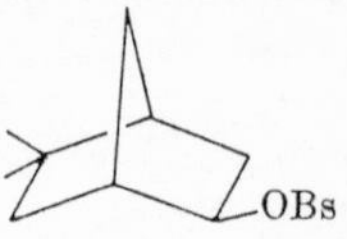   (73)

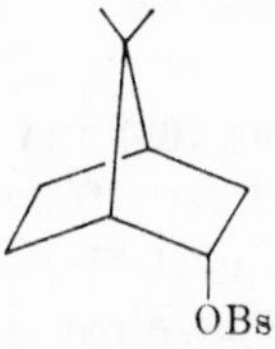

OBs   OBs   OBs
(74) 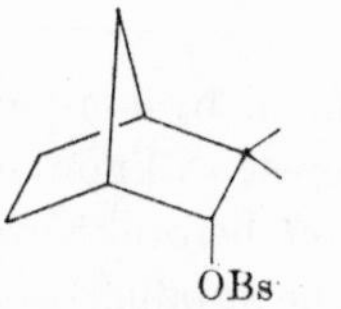   (75) 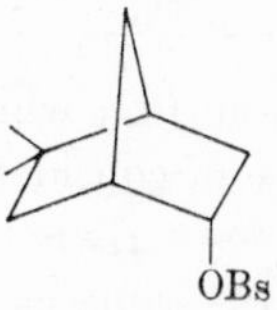   (76)

from the reaction mixture of the more slowly reacting 3,3-dimethyl-*exo*-norbornyl *p*-bromobenzenesulphonate (72). Ion-pair return would, of course, remain undetected in this way in the solvolysis of the latter, and also in the solvolysis of 5,5-dimethyl-*exo*-norbornyl *p*-bromobenzenesulphonate (73) because of the symmetry of the ion. All three of these compounds underwent solvolysis much more rapidly than the corresponding *endo*-isomers (Table 4);

Table 4. The rates of acetolysis of some *gem*-dimethyl substituted norbornyl *p*-bromobenzenesulphonates at 25°.

Norbornyl *p*-bromobenzene-sulphonate	10^6k (sec^{-1}) *exo*-isomer	10^6k (sec^{-1}) *endo*-isomer
7,7-Dimethyl	770	0.188
3,3-Dimethyl	32.9	0.0266
5,5-Dimethyl	26.1	0.039
Unsubstituted	88.2	

and the 7,7- and 3,3-dimethyl compounds, which would be interconverted by a Wagner–Meerwein shift, yielded almost identical products (presumably isolated at 100% reaction) on solvolysis in three solvents (Table 5). These are exclusively *exo*-products, and no *endo*-product (< 0.5%) was observed even when the *gem*-dimethyl group was at position 7.[48] Unfortunately, it is

[47] A. Colter, E. C. Friedrich, N. J. Holness, and S. Winstein, *J. Am. Chem. Soc.*, **87**, 378 (1965).
[48] For the significance of this point see p. 12.

Table 5. Products from the solvolyses of *gem*-dimethylnorbornyl
p-bromobenzenesulphonates.

Norbornyl p-bromobenzene-sulphonate		7,7-Dimethyl *exo*	3,3-Dimethyl *exo*	5,5-Dimethyl *exo*
At 25°			AcOH, 0.049м-NaOAc	
7,7-Me$_2$ *exo*-	**(71)**	47.0	4.5	48.5
3,3-Me$_2$ *exo*-	**(72)**	47.0	4.0	49.0
At 75°				
7,7-Me$_2$ *exo*-	**(71)**	43.0	6.0	51.0
7,7-Me$_2$ *endo*-	**(74)**	43.0	6.0	51.0
			72.4% Aqueous dioxan at 25°	
7,7-Me$_2$ *exo*-	**(71)**	(70.9)[a]	(11.3)[a]	(18.5)[a]
3,3-Me$_2$ *exo*-	**(72)**	71.5	10.5	18.0
5,5-Me$_2$ *exo*-	**(73)**	12.5	2.0	85.5
			70% Aqueous acetone at 75°	
7,7-Me$_2$ *exo*-	**(71)**	64.0	11.5	24.5
7,7-Me$_2$ *endo*-	**(74)**	63.5	10.5	25.0
3,3-Me$_2$ *exo*-	**(72)**	62.5	12.0	25.5
3,3-Me$_2$ *endo*-	**(75)**	59.0	13.5	27.5

[a] Infrared analysis: others by gas–liquid chromatography.

not clear if the small differences in ratios for the 7,7- and 3,3-dimethyl compounds in 70% aqueous acetone are significant. The 5,5-dimethyl system is interconverted with the other two through a $6 \rightarrow 2$ (or $6 \rightarrow 1$) hydride shift and the observation that there is less of this product in the more nucleophilic aqueous solvents than in acetic acid led the authors to conclude this occurs subsequent to, rather than concurrently with, the Wagner–Meerwein shift. It was also suggested that on theoretical grounds the $6 \rightarrow 2$ hydride shift probably involves an edge-protonated rather than a

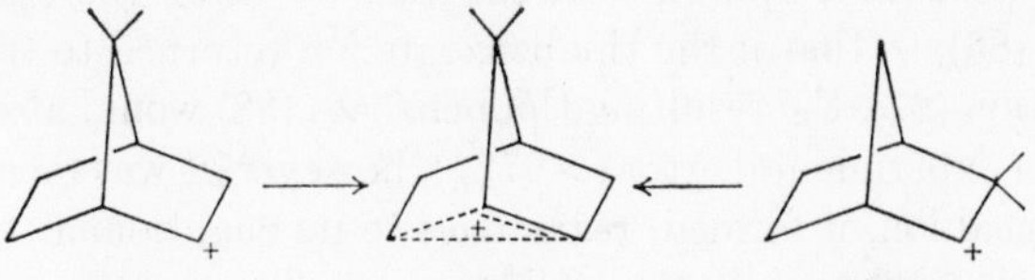

(77)   (79)   (78)

face-protonated transition state (cf. p. 22). The *endo-p*-bromobenzenesulphonates **(74)** and **(75)** yield products in a very similar ratio to those from their *exo*-isomers, a result which was interpreted as indicating efficient "leakage" from the classical ions **(77)** and **(78)** to the non-classical ion **(79)**.

Winstein and his co-workers have also forcibly restated the case for re-
garding *exo*-norbornyl and substituted *exo*-norbornyl cations as having non-
classical structures.[49,50]

The products of the solvolysis of *exo*-norbornyl *p*-bromobenzenesulphonate
have been re-investigated and the high stereospecificity of the reaction has
been even more strikingly demonstrated.[51] It is claimed that in 75% aqueous
acetone the product of substitution contains less than 0.02% of *endo*-
norbornanol, and that the *exo*-norbornanol obtained from optically active
p-bromobenzenesulphonate retains less than 0.05% of the optical activity.

Ethanolysis of camphene hydrochloride (**80**) and isobornyl chloride (**81**)
has been shown to give mainly camphene (**82**) and the tertiary ether (**83**)
with small amounts of the secondary ether (**84**) and tricyclene (**85**).[52] It is
not clear though whether the slightly smaller proportion of secondary ether
formed from the tertiary camphene hydrochloride (see Table 6) is significant
or not.

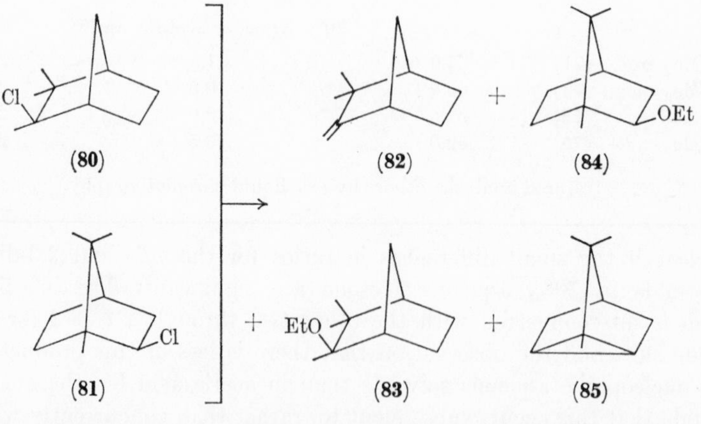

(**80**)   (**82**)   (**84**)

(**81**)   (**83**)   (**85**)

Ion-pair return occurs concurrently with acetolysis of *O-exo*-norbornyl *p*-
trifluoromethylthiobenzoate (**86**) to yield the thiol ester (**88**).[53] Positions 2
and 1 of the norbornyl system were labelled by using the optically active
thiobenzoate (**86**), so that if the thiobenzoate ion returned to trap a classical
norbornyl cation (**87**), the resulting thiobenzoate (**88**) would also be optically
active. Within experimental error (> 97%), however, it was racemic, showing
that the classical ion, if formed, rearranges to its enantiomer (by movement
of the 1,6-bond) more rapidly than it is trapped. The thiobenzoate ion is not,

[49] R. Howe, E. C. Friedrich, and S. Winstein, *J. Am. Chem. Soc.*, **87**, 379 (1965).
[50] S. Winstein, *J. Am. Chem. Soc.*, **87**, 381 (1965).
[51] S. Winstein, E. Clippinger, R. Howe, and E. Vogelfanger, *J. Am. Chem. Soc.*, **87**, 376 (1965).
[52] C. A. Bunton and C. O'Connor, *Chem. Ind.* (*London*), **1965**, 1182.
[53] S. G. Smith and J. P. Petrovich, *J. Org. Chem.*, **30**, 2882 (1965).

Table 6. Products of ethanolysis of camphene hydrochloride
and isobornyl chloride.

Substrate	[OEt$^-$] (M)	Camphene (moles %)	Ratio of ethers $\dfrac{\text{sec } (84)}{\text{tert } (83)}$
Camphene hydrochloride	0.1	68	0.02
	0.4	81	0.04
Isobornyl chloride	0.2	76	0.05
	0.5	84	0.06

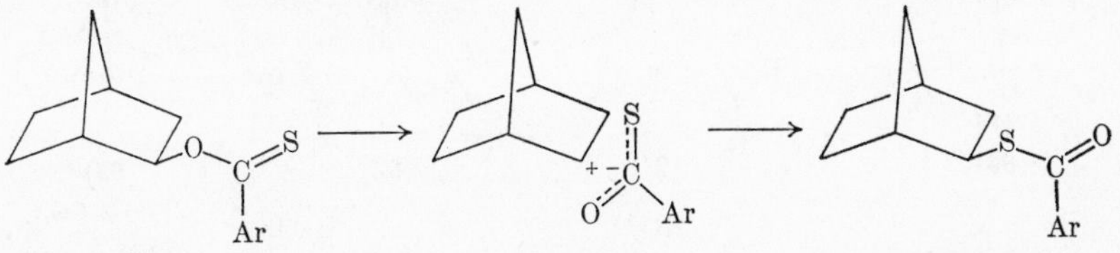

(86)   (87)   (88)

however, a very efficient trap since only 16% of thiol ester is formed compared with 83% of acetate.

The volumes of activation for the solvolysis of *exo*-norbornyl, *endo*-norbornyl, and cyclopentyl *p*-bromobenzenesulphonates in aqueous acetone at 40.0° are 14.3, 17.7, and 17.8 cm^3 mole^{-1}, respectively.[54] The low value for the *exo*-norbornyl derivative may result from its reacting via a non-classical transition state, with a more diffuse charge and smaller attraction from solvent molecules, but it could possibly be explained by a classical transition state in which solvation from the *endo*-side was rather difficult. The entropies of activation for the acetolyses of *exo*- and *endo*-norbornyl *p*-bromobenzenesulphonate are -3.5 ± 1.2 and -2.0 ± 0.3 cal deg^{-1} mole^{-1}, respectively,[55] though the significance of this similarity is not clear.

The secondary deuterium isotope effect on the polarimetic rate constant for the solvolysis of *exo*-[3,3-^{2}H]norbornyl bromide in aqueous acetic acid at 51.25° is $k_H/k_D = 1.087 \pm 0.005$.[56] This result was about 40% less than expected for a reaction involving a classical transition state and was interpreted as indicating a non-classical one. On dideuteration at position 7 of norbornyl bromide, zero isotope effect was observed on the polarimetric rate constant for the solvolysis in aqueous acetic acid. This was interpreted as indicating that the correct stereospecificity for electron release by the 7-hydrogen atoms was lacking, rather than that there was no electron

[54] W. J. le Noble and B. L. Yates, *J. Am. Chem. Soc.*, **87**, 3515 (1965).
[55] C. C. Lee and E. W. C. Wong, *Can. J. Chem.*, **43**, 2254 (1965).
[56] J. P. Schaefer and D. S. Weinberg, *Tetrahedron Letters*, **1965**, 2491.

deficiency at $C_{(1)}$ in the transition state. The α-deuterium effect on the acetolysis of *endo*-2-norbornyl toluene-*p*-sulphonate has also been measured.[55]

The question whether the $6 \rightarrow 2$ hydride shifts of the norbornyl cation involve nortricyclonium ions as (89) (i.e. a face-protonated cyclopropane) or edge-proponated cyclopropanes as (90) has been studied experimentally by Berson and Grubb.[57] Cation (89) is three-fold symmetric and, on collapse to ions (91), (92), and (93) (written here as non-classical) or products therefrom, the migrating hydride should show no preference for *exo*- or *endo*-attachment.

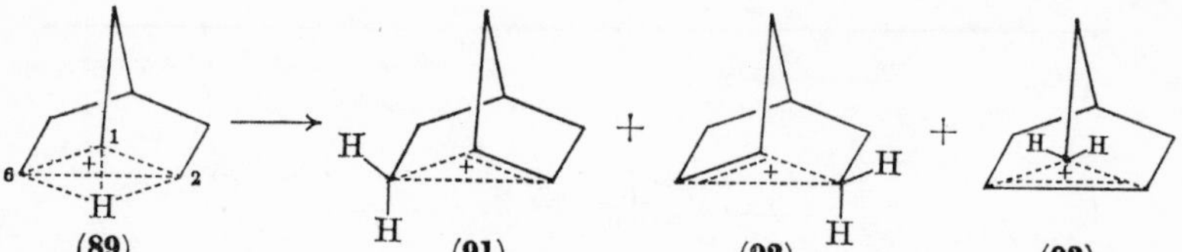

(89)   (91)   (92)   (93)

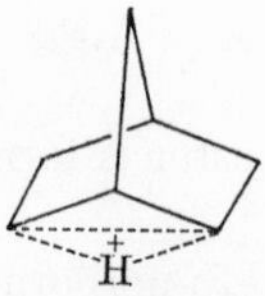

(90) 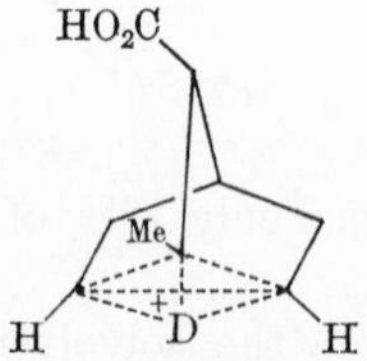   (94) 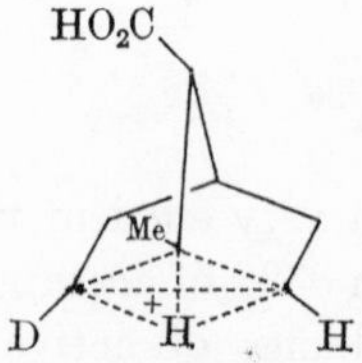   (95)

With (90), however, only *endo*-to-*endo* migrations should occur. The conversion of 3-*endo*-deuterio-3-*exo*-methyl- (96) and 3-*exo*-deuterio-3-*endo*-methyl-norborn-5-ene-2-*exo*-carboxylic acid (97) by 50% aqueous sulphuric acid into lactone (103) was investigated. These reactions involve the shifts (98) to (99), (101) to (102), and (100) to (99), which must occur almost exclusively in *endo* to *endo* fashion since the lactone (103) contains less than 3% of the 2-protio-species. The two nortricyclonium ions which would be involved in the conversion of (96) into (103) are (94) and (95). Ion (94), which would be intermediate in the step from (98) to (99), would lead to non-stereospecific labelling of (99) and hence to (103) with protium at $C_{(2)}$, and therefore it cannot be formed. Since ion (95) differs from (94) only in the position of the deuterium it too is presumably not involved. Migration of hydride ion in an *endo,endo* sense would also be expected, of course, if the intermediate ions were classical.

The stereochemical course of the formation of 2-*endo*-methyl-2-*exo*-norbornyl acetate on acetolysis of 3-*exo*-methyl-2-*endo*-norbornyl *p*-bromo-benzenesulphonate (104) has also been elucidated by Berson and his co-

[57] J. A. Berson and P. W. Grubb, *J. Am. Chem. Soc.*, **87**, 4016 (1965).

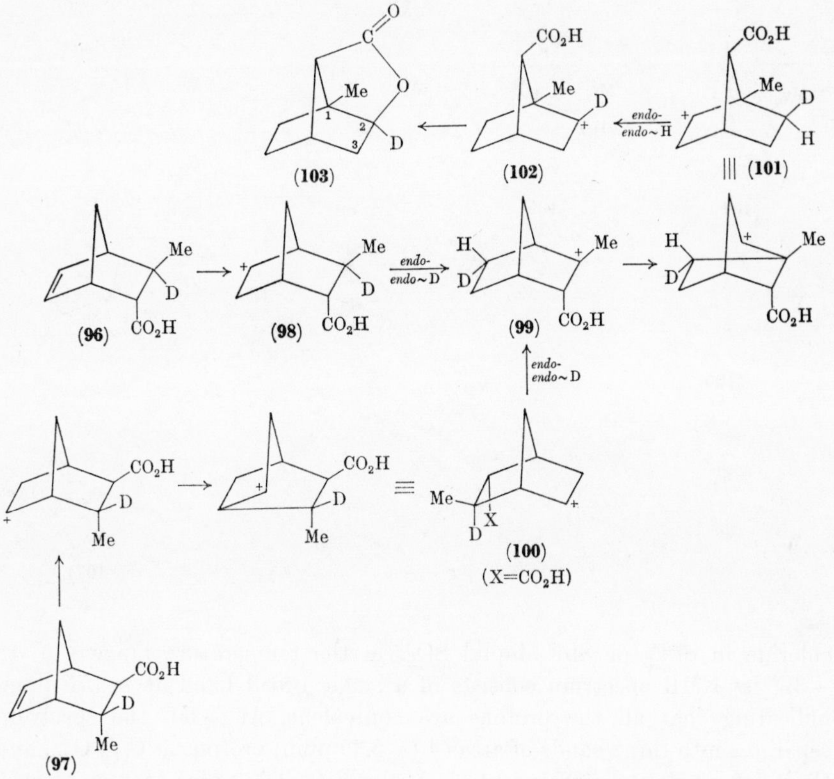

(**103**)   (**102**) ||| (**101**)

(**96**)   (**98**)   (**99**)

(**100**)
(X=CO$_2$H)

(**97**)

workers.[58] The reaction does not involve an *endo*-hydride migration in the cation, (**105**) → (**106**), since at least 90% of acetate is of opposite configuration to that of the starting *p*-bromobenzenesulphonate and hence must be formed by a more circuitous route such as (**104**) to (**107**). This non-occurrence of an *endo*-hydride shift was interpreted as indicating that the intermediate ion (**105**) had a non-classical structure. The analogous *exo*-hydride shift occurs quite readily in the acetolysis of 3-*endo*-methyl-2-norbornyl *p*-bromobenzenesulphonate (**108**) to yield 2-*endo*-methyl-2-*exo*-norbornyl acetate (**109**) of the same configuration.[59]

In an important communication Schleyer, Watts, Fort, Comisarow, and Olah have reported direct observation of the nuclear magnetic resonance of the 2-norbornyl cation.[60] This ion was prepared by dissolving *exo*-norbornyl

[58] J. A. Berson, J. H. Hammons, A. W. McRowe, R. G. Bergman, A. Remanick, and D. Houston, *J. Am. Chem. Soc.*, **87**, 3248 (1965).

[59] J. A. Berson, R. G. Bergman, J. H. Hammons, and A. W. McRowe, *J. Am. Chem. Soc.*, **87**, 3247 (1965).

[60] P. von R. Schleyer, W. E. Watts, R. C. Fort, M. B. Comisarow, and G. A. Olah, *J. Am. Chem. Soc.*, **86**, 5679 (1964).

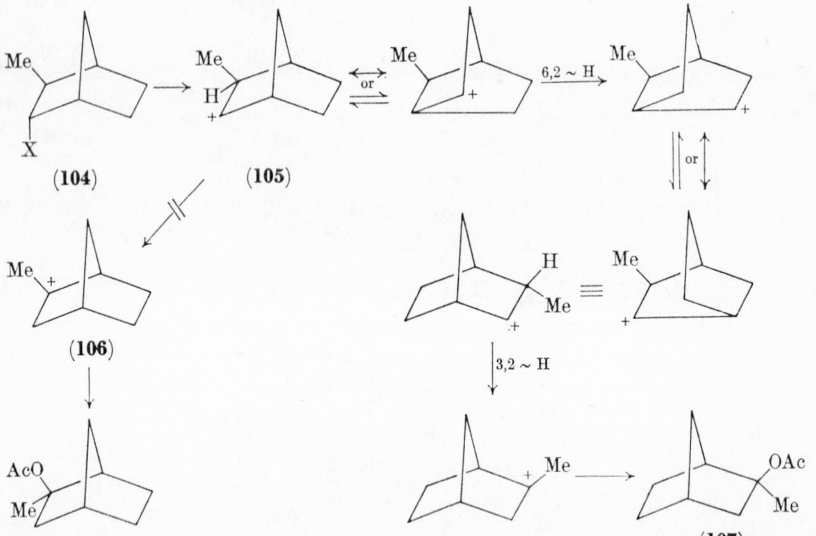

(104)   (105)

(106)

(107)

chloride in SbF_5 or SbF_5–liquid SO_2; in the temperature range $-5°$ to $+37°$ its NMR spectrum consists of a single broad band at -3.75 ppm, indicating that all the protons are equivalent. At $-60°$ the spectrum separates into three bands of areas 4 (-5.35 ppm; protons at $C_{(1)}$, $C_{(2)}$, and $C_{(6)}$), 1 (-3.15 ppm; proton at $C_{(4)}$), and 6 (-2.20 ppm; protons at $C_{(3)}$, $C_{(5)}$, and $C_{(7)}$); and on stronger cooling to $-120°$ no further change in the

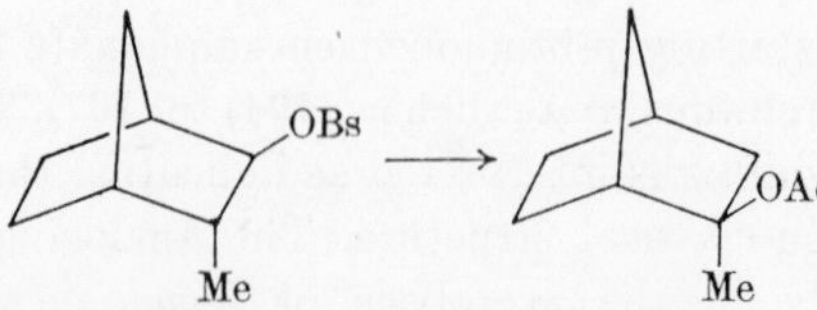

Me   Me
(108)   (109)

spectrum occurs. These results can be interpreted in terms of a Wagner–Meerwein rearrangement (110) → (111) and a $6 \rightarrow 2$ hydride shift (110) → (112), which occur rapidly over the temperature range studied, and a $3 \rightarrow 2$ hydride shift (110) → (113), which is fast at room temperature but slow below $-23°$. The rate constant for the $3 \rightarrow 2$ hydride shift was estimated to be 10^4 sec^{-1} at $+12.9°$ with $E_a = 10.8 \pm 0.6$ kcal mole^{-1} and $A = 10^{12.3}$ sec^{-1}.[61] Even at $-143°$, with SbF_5–SO_2ClF–SO_2 as solvent, any line

[61] M. Saunders, P. von R. Schleyer, and G. A. Olah, *J. Am. Chem. Soc.*, **86**, 5680 (1964).

broadening of the low field peak over that ascribable to the high viscosity
of the medium was less than 5 cps. From this it was concluded that the rate
constants for the $6 \to 2$ hydride shift and the Wagner–Meerwein rearrangement were at least 10^9 times greater than for the $3 \to 2$ shift, and hence the
former would appear to have values of at least 10^{13} sec^{-1} at room temperature.
According to Eigen and De Maeyer[62] the upper limit for the rate constant
for an intramolecular transformation which involves mass transfer is 10^{11} to
10^{13} sec^{-1}; if it involves only changes in electronic configuration higher
values are possible, but in any case the rate constant will remain considerably
below the electronic resonance frequencies ($\sim 10^{15}$ sec^{-1}). It is clear then

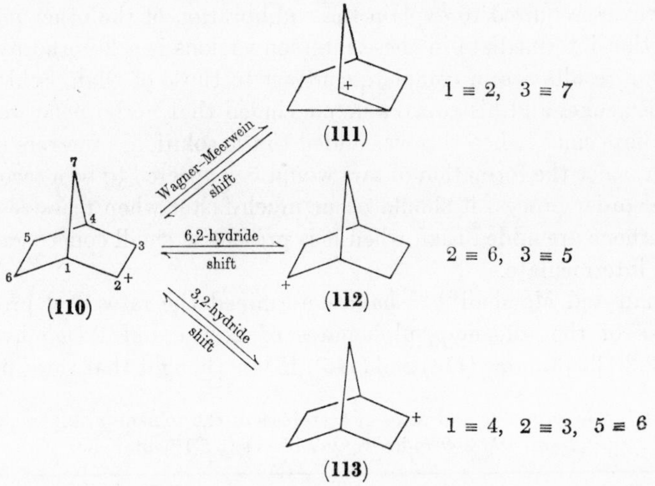

$$1 \equiv 2, \quad 3 \equiv 7 \quad \textbf{(111)}$$

$$2 \equiv 6, \quad 3 \equiv 5 \quad \textbf{(112)}$$

$$1 \equiv 4, \quad 2 \equiv 3, \quad 5 \equiv 6 \quad \textbf{(113)}$$

that if the ion observed by Olah, Schleyer, and their co-workers is not to be
regarded as a single completely delocalized species it is not far removed from
this. However, the result must clearly not be extrapolated to other more
highly solvating and ionizing solvents in which specific solvation effects
could cause charge to become localized on a single carbon atom.

It would be of considerable interest if the stability of the norbornyl cation
relative to that of the parent alcohol were known. In sulphuric acid–trifluoroacetic acid mixtures the alcohol is the stable species, but since the
NMR spectrum indicates that it is undergoing rapid rearrangement a small
amount of the ion or some other intermediate is presumably present. In pure
trifluoroacetic acid the spectrum consists of a signal at -4.96 ppm due to the
proton at $C_{(2)}$ and multiplets at -2.25 and -1.55 ppm ascribable to the

[62] M. Eigen and L. De Maeyer in "Investigation of Rates and Mechanisms of Reactions",
Interscience, New York, 1963, Part II, p. 1047.

protons at the bridgehead and $C_{(3)}$–$C_{(5)}$–$C_{(6)}$–$C_{(7)}$, respectively.[63] On addition
of sulphuric acid the spectrum changes, and with 60% sulphuric acid at 35°
the signal due to the proton at $C_{(2)}$ has disappeared and the spectrum
consists of two broad lines, at −2.2 and −1.2 ppm of relative areas 2:9.
These changes are reversible and *exo*-norbornanol can be recovered unchanged.
It was concluded that the line at −2.2 ppm was the signal of the bridgehead
protons, while that at −1.2 ppm was the time-averaged signal of all the other
protons. The fact that the position of the signal due to the bridgehead
protons is almost unaltered was taken to indicate that they are not undergoing exchange and must mean, therefore, that a fast Wagner–Meerwein
shift is not occurring in any intermediate. Rapid $6 \rightarrow 2$ and $3 \rightarrow 2$ shifts
are, however, required to explain the equilibration of the other protons. If
the reaction intermediate in these interconversions is a 2-norbornyl cation,
then these results are in complete contrast to those of Olah, Schleyer, and
others. Fraenkel and his co-workers concluded that norbornene was not an
intermediate since, when this was added to the solution, tars were obtained.
However, since the formation of tars would be expected to be a second-order
or higher-order process it should occur much faster when massive amounts
of norbornene are added than when it is present in small concentration as a
reaction intermediate.

Gassman and Marshall[64,65] have determined the rates and products of
acetolysis of the toluene-*p*-sulphonates of 2-*exo*- and 2-*endo*-hydroxy-7-
bicyclo[2.2.1]heptanone (114) and (115). It was thought that these norbornyl

Table 7. The products and rates of acetolysis of the toluene-*p*-sulphonates of
2-*exo*- and 2-*endo*-hydroxy-7-bicyclo[2.2.1]heptanone.

Starting material	$10^4 k$ (sec.$^{-1}$)	Product (%)	
		exo-acetate (118)	*endo*-acetate (119)
exo- (114)	5.96	60	40
endo- (115)	17.9	98.3	1.7

derivatives should be unable to react via non-classical ions, and it was concluded that this is so, since the *endo*-compound reacted faster than the *exo*-
isomer and since they both gave mixtures of the *exo*- and the *endo*-acetate
(see Table 7). It therefore appears that, contrary to Brown's proposal based
on the reactivity of and products from tertiary norbornyl derivatives,[66] re-

[63] G. Fraenkel, P. D. Ralph, and J. P. Kim, *Can. J. Chem.*, **43**, 674 (1965).
[64] P. G. Gassman and J. L. Marshall, *Tetrahedron Letters*, **1965**, 4073.
[65] P. G. Gassman and J. L. Marshall, *J. Am. Chem. Soc.*, **87**, 4648 (1965).
[66] H. C. Brown, F. J. Chloupek, and Min-Hon Rei, *J. Am. Chem. Soc.*, **86**, 1246, 1247, 1248
(1964); H. C. Brown and H. M. Bell, *ibid.*, pp. 5003, 5006, 5007; H. C. Brown and Min-
Hon Rei, *ibid*, pp. 5004, 5008.

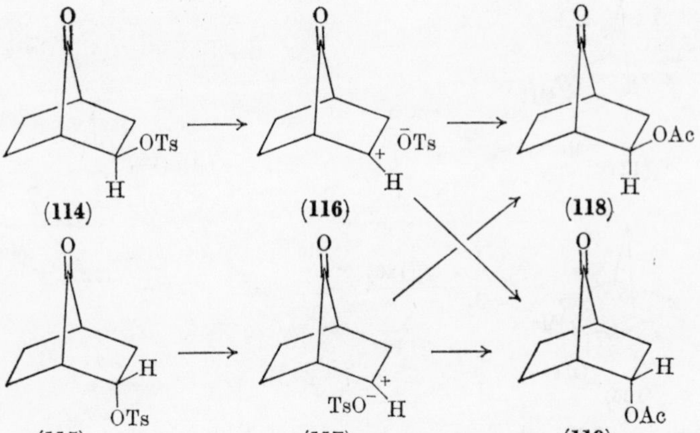

(114) → (116) → (118)

(115) → (117) → (119)

actions of the *exo*-isomer which involve classical norbornyl cations are not faster than those of the *endo*-isomer and do not yield exclusively an *exo*-product. Unfortunately, it is difficult to assess the validity of this conclusion because it is uncertain to what extent the carbonyl group is implicated and because the structures of a number of minor products were undetermined. Also of interest is the very large difference in the product ratios from the two reactions, which suggests that the acetates are formed from ion pairs (116) and (117) rather than from free ions.

The esterification of several bicyclic alcohols in formic acid has also been investigated.[67] Alcohols (120) to (123) react at similar rates, to yield the unrearranged formate, and the reactions must therefore involve acyl–oxygen fission. The corresponding tertiary alcohols, however, react at different rates and yield rearranged formates, so the reactions probably involve carbonium ions. *endo*-2-Methyl-*exo*-norbornen-2-ol (124) reacts 210 times as

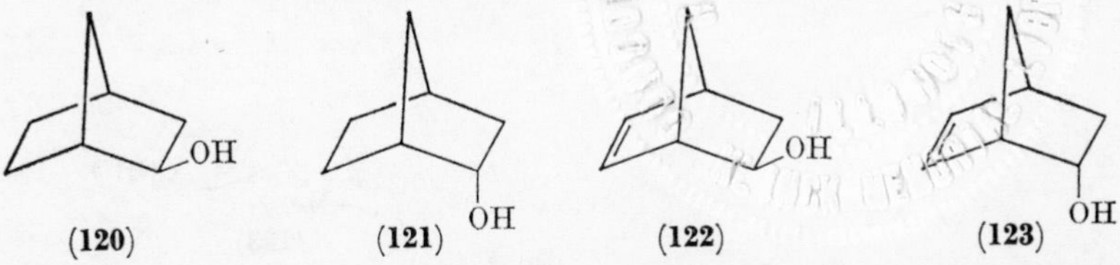

(120)   (121)   (122)   (123)

fast as its epimer (125), and both yield 1-methyl-3-nortricyclyl formate (127), presumably by way of the homoallylic cation (126). The saturated *exo*-alcohol (128) is esterified 5800 times faster than its *endo*-isomer (129); it yields first the unrearranged formate (131), but this rearranges to the *endo*-formate (132) and finally to the secondary formate (133). In the Reviewers' opinion these results are explained most simply in terms of a classical

[67] J. Paasivirta, *Ann. Chem.* **686**, 1 (1965).

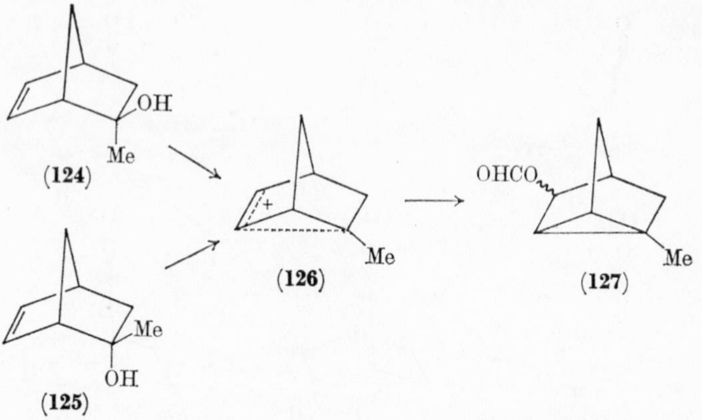

(124)

(125)

(126)

(127)

carbonium ion (130) which reacts, in order of decreasing rate, by *exo*-attack, *endo*-attack, and 1,6-bond migration, the last of these processes yielding the stable secondary formate (133).

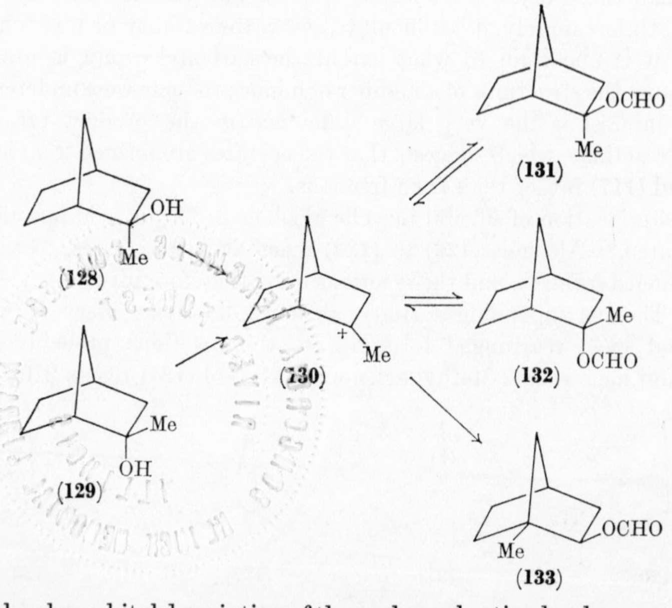

(131)

(128)

(130)

(132)

(129)

(133)

A molecular-orbital description of the norbornyl cation has been reported.[68]
The rates and products of the acetolysis of *exo*- and *endo*-6-bicyclo[3.1.0]-hexanylmethyl toluene-*p*-sulphonate (134) and (135) have been determined[69] and compared with those for the isomeric 6-bicyclo[3.2.0]heptyl derivatives (136) and (137).[70] No 2-vinylcyclopentyl toluene-*p*-sulphonate was obtained

[68] W. S. Tranhanovsky, *J. Org. Chem.*, **30**, 1666 (1965).
[69] K. B. Wiberg and A. J. Ashe, *Tetrahedron Letters*, **1965**, 1553.
[70] H. L. Goering and F. F. Nelson; F. F. Nelson, Ph.D. thesis, University of Wisconsin, 1960.

from (**136**), but this would have undergone solvolysis under the reaction conditions and thus the products from (**134**) and (**136**) are very similar and the reactions probably involve the same ion or ions. If the toluene-*p*-sulphonyloxy

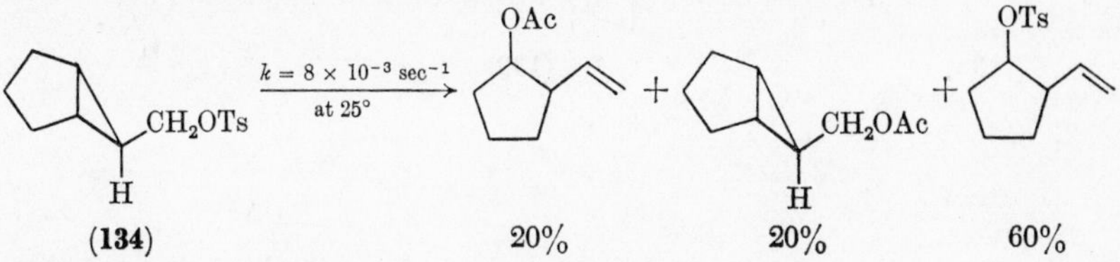

$k = 8 \times 10^{-3} \text{ sec}^{-1}$ at 25°

CH$_2$OTs
H
(**134**)

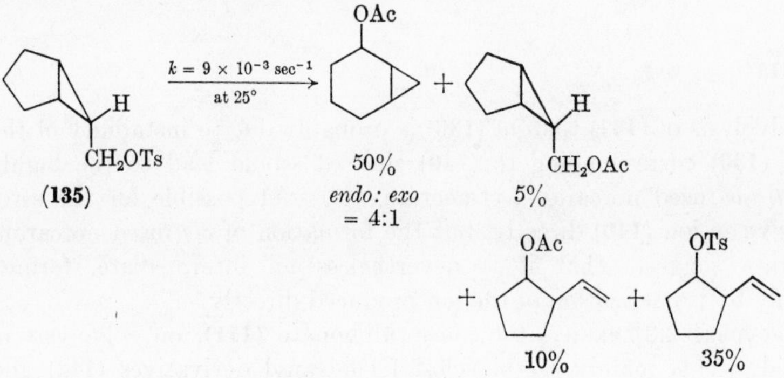

OAc
20%

CH$_2$OAc
H
20%

OTs
60%

$k = 9 \times 10^{-3} \text{ sec}^{-1}$ at 25°

H
CH$_2$OTs
(**135**)

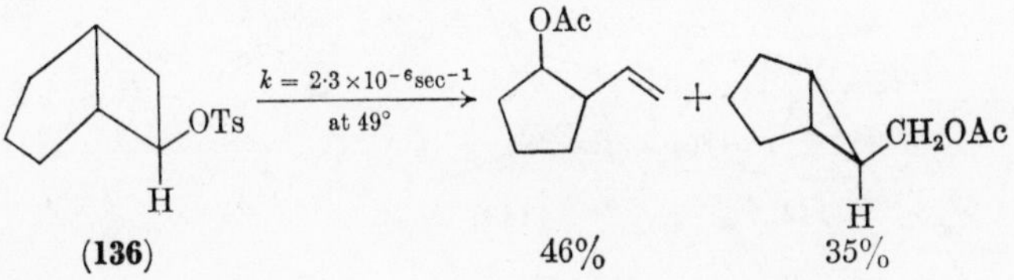

OAc
50%
endo : exo = 4:1

H
CH$_2$OAc
5%

OAc
10%

OTs
35%

$k = 2 \cdot 3 \times 10^{-6} \text{sec}^{-1}$ at 49°

OTs
H
(**136**)

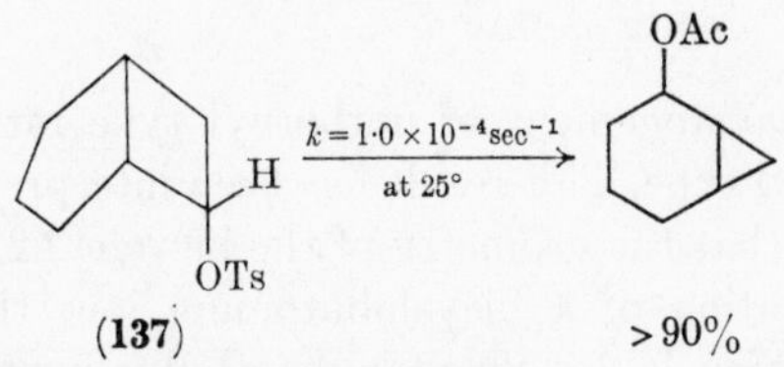

OAc
46%

CH$_2$OAc
H
35%

$k = 1 \cdot 0 \times 10^{-4} \text{sec}^{-1}$ at 25°

H
OTs
(**137**)

OAc
>90%

group of (**136**) takes up a pseudo-equatorial position (see **138**) the ion may then be written in the non-classical convention as (**139**). The products from (**135**) and (**137**) are, however, quite different and the reactions must involve different ions. That from (**137**) can be formulated as (**140**); and the higher

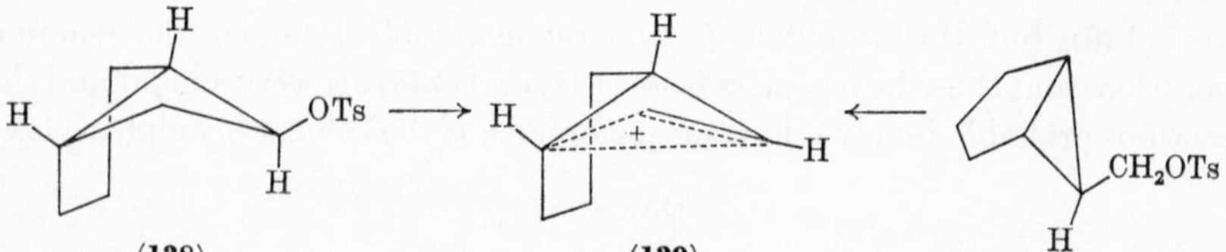

(138)   (139)

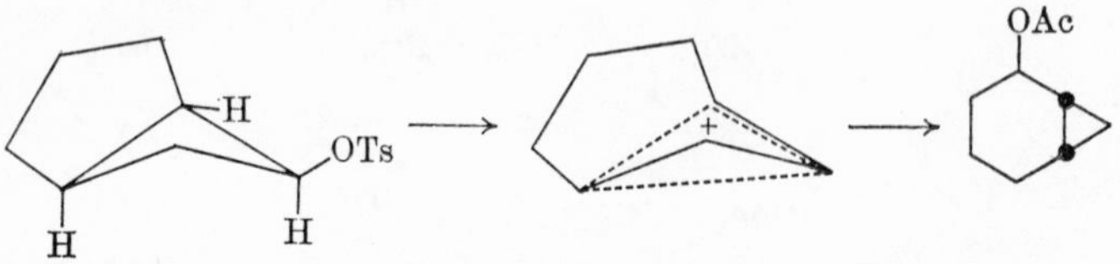

(137)   (140)

rate of solvolysis of (137) than of (136) is probably due to instability of the
ion from (136) corresponding to (140) since it would lead to the highly
strained *trans*-fused norcaran-2-yl acetate. It is not possible for the ester
(135) to give an ion (140) directly, but the formation of *cis*-fused norcaran-
2-yl acetate suggests that it is nevertheless an intermediate, formed
presumably by isomerization of the ion produced directly.

exo-2-Bicyclo[2.2.0]hexanyl toluene-*p*-sulphonate (141), on solvolysis in
acetic acid, yields mainly *exo*-bicyclo[2.1.1]hexanyl derivatives (142) and

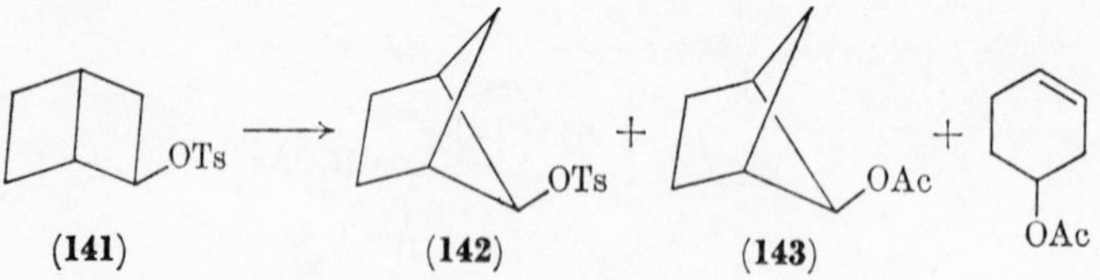

(141)   (142)   (143)

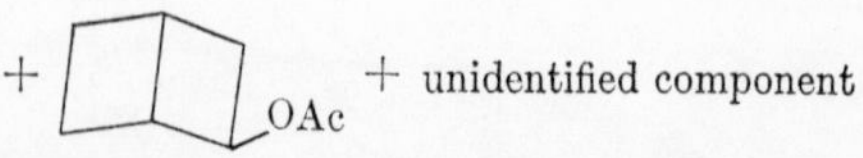

(143),[71] i.e. the reaction involves rearrangement of norbornyl type rather
than of cyclobutyl–cyclopropylmethyl type. This result has been interpreted
by Wiberg and Ashe[72] as indicating that the geometry of the bicyclo[2.2.0]-
hexanyl system precludes the formation of a bicyclobutonium ion; they
conclude also that the isomeric *exo*-5-bicyclo[2.1.0]pentanylmethyl toluene-*p*-
sulphonate (144) does not react via this ion, in agreement with their observation that this compound undergoes acetolysis without rearrangement.

[71] R. N. McDonald and C. E. Reineke, *J. Am. Chem. Soc.*, **87**, 3020 (1965).
[72] K. B. Wiberg and A. J. Ashe, *Tetrahedron Letters*, **1965**, 4245.

The reaction is, however, strongly accelerated, and Wiberg and Ashe conclude that the reaction involves a carbonium ion stabilized by the interaction of the p orbital at the methyl-carbon with the bent ring orbitals. The geometry of the *exo*-5-bicyclo[2.1.0]pentanylmethyl system should be ideal for this kind of interaction (see p. 43). In the Reviewers' opinion it is highly likely that the driving force for the rapid solvolysis of other cyclopropylmethyl derivatives also results from this type of interaction rather than from the formation of a bicyclobutonium ion.[73]

Preliminary details of an investigation of the solvolysis of the methoxyl-substituted bicyclic toluene-p-sulphonate (145) in 98% formic acid have been reported;[74] the reaction involves a series of complex rearrangements.

Other reactions of bicyclic and polycyclic systems which have received attention include acetolyses of *exo*- and *endo*-3-bicyclo[3.2.1]octanyl toluene-p-sulphonate,[75] acetolysis of the p-bromobenzenesulphonate of the bridgehead birdcage alcohol,[76] reaction of *endo*-norbornanol and norbornan-7-ol with triphenylphosphine and bromine,[77] rearrangement of quadricyclone,[78] ring expansion of nortricyclane,[79] solvolyses of 2-bromobicyclo[2.2.1]heptane-2-bromobicyclo[2.2.2]octane-, and 2-bromobicyclo[3.2.1]octane-1-carboxylic acid.[80]

Phenonium Ions

The cases for[81] and against[82] phenonium ions as discrete intermediates in certain Wagner–Meerwein rearrangements have been stated.

Brown's postulate of the intervention of a pair of equilibrating classical ions in the acetolysis of D-*erythro*-$\alpha\beta$-dimethylphenethyl toluene-p-sulphonate

[73] For previous suggestions that this is so see E. S. Gould, "Mechanism and Structure in Organic Chemistry", Henry Holt and Company, New York, 1959, p. 588, and B. Capon, *Quart. Rev. (London)*, **18**, 100 (1964).

[74] K. L. Rabone and N. A. J. Rogers, *Chem. Ind. (London)*, **1964**, 1838.

[75] C. W. Jefford, J. Gunsher, and B. Waegell, *Tetrahedron Letters*, **1965**, 3405.

[76] P. Carter, R. Howe, and S. Winstein, *J. Am. Chem. Soc.*, **87**, 914 (1965).

[77] J. P. Schaefer and D. S. Weinberg, *J. Org. Chem.*, **30**, 2635, 2639 (1965).

[78] P. R. Story and S. R. Fahrenholtz, *J. Am. Chem. Soc.*, **87**, 1623 (1965).

[79] J. T. Lumb and G. H. Whitham, *Tetrahedron*, **21**, 499 (1965).

[80] W. R. Vaughan, R. Caple, J. Csapilla, and P. Scheiner, *J. Am. Chem. Soc.*, **87**, 2204 (1965).

[81] D. J. Cram, *J. Am. Chem. Soc.*, **86**, 3767 (1964).

[82] H. C. Brown, K. J. Morgan, and F. J. Chloupek *J. Am. Chem. Soc.*, **87**, 2137 (1965).

has been explored mathematically by Collins, Benjamin, and Lietzke,[83] who have shown that the observed product ratios are consistent with reasonable values of the rate constants for phenyl migration, rotational isomerization, and reaction of the ions with solvent.

The solvolyses of triphenylethyl chloride and toluene-p-sulphonate proceed at enhanced rates and, since they involve migration of a phenyl group, must involve transition states such as (146). Three views have been taken as to the origin of these high rates: (i) that they are exclusively the results of "a release of bonding energy in the transition state" (i.e. the reactions are

$$\text{Ph}_3\text{C—CH}_2\text{X} \longrightarrow \text{Ph}_2\overset{\overset{\text{Ph}}{\frown}}{\underset{+}{\text{C}}}\!\!\!\!=\!\!\!\text{CH}\cdots\text{X} \longrightarrow \text{Ph}_2\text{C}\!=\!\text{CHPh}$$

(146)

$$\text{Ph}_2\text{C}\!\!\overset{\delta+}{\underset{\overset{\displaystyle|}{\text{X}^{\delta-}}}{\text{—CH}_2}}$$

(147)

anchimerically assisted);[84] (ii) that they are largely the result of release of compression energy in the transition state (i.e. are sterically assisted);[85] and (iii) that both these factors are important.[86] By using ketones as models for carbonium ions Brown, Bernheimer, and Morgan[87] have now provided evidence for the operation of steric factors in the solvolyses of these and related compounds. The argument runs that the formation of a carbonium ion involves going from a tetra-coordinated to a tri-coordinated state while the reduction involves the reverse of this, so that steric factors should operate in opposite senses in the two reactions. That this is qualitatively and even semiquantitatively so is shown by the results given in Table 8. It is, therefore, likely that the same factors are important in controlling both sets of reactivities and, since it seems improbable that bridging by the phenyl group could be important in the reduction, this common factor must be the steric one as outlined above. Since there is little difference in the rates of solvolysis of 1-methyl-2,2,2-triphenylethyl (149) and $\alpha\beta\beta$-trimethylphenethyl (148) p-bromobenzenesulphonate (see Table), it seems unlikely that phenyl migration is well developed in the transition state, for after migration these two compounds should give ions of quite different stabilities. Also, since the effects of substituents on the rate of solvolysis of 2-methyl-2-phenylpropyl p-bromo-benzenesulphonate are only moderate, it seems likely that bridging is

[83] C. J. Collins, B. M. Benjamin, and M. H. Lietzke, *Ann. Chem.*, **687**, 150 (1965).
[84] C. K. Ingold, "Structure and Mechanism in Organic Chemistry", Cornell University Press, Ithaca, N.Y., 1953, p. 514.
[85] H. C. Brown in "The Transition State", *Chem. Soc. Special Publ.*, No. 16, p. 152 (1962).
[86] S. Winstein, B. K. Morse, E. Grunwald, K. C. Schreiber, and J. Corse, *J. Am. Chem. Soc.*, **74**, 1113 (1952).
[87] H. C. Brown, R. Bernheimer, and K. J. Morgan, *J. Am. Chem. Soc.*, **87**, 1280 (1965).

Table 8. Relative rates of acetolysis of some arenesulphonates and of reduction of the corresponding ketones by borohydride.

Primary toluene-p-sulphonate	Relative rate of acetolysis at 50°	Secondary p-bromo-benzenesulphonate at 50°	Relative rate of acetolysis at 50°	Ketone	Relative rate of reduction by borohydride at 50°
CH$_3$CH$_2$OTs	1.00	CH$_3$CH(CH$_3$)OBs	1.00	CH$_3$COCH$_3$	1.00
PhCH$_2$CH$_2$OTs	0.35	PhCH$_2$CH(CH$_3$)OBs	0.32	PhCH$_2$COCH$_3$	1.6
		PhCH(CH$_3$)CH(CH$_3$)OBs	1.1 (threo) 1.3 (erythro)	PhCH(CH$_3$)COCH$_3$	0.23
Ph$_2$CHCH$_2$OTs	2.6	Ph$_2$CHCH(CH$_3$)OBs	1.8	Ph$_2$CHCOCH$_3$	0.19
		(CH$_3$)$_3$CCH(CH$_3$)OBs	3.5	Me$_3$CCOCH$_3$	0.10
		PhC((CH$_3$)$_2$CH(CH$_3$)OBs (148)	73	PhC(CH$_3$)$_2$COCH$_3$	0.034
Ph$_3$CCH$_2$OTs	378	Ph$_3$CCH(CH$_3$)OBs (149)	417	Ph$_3$CCOCH$_3$	0.0051

only slightly developed in the transition state. Brown therefore concludes that this is best represented by (147) and that "the enhanced rates in β,β,β-triphenylethyl and related derivatives are largely to be attributed to relief of steric strain with phenyl participation providing a portion of the mechanism for relief of that strain."

When dissolved in SO_2–SbF_5–FSO_3H at $-60°$, *erythro-* and *threo-$\alpha\beta$-*dimethylphenethyl alcohol (150) and (151) give the same NMR spectrum, a fact that was attributed to formation of equimolar amounts of the phenonium ions (152) and (153).[88] Since stereochemistry is not preserved, classical

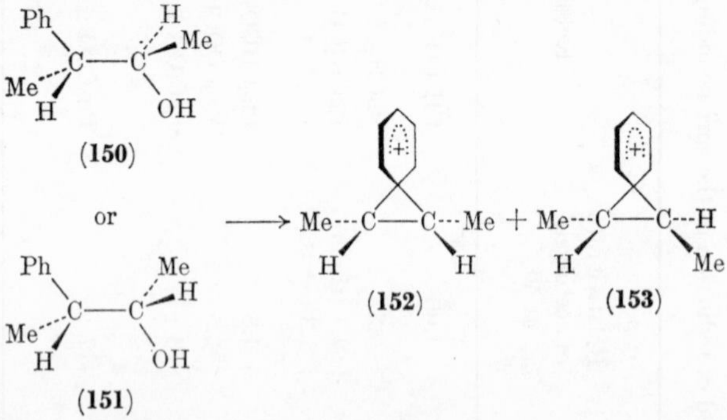

(150)

or

(151)

$\longrightarrow$ (152) + (153)

$\alpha\beta$-dimethylphenethyl cations presumably intervene at some stage in the formation of these bridged ions. It was emphasized by Olah and Pittman that their results "should not be extrapolated to previous stereochemical and kinetic solvolyses studies since in our highly acidic and ionizing solvent a shift from possible equilibrating ions, in more basic solvents, to static bridged ions would be favoured."[88]

A similar bridged anthryl ion (155) has been prepared by Eberson and Winstein[89] by dissolving the spiro alcohol (154) in SO_2–SbF_5 at $-80°$. The signals of the protons of the cyclopropane ring occur at much lower field (2 ppm) than those of the original spiro alcohol. This deshielding is appreciably larger than occurs with the signals of the methyl groups on conversion of (158) into (159). For this reason Eberson and Winstein concluded that there was appreciable delocalization of charge into the cyclopropane-methylene groups and preferred to represent the bridged ion as (155) rather than as (156).[90] This ion has also been shown to intervene in the solvolysis of 2-(9-anthryl)ethyl toluene-p-sulphonate (157) in a variety of solvents;[91] its

[88] G. A. Olah and C. U. Pittman, *J. Am. Chem. Soc.*, **87**, 3509 (1965).

[89] L. Eberson and S. S. Winstein, *J. Am. Chem. Soc.*, **87**, 3506 (1965).

[90] It is perhaps significant that the signals of the protons on the cyclopropane ring of ions (152) and (153) occur at similar positions (-3.0 and -3.50 ppm) to those of ion (155) (-3.44 ppm).

[91] L. Eberson, J. P. Petrovich, R. Baird, D. Dyckes, and S. Winstein, *J. Am. Chem. Soc.*, **87**, 3504 (1965).

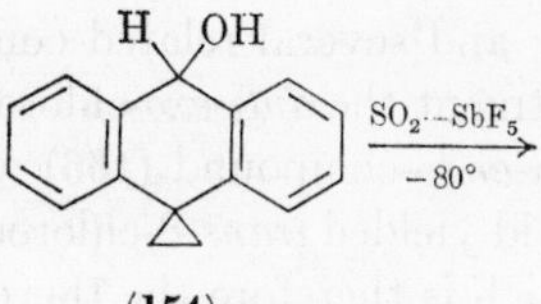

(154) 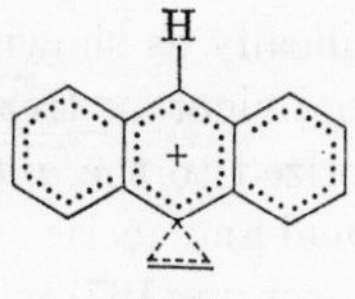   (155) 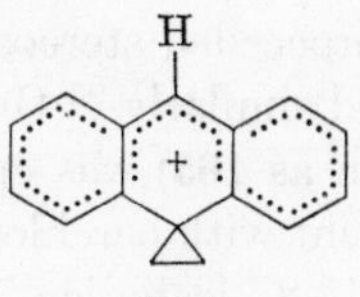   (156)

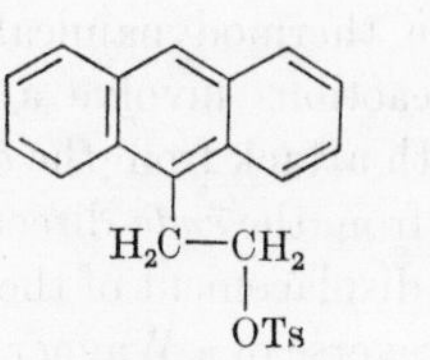

(157) 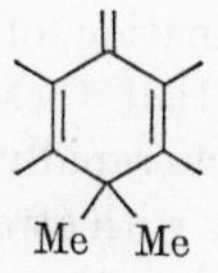   (158) 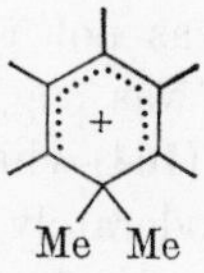   (159)

formolysis, for instance, is 100 times faster than that of phenethyl toluene-*p*-sulphonate, and hydrolysis in 60% aqueous dioxan in the presence of sodium hydrogen carbonate yields 2-(9-anthryl)ethanol (15%) and the spiro alcohol (154) (85%).

The NMR spectra of a number of 2-phenylalkyl cations in which the charge can reside on tertiary carbon were also measured.[92] The results either indicated a rapid equilibration between classical structures as with (160) or did not allow a clear choice to be made between the classical and non-classical structures as with (161).

$$Ph_3C \overset{+}{-} \overset{+}{C}Ph_2 \qquad PhMe_2C \overset{+}{-} \overset{+}{C}Me_2$$
(160)   (161)

The acetolyses of the *cis*- and *trans*-chloro-toluene-*p*-sulphonates (162) and

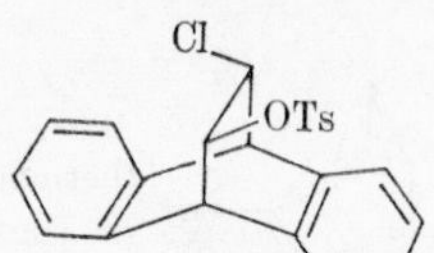

(162) 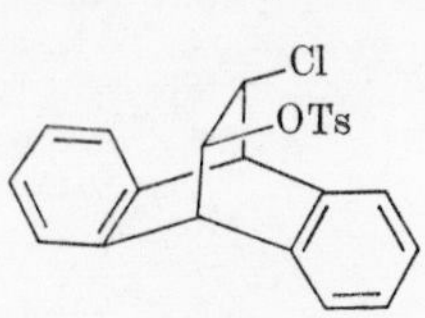   (163)

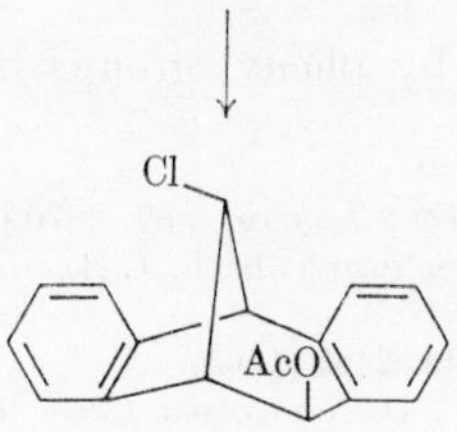

(165) 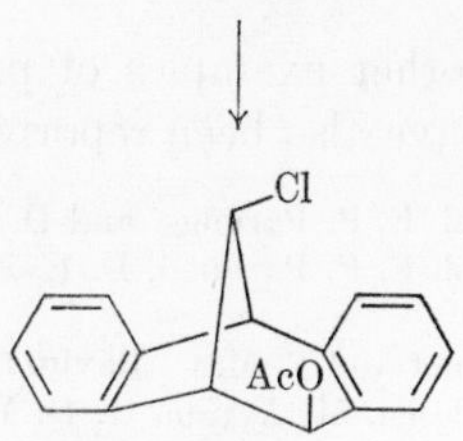   (166)

[92] G. A. Olah and C. U. Pittman, *J. Am. Chem. Soc.*, **87**, 3507 (1965).

(**163**) proceeded stereospecifically as shown, and several related compounds behaved similarly.[93] On more vigorous treatment the *anti-exo*-chloro acetate (written as **165**) was epimerized to the *anti-endo*-compound (**166**) which on treatment with perchloric acid and acetic acid yielded *trans*-8-chlorodibenzobicyclo[2.2.2]octadien-7-yl acetate (**167**) which is therefore the thermodynamically most stable of these chloro-acetates.[94] It was therefore concluded that the driving force for the rearrangement occurring in the original solvolyses was not the formation of the thermodynamically most stable product. It was suggested that these reactions involve a single classical benzylic ion (**164**) which reacts rapidly with attack from the *exo*(quasi axial)-direction, moderately rapidly with attack from the *endo*-direction, and slowly with attack at the bridgehead carbon and displacement of the carbon–carbon bond. The last process is the microscopic reverse of a Wagner–Meerwein shift and it was suggested that such processes should be termed "geitonodesmic".

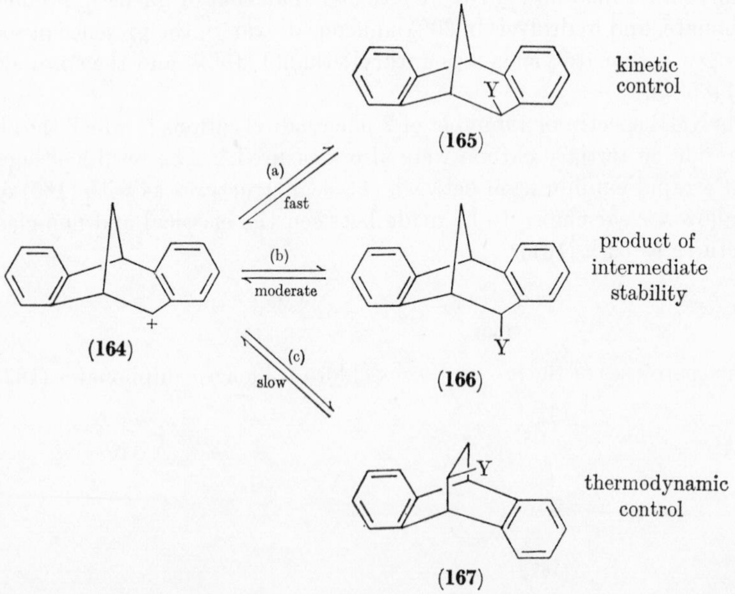

Several other examples of participation by phenyl groups in solvolysis reactions have also been reported.[95,96,97]

[93] S. J. Cristol, F. P. Parungo, and D. E. Plorde, *J. Am. Chem. Soc.*, **87**, 2870 (1965).

[94] S. J. Cristol, F. P. Parungo, D. E. Plorde, and K. Schwarzenbach, *J. Am. Chem. Soc.*, **87**, 2879 (1965).

[95] F. A. L. Anet and P. M. G. Bavin, *Can. J. Chem.*, **43**, 2465 (1965).

[96] I. G. Dinulescu, M. Avram, G. D. Mateescu, and C. D. Nenitzescu, *Chem. Ind. (London)*, **1964**, 2023.

[97] E. Ciorănescu, A. Bucur, M. Elian, M. Banciu, M. Voicu, and C. D. Nenitzescu, *Tetrahedron Letters*, **1964**, 3835.

Participation by Double[98] and Triple Bonds

A detailed investigation of participation by carbon–carbon double bonds has been reported by Bartlett and his co-workers. Included are full details of the solvolyses of 2-(cyclopent-3-enyl)ethyl arenesulphonates.[99] The *p*-nitrobenzenesulphonate (168a) undergoes formolysis 640 times faster, and the toluene-*p*-sulphonate (168b) undergoes acetolysis 74 times and solvolysis in 50% aqueous ethanol 5.7 times faster, than their saturated analogues. High yields of *exo*-norbornyl derivatives were obtained. Introduction of successive methyl groups at positions 3 and 4 (cf. 171, 172, and 173) results in approximately equal rate enhancements, which suggests that the transition state is symmetrical, as (169).[100] It was argued that since the transition state lies on the pathway from starting material to the first-formed intermediate ion this must also be symmetrical and hence have the bridged structure (170).

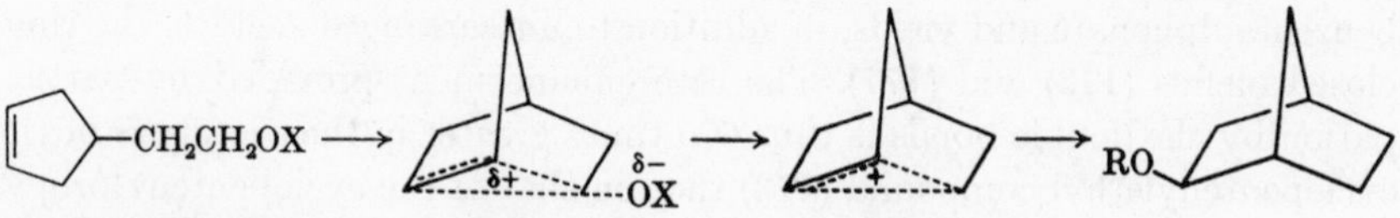

(168a)
X = *p*-NO$_2$.C$_6$H$_4$.SO$_2$
(168b)
X = *p*-Me.C$_6$H$_4$.SO$_2$

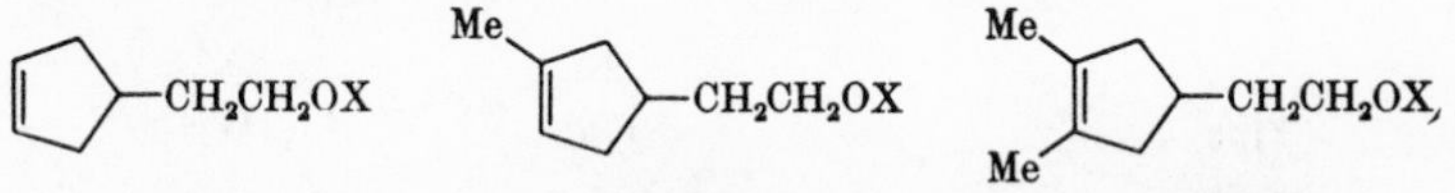

(169)   (170)

(171)   (172)   (173)
1·10   7·69   42·3

10^4 k (sec^{-1}) for acetolysis at 60°
X = *p*-NO$_2$.C$_6$H$_4$.SO$_2$

A 1,3-hydride shift analogous to that observed in the solvolysis of *exo* norbornyl derivatives (see p. 3) was also observed in the solvolysis of 2-(cyclopent-3-enyl)[1,1-^{2}H$_2$]ethyl toluene-*p*-sulphonate (174).[101] After conversion of the product of solvolysis in 80% aqueous acetic acid to *exo*-norbornanol, analysis by NMR spectroscopy indicated 10—15% migration of

[98] For reviews see B. Capon, *Quart. Rev.* (*London*), 18, 97 (1964); B. Capon and C. W. Rees, *Ann. Rept. Progr. Chem.* (*Chem. Soc. London*), 61, 234 (1964).
[99] P. D. Bartlett, S. Bank, R. J. Crawford, and G. H. Schmid, *J. Am. Chem. Soc.*, 87, 1288 (1965).
[100] P. D. Bartlett and C. D. Sargent, *J. Am. Chem. Soc.*, 87, 1297 (1965).
[101] K. Humski, S. Borcic, and D. E. Sunko, *Croat. Chem. Acta*, 37, 3 (1965).

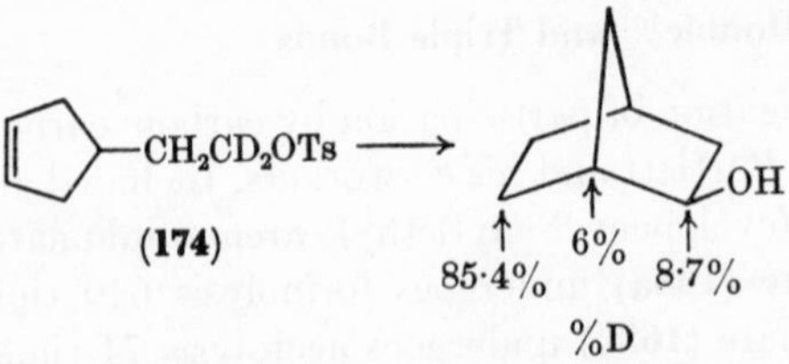

CH₂CD₂OTs → ... OH

(174)

85·4%   6%   8·7%

%D

deuterium to positions 1 and 2. The α-deuterium isotope effect for the solvolysis of 2-(cyclopent-3-enyl)[1,1-^{2}H$_2$]ethyl p-nitrobenzenesulphonate has
also been measured and discussed.[102]

In contrast to (168), 3-(cyclopent-3-enyl)propyl p-nitrobenzenesulphonate
undergoes acetolysis without appreciable acceleration by the double bond.[103]
However, 3-(3,4-dimethylcyclopent-3-enyl)propyl p-nitrobenzenesulphonate
(175) undergoes acetolysis 3.2 times as fast as 3-cyclopentylpropyl p-nitro-
benzenesulphonate and yields, in addition to unrearranged acetate, the ring-
closed olefins (176) and (177). The rate enhancement provided by participation by the double bonds is thus 600 times greater in the acetolysis of the
cyclopentenylethyl compound (173) than in that of the cyclopentenylpropyl
compound (175). This difference was reflected in both the energy and the

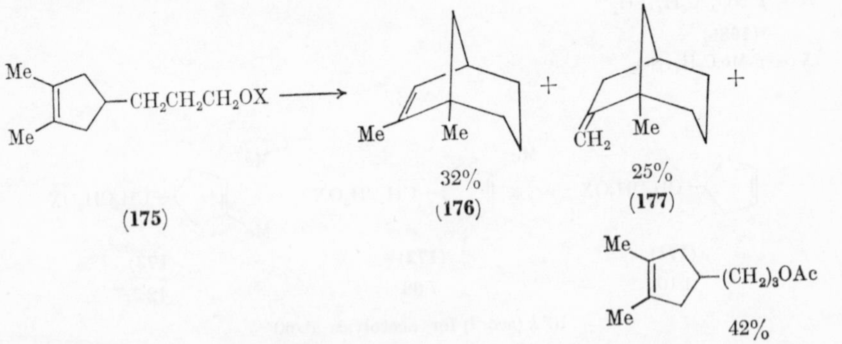

Me — CH₂CH₂CH₂OX →

Me   Me   +   CH₂   Me   +

32%   25%

(176)   (177)

(175)

Me
|
—(CH₂)₃OAc
|
Me   42%

entropy of activation. The less favourable entropy of activation was attributed
to the larger number of conformations of the initial state of (175) which are
unsuitable for reaction. The more favourable energy of activation for the
reaction of (173) was thought to arise from the fact that attack by the centre
of the double bond on C$_{(1)}$ can occur in this compound in a conformation with
the 1- and 2-hydrogen atoms all staggered, but with (175) this is not possible.

Participation by the double bond has been shown to occur in the acetolysis
of 2-(cyclopent-1-enyl)ethyl p-bromobenzenesulphonate (178) which pro-

[102] C. C. Lee and E. W. C. Wong, *Tetrahedron*, 21, 539 (1965).
[103] P. D. Bartlett, W. S. Trahanovsky, D. A. Bolon, and G. H. Schmid, *J. Am. Chem. Soc.*,
87, 1314 (1965).

ceeds about forty times as fast as that of 2-cyclopentylethyl p-bromobenzene-sulphonate and yields the complex mixture of products shown.[104,105]

Acetolysis of 5-hexenyl p-nitrobenzenesulphonate (**180**; Ns = p-NO$_2$·C$_6$H$_4$·SO$_2$ is 1.7 times as fast as that of hexyl p-nitrobenzenesulphonate and in the presence of sodium acetate yields 83.7% of 5-hexenyl acetate, 4.7% of

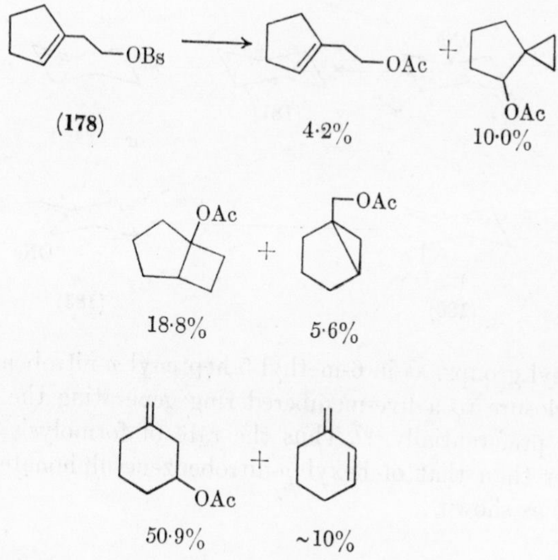

(**178**)   4·2%   10·0%

18·8%   5·6%

50·9%   ~10%

cyclohexene, and 11.6% of cyclohexyl acetate;[106] participation by the double bond therefore occurs. The cationic intermediate cannot, however, be any of those involved in the acetolyses of cyclopentyl methyl p-nitrobenzenesul-phonate (**179**) and cyclohexyl p-nitrobenzenesulphonate, which also yield cyclohexene and cyclohexyl acetate but in different ratios. It was suggested that this would be explained if the cyclohexyl compound gave the cyclohexyl cation (**181**) directly, and the cyclopentylmethyl and 5-hexenyl compounds gave non-classical ion pairs (**182**) and (**183**) in which the p-nitrobenzene-sulphonate ion is oriented differently in respect of the carbonium ion. In the ion pair (**182**) from the cyclopentylmethyl compound, the orientation is highly suitable for ion-pair return to give cyclohexyl p-nitrobenzenesul-phonate, which is solvolysed rapidly. The high proportion of cyclohexene in the products from the cyclopentylmethyl and cyclohexyl compounds can then be explained if the cyclohexyl cation gives cyclohexene much more readily than the non-classical ion does.

When the terminal methylene group of the 5-hexenyl system is substituted

[104] W. D. Closson and G. T. Kwiatkowski, *Tetrahedron Letters*, **1964**, 3831.
[105] W. D. Closson and G. T. Kwiatkowski, *Tetrahedron*, **21**, 2779 (1965).
[106] P. D. Bartlett, W. D. Closson, and T. J. Cogdell, *J. Am. Chem. Soc.*, **87**, 1308 (1965).

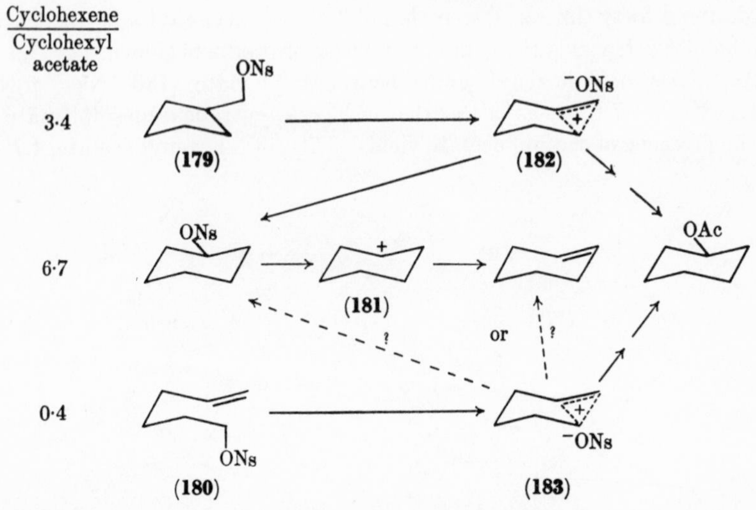

Cyclohexene
——————
Cyclohexyl
acetate

3·4

(179)   (182)

6·7

(181)

or

0·4

(180)   (183)

by two methyl groups, as in 6-methyl-5-heptenyl p-nitrobenzenesulphonate
(184), ring closure to a five-membered ring generating the tertiary cation
(185) occurs preferentially.[107] Thus the rate of formolysis of (184) is 20
times greater than that of hexyl p-nitrobenzenesulphonate, and the final
products are as shown.

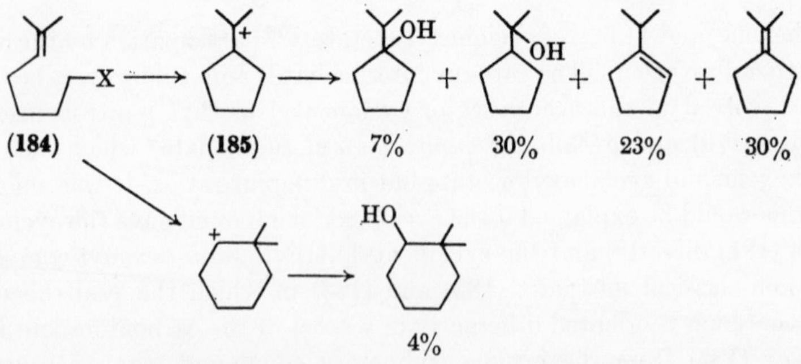

(184)   (185)   7%   30%   23%   30%

4%

Participation by the double bond of the ester (186), but not of its epimer
(188), occurs readily with the formation of 1-adamantanol (187).[108] Participation by an exocyclic methylene group has also been shown to occur in the
solvolysis of 6-methylenecyclodecyl toluene-p-sulphonate (189), which in
buffered aqueous acetone proceeds rapidly with the formation of *cis*-
bicyclo[4.4.1]undecan-1-ol (190) and bicyclo[4.4.1]undec-1-ene (191).[109]

107 W. S. Johnson and R. Owyang, *J. Am. Chem. Soc.*, **86**, 5593 (1964).
108 M. Eakin, J. Martin and W. Parker, *Chem. Comm.* **1965**, 206.
109 T. L. Westman and R. D. Stevens, *Chem. Comm.* **1965**, 459.

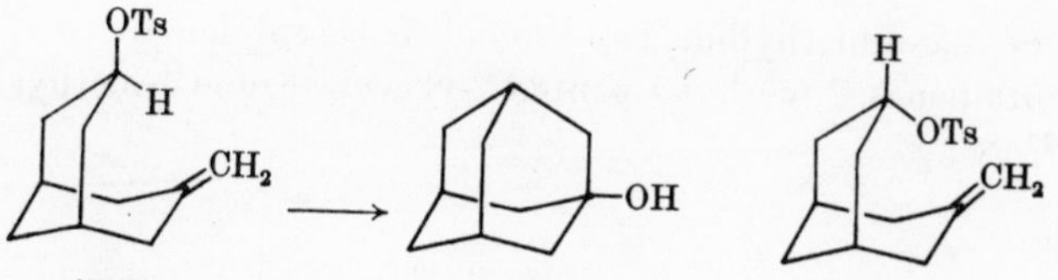

(186)   (187)   (188)

Acetolysis of 4-cyclooctenylmethyl *p*-bromobenzenesulphonate (192) yields 75% of *endo*bicyclo[3.3.1]nonan-1-yl acetate (193), 14% of bicyclo-[3.3.1]non-2-ene (194), 3% *exo*-bicyclo[3.3.1]nonan-2-yl acetate (195), and

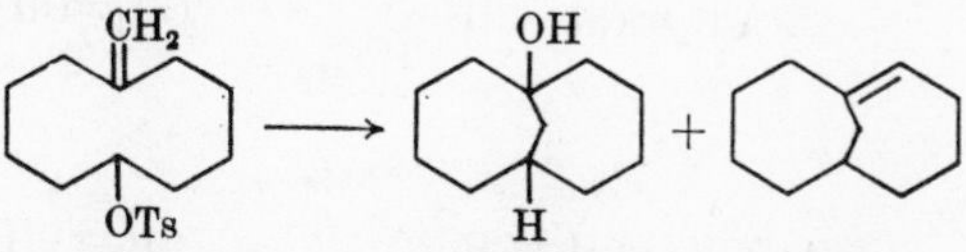

(189)   (190)   (191)

7% of a compound tentatively identified as bicyclo[4.2.1]nonan-2-yl acetate (196) of unspecified stereochemistry.[110] On the usual assumption, then, that non-classical ions cannot react with nucleophiles on the same side as the

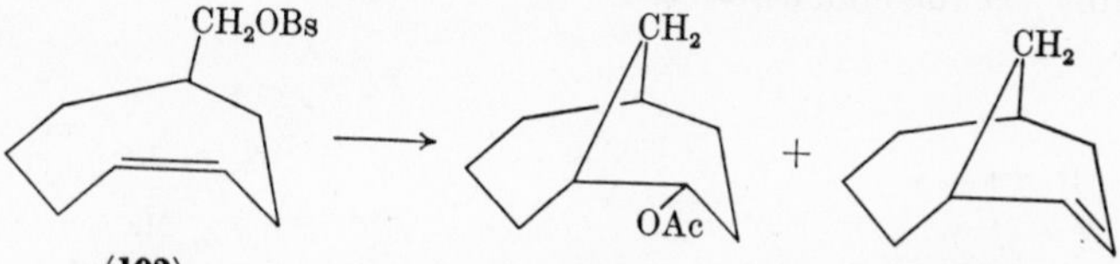

(192)   (193)   (194)

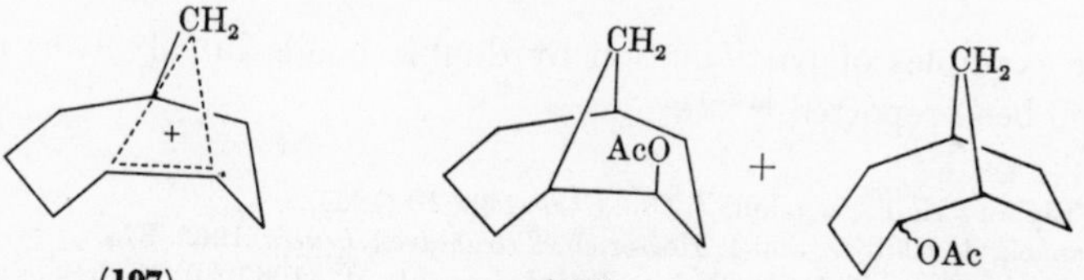

(197)   (195)   (196)

bridge, formation of (195) indicates that (197) cannot be the sole intermediate.

Homoallylic participation occurs in the acetolysis of 2,2-dimethyl-3-butenyl *p*-bromobenzenesulphonate (198), which is 60 times as fast at 100°

---

[110] A. C. Cope, D. L. Nealy, P. Scheiner, and G. Wood, *J. Am. Chem. Soc.*, **87**, 3130 (1965); see also M. Hanack and W. Kaiser, *Angew. Chem. Internt. Ed. Engl.*, **3**, 583 (1964); W. F. Erman and H. C. Kretschmar, *Tetrahedron Letters*, **1965**, 1717.

2*

as that of 2,2-dimethylbutyl *p*-bromobenzenesulphonate and yields 2-methylpentadienes, 2-methyl-4-penten-2-yl acetate, and 4-methyl-3-pentenyl acetate.[111]

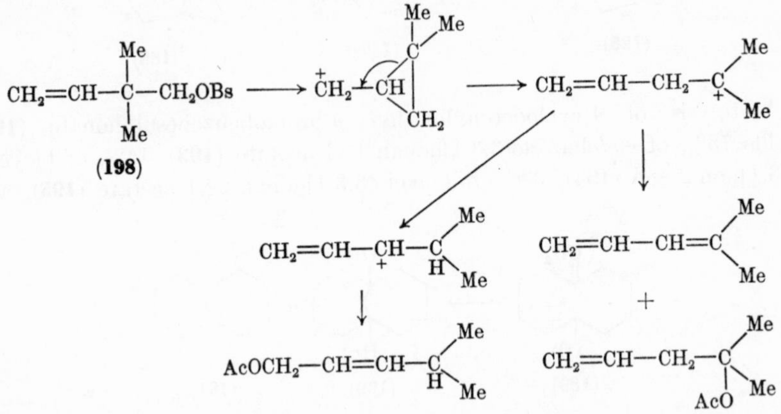

$$CH_2=CH-\underset{\underset{Me}{|}}{\overset{\overset{Me}{|}}{C}}-CH_2OBs \longrightarrow \overset{+}{C}H_2-CH \longrightarrow CH_2=CH-CH_2-\overset{+}{C}$$

(198)

Participation by a 2,3-triple bond has been reported to occur in the formolysis of 3-pentynyl toluene-*p*-sulphonate and 2-naphthalenesulphonate (199), which yields about 20% of 2-methylcyclobutanone (201) formed presumably via the enol ester (200).[112]

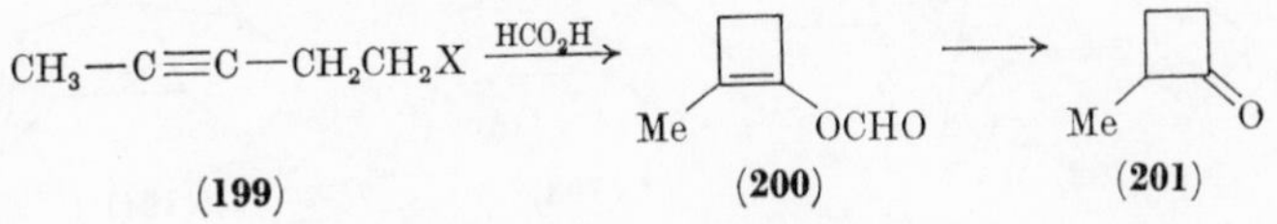

$$CH_3-C\equiv C-CH_2CH_2X \xrightarrow{HCO_2H}$$

(199)   (200)   (201)

Other examples of participation by double bonds in solvolysis reactions have also been reported.[113–118]

[111] R. S. Bly and R. T. Swindell, *J. Org. Chem.*, **30**, 10 (1965).

[112] M. Hanack, J. Häffner, and I. Herterich, *Tetrahedron Letters*, **1965**, 875.

[113] R. M. Moriarty and T. D. J. D'Silva, *Tetrahedron*, **21**, 547 (1965); R. M. deSousa and R. M. Moriarty, *J. Org. Chem.*, **30**, 1509 (1965).

[114] D. N. Gupta, G. Schilling, and G. Just, *Can. J. Chem.*, **43**, 792 (1965); J. B. Jones and D. C. Wigfield, *Tetrahedron Letters*, **1965**, 4103.

[115] J. Tadanier, *Experientia*, **21**, 563 (1965).

[116] H. Tanida and Y. Hata, *J. Org. Chem.*, **30**, 977 (1965).

[117] W. S. Johnson and J. K. Crandall, *J. Org. Chem.*, **30**, 1785 (1965).

[118] W. Rittersdorf, *Angew. Chem. Intern. Ed. Engl.*, **4**, 444 (1965).

[119] Deno (ref. 1) terms ions of the type ▷— + cyclopropyl carbonium ions and those of the type ▷ + cyclopropyl cations.

[120] See also the discussion on p. 31.

Cyclopropyl Carbonium Ions[119, 120]

The direct observation of a series of cyclopropyl carbonium ions (202) and (203) by NMR spectroscopy has been reported.[121, 122] These ions were

(202)   (203)   (204)

prepared by dissolving the corresponding alcohols in FSO_3H–SO_2–SbF_5 at $-50°$ to $-78°$. The methyl groups of the dimethylcyclopropyl carbonium ion are not equivalent since their signals are separated by 0.54 ppm. This was taken to indicate that the cyclopropyl ring lies in a plane which is perpendicular to the plane of the $\overset{+}{-C}Me_2$ system (204). If this is so, one methyl group lies *cis*- and the other *trans*- to the cyclopropyl ring. The *cis*-methyl group will experience the diamagnetic anisotropy of the cyclopropyl ring and show resonance at a higher field than the *trans*-methyl group. In this conformation there is presumably maximum overlap between the "banana" bonds of the cyclopyl ring and the vacant p orbital. A theoretical discussion of this conjugation has been given.[123]

Deno and his co-workers[124] consider that this result would not be inconsistent with a bicyclobutonium structure (see p. 6) for the cyclopropyl-methyl cation if it were assumed that rapid oscillation occurs between two such structures, with a structure such as (204) as the symmetrical mid-point.

The ability of the cyclopropyl group to stabilize carbonium ions is striking and, as measured by the equilibrium constant for the ion–alcohol equilibrium in sulphuric acid, the tricyclopropyl carbonium ion (pK -2.34) is more stable than the triphenylcarbonium ion (pK -6.63).[124]

Evidence that phenylcyclopropane preferentially takes up a conformation similar to that of the cyclopropyl carbonium ion has been reported by Closs and Klinger.[125] The NMR spectrum of *p*-deuteriophenylcyclopropane shows an A_2B_2 pattern for the protons of the aryl ring. The signals of the *ortho*-protons are highly temperature-dependent and move to abnormally high field with decreasing temperature. This was interpreted in terms of an equilibrium between a low-energy (205) and high-energy (206) conformation.

[121] C. U. Pittman and G. A. Olah, *J. Am. Chem. Soc.*, 87, 2999, 5123 (1965).
[122] N. C. Deno, J. S. Liu, J. O. Turner, D. N. Lincoln, and R. E. Fruit, *J. Am. Chem. Soc.*, 87, 3001 (1965).
[123] R. Hoffmann, *Tetrahedron Letters*, 1965, 3819; *J. Chem. Phys.*, 40, 2480 (1964).
[124] N. C. Deno, H. G. Richey, J. S. Liu, D. N. Lincoln, and J. O. Turner, *J. Am. Chem. Soc.*, 87, 4533 (1965).
[125] G. L. Closs and H. B. Klinger, *J. Am. Chem. Soc.*, 87, 3265 (1965).

In the low-energy conformation (205) one of the *ortho*-protons will experience the anisotropic deshielding of the cyclopropane ring. The observed chemical shifts are the averages for both *ortho*- and both *meta*-protons, because even below $-100°$ the A_2B_2 pattern is preserved. It was estimated that the energy separating the conformations was about 1.4 kcal mole^{-1}.

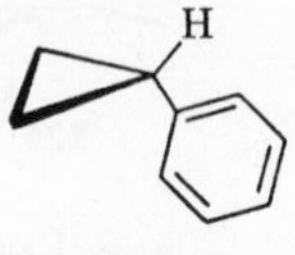

(205) 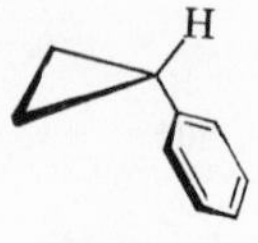   (206)

The solvolyses of some 1-substituted cyclopropylmethyl toluene-*p*-sulphonates have been investigated.[126]

Cyclopropyl Cations[119]

Reactions which involve the cyclopropyl cation (207) frequently result in ring opening through cleavage of the bond opposite the cationic centre, to give an allyl cation (208). This could occur by conrotatory or disrotatory

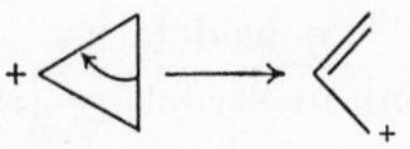

(207)   (208)

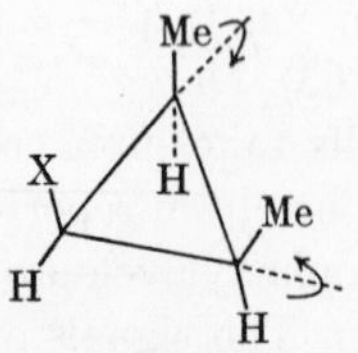

(209) 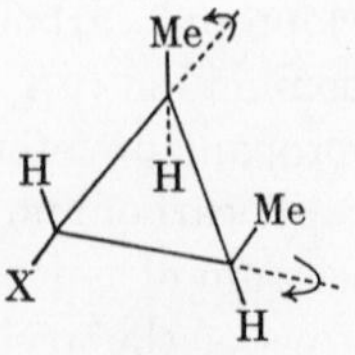   (210) 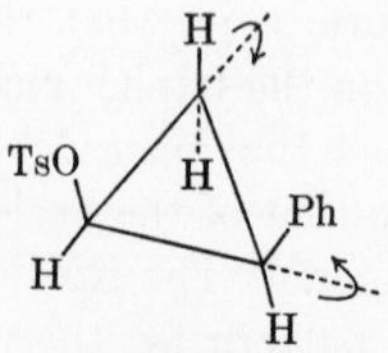   (211)

processes (see p. 127), and calculations by Woodward and Hoffmann[127] indicate that when the cyclopropyl cation is produced by ionization of a group X with concerted electrocyclic transformation to an allyl cation the stereoelectronically favoured processes are the disrotatory ones, (209) and (210). Experimental support for this has come from the work of DePuy and

[126] D. D. Roberts, *J. Org. Chem.*, **30**, 23 (1965).
[127] R. B. Woodward and R. Hoffmann, *J. Am. Chem. Soc.*, **87**, 395 (1965).

his co-workers who have shown that the solvolysis of *trans*-2-phenylcyclo-propyl toluene-*p*-sulphonate occurs 15 times faster than that of its *cis*-isomer.[128] This difference was ascribed to the fact that for the latter the phenyl group has to be rotated inwards (211) and this leads to a less stable, sterically hindered cation. A similar explanation was offered for the observation that, whereas (212) undergoes acetolysis at 124.6° with $k = 1.4 \times 10^{-6}$ sec^{-1}, (213) was recovered unchanged after 692 hours ($k < 8 \times 10^{-9}$ sec^{-1}).[129a]

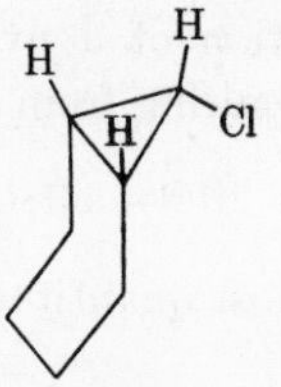

(212) 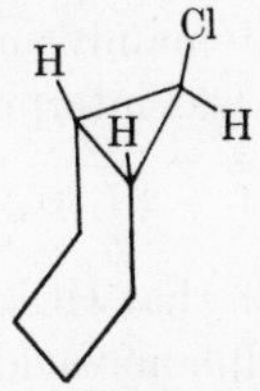   (213)

Solvolytic ring-opening of 1,1-dibromobicyclo[3.1.0]hexane has also been investigated.[129b] So has acetolysis of a series of 1-arylcyclopropyl toluene-*p*-sulphonates; the 1-phenyl compound reacts at 108° as shown in Scheme 2.[128] The ρ values for the processes corresponding to k_1 and k_2 are -4.4 and -4.0, respectively.

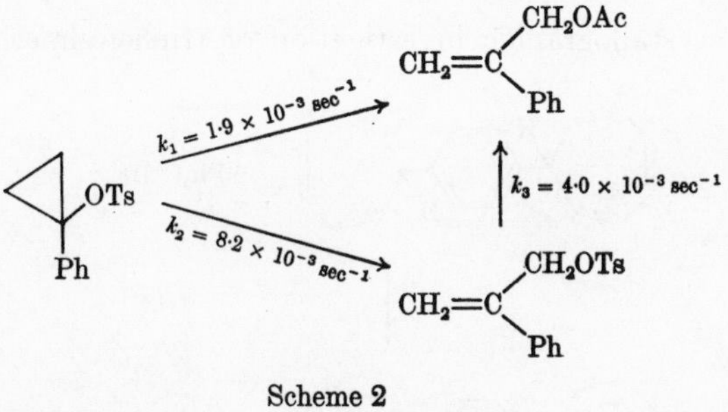

Scheme 2

Other Stable Carbonium Ions and Their Reactions

The claim by Hart and Fish[130] that trichloromethylmesitylene, pentamethyl-trichloromethylbenzene, and trichloromethylprehnitene ionize in sulphuric

---

[128] C. H. DePuy, L. G. Schnack, J. W. Hausser, and W. Wiedemann, *J. Am. Chem. Soc.*, **87** 4006 (1965).

[129a] S. J. Cristol, R. M. Sequeira, and C. H. DePuy, *J. Am. Chem. Soc.*, **87**, 4007 (1965).

[129b] L. Gatlin, R. E. Glick, and P. S. Skell, *Tetrahedron*, **21**, 1315 (1965).

[130] H. Hart and R. W. Fish, *J. Am. Chem. Soc.*, **80**, 5894 (1958); **82**, 5419 (1960); **83**, 4460 (1961).

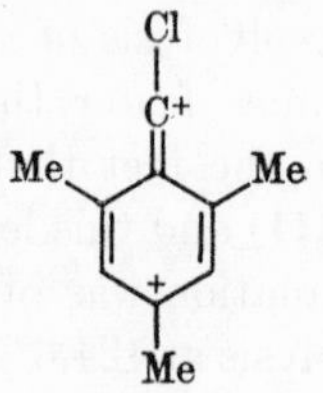

(214) 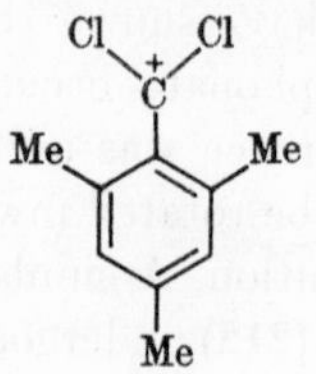   (215)

acid to give dicarbonium ions (as **214**) has been shown to be incorrect.[131-135] The claim was based mainly on observation of a five-fold depression of freezing point, which was interpreted as resulting from an ionization:

$$RCCl_3 + 2H_2SO_4 \rightarrow RCCl^{2+} + 2HCl + 2HSO_4^-$$

It has now been shown that HCl reacts almost quantitatively with sulphuric acid to form chlorosulphonic acid:

$$HCl + 2H_2SO_4 = HClSO_3 + H_3O^+ + HSO_4^-$$

It is thought that the ionic species formed is, in fact, the dichloromonocarbonium ion (**215**), and this was confirmed by its NMR spectrum. This ion has also been prepared by treating trichloromethylmesitylene with antimony pentachloride in carbon tetrachloride.[136a]

The ditropylium ion (**216**) has been prepared by the reaction shown in Scheme 3.[136b]

An X-ray crystallographic investigation of triphenylmethyl perchlorate

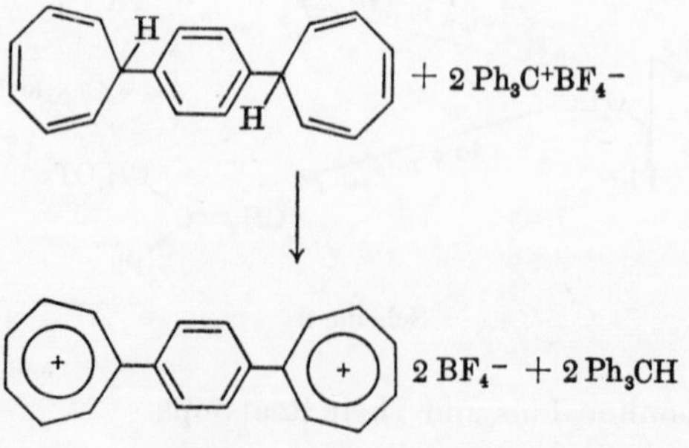

Scheme 3

(216)

[131] R. J. Gillespie and E. A. Robinson, *J. Am. Chem. Soc.*, **86**, 5676 (1964).
[132] N. C. Deno, N. Friedman, and J. Mockus, *J. Am. Chem. Soc.*, **86**, 5676 (1964).
[133] E. A. Robinson and J. A. Ciruna, *J. Am. Chem. Soc.*, **86**, 5677 (1964).
[134] R. J. Gillespie and E. A. Robinson, *J. Am. Chem. Soc.*, **87**, 2428 (1965).
[135] J. A. Ciruna, K. Ng, E. A. Robinson, S. A. A. Zaidi, R. J. Gillespie, and J. S. Hartman, *Tetrahedron Letters*, **1965**, 1101.
[136a] H. Volz and M. J. Volz de Lecca, *Tetrahedron Letters*, **1965**, 3413.
[136b] R. W. Murray and M. L. Kaplan, *Tetrahedron Letters*, **1965**, 2903.

has shown the triphenylmethyl cation to have a propellor shape with adjacent phenyl rings at angles of 54°.[137]

The proton-decoupled ^{19}F NMR spectra of the 3,3′-difluoro- and 3,3′,3″-trifluoro-triphenylmethyl cations in liquid HF show, respectively, four and two lines of equal intensity from $-60°$ to $-80°$. The spectra of the 3-fluoro-, 4-fluoro-, 4,4′-difluoro-, and 4,4′,4″-trifluoro-triphenylmethyl cations, however, show only one line down to $-80°$.[138] These results are consistent with propellor conformations, and for the 3,3′-difluoro- and 3,3′,3″-trifluoro-ions those in the annexed formulae are possible:

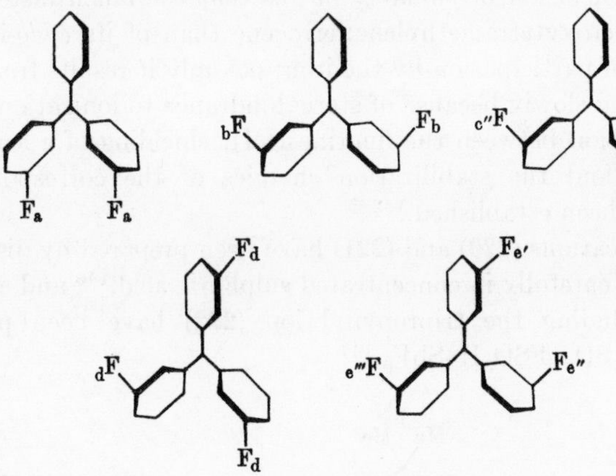

Four different environments are, therefore, possible for the fluorine atoms in the two ions, the presence of only two lines in the 3,3′,3″-trifluoro-ion being ascribed to the accidental coincidence of lines from fluorine atoms a and e‴ and from e′ and e″. It was calculated from the variation of line position with temperature that the energy of activation for interconversion of the different propellor conformations was 9.1 kcal.

Salts of the diphenylferrocenyl carbonium ion have been reported.[139] It has been suggested[140] that the powerful stabilizing effect of the α-ferrocenyl group on carbonium ions is due to a resonance effect, (217)↔(218), rather than to participation of the non-bonding electrons of the iron, as (219). The basis for this is that methoxyl and ferrocenyl groups, which have similar rate-enhancing effects in solvolyses when attached to the α-carbon of alkyl

[137] A. H. Gomes de Mesquita, C. H. MacGillavry, and K. Eriks, *Acta Cryst.*, 18, 437 (1965); see also P. Andersen and B. Klewe, *Acta Chem. Scand.*, 19, 791 (1965).
[138] A. K. Colter, I. I. Schuster, and R. J. Kurland, *J. Am. Chem. Soc.*, 87, 2278, 2279 (1965).
[139] A. N. Nesmeyanov, V. A. Sazonova, V. N. Drozd, and N. A. Rodionova, *Dokl. Akad. Nauk SSSR*, 160, 355 (1965).
[140] J. C. Ware and T. G. Traylor, *Tetrahedron Letters*, 1965, 1295.

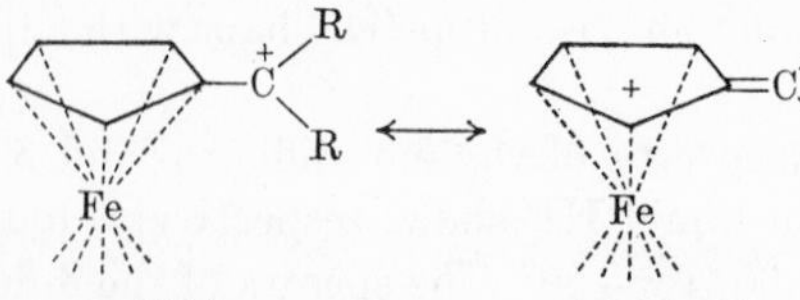

(217)   (218) 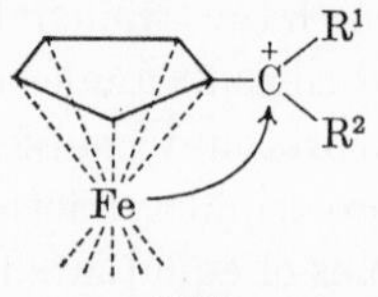   (219)

chlorides, have also similar electron-releasing resonance effects as measured by their σ^+_{para} constants. The rate of solvolysis of 1-ferrocenylethyl chloride can also be well correlated with the stretching frequency of the corresponding ketone. This conclusion presumably means that the much faster solvolysis of 1,2-(1-*exo*-acetoxytetramethylene)ferrocene than of its *endo*-isomer[141] is not the result of participation by the iron; possibly it results from the *endo*-isomer's reacting slowly because of steric hindrance to ionization.

A linear relation between the fluorine NMR shielding of a series of ions, $FC_6H_4–C^+Ar_2$ and the stabilization energies of the corresponding ions, $Ph–C^+Ar_2$ has been established.[142]

The alkynyl cations (220) and (221) have been prepared by dissolving the parent alcohols carefully in concentrated sulphuric acid,[143] and a number of other ions including the tripropynyl ion (222) have been prepared in $SO_2–FSO_3H$ or $SO_2–FSO_3H–SbF_5$.[144]

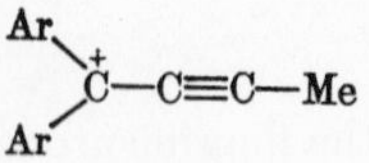

(220)   (221)   (222)

Ar = p-MeO . C_6H_4—

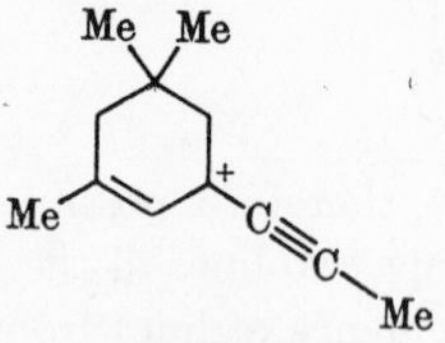

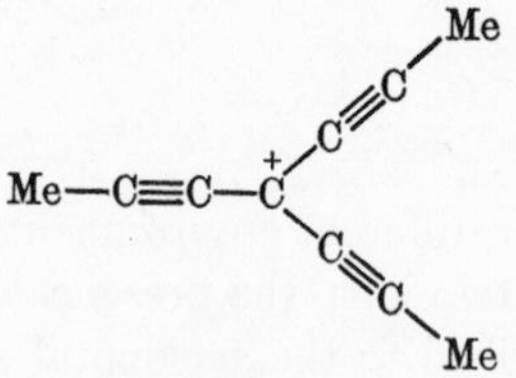

The preparation and reactions of a series of dienylic, trienylic and polyenylic cations have been reported.[145,146]

Electron-transfer reactions of the tropylium cation have been investigated.[147,148]

[141] E. A. Hill and J. H. Richards, *J. Am. Chem. Soc.*, **83**, 4216 (1961).

[142] R. W. Taft and L. D. McKeever, *J. Am. Chem. Soc.*, **87**, 2489 (1965).

[143] H. G. Richey, J. C. Philips and L. E. Rennick, *J. Am. Chem. Soc.*, **87**, 1381 (1965).

[144] H. G. Richey, L. E. Rennick, A. S. Kushner, J. M. Richey, and J. C. Philips, *J. Am. Chem. Soc.*, **87**, 4017 (1965).

[145] T. S. Sorensen, *Can. J. Chem.*, **43**, 2744, 2768 (1965); *J. Am. Chem. Soc.*, **87**, 5075 (1965).

[146] N. C. Deno, C. U. Pittman, and J. O. Turner, *J. Am. Chem. Soc.*, **87**, 2153 (1965); N. C. Deno and E. Sacher, *ibid.*, **87**, 5120 (1965).

[147] C. E. H. Bawn, C. Fitzsimmons, and A. Ledwith, *Proc. Chem. Soc.*, **1964**, 391.

[148] A. Ledwith and M. Sambhi, *Chem. Comm.*, **1965**, 64.

Similarities in the NMR spectra of acyl cations and *N*-nitrosoamines have been reported.[149]

Other stable carbonium ions which have been investigated include the allyl and 2-methylallyl cations,[150] arylalkyl cations,[151,152] the methylbenzenium ion,[153] the heptamethylbenzenium ion,[154] dioxenium ions,[155,156,157] the triphenylmethyl cations[158] the tropylium ion,[159,160] the hydroxytropylium ion,[161] the pentachlorocyclopentadienyl cation,[162] the 1,3-dithiolium cation,[163] the thienotropylium ion,[164] the 2,2,4-trimethylpentyl cation,[165] substituted cyclopropenium ions,[166] the bicyclo[5.4.1]dodecapentaenylium cation,[167] the benzyl cation,[168] and carbonium ions derived from aromatic olefins.[169] Cyclopropenium ions have been reviewed.[170]

[149] J. G. Traynham and M. T. Yang, *Tetrahedron Letters*, **1965**, 575.
[150] G. A. Olah, M. B. Comisarow, C. A. Cupas, and C. U. Pittman, *J. Am. Chem. Soc.*, **87**, 2997 (1965).
[151] G. A. Olah and M. B. Comisarow, *J. Am. Chem. Soc.*, **86**, 5682 (1964).
[152] H. Volz and H. W. Schnell, *Angew. Chem. Intern. Ed. Engl.*, **4**, 873 (1965).
[153] G. A. Olah, *J. Am. Chem. Soc.*, **87**, 1103 (1965).
[154] V. A. Koptyug, V. G. Shubina, and A. I. Resvukhin, *Izv. Akad. Nauk SSSR, Ser. Khim.*, **1965**, 201.
[155] C. F. Wilcox and D. L. Nealy, *J. Org. Chem.*, **29**, 3668 (1964).
[156] J. A. Magnuson, C. A. Hirt, and P. J. Launer, *Chem. Ind.* (*London*), **1965**, 691.
[157] G. Schneider and O. K. J. Kovacs, *Chem. Comm.*, **1965**, 202.
[158] K. M. Harmon and F. E. Cummings, *J. Am. Chem. Soc.*, 1965, **87**, 539.
[159] K. M. Harmon, A. B. Harmon, and F. E. Cummings, *J. Am. Chem. Soc.*, **86**, 5511 (1964).
[160] K. M. Harmon, S. D. Alderman, K. E. Benker, D. J. Diestler, and P. A. Gebauer, *J. Am. Chem. Soc.*, **87**, 1700 (1965).
[161] K. M. Harmon and T. T. Coburn, *J. Am. Chem. Soc.*, **87**, 2499 (1965).
[162] R. Breslow, R. Hill, and E. Wasserman, *J. Am. Chem. Soc.*, **86**, 1964, 5349 (1964).
[163] E. Klingsberg, *J. Am. Chem. Soc.*, **86**, 5290 (1964).
[164] R. G. Turnbo, D. L. Sullivan, and R. Pettit, *J. Am. Chem. Soc.*, **86**, 5630 (1964).
[165] J. E. Hofmann, *J. Org. Chem.*, **29**, 3627 (1964).
[166] B. Fohlisch and P. Burgle, *Tetrahedron Letters*, **1965**, 2661.
[167] W. Grimme, H. Hoffmann, and E. Vogel, *Angew. Chem. Intern. Ed. Engl.*, **4**, 354 (1965).
[168] I. Hanazaki and S. Nagakura, *Tetrahedron*, **21**, 2441 (1965).
[169] A. Gandini and P. H. Plesch, *J. Chem. Soc.*, **1965**, 4765, 6019.
[170] A. W. Krebs, *Angew. Chem. Intern. Ed. Engl.*, **4**, 10 (1965).

CHAPTER 2

Nucleophilic Aliphatic Substitution

Ion-pair Return and Related Phenomena

An extensive summarizing paper on their work on ion-pair return has been published by Winstein and his co-workers.[1]

Another significant addition has been made this year by Goering and his co-workers to their penetrating study of the stereochemistry of ion-pair return.[2,3] In solvolytic media certain arylalkyl p-nitrobenzoates may undergo three processes involving alkyl–oxygen fission. These are solvolysis (1), randomization of the ester–oxygen atoms (2), and racemization (3):

$$\text{ROOCC}_6\text{H}_6\text{NO}_2 \xrightarrow{k_t} \text{ROH} + \text{HOOCC}_6\text{H}_4\text{NO}_1 \tag{1}$$

$$\underset{{}^{18}\text{O}}{\text{ROOCC}_6\text{H}_4\text{NO}_2} \xrightarrow{k_{eq}} \underset{{}^{18}\text{O}}{\text{R}^{18}\text{O}-\text{CC}_6\text{H}_4\text{NO}_2} \tag{2}$$

$$(+ \text{ or } -)\text{—ROOCC}_6\text{H}_4\text{NO}_2 \xrightarrow{k_{rac}} (\pm)\text{—ROOCC}_6\text{H}_4\text{NO}_2 \tag{3}$$

The kinetics of these transformations are of the first order and the last two are intramolecular. It is thought that k_{eq} is the rate constant for total ionization and k_{eq}/k_t is the ratio of ion-pair return to solvolysis. Racemization associated with return is measured by k_{rac}, and the rest of the return, $k_{eq} - k_{rac}$, proceeds with preservation of configuration.

Table 1. Kinetic data for the reactions of alkyl p-nitrobenzoates ($\text{ROOCC}_6\text{H}_4\text{NO}_2\text{-}p$) in 90% aqueous acetone.

R	Temp.	k_{eq}/k_t	k_{rac}/k_{eq}
p-ClC$_6$H$_4$CHPh—	100°	2.5	0.38
p-MeC$_6$H$_4$CHPh—	80°	3.2	0.37
p-MeOC$_6$H$_4$CHMe—	80°	2.2	0.53
PhCMeEt—	78.6°	0.60	0.079

[1] S. Winstein, B. Appel, R. Baker, and A. Diaz, "Organic Reaction Mechanisms", *Chem, Soc. Special Publ. No.* 19, p. 109.

[2] H. L. Goering and S. Chang, *Tetrahedron Letters*, **1965**, 3607.

[3] For earlier work see H. L. Goering, R. G. Briody, and J. F. Levy, *J. Am. Chem. Soc.*, **85**, 3059 (1963); H. L. Goering and J. F. Levy, *ibid.*, **86**, 120 (1964).

With the three secondary compounds listed in Table 1, ion-pair return is associated with substantial racemization (k_{rac}/k_{eq} = 0.37—0.53), but with the fourth, tertiary compound the amount of racemization is substantially less (k_{rac}/k_{eq} = 0.079). It was suggested that this was the result of the greater distance separating the anion from the opposite side of the cation (see **1 → 2**; PNB = p-nitrobenzoyl). The solvolysis proceeds with 31% substitution and 69% elimination, and the product of substitution is formed

(1)   (2)

with 38% excess retention of configuration. This result is of considerable significance since it indicates that overall retention of configuration may result in a reaction which does not involve neighbouring-group participation.

Fava and his co-workers have continued their investigations on the mechanism of the isomerization of benzhydryl thiocyanates to isothiocyanates.[4] This year they have reported a detailed investigation of the reaction of 4,4'-dimethylbenzhydryl thiocyanate in acetonitrile. They also studied the rate of exchange with ^{35}S-labelled sodium thiocyanate, and it was concluded that this first-order process involved dissociation of the benzhydryl thiocyanate to free ions. Isomerization is faster than exchange, so it was concluded that the former proceeds via an ion pair; since "normal" salt effects were observed this was thought to be an intimate ion pair. This was shown to collapse to thiocyanate and isothiocyanate in the ratio 5:1 by measuring the relative initial rates of uptake of radioactivity by the benzhydryl thiocyanate and isothiocyanate in the presence of ^{35}S labelled sodium thiocyanate. It was thus concluded that for 100 intimate ion pairs formed, about 5 undergo further ionization and 95 return to the covalent state. Of the latter, about 79 return to thiocyanate and the rest to isothiocyanate.

The stereochemical course of this isomerization has been investigated for optically active 4-chlorobenzhydryl thiocyanate, again in acetonitrile.[5] Racemization occurs at a rate comparable with that of isomerization. The isothiocyanate is formed with a slight stereospecificity and it was estimated "that of one hundred isothiocyanate molecules formed, 48 lose and 52 retain the configuration of the parent thiocyanate." By carrying out the reaction in the presence of labelled sodium thiocyanate, it was estimated that less

[4] A. Fava, A. Iliceto, A. Ceccon, and P. Koch, *J. Am. Chem. Soc.*, **87**, 1045 (1965).
[5] A. Fava, U. Tonellato, and L. Congui, *Tetrahedron Letters*, **1965**, 1657.

than 4% of the racemization and less than 2% of the isomerization of the thiocyanate occurred by way of free ions.

The acetolysis of 1-methylheptyl toluene-p-sulphonate yields mainly the acetate of inverted configuration.[6] The racemized product was shown to come from the reaction of 1-methylheptyl acetate with the toluene-p-sulphonic acid also formed, from the addition of acetic acid to octene, and from racemization of starting toluene-p-sulphonate; the acetolysis reaction thus proceeds with complete inversion. The degree of racemization of the product was increased in the presence of lithium toluene-p-sulphonate and this was shown to be due to increased racemization of the starting material.[7] The acetolysis of 1-methylheptyl p-nitrobenzenesulphonate in the presence of lithium toluene-p-sulphonate was accompanied by the formation of the toluene-p-sulphonate ester, this reaction being visualized as an ion-pair interchange:

$$\text{RONs} \rightleftharpoons \text{R}^+ \ \bar{\text{O}}\text{Ns} \rightleftharpoons \text{Li}^+ \ \bar{\text{O}}\text{Ts} \ \text{R}^+ \ \bar{\text{O}}\text{Ns}$$
$$\text{ROTs} \rightleftharpoons \text{R}^+ \ \bar{\text{O}}\text{Ts} \rightleftharpoons \bar{\text{O}}\text{Ts} \ \text{R}^+ \ \bar{\text{O}}\text{Ns} \ \text{Li}^+$$

$$(\text{Ts} = p\text{-Me} \cdot \text{C}_6\text{H}_4 \cdot \text{SO}_2; \ \text{Ns} = p\text{-NO}_2 \cdot \text{C}_6\text{H}_4 \cdot \text{SO}_2)$$

and the racemization of the toluene-p-sulphonate was thought to involve a similar mechanism.

Addition of lithium perchlorate in the acetolysis of 1-methylheptyl toluene-p-sulphonate also increased the racemization of the product, owing to increase in the proportion of olefin formed and in the racemization of starting material. This effect was also thought to involve ion-pairs:

$$\text{R}^+ \ ^-\text{OTs} \ \underset{}{\overset{\text{LiClO}_4}{\rightleftharpoons}} \ \text{R}^+ \ \text{ClO}_4^- \longrightarrow \text{Products}$$

The perchlorate ion pair must, therefore, produce a higher proportion of olefin than the toluene-p-sulphonate ion pair. It was thought that the former might tend to be more solvent-separated, so that the carbonium ion has to distribute a greater positive charge internally, and that this might lead to an increased proportion of olefin.[7]

Optically active 1-methylheptyl p-bromobenzenesulphonate yields inverted alcohol of 76.8% optical purity on solvolysis in 75% aqueous dioxan.[8] In the presence of sodium azide, however, 100% optically pure, inverted alcohol is obtained. This suggests that the solvolysis proceeds partly by way of an S_N2 displacement, to yield inverted alcohol, and partly by way of an intermediate, that can be captured by azide ion, to yield alcohol which is

[6] A. Streitwieser, T. D. Walsh, and J. R. Wolfe, *J. Am. Chem. Soc.*, **87**, 3682 (1965).
[7] A. Streitwieser and T. D. Walsh, *J. Am. Chem. Soc.*, **87**, 3686 (1965).
[8] H. Weiner and R. A. Sneen, *J. Am. Chem. Soc.*, **87**, 287 (1965).

racemized or is of the same configuration as starting material. Since the amount of racemization increases with increasing dioxan content of the solvent and since 1-methylheptyl methanesulphonate, which behaves similarly, yields 100% inverted alcohol in pure water, this intermediate was identified as the oxonium ion (3). Acetone is also thought to act as a nucleo-

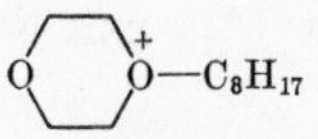

$$
\text{Me}_2\text{C}=\overset{+}{\text{O}}\overset{\curvearrowright}{}\text{C}_8\text{H}_{17}\overset{\curvearrowleft}{}\text{OBs} \longrightarrow \text{Me}_2\text{C}=\overset{+}{\text{O}}-\text{C}_8\text{H}_{17} \longrightarrow \text{Me}_2\text{C}\overset{\text{OC}_8\text{H}_{17}}{\underset{\text{OMe}}{\diagdown}}
$$

(3)

(4a)

$$\Big\downarrow \text{H}^+$$

$$\longleftarrow \quad \text{Me}_2\overset{+}{\text{C}}=\overset{}{\text{OMe}} + \text{C}_8\text{H}_{17}\text{OH}$$

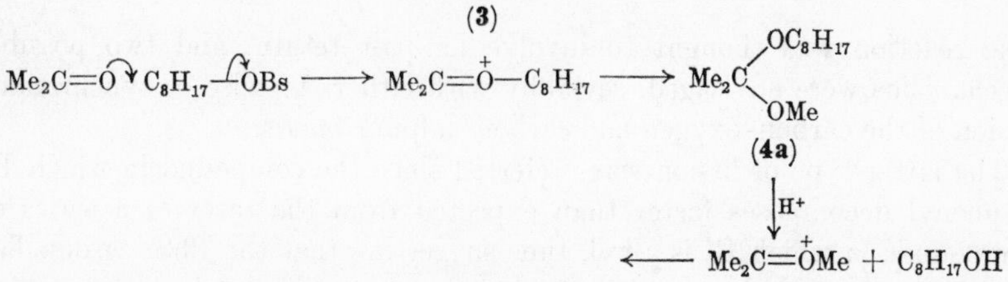

Scheme 1

phile in the solvolysis of 1-methylheptyl *p*-bromobenzenesulphonate in 80% methanolic acetone.[8] This would account for the formation of octan-2-ol in this reaction (see Scheme 1), and evidence was provided for the presence of the mixed ketal (4a) in reaction mixtures which were buffered with 2,6-lutidine. *s*-Butyl chloride does not, however, participate in the solvolysis of isopropyl *p*-nitrobenzenesulphonate in mixtures of acetic acid and butyl chloride.[9]

In 25% aqueous dioxan, the addition of sodium azide (0.0462M) does not affect the rate of solvolysis of 1-methylheptyl *p*-bromobenzenesulphonate (0.018M), but diverts 31% of the product to the azide ester of inverted configuration.[10] This is strong evidence for an ion-pair mechanism and it was suggested that a similar mechanism may also occur in 75% aqueous dioxan, for which the addition of sodium azide diverts the product to 1-methyl-heptyl azide and, in addition, increases the reaction rate. The latter observation could be accommodated if the reaction involved a rate-determining attack by azide ion on a reversibly formed ion pair. Indeed it was suggested that other reactions, that have been allocated an S_N2 mechanism from their kinetic behavior, may also proceed by this mechanism.

A discrepancy in the ratio of the rate constants for the trapping of the triphenylmethyl cation in aqueous acetone by azide ion and water has been noted.[11] The system has been re-investigated and results in approximate

[9] D. E. Appelquist and G. W. Burton, *J. Org. Chem.*, **30**, 2460 (1965).
[10] H. Weiner and R. A. Sneen, *J. Am. Chem. Soc.*, **87**, 292 (1965).
[11] E. A. Hill, *Chem. Ind.* (*London*), **1965**, 1696.

agreement with those of Swain and his co-workers,[12] but at variance with those of Golomb[13] and of Pocker,[14] have been obtained. This was attributed to inefficient mixing of the reactant solutions and possibly to prior hydrolysis of the triphenylmethyl chloride in some of the work.

Arylalkyl thiocarbonates undergo an S_Ni reaction when heated in inert solvents, yielding the organic sulphide:[15]

$$RO—CSR' \longrightarrow RSR' + CO_2$$
$$\underset{O}{\overset{\|}{}}$$

The reaction was thought to involve ion-pair return, and two possible mechanisms were envisaged, severally with either stepwise or synchronous fission of the carbon–oxygen and carbon–sulphur bonds.

The latter type of fission was preferred since the compound in which R' is phenyl decomposes faster than expected from the rates of a series of compounds in which R' is alkyl, thus suggesting that the PhS$^-$ group has started to ionize in the transition state.

In contrast to hydrolysis of a series of straight-chain alkyl chlorosulphates, $ROSO_2Cl$, which is thought to proceed by a bimolecular mechanism, the hydrolysis of neopentyl chlorosulphate is thought to proceed via a carbonium ion since it yields t-pentylalcohol.[16] A change in mechanism is also indicated by a change in the entropy of activation from ~ -20 e.u. to $+0.1$ e.u. It is not thought, however, that the reaction of the neopentyl compound is a simple S_N1 reaction with $-SO_3Cl$ as leaving group since the value of 0.1 e.u. for the entropy of activation is 10 to 15 e.u. more positive than is normally associated with such a process. Instead it was thought that there is here, also, a synchronous fission of the C–O and S–Cl bonds leading to a transition state with greater charge separation, stronger solvation, and a more positive value of $\Delta S^{\ddagger}$.

The steric course of the thermal decomposition of thiosulphonates:

$$RSO_2SR' \longrightarrow RSR' + SO_2$$

has been investigated with optically active S-phenyl [α-^{2}H]toluene-α-thio-sulphonate, $PhCHD \cdot SO_2SPh$, and shown to involve almost complete racemization.[17]

[12] C. G. Swain, C. B. Scott, and K. H. Lohmann, *J. Am. Chem. Soc.*, **75**, 136 (1953).

[13] D. Golomb, *J. Chem. Soc.*, **1959**, 1334.

[14] Y. Pocker, *Chem. Ind.* (*London*), **1957**, 1599.

[15] J. L. Kice, R. A. Bartsch, M. A. Dankleff, and S. L. Schwartz, *J. Am. Chem. Soc.*, **87**, 1734 (1965).

[16] E. Buncel and J. P. Millington, *Proc. Chem. Soc.*, **1964**, 406; *Can. J. Chem.*, **43**, 547, 556 (1965).

[17] J. L. Kice, R. H. Engebrecht, and N. E. Pawlowski, *J. Am. Chem. Soc.*, **87**, 4131 (1965).

Neighbouring-group Participation[18]

A detailed investigation of neighbouring-group participation by the carbonyl group has been reported by Pasto and Serve.[19] The relative rates of solvolyses of a series of chloro ketones, $Cl \cdot [CH_2]_{n-2} \cdot COPh$, promoted by silver ion, in 80% aqueous ethanol were: 1.3, $n = 3$; 7.9, $n = 4$; 759, $n = 5$; 21.3, $n = 6$; and 2.7, $n = 7$; where the rate for n-butyl chloride was taken as 1.0. The reactions of the compounds with $n = 5$ or 6 clearly involve participation which was written as follows:

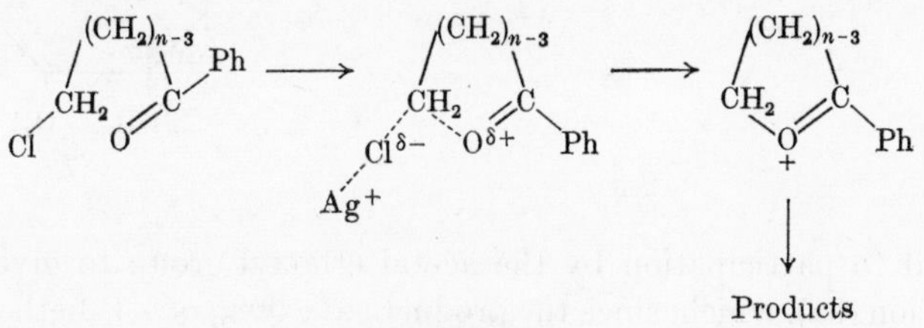

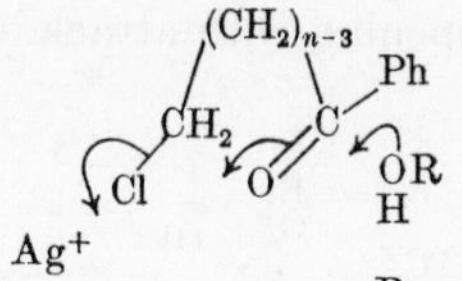

Products

(4b) R = H, Et

It was considered that the p–π bonding electrons of the carbonyl group were not involved in the formation of the transition state since the introduction of electron-releasing substituents into the *para*-position of the phenyl group caused a rate decrease. In the Reviewers' opinion, however, a mechanism in which attack by solvent on the carbon of the carbonyl group occurred synchronously with, or before, participation by the carbonyl group (see **4b**) would also account for this observation. This is also consistent with a large proportion of the product's being the cyclic ketal.

Neighbouring-group participation by acetal groups has also been reported. Thus treatment of the ribose derivative (5) with tetrabutylammonium benzoate in 1-methylpyrrolidone, instead of giving the expected 4-*O*-benzoyl-L-lyxose derivative, gave 2,3,5-tri-*O*-benzyl-4-*O*-methyl-L-lyxose methyl hemiacetal 1-benzoate (7), presumably via the cyclic oxonium ion (6).[20] It has also been reported that 4-chloro-1,1-diethoxybutane (8) liberates chloride ion in the presence of potassium hydroxide in diethylene glycol much more rapidly than does 5-chloro-1,1-diethoxypentane.[21] This was

---

[18] For a review see B. Capon, *Quart. Rev. (London)*, **18**, 45 (1964).

[19] D. J. Pasto and M. P. Serve, *J. Am. Chem. Soc.*, **87**, 1515 (1965).

[20] N. A. Hughes and P. R. H. Speakman, *Chem. Comm.*, **1965**, 199.

[21] J. P. Ward, *Tetrahedron Letters*, **1965**, 3905.

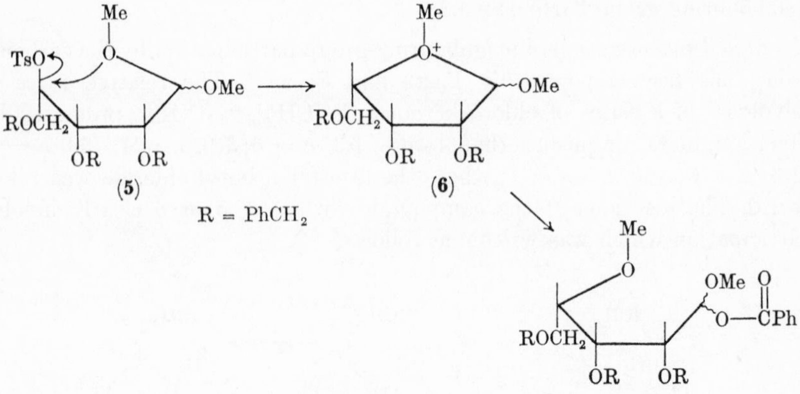

(5)   (6)

R = PhCH$_2$

(7)

attributed to participation by the acetal ethoxyl group to give the cyclic oxonium ion (9a), which, since the product gave 90% of 4,4-diethoxybutan-1-ol (9b) must undergo ring opening with attack at the original position 4.

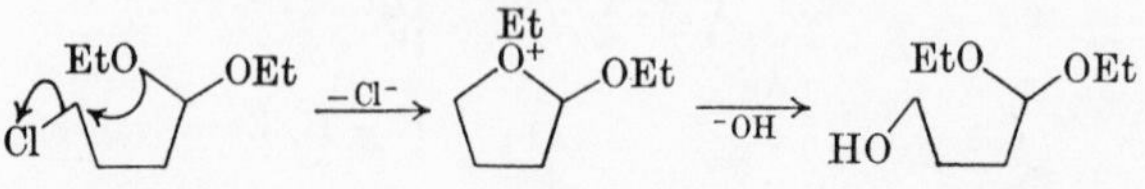

(8)   (9a)   (9b)

The detailed mechanism of the ring closure of 4-chlorobutanol to tetrahydrofuran has been discussed by Swain, Kuhn, and Schowen.[22] The reaction is catalysed by water and borate ion as well as by hydroxide ion, in contrast to the ring closure of 2-chloroethanol which is catalysed only by hydroxide ion. It is also of interest that the rate of reaction apparently increases linearly with borate concentration up to 0.528M and is thus unaffected by the monomer–polymer equilibria of the borate. The slope of the three-point Brønsted plot of these results was 0.25, which was interpreted as indicating that transfer of the alcohol proton was about 25% complete in the transition state. The solvent isotope effects were shown to be $k_{H_2O}/k_{D_2O} = 1.28$ and $k_{OD^-}/k_{OH^-} = 1.07$. These were considered to be secondary isotope effects, showing that alcoholic hydrogen is excluded from the reaction coordinate so that the symmetric stretch of the OHO bond in the transition state is a real vibration, with zero-point energy, and not a translation. The transition state for this reaction was then visualized as (10) with a long, loose bond between carbon and oxygen.

[22] C. G. Swain, D. A. Kuhn, and R. L. Schowen, *J. Am. Chem. Soc.*, **87**, 1553 (1965).

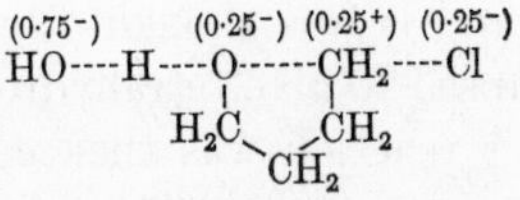

$$(0.75^-) \qquad (0.25^-) \; (0.25^+) \quad (0.25^-)$$
$$\text{HO----H----O-----CH}_2\text{----Cl}$$

(10)

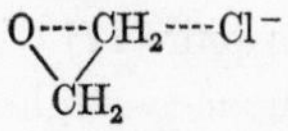

(11)

The reaction of 2-chloroethanol to give ethylene oxide is catalysed only by hydroxide ion, and the transition state was visualized as carrying a full negative charge, as (11). The reason for this difference was discussed in terms of two factors. The first was the greater acidity of the alcoholic group in 2-chloroethanol than in 4-chlorobutan-1-ol, which would give a higher concentration of $^-\text{OCH}_2\text{CH}_2\text{Cl}$. The second may be visualized by considering the microscopic reverse of the ring closure, namely the reaction of ethylene oxide with chloride ions. In this reaction the oxygen of the oxide bridge behaves as a much better leaving group than even the fully protonated oxygen of tetrahydrofuran. This means that it is a poor nucleophile in the ring closure, which causes the O–C bond to be shorter in this transition state than in that of the corresponding reaction of 4-chlorobutan-1-ol, and this is accompanied by greater transfer of the proton to the attacking base.

The oxide migrations (12) → (13)[23] and (14) → (15)[24] have been investigated.

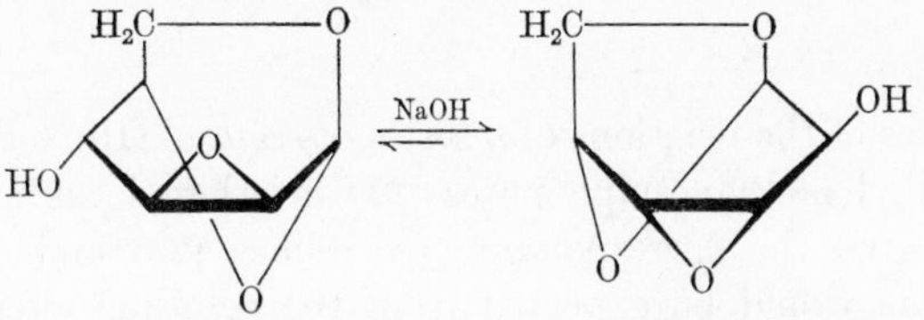

(12)

(13)

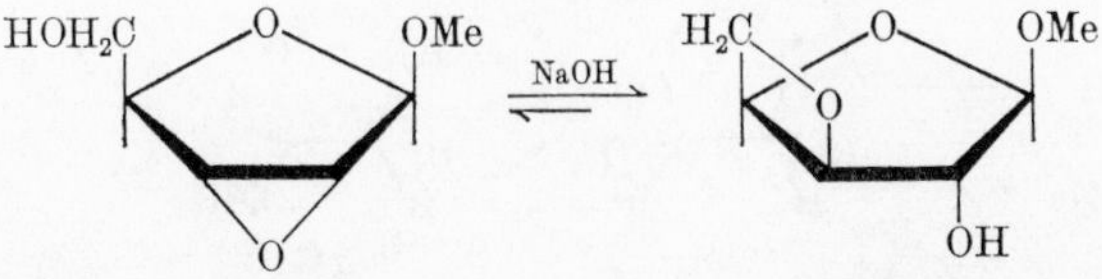

(14)

(15)

Neighbouring-group participation by the benzyloxy-group has been investigated by Barker and his co-workers.[25] 4-Benzyloxypentyl toluene-*p*-sulphonate (16) in ethanol at 75° gives tetrahydro-2-methyl furan, toluene-*p*-sulphonic acid, and benzyl ethyl ether. The kinetics determined by the rate

[23] M. Černý, J. Pacák, and J. Staněk, *Collection Czech. Chem. Commun.*, **30**, 1151 (1965).
[24] P. W. Austin, J. G. Buchanan, and E. M. Oakes, *Chem. Comm.*, **1965**, 374, 472.
[25] G. R. Gray, F. C. Hartman, and R. Barker, *J. Org. Chem.*, **30**, 2020 (1965).

of formation of acid were complex, and it was thought that the reaction involved the ion pair (17) which collapsed to give tetrahydro-2-methylfuran and benzyl toluene-p-sulphonate; this reaction was then considered to be followed by a slightly faster solvolysis of the ester. The rates of the corres-

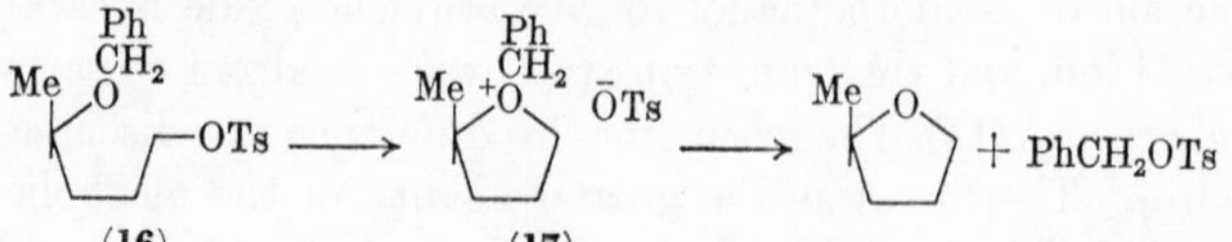

$$\text{(16)} \qquad\qquad \text{(17)}$$

ponding reactions of 2,3,4-tri-O-benzyl-1,5-di-O-toluene-p-sulphonyl-ribitol, -arabinitol, and -xylitol were also measured. The rate constant (after symmetry correction) for the ribitol compound is about 10 times smaller than those for the other two. The conformational reasons for this were discussed.

Participation by the oxygen of the tetrahydrofuran ring has been shown to occur in the acetolyses of compounds (18) when Bs = p-bromobenzene-sulphonyl and n = 5 or 6.[26]

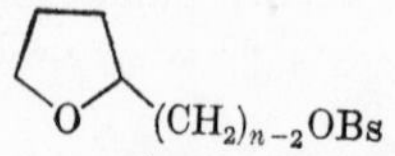

$$\text{(18)}$$

The kinetics of the reactions of a series of *erythro-* (19) and *threo-*2-p-nitro-benzamidoalkyl methanesulphonates (21) in ethanol containing potassium acetate, to give the corresponding oxazolines (20) and (22), have been reported.[27] As would be expected from the eclipsing effects, the *erythro-*compounds all react faster than their *threo-*isomers. The *erythro:threo* rate

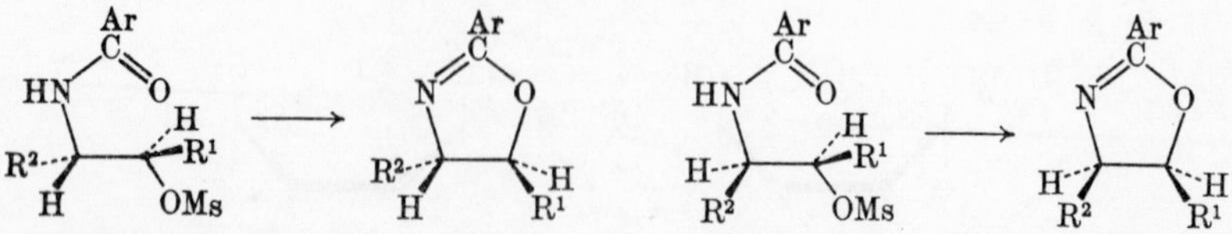

$$\text{(19)} \qquad\qquad \text{(20)} \qquad\qquad \text{(21)} \qquad\qquad \text{(22)}$$

ratio varies from 1.7 when R^1 and R^2 are cyclohexylmethyl to 611 when they are cyclohexyl, mainly as a result of variation in the rate for the *threo-*isomer. A series of *cis-* and *trans-*2-benzamidocycloalkyl methanesulphonates was also investigated.[28] Among the lower members, the *trans-* reacted faster

[26] G. T. Kwiatkowski, S. J. Kavarnos, and W. D. Closson, *J. Heterocyclic Chem.*, **2**, 11 (1965).
[27] M. Pankova and J. Sicher, *Collection Czech. Chem. Commun.*, **30**, 388 (1965).
[28] M. Svoboda and J. Sicher, *Collection Czech. Chem. Commun.*, **30**, 2948 (1965).

than the *cis*-compound, but with the twenty-six membered ring, the largest studied, the *cis*-compound reacted the faster. For ring size greater than fifteen the *trans*-compounds reacted approximately as fast as an analogous acyclic compound, *threo*-6-benzamido-5-methanesulphonyloxydecane, but the *cis*-compounds all reacted more slowly than *erythro*-6-benzamido-5-methanesulphonyloxydecane, even when the ring size was twenty-six. These results were discussed in terms of the preferred conformations for ring closure (23) (*trans*) and (24) (*cis*). In the absence of the constraint of the ring the *cis*-compound, which resembles an acyclic *erythro*-isomer, would be expected to react faster, but this becomes possible only when the ring has 26 members.

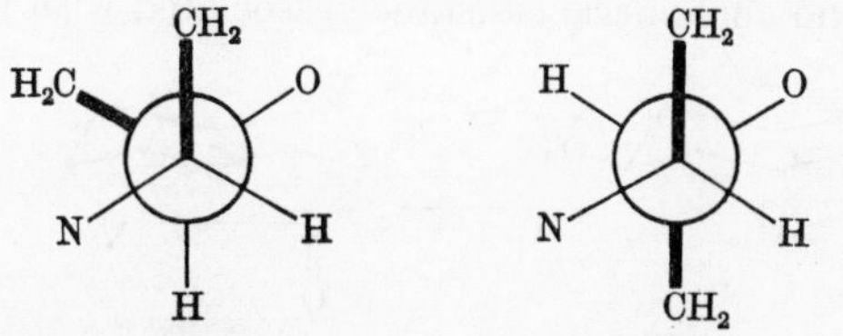

(23)   (24)

However, to afford conformation (24) the amido-group has to take up an unfavourable intra annular position, so that the rate of reaction is always less than that of the corresponding acyclic compound. With the *trans*-compounds the amido- and methanesulphonyloxy-groups can be antiperiplanar and outside the ring, but then the methylene groups of the ring are in the less favourable synclinal arrangement.

A kinetic study of the ethoxide-promoted cyclization of a series of iodo amides (25) has been reported.[29]

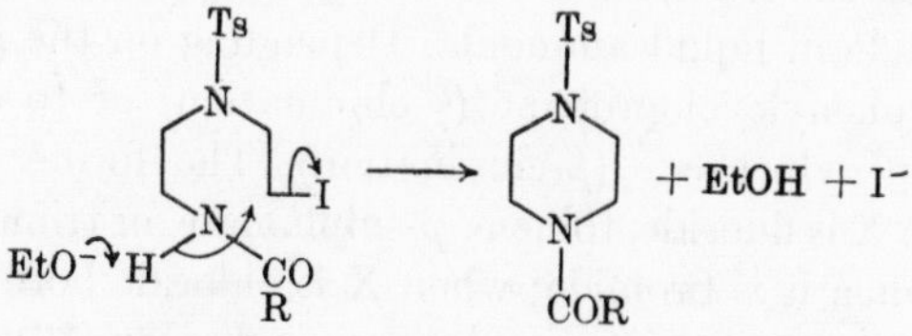

(25)

The reaction of iodo carbamates of general structure (26) with a base, to give the aziridine (28), has been shown to proceed via the aziridine-1-carboxylate (27).[30]

Participation by the neighbouring PhCH$_2$OOCNH group has been shown

[29] A. Chambers and C. J. M. Stirling, *J. Chem. Soc.*, **1965**, 4558.
[30] A. Hassner and C. Heathcock, *J. Org. Chem.*, **29**, 3640 (1964).

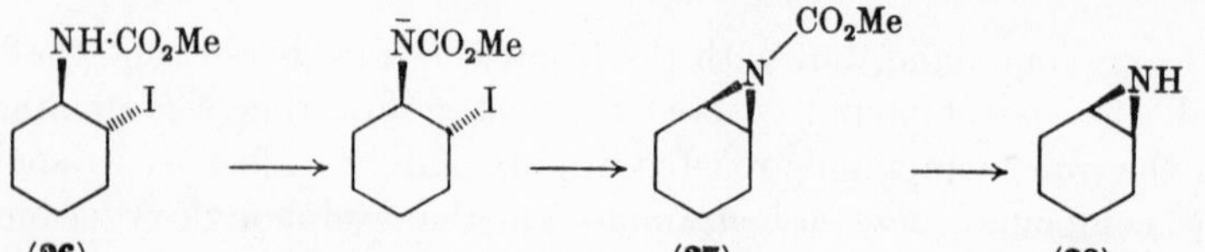

to occur in reactions of 3,4-anhydro-*N*-benzyloxycarbonyl-D-galactosamine derivatives.[31]

A small yield (4.4%) of the lactone (30) was obtained on formolysis of the toluene-*p*-sulphonate of methyl *trans*-5-hydroxycyclooctanecarboxylate (29), indicating neighbouring-group participation by the ester group.[32] There appeared to be no anchimeric assistance associated with this process.

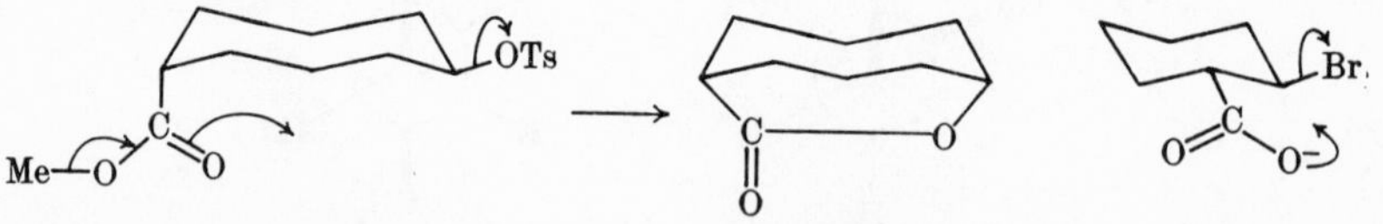

Participation by the carboxylate group has been reported to occur in the hydrolysis of *trans*-2-bromocyclohexanecarboxylic acid (31), which is considerably faster than that of its *cis*-isomer in alkaline media.[33] The possibility of neighbouring-group participation by the carboxyl group in the solvolysis of some trialkylsulphonium salts has also been discussed.[34]

Various examples of participation by amino groups in reactions of β-halogeno amines, to give aziridines, and in aziridine ring-opening have been reported.[35]

Participation by neighbouring carbanion was investigated by Bumgardner[36] who studied the reactions of a series of compounds, PhCH₂CH₂CH₂X, with sodium amide in liquid ammonia. Depending on the nature of X this leads either to phenylcyclopropane (γ-elimination) or to diphenylhexenes formed via propenylbenzene (β-elimination). The former reaction occurs exclusively when X is fluoride, toluene-*p*-sulphonate, or trimethylammonium, and the latter when it is bromide; when X is chloride both reactions occur.

Attempts to observe participation by a radical anion (33) generated by the

[31] P. H. Gross, K. Brendel, and H. K. Zimmerman, *Ann. Chem.*, **680**, 159 (1964).

[32] A. C. Cope and D. L. Nealy, *J. Am. Chem. Soc.*, **87**, 3122 (1965).

[33] W. R. Vaughan, R. Caple, J. Csapilla, and P. Scheiner, *J. Am. Chem. Soc.*, **87**, 2204 (1965).

[34] J. B. Hyne and J. H. Jensen, *Can. J. Chem.*, **43**, 57 (1965).

[35] N. J. Leonard and J. Paukstelis, *J. Org. Chem.*, **30**, 821 (1965); D. L. Klayman, J. W. Lown, and T. R. Sweeney, *ibid.*, **30**, 2275 (1965); A. Hassner and C. Heathcock, *ibid.*, **30**, 1748 (1965); A. T. Bottini, B. F. Dowden, and L. Sousa, *J. Am. Chem. Soc.*, **87**, 3249 (1965).

[36] C. L. Bumgardner, *Chem. Comm.*, **1965**, 374.

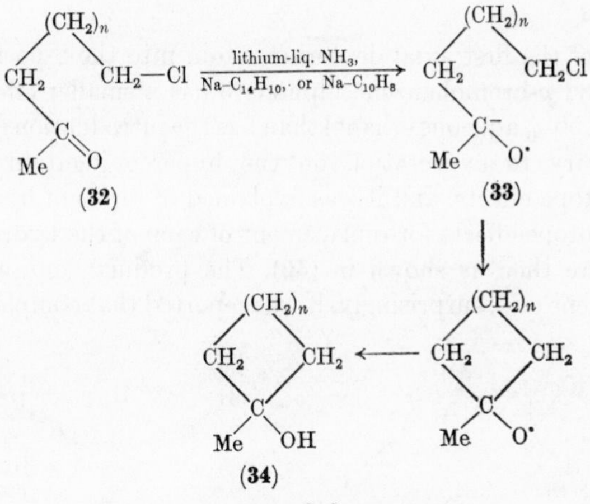

(32)   (33)

(34)

$n = 2, 3$

Scheme 2

reduction of the ketones (32), as shown in Scheme 2, was not very successful since only poor yields of the cyclized products (34) were obtained.[37]

Participation by neighbouring carbon has been found on solvolysis of methyl α-D-glucoside 3-nitrobenzenesulphonate (35), which in acetate buffer at 100° and pH 5 yields the hemiacetal (36) of methyl 3-deoxy-3-formyl-α-D-xylofuranoside.[38] Another example is found in the conversion of the 5β,6β-

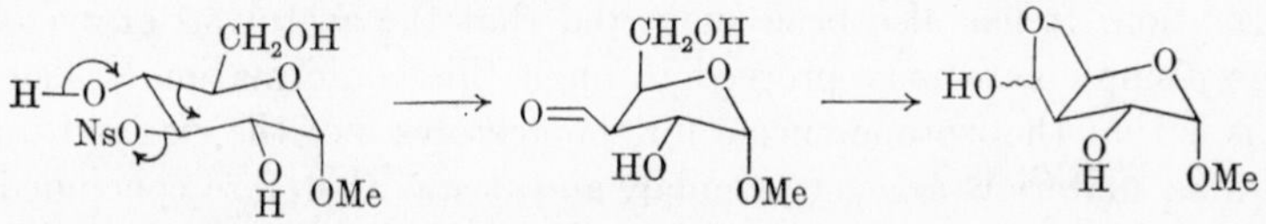

(35)   (36)

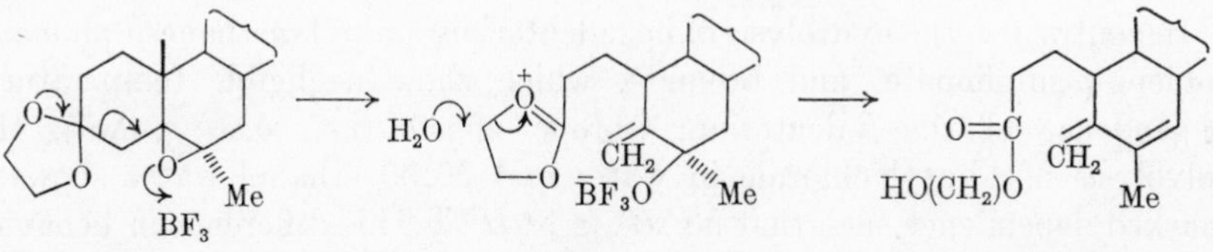

(37)   (38)   (39)

epoxy-6α-methylcholestane derivative (37) to the diene ester (39), which was envisaged as proceeding via the oxonium ion (38) with subsequent ring opening and dehydration on isolation.[39]

[37] H. O. House, J. J. Riehl, and C. G. Pitt, *J. Org. Chem.*, **30**, 650 (1965).
[38] P. W. Austin, J. G. Buchanan, and R. M. Saunders, *Chem. Comm.*, **1965**, 146.
[39] J. W. Blunt, M. P. Hartshorn, and D. N. Kirk, *Chem. Comm.*, **1965**, 545.

Isotope Effects

Introduction of the first axial deuterium atom into the 2-position of *cis*-4-*t*-butylcyclohexyl *p*-bromobenzenesulphonate has a smaller effect on the rate of solvolysis in 50% aqueous ethanol than has the introduction of the second.[40] This is contrary to expectation on the hyperconjugation model for *β*-deuterium isotope effects, and it was explained in terms of hydrogen participation. The isotope effects for replacement of each of the hydrogen atoms by deuterium were then as shown in (40). The product contained 86% of *t*-butylcyclohexene and, surprisingly, it was reported that complete deuteration

of the *β*-positions did not substantially lower the olefin fraction of the product. Isotope effects on the solvolysis of *trans*-4-*t*-butylcyclohexyl *p*-bromobenzene-sulphonate were also reported (see 41).[41] The reaction of undeuterated material yielded 67% of 4-*t*-butylcyclohexene, 27% of *t*-butylcyclohexanols (97% *cis*), and 6% of ethers. It was suggested that the equivalence of the equatorial *trans*-isotope effects and the non-equivalence of the axial *cis*-isotope effects could be explained if the transition state had a twist boat conformation. It has also been suggested that the acetolysis of cyclohexyl toluene-*p*-sulphonate may proceed through this conformation.[42] The basis for this is that the isotope effects for the acetolyses of the *cis*- and *trans*-2-deuterated materials are very similar, and it was therefore concluded that there is little average difference in the dihedral angle between C–OTs and C–D bonds in the transition states for the *cis*- and *trans*-isomers.

In contrast to the hydrolysis of hexadeuterioisopropyl methanesulphonate, toluene-*p*-sulphonate, and bromide, which show negligible temperature-dependence of the *β*-deuterium isotope effects (i.e., $\delta\Delta G^{\ddagger} \approx \Delta S^{\ddagger}$), the solvolysis of *t*-butyl chloride in water and 50:50 ethanol–water shows a marked dependence such that $\delta S\Delta G^{\ddagger} \approx \delta\Delta H^{\ddagger}$.[43] This difference in behavior was attributed to nucleophilic interaction between water and the *α*-carbon atom of the secondary compounds. According to an analysis by Wolfsberg and Stern[44] this would lead to an increase in torsional and bending frequencies

[40] V. J. Shiner and J. G. Jewett, *J. Am. Chem. Soc.*, **87**, 1382 (1965).
[41] V. J. Shiner and J. G. Jewett, *J. Am. Chem. Soc.*, **87**, 1383 (1965).
[42] W. H. Saunders and K. T. Finley, *J. Am. Chem. Soc.*, **87**, 1384 (1965).
[43] L. Hakka, A. Queen, and R. E. Robertson, *J. Am. Chem. Soc.*, **87**, 161 (1965).
[44] M. Wolfsberg and M. J. Stern, *Pure Appl. Chem.*, **8**, 325 (1964).

of the methyl group about the C–C axis which would compensate for a decrease in "high" frequencies, such as bending, owing to electronic changes associated with partial bonding to O and to Cl. In the solvolysis of the *t*-butyl chloride there is no nucleophilic participation and hence this compensation does not occur.

A kinetic isotope effect, $k_H/k_D = 1.20$, was observed in the ethanolysis of compound (42), but not of compound (43).[45] This result was thought to

$$Me_2C=CH-CD_2Cl \qquad Me_2C-CH=CD_2$$
$$\qquad\qquad\qquad\qquad\qquad\qquad |$$
$$\qquad\qquad\qquad\qquad\qquad\qquad Cl$$

$$\text{(42)} \qquad\qquad\qquad \text{(43)}$$

accord with the view that the α-deuterium isotope effect is the result of vibrational changes in the carbon–deuterium bonds occurring in sp^3–sp^2 rehybridization, since the deuterium atoms of compound (43) are already bonded to sp^2-hybridized carbon in the initial state.

The α-secondary deuterium isotope effect has been measured for the isotopic exchange reaction of benzyl and thienyl chloride[46] and of the reactions of trideuteriomethyl iodide with tertiary amines in benzene.[47] The effect of ring deuteration on the rate of solvolysis of benzhydryl chloride has also been determined.[48]

The carbon-13 isotope effect of the ionization of triphenylmethyl chloride to the triphenylmethyl carbonium ion has been determined and analysed theoretically.[49]

Deaminations

In 1953 it was reported that deamination of [1-^{14}C]-n-propylammonium perchlorate by nitrous acid gave 1-propanol in which 8% of the label was rearranged to $C_{(2)}$ and $C_{(3)}$.[50] This was interpreted as involving migration of a

$$CH_3$$
$$H_2\overset{+}{C}\!=\!=\!=\!\overset{}{C}H_2$$
$$\text{(44)}$$

methyl group via a protonated cyclopropane intermediate (44). In 1962 this view was criticized by Reutov and Shatkina,[51] who reported that ^{14}C was found only at positions 1 and 3 and suggested that this could result from two

[45] V. Belanić-Lipovac, S. Borčič, and D. E. Sunko, *Croat. Chem. Acta*, **37**, 61 (1965).
[46] B. Östman, *J. Am. Chem. Soc.*, **87**, 3163 (1965).
[47] K. T. Leffek and J. W. Maclean, *Can. J. Chem.*, **43**, 40 (1965).
[48] A. Streitwieser and H. S. Klein, *J. Am. Chem. Soc.*, **86**, 5170 (1964).
[49] A. J. Kresge, N. N. Lichtin, K. N. Rao, and R. E. Weston, *J. Am. Chem. Soc.*, **87**, 437 (1965).
[50] J. D. Roberts and M. Halmann, *J. Am. Chem. Soc.*, **75**, 5759 (1953).
[51] O. A. Reutov and T. N. Shatkina, *Tetrahedron*, **18**, 237 (1962).

1,2- or one 1,3-hydride shift. Support for the occurrence of a 1,3-hydride shift was provided by Karabatsos and Orzech[52] who showed that 1,1,2,2-tetradeuteriopropylamine yielded propanol with most of the protons at positions 1 and 3, as determined by NMR spectroscopy. This reaction has now been re-investigated by two groups of workers (using different techniques); both report that the amount of rearrangement is only 2–4%; and, in experiments with 1-tritiopropylamine, tritium was found in small amounts at positions 2 (2.9%) and 3 (1.6%).[53,54] Re-investigation of the reaction by means of a ^{14}C label also indicated about 4% of rearrangement, and the label was found to be equally distributed between positions 2 and 3.[55] These results are consistent with the intervention, in the rearrangement, of a protonated cyclopropane in which the three carbon atoms approach equivalence. This could be a series of edge-protonated cyclopropanes (45) to (47), or a face-protonated cyclopropane (48).

A definitive study of conformational effects on the deamination of 2-aminocyclohexanols has been reported.[56] The four epimeric 2-amino-4-*t*-butylcyclohexanols undergo deamination as shown in Scheme 3, the major products being those expected on conformational grounds. The formation of small amounts of the aldehyde (49) from (50) and (51) indicates stereochemical leakage at the carbonium-ion stage, which was attributed to solvation of the carbonium ion from the equatorial (rear) side. However, the ratio of the products of the deamination of *cis*-2-aminocyclohexanol does not reflect the expected ratio of the conformers. On the basis that the *A*-value of the amino group is 1.8 and that of the hydroxyl group 0.8, it was estimated that the ratios of the concentrations of conformation (52) and (53) should be 3:1, and this was confirmed experimentally. The products, (54) and (55), expected to be formed uniquely from them were present, however, in almost equal amounts. To explain this it was suggested that there was a conformational re-equilibration at the diazo-hydroxide stage of the reaction.

The deamination of *cis*- and *trans*-2-amino-cyclooctanols by nitrous acid has also been investigated.[57]

[52] G. J. Karabatsos and C. E. Orzech, *J. Am. Chem. Soc.*, 84, 2838 (1962).
[53] C. C. Lee, J. E. Kruger, and E. W. C. Wong, *J. Am. Chem. Soc.*, 87, 3985 (1965).
[54] G. J. Karabatsos, C. E. Orzech, and S. Meyerson, *J. Am. Chem. Soc.*, 87, 4394 (1965).
[55] C. C. Lee and J. E. Kruger, *J. Am. Chem. Soc.*, 87, 3986 (1965).
[56] M. Chérest, H. Felkin, J. Sicher, F. Šipoš, and M. Tichý, *J. Chem. Soc.*, 1965, 2513.
[57] J. G. Traynham and M. T. Yang, *J. Am. Chem. Soc.*, 87, 2394 (1965); see also J. G. Traynham and J. Schneller, *ibid.*, 87, 2398 (1965).

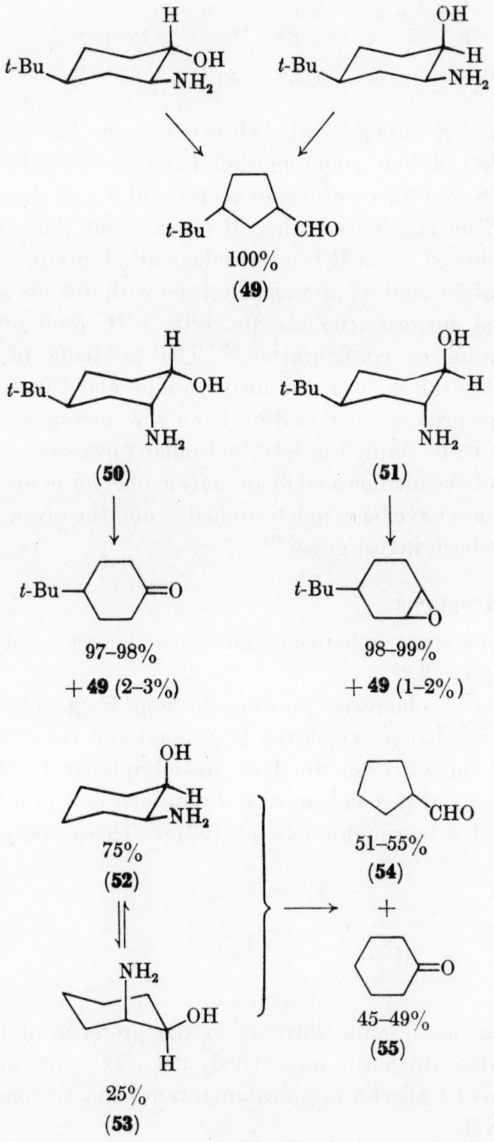

Scheme 3

The relative importance of a stepwise (reaction 1) and synchronous (reaction 2) mechanism for the decomposition alkyldiazonium compounds (X = OH, OAc, $^+$NH$_2$Ar, etc.) has been discussed. It was considered that with the synchronous

3+

$$R—N{=}N—X \longrightarrow R—^+N{\equiv}N + X^- \\ R—^+N{\equiv}N \longrightarrow R^+ + N_2 \longrightarrow Products \Bigg\} \tag{1}$$

$$R—N{=}N—X \longrightarrow R^+ + N_2 + X^- \longrightarrow Products \tag{2}$$

mechanism R^+ and X^- are generated close to one another, so there should be a higher probability of their recombination than with the stepwise mechanism. By using this criterion for reactions in acetic acid it was considered that the stepwise mechanism is followed when R is octyl but that the synchronous mechanism is followed when R is a secondary alkyl group.[58]

Deamination of 3α- and 3β-aminocholestane with nitrous acid and via the nitrosoamide and nitrosocarbonate proceeds with predominant, but not exclusive, retention of configuration.[59] The products of retention and inversion of configuration from the nitrosoamide and the nitrosocarbamate reaction, and the product of retention from the nitrous acid deamination, were all thought to be formed in intramolecular processes.

The products of the nitrous acid deamination of *erythro*- and *threo*-2-amino-1,2-diphenylethanol have been redetermined,[60] and the deamination of cyclopropylamine has been investigated.[61]

Fragmentation Reactions

A brief review of the mechanism and stereochemistry of fragmentation reactions has appeared.[62]

The preferred stereochemistry for a synchronous fragmentation (reaction 3) is considered to be that in which the C–X bond and the orbital of the lone electron pair of the nitrogen are both antiperiplanar to the fragmenting C_β–C_γ bond. This ideal stereochemistry is obtained in 3-bromoadamantan-1-amine (**56**)[63] and 4-bromoquinuclidine (**57**).[64] These compounds undergo

$$R_2\overset{\curvearrowright}{N}—\underset{\gamma}{C}—\underset{\beta}{C}—\underset{\alpha}{C}—X \longrightarrow R_2N{=}C\big< + \big>C{=}C\big< + X^- \tag{3}$$

solvolysis in aqueous organic solvents in the presence of base with fragmentation, at rates which are, respectively, 30—500 times and 50,000 times faster than those of 1-alkyl-3-bromoadamantanes and 1-bromobicyclo[2.2.2]-octane, respectively.

[58] H. Maskill, R. M. Southam, and M. C. Whiting, *Chem. Comm.*, **1965**, 496.
[59] E. H. White and F. W. Bachelor, *Tetrahedron Letters*, **1965**, 77.
[60] J. W. Huffman and R. P. Elliott, *J. Org. Chem.*, **30**, 365 (1965).
[61] E. J. Corey and R. F. Atkinson, *J. Org. Chem.*, **29**, 3703 (1964).
[62] C. A. Grob, *Angew. Chem. Intern. Ed. Engl.*, **4**, 440 (1965).
[63] C. A. Grob and W. Schwarz, *Helv. Chim. Acta*, **47**, 1870 (1964).
[64] P. Brenneisen, C. A. Grob, R. A. Jackson, and M. Ohta *Helv. Chim. Acta*, **48**, 146 (1965).

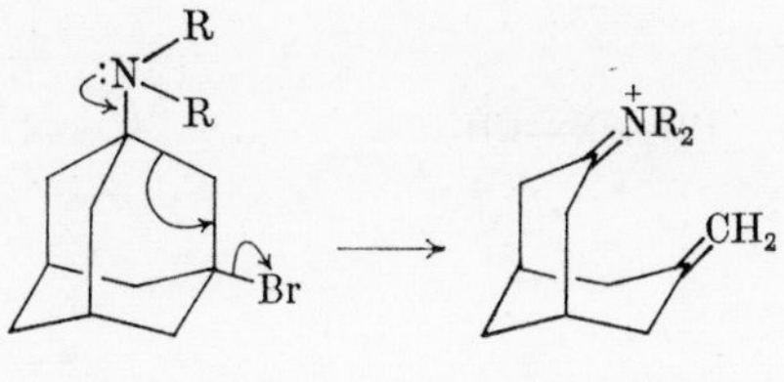

(56) 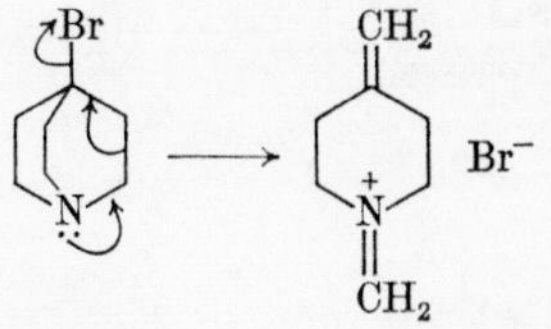   (57)

This stereochemical requirement is also illustrated by the results shown in Scheme 4.[65]

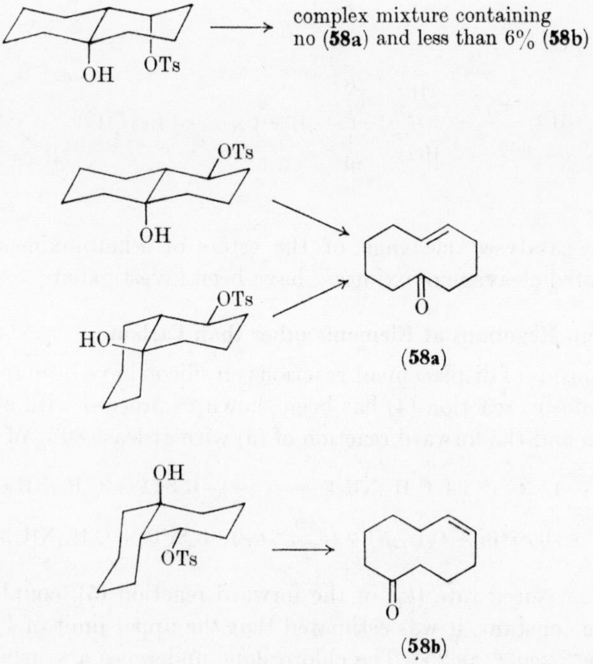

complex mixture containing
no (58a) and less than 6% (58b)

(58a)

(58b)

Scheme 4

The fragmentation of (59),[66] (60a),[67] and (60b)[68] have also been investigated.

[65] P. S. Wharton and G. A. Hiegel, *J. Org. Chem.*, **30**, 3254 (1965).
[66] C. A. Grob and V. Krasnobajew, *Helv. Chim. Acta*, **47**, 2145 (1964).
[67] S. Beckmann and H. Geiger, *Chem. Ber.*, **98**, 516 (1965).
[68] F. Nerdel, D. Frank, and H. J. Lengert, *Chem. Ber.*, **98**, 728 (1965).

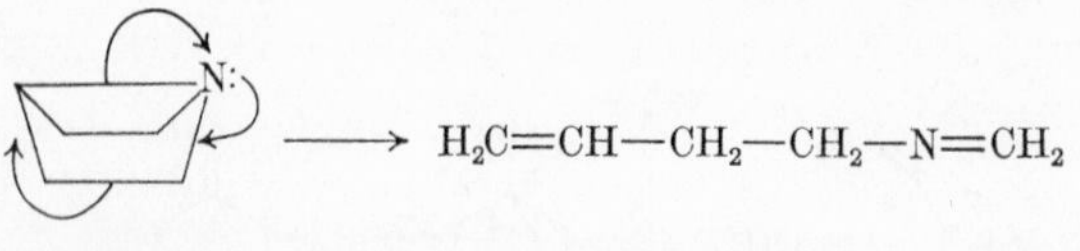

(59)

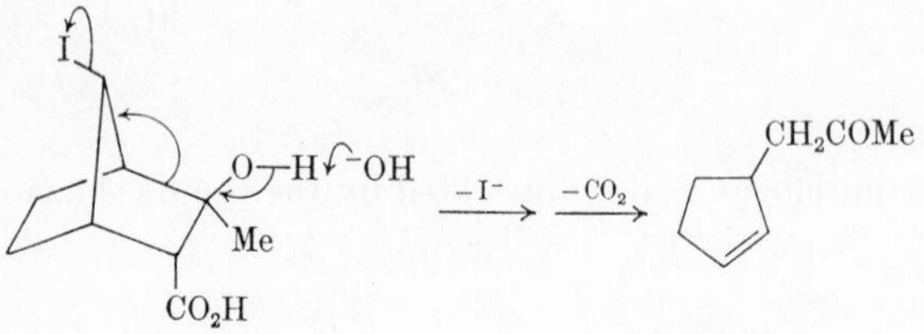

(60a)

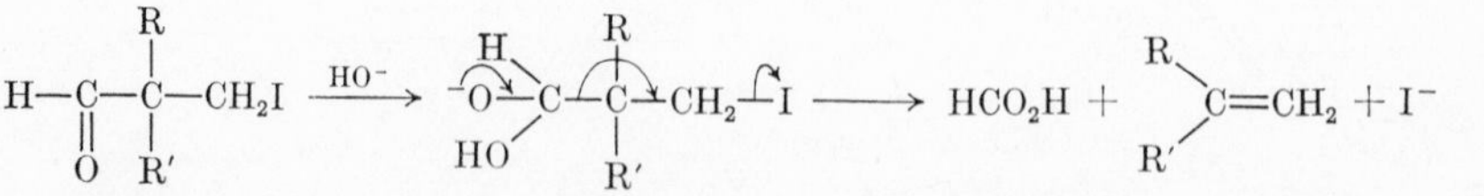

(60b)

The base-catalysed cleavage of the esters of α-keto-oximes[69] and the PCl₅-promoted cleavage of oximes[70] have been investigated.

Displacement Reactions at Elements other than Carbon

The mechanisms of displacement reactions on silicon have been reviewed.[71]

In chloroform, reaction (4) has been shown to proceed with at least 90% of inversion and the forward reaction of (5) with at least 80% of inversion.[72]

$$(-)\text{—}R_3Si^*Cl + C_6H_{11}NH_3F \longrightarrow (-)\text{—}R_3Si^*F + C_6H_{11}NH_3Cl \qquad (4)$$

$$(-)\text{—}R_3Si^*Br + C_6H_{11}NH_3Cl \underset{k_r}{\overset{k_f}{\rightleftharpoons}} (+)\text{—}R_3Si^*Cl + C_6H_{11}NH_3Br \qquad (5)$$

From the measured rate (k_f) of the forward reaction (5), coupled with the equilibrium constant, it was estimated that the upper limit of k_r was 2.8×10^{-3} l mole^{-1} sec^{-1} at 25°. The chlorosilane undergoes a salt-induced racemization which is approximately of the first order in added salt, the rate constant for cyclohexylammonium bromide as the salt being at least 15 times the value of k_r estimated as above. This result was interpreted as

[69] A. Hassner and W. A. Wentworth, *Chem. Comm.*, 1965 44; see also M. S. Newman and C. Courduvelis, *J. Org. Chem.*, **30**, 1795, footnote 7 (1965).

[70] A. Hassner and E. G. Nash, *Tetrahedron Letters*, **1965**, 525.

[71] L. H. Sommer, "Stereochemistry, Mechanism and Silicon", McGraw-Hill Book Co., New York, N.Y., 1965.

[72] L. H. Sommer, F. O. Stark, and K. W. Michael, *J. Am. Chem. Soc.*, **86**, 5683 (1964).

indicating that the racemization involves the formation of a siliconium ion-pair. In the presence of cyclohexylammonium [[36]Cl]chloride the rates of racemization and of chloride exchange are, within experimental error, identical. It was proposed that there was a rate-determining formation of a siliconium ion pair which undergoes equally fast exchanges by retention or inversion.

Stereochemical studies of the acid-catalysed cleavage of the silicon–nitrogen bond,[73] and of the cleavage of an optically active disilane by metallic lithium[74] have also been reported.

The ferrocenyl group appears to be able to stabilize a siliconium ion as well as a carbonium ion (see p. 47).[75] Thus the silicon–silicon bond of pentamethyldisilanylferrocene (61) undergoes extremely ready cleavage on treatment with 10^{-3}M ethanolic hydrogen chloride. The product, compound (63), is formed in 47% yield, presumably via the siliconium ion (62).

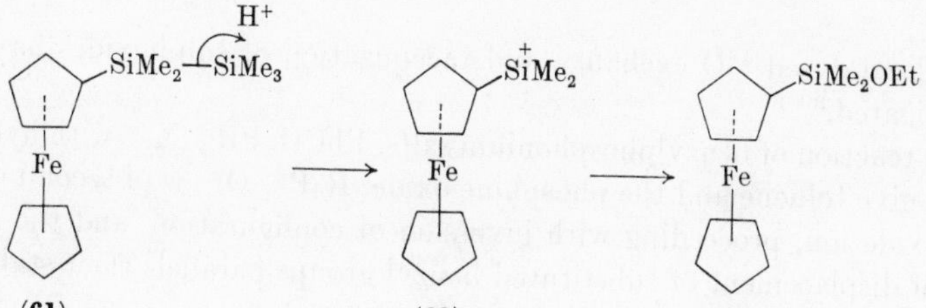

(61)   (62)   (63)

The kinetics of the methanolysis of methyl 2,3,4,6-tetrakis-O-trimethylsilyl α-D-glucopyranoside[76] and of the alkaline cleavage of organosilicon hydrides,[77] phenylethynyl-silanes and -germanes,[78a] and benzyl trimethyl silanes[78b] have also been investigated.

The following reactions have been studied by Day and Cram:[79]

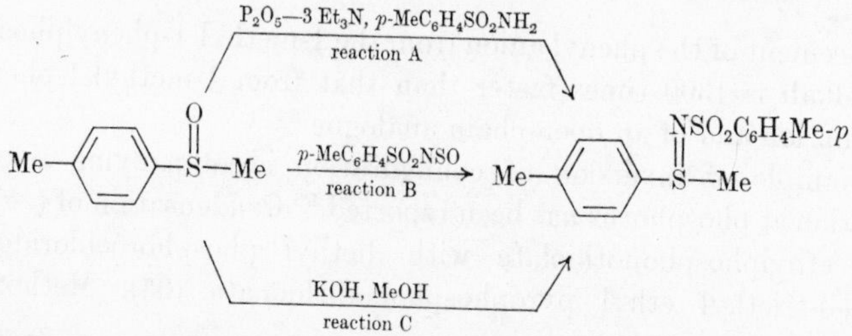

[73] L. H. Sommer, J. D. Citron, and C. L. Frye, *J. Am. Chem. Soc.*, **86**, 5684 (1964).
[74] L. H. Sommer and R. Mason, *J. Am. Chem. Soc.*, **87**, 1619 (1965).
[75] M. Kumada, K. Mimura, M. Ishikawa, and K. Shiina, *Tetrahedron Letters*, **1965**, 83.
[76] A. G. McInnes, *Can. J. Chem.*, **43**, 1998 (1965).
[77] J. Hetflys, F. Mares, and V. Chvolovsky, *Collection Czech. Chem. Commun.*, **30**, 1643 (1965).
[78a] C. Eaborn and D. R. M. Walton, *J. Organomet. Chem.*, **4**, 217 (1965).
[78b] R. W. Bott, C. Eaborn, and B. M. Rushton, *J. Organomet. Chem.*, **3**, 448 (1965); R. W. Bott, B. F. Dowden, and C. Eaborn, *J. Chem. Soc.*, **1965**, 4994.
[79] J. Day and D. J. Cram., *J. Am. Chem., Soc.*, **87**, 4398 (1965).

All three involved inversion of configuration and reactions B and C yielded products of 98% and 96% optical purity, respectively. Since reaction B gave also sulphur dioxide it was concluded that the entering and the leaving nucleophile must be coupled in a ring system formally similar to that involved in the Wittig reaction. This was formulated (see **64**) as involving two molecules of N-sulphinyl-toluene-p-sulphonamide, since preliminary kinetic results indicated that the reaction was of the second order in this reactant.

(**64**)

Acid-catalysed ^{18}O exchange and racemization of sulphoxides have been investigated.[80-82]

The reaction of benzylphosphonium salts, $PhCH_2PR_3^+$ X^-, with hydroxide ion to give toluene and the phosphine oxide, R_3P^+-O^-, is of second order in hydroxide ion, proceeding with inversion of configuration, and the relative ease of displacement of substituted benzyl groups parallels their stability as anions.[83] The following mechanism was proposed:

$$R_4P^+ + {}^-OH \longrightarrow R_4P-OH$$

$$R_4P-OH + {}^-OH \longrightarrow R_4P-O^- + H_2O$$

$$R_4P-O^- \longrightarrow R_3P{=}O + R^-$$

$$R^- + H_2O \longrightarrow RH + OH^-$$

Displacement of the phenyl anion from the 1-methyl-1-phenylphospholium ion by alkali is 1300 times faster than that from 1-methyl-1-phenylphosphinanium ion and of an open-chain analogue.[84]

An example of inversion of configuration accompanying nucleophilic substitution at phosphorus has been reported.[85] Condensation of ($-$)-sodium O-ethyl ethylphosphonothiolate with diethyl phosphorochloridothionate gave ($+$)-triethyl ethyl pyrophosphonodithionate (**65**). Methoxide ion

[80] S. Oae, T. Kitao, Y. Kitaoka, and S. Kawamura, *Bull. Chem., Soc. Japan*, **38**, 546 (1965).

[81] S. Oae, T. Kitao, and Y. Kitaoka, *Bull. Chem. Soc. Japan*, **38**, 543 (1965).

[82] S. Oae, N. Kunieda, and W. Tagaki, *Chem. Ind.* (*London*), 1965, 1790.

[83] W. E. McEwen, G. Axelrad, M. Zanger, and C. A. VanderWerf, *J. Am. Chem. Soc.*, **87**, 3948 (1965); W. E. McEwen, K. F. Kumli, A. Blade-Font, M. Zanger, and C. A. Vander-Werf, *ibid.*, **86**, 2378 (1964).

[84] G. Aksnes and K. Bergesen, *Acta Chem. Scand.*, **19**, 931 (1965).

[85] J. Michalski, M. Mikolajczyk, and J. Omelańczuk, *Tetrahedron Letters*, **1965**, 1779; see also J. Michalski and M. Mikolajczyk, *Chem. Comm.*, **1965**, 35.

attacks this compound at P_I and hence it was thought that hydroxide ion would do the same. The latter reaction yields (+)-O-ethyl ethylphosphono-thionic acid and hence must proceed with inversion of configuration.

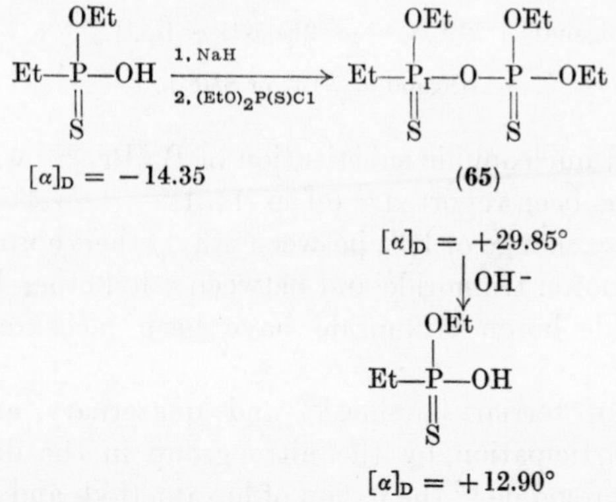

$$
\underset{\underset{\textstyle S}{\|}}{\overset{\textstyle OEt}{\overset{|}{Et-P-OH}}} \xrightarrow[\text{2, (EtO)}_2\text{P(S)Cl}]{\text{1, NaH}} \underset{\underset{\textstyle S}{\|}}{\overset{\textstyle OEt}{\overset{|}{Et-P_I-O}}}\underset{\underset{\textstyle S}{\|}}{\overset{\textstyle OEt}{\overset{|}{-P-OEt}}}
$$

$$[\alpha]_D = -14.35 \qquad\qquad (65)$$

$$[\alpha]_D = +29.85°$$
$$\downarrow OH^-$$

$$
\underset{\underset{\textstyle S}{\|}}{\overset{\textstyle OEt}{\overset{|}{Et-P-OH}}}
$$

$$[\alpha]_D = +12.90°$$

The kinetics of chloride exchange of a series of compounds, $RPOCl_2$, in 1,2-dichloroethane have been measured. All the reactions are of the second order, and the rates decrease in the order $R = Ph > Me > OMe \sim Cl > OPh$.[86a]
The reaction of triethylboron with carboxylic acids:

$$Et_3B + RCOOH \xrightarrow[\text{Diglyme}]{} Et_2BO_2CR + EtH$$

has been investigated.[86b] It is of the first order in each reactant and log k_2 is an inverse function of the pK_a of the acid. It was thought, therefore, that the reaction involves nucleophilic co-ordination of the organoborane by the oxy-function of the acid and that this increases the electrophilic character of the proton of the acid and weakens the C–B bond, the reaction probably proceeding as illustrated.

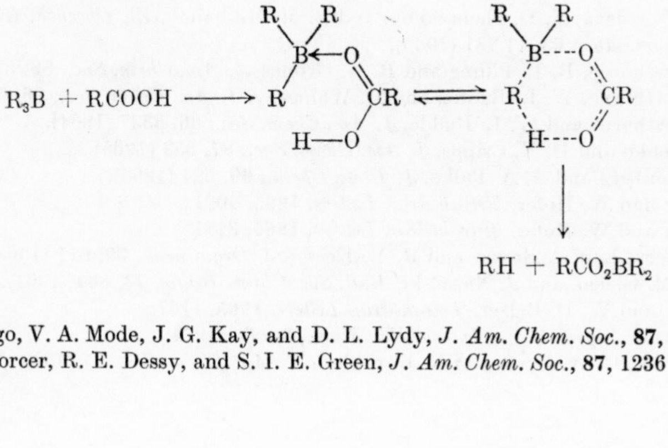

$$RH + RCO_2BR_2$$

[86a] R. S. Drago, V. A. Mode, J. G. Kay, and D. L. Lydy, *J. Am. Chem. Soc.*, **87**, 5010 (1965).
[86b] L. H. Toporcer, R. E. Dessy, and S. I. E. Green, *J. Am. Chem. Soc.*, **87**, 1236 (1965).

The alcoholysis of bisdimethylaminoboronium salts has been studied.[87] Also, the following reactions have been investigated kinetically.[88-90]

$$Bu_3P + R\bar{B}H_2\overset{+}{N}Me_3 \longrightarrow Bu_3\overset{+}{P}-\bar{B}H_2R + NMe_3$$

$$B_{10}H_{12}(ligand)_2 + 2Et_3N \longrightarrow 2Et_3\overset{+}{N}H + B_{10}H_{10}{}^{2-} + 2\ ligand$$

$$(ligand = NEt_3\ or\ SMe_2)$$

A photoinduced nucleophilic substitution of $B_{12}Br_{12}{}^{2-}$ with KCN to give $B_{12}Br_3(CN)_9{}^{2-}$ has been reported[91] (cf. p. 145).

The kinetics of exchange of BF_3 between ethyl ether–boron trifluoride and tetrahydrofuran–boron trifluoride and between ethyl ether–boron trifluoride and ethyl sulphide–boron trifluoride have been measured by an NMR method.[92]

Racemization of tertiary arsines[93] and quaternary arsonium salts,[94] intramolecular participation by the nitro group in the dissociation of *o*-nitrophenyl dichloroiodide,[95] the fission of hexamethyl- and hexabutyl- ditin by iodine,[96] and the steric course of deoxygenation of tertiary phosphine oxides to tertiary phosphines with trichlorosilane[97] have been investigated.

Ambident Nucleophiles

The many factors influencing the reactions of ambident nucleophiles were comprehensively reviewed last year.[98] One of the classic examples of an ambident nucleophile is the anion of an aliphatic nitro-compound; this can be alkylated on carbon or on oxygen (Scheme 4), and the latter process, leading to a carbonyl compound and an oxime, usually predominates. Kerber, Urry, and Kornblum[99] have now shown that *C*- and *O*-alkylation of the lithium salt of 2-nitropropane by the nitrobenzyl halides in dimethyl-formamide involve fundamentally different mechanisms. Thus the simple concept of competing nucleophilic attack by either end of a delocalized anion needs modification, at least for these reactions and possibly for others.

[87] T. A. Shchegoleva, V. D. Sheluyakov, and B. M. Mikhailov, *Zh. Obschch. Khim.*, **35**, 1066 (1965); *Chem. Abs.*, **63**, 11,281 (1965).
[88] M. F. Hawthorne, R. L. Pilling and R. N. Grimes, *J. Am. Chem. Soc.*, **86**, 5338 (1964).
[89] M. F. Hawthorne, W. L. Budde, and D. Walmsley, *J. Am. Chem. Soc.*, **86**, 5337 (1964).
[90] M. F. Hawthorne and W. L. Budde, *J. Am. Chem. Soc.*, **86**, 5337 (1964).
[91] S. Trofimenko and H. N. Cripps, *J. Am. Chem. Soc.*, **87**, 653 (1965).
[92] A. C. Rutenberg and A. A. Palko, *J. Phys. Chem.*, **69**, 527 (1965).
[93] L. Horner and W. Hofer, *Tetrahedron Letters*, **1965**, 4091.
[94] L. Horner and W. Hofer, *Tetrahedron Letters*, **1965**, 3281.
[95] E. A. Jeffery, L. J. Andrews, and R. M. Keefer, *J. Org. Chem.*, **30**, 617 (1965).
[96] S. Boué, M. Gielen, and J. Nasielski, *Bull. Soc. Chim. Belges*, **73**, 864 (1964).
[97] L. Horner and W. D. Balzer, *Tetrahedron Letters*, **1965**, 1157.
[98] R. Gompper, *Angew. Chem. Intern. Ed., Engl.*, **3**, 560 (1964).
[99] R. C. Kerber, G. W. Urry, and N. Kornblum, *J. Am. Chem. Soc.*, **87**, 4520 (1965).

It is proposed that O-alkylation is a straightforward bimolecular displacement by oxygen and that C-alkylation is a radical-ion process as shown in Scheme 5. There were three lines of evidence for this. On passing from

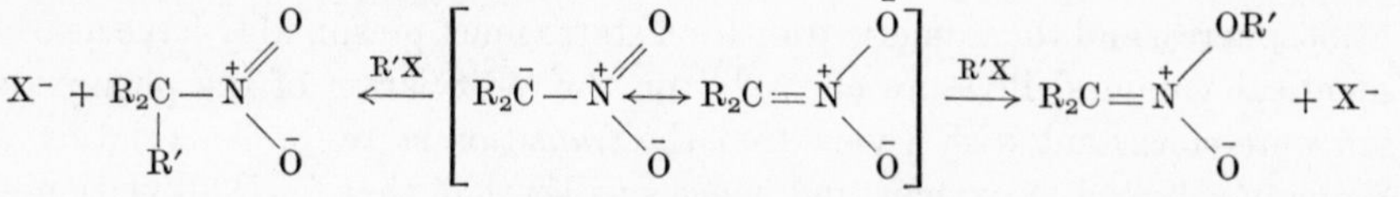

Scheme 4

chloride to bromide to iodide the rate of O-alkylation increased about 1000-fold, as expected for an S_N2 reaction in a dipolar aprotic solvent, but the C-alkylation rate increased only 7-fold. Electron paramagnetic resonance results showed that electron transfer from the nitropropane anion to the nitro-compound does occur. It is of even greater significance that electron

$$p\text{-}O_2NC_6H_4CH_2Cl + Me_2\bar{C}NO_2 \longrightarrow [O_2NC_6H_4CH_2Cl]^{\cdot} + Me_2CNO_2$$
$$\qquad\qquad\qquad\qquad\qquad\qquad\qquad\qquad (66) \qquad\qquad\qquad (67)$$

$$66 \longrightarrow O_2NC_6H_4CH_2^{\cdot} + Cl^-$$
$$(68)$$

$$67 \text{ and } 68 \longrightarrow O_2NC_6H_4CH_2CMe_2NO_2$$

Scheme 5

acceptors such as p-dinitrobenzene inhibit C-alkylation but not O-alkylation. Dimers of the relatively stable radicals (67) and (68) were not detected and it has to be assumed that these radicals were never free.[99] However, in the reactions of benzhydryl halides with hindered nucleophiles such as potassium

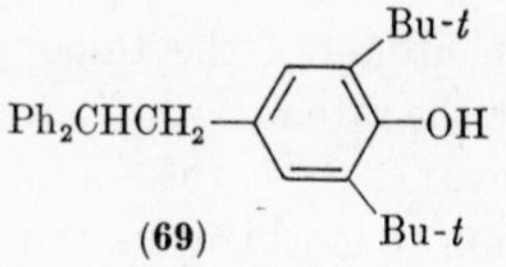

(69) 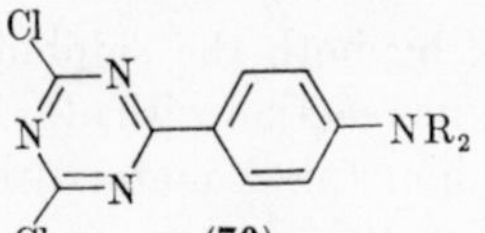   (70)

(71)   (72)

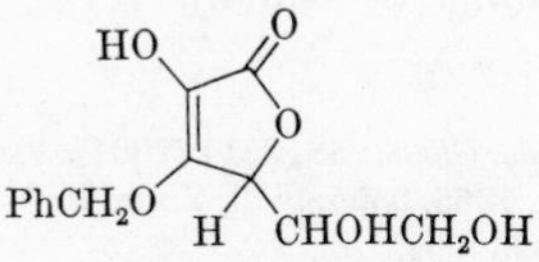

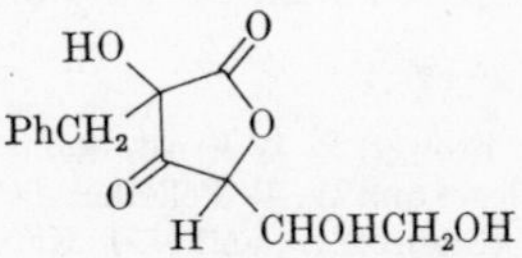

4-methyl-2,6-di-t-butyl phenoxide products derived from the benzhydryl radical were obtained, namely, *sym*-tetraphenylethane, diphenylmethane, and 1-(4-hydroxy-3,5-di-t-butylphenyl)2,2-diphenylethane (69).[100]

[100] K. Okamoto, Y. Matsui, and H. Shingu, *Bull. Chem. Soc. Japan*, **38**, 153 (1965); *Chem. Abs.*, **62**, 10305 (1965).

3*

Volumes of activation have been determined for several alkylations of sodium nitrite and of the ambident anions of acetylacetone, methyl acetoacetate, phenol, and 2-pyridone. In all save the phenoxide reactions the proportion of isomeric products was not influenced by pressure, up to 1360 atmospheres, and the isomeric transition states must presumably have nearly identical volumes. Pressure effects[101] on the benzylation of the phenoxide ion were consistent with a view that the transition state for C-alkylation is strongly solvated at oxygen, and hence smaller than that for O-alkylation— the latter transition state would be largely desolvated at oxygen as the new bond is being formed. A preliminary account of the effects of solvent, halide, base, and temperature on butylation of the anion of ethyl acetoacetate has been published; in dipolar aprotic solvents, such as 1-methylpyrrolidone and dimethyl sulphoxide, and with potassium carbonate as base, extensive O-alkylation occurs.[102]

N,N-Dialkyl-anilines, -toluidines, and -xylidines have been found to react as neutral ambident nucleophiles with cyanuric chloride, giving C-substituted (e.g. **70**) as well as N-substituted products.[103] Other examples of compounds reacting as ambident nucleophiles include: the phosphoranes, $Ph_3P = CHCOR$, which are normally alkylated and acylated on oxygen but are acylated on carbon by acid anhydrides (an example of kinetic vs. thermodynamic control);[104] the monoanion of L-ascorbic acid, which with benzyl chloride gives compounds **(71)** and **(72)**;[105] the anion of 3-phenylindene, either as the sodium salt or derived from 3-phenylindenylmagnesium bromide, in its reactions with 2-dimethylaminoethyl chloride;[106] the O,O-diethylthiophosphate anion, which is alkylated on oxygen and sulphur;[107] and p-alkylphenoxide ions, which with allyl bromide give the corresponding cyclohexa-2,5-dienones.[108a]

Attack by both the sulphur and nitrogen atoms of the thiocyanate ion occurs in its reaction with alkyl thiocyanates; the rate ratio k_S/k_N for reaction of potassium thiocyanate with benzyl thiocyanate[108b] was found to range from 10^2 to 10^3 when the rate of isomerization to isothiocyanate (N-attack) was compared with the rate of isotopic exchange (S-attack).

[101] K. R. Brower, R. L. Ernst, and J. S. Chen, *J. Phys. Chem.*, **68**, 3814 (1964).
[102] G. Brieger and W. M. Pelletier, *Tetrahedron Letters*, **1965**, 3555; S. T. Yoffe, K. V. Vatsuro, E. E. Kugutcheva, and M. I. Kabachnik, *ibid.*, **1965**, 593.
[103] R. A. Shaw and P. Ward, *J. Chem. Soc.*, **1965**, 5446.
[104] P. A. Chopard, R. J. G. Searle, and F. H. Devitt, *J. Org. Chem.*, **30**, 1015 (1965).
[105] E. Buncel, K. G. A. Jackson, and J. K. N. Jones, *Chem. Ind.* (*London*), **1965**, 89.
[106] C. R. Ganellin, J. M. Loynes, and M. F. Ansell, *Chem. Ind.* (*London*), **1965**, 1256.
[107] G. Kuhlow, H. Teichmann, and G. Hilgetog, *Z. Chem.*, **5**, 179 (1965).
[108a] R. Barner, A. Boller, J. Borgulya, E. G. Herzog, W. von Philipsborn, C. von Planta, A. Fürst, and H. Schmid, *Helv. Chim. Acta*, **48**, 94 (1965).
[108b] A. Fava, A. Iliceto, and S. Bresadola, *J. Am. Chem. Soc.*, **87**, 4791 (1965).

Other Reactions

An important discussion of the effect of solvent composition on the activation parameters for the solvolysis of t-butyl chloride has appeared.[109] It has been known[110] for a number of years that the plot of $\Delta H^{\ddagger}$ for this reaction against solvent composition shows a minimum when the mole fraction of the water is 0.8—0.9.[110]

The partial molar heats of solution of a number of electrolytes and non-electrolytes have now been shown to have maximum values at the same solvent composition as that for which $\Delta H^{\ddagger}$ for the solvolysis reaction shows a minimum. The size of these maxima are greater for non-electrolytes than for electrolytes of comparable size. It was, therefore, concluded that the major part of the variation of $\Delta H^{\ddagger}$ for the solvolysis reaction can be attributed to variation of the enthalpy of the initial state rather than of the transition state (which will be salt-like in character). The reason for these maxima was considered to be that the addition of alcohol to water causes, at first, an increase in the "degree of solvent structuredness" up to an alcohol mole fraction of 0.1—0.2; thus addition of a third component appears to result in less structure formation than when addition is made to pure water. In a later communication,[111] a complete dissection of the solvent effect into the free energy, enthalpy, and entropy of solution of ground and transition states was made for the range 100% to 40% ethanol. The plot of entropy of solution against enthalpy of solution for the transition state is almost linear, but that for the initial state shows a maximum and a minimum. It was, therefore, concluded that the "serpentine wanderings" of the plot of entropy of activation against enthalpy of activation was largely a reflection of the variation of the properties of the initial state. Interestingly, the one point that did not fall on the linear plot of entropy against enthalpy of solution for the transition state was that for 100% ethanol; it was suggested that here the solvent was assuming a special role (presumably nucleophilic participation).

Careful measurements of the dependence of rate constant on temperature have been made for the solvolyses of t-butyl chloride,[112] t-pentyl chloride,[113] 2-chloro-2-methylpropyl methyl ether,[113] and the dimethyl-t-butylsulphonium ion[113] in water.

Swain and Thornton's criticism[114] of the view of Robertson and his

[109] E. M. Arnett, W. G. Bentrude, J. J. Burke, and P. McC. Duggleby, *J. Am. Chem. Soc.*, **87**, 1541 (1965); cf. E. M. Arnett, J. J. Burke, and P. McC. Duggleby, *ibid.*, **85**, 1350 (1963); E. M. Arnett and D. R. McKelvey, *Rec. Chem. Progr.*, **26**, 184 (1965).

[110] S. Winstein and A. H. Fainberg, *J. Am. Chem. Soc.*, **79**, 5937 (1957).

[111] E. M. Arnett, W. G. Bentrude, and P. McC. Duggleby, *J. Am. Chem. Soc.*, **87**, 2049 (1965).

[112] E. A. Moelwyn-Hughes, R. E. Robertson, and S. Sugamori, *J. Chem. Soc.*, **1965**, 1965.

[113] K. T. Leffek, R. E. Robertson, and S. Sugamori, *J. Am. Chem. Soc.*, **87**, 2097 (1965).

[114] C. G. Swain and E. R. Thornton, *J. Am. Chem. Soc.*, **84**, 822 (1962).

co-workers that the solvent isotope effect k_{D_2O}/k_{H_2O} on the solvolysis of methyl halides results mainly from an initial-state difference has been negated by Laughton and Robertson.[115] The basis for Swain and Thornton's criticism was that the solubility of each halide in H_2O and D_2O is almost the same, so that the standard free-energy differences between the initial states are never greater than 60 cal mole^{-1}. This, however, ignores differences in the free-energies of the D_2O and H_2O, which are large and to which the isotope effect can be ascribed. This view is also supported by the observation that the solvolyses of t-butyl fluoride and phenethyl fluoride have isotope effects similar to those of other halides. This is contrary to expectation of Swain and Thornton's view that the isotope effect is dependent on transition-state solvation since fluoride ion, unlike the other halide ions, has a lower free energy in D_2O than in H_2O.

Other investigations of the solvolyses of tertiary chlorides have been reported.[116]

The effect of electrolytes on the activity coefficient of initial and transition states of the solvolysis of 4-nitro-4'-phenyldiphenylmethyl chloride have been reported.[117] The variations in rate were shown to be due mainly to effects on the stability of the initial state.

The effects of substituents on the rate of solvolysis of benzyl bromide in formic acid containing 0.5% of water have been compared with their effects on the rate of the S_N1 solvolysis of 4'-nitrodiphenylmethyl chloride in 85% aqueous acetone.[118] Electron-releasing substituents increase the rates of both reactions similarly, but electron-withdrawing substituents retard the latter reaction much more markedly. Also, the plot of log k for the solvolyses of the benzyl bromides in the aqueous formic acid against σ^+ shows marked curvature. It was, therefore, concluded that benzyl bromides undergo solvolysis in aqueous formic acid by an S_N1 mechanism only when a 4-phenyl substituent, or some better electron donor, is present; this was supported by the entropies of activation.

Solvolyses of *trans,cis*-3,5-di-t-butylcyclohexyltoluene-p-sulphonate, which probably exists in a twist conformation, are much faster than those of its *cis,cis*- and *trans,trans*-isomers.[119] Extensive investigations have also been reported on the solvolyses of menthyl, neomenthyl, and isomenthyl toluene-p-sulphonates,[120] the toluene-p-sulphonates of the four isomeric carvo-

[115] P. M. Laughton and R. E. Robertson, *Can. J. Chem.*, **43**, 154 (1965).

[116] C. Prevost and J. Landais, *Bull. Soc. Chim. France*, **1965**, 361; J. Landais and C. Prevost, *ibid.*, **1965**, 2982; E. M. Kosower, *Reaktsionnaya Sposobniost Organ. Soedin. Tartusk. Zos. Univ.*, **1**, 238 (1964); *Chem. Abs.*, **62**, 10,307 (1965).

[117] E. Jackson and G. Kohnstam, *Chem. Comm.*, **1965**, 279.

[118] J. R. Fox and G. Kohnstam, *Chem. Comm.*, **1965**, 249.

[119] M. Hanack and K. W. Heinz, *Ann. Chem.*, **682**, 75 (1965).

[120] W. Hückel and C. M. Jennewein, *Ann. Chem.*, **683**, 100 (1965).

menthols[121] and of the four *t*-butylmenthols,[122] *cis*- and *trans*-2-isopropyl-cyclopentyl toluene-*p*-sulphonate,[123a] and 2,2-dimethylcyclohexyl toluene-*p*-sulphonate.[123b]

Trifluoroacetolysis of secondary alkyl toluene-*p*-sulphonates has been shown to be highly sensitive to electron release by the alkyl group. Possibly this is due to the poor nucleophilicity of the solvent, which results in little effective solvation of the developing cation in the transition state.[123c]

Solvolyses of some glucofuranosyl halides have been investigated.[124] Anomerization of the tetra-*O*-acetyl-D-glucopyranosyl chlorides in aceto-nitrile is promoted by tetraethylammonium chloride and involves nucleo-philic attack by chloride ion on the anomeric carbon.[125]

A theoretical discussion of leaving-group orders in S_N2 reactions has been presented.[126]

The rates of reaction of 39 amines with 15 substrates have been correlated by the Swain–Scott equation.[127]

The rates of the S_N2 reactions of a series of 4-substituted bicyclo[2.2.2]-octylmethyl toluene-*p*-sulphonates with sodium thiophenoxide in ethanol have been measured.[128] Electron-withdrawing substituents accelerate the

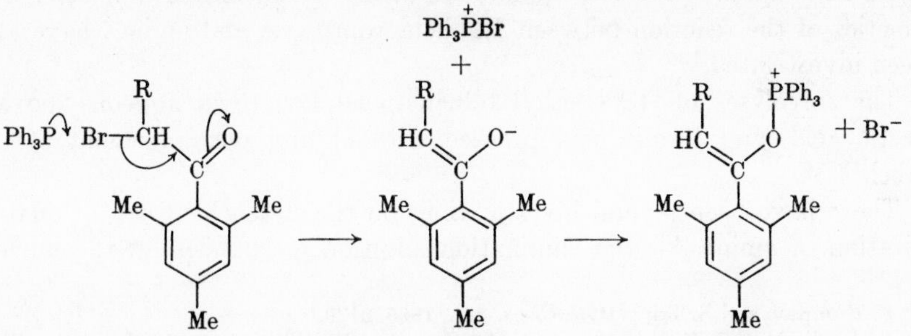

(73) **(74)**

reaction and this was discussed in terms of the work necessary to locate the partial negative charges of the transition state in the field of the substituent dipole.

The reaction of triphenylphosphine with the bromo ketones **(73)**, to give

[121] W. Hückel and P. Heinzelmann, *Ann. Chem.*, **687**, 82 (1965).
[122] W. Hückel and W. Sommer, *Ann. Chem.*, **687**, 102 (1965).
[123a] W. Hückel and R. Bross. *Ann. Chem.*, **685**, 118 (1965).
[123b] W. Hückel and S. K. Gupté, *Ann. Chem.*, **685**, 105 (1965).
[123c] P. E. Peterson, R. E. Kelley, R. Belloli, and K. A. Sipp, *J. Am. Chem. Soc.*, 87, 5169 (1965).
[124] C. P. J. Glaudemans and H. G. Fletcher, *J. Am. Chem. Soc.*, **87**, 2456, 4636 (1965).
[125] R. U. Lemieux and J. Hayami, *Can. J. Chem.*, **43**, 2162 (1965).
[126] R. E. Davis, *J. Am. Chem. Soc.*, **87**, 3010 (1965).
[127] H. K. Hall, *J. Org. Chem.*, **29**, 3539 (1964).
[128] H. D. Holtz and L. M. Stock, *J. Am. Chem. Soc.*, **87**, 2404 (1965).

the enolphosphonium salts (74), has been investigated.[129] It was considered unlikely that these sterically hindered ketones would react with attack on carbon and from the observation that the entropies of activation are highly negative it was concluded that the transition state is dipolar. Since the reaction is slower when R is Me and much faster when it is Br it appears that the halogen-bearing carbon atom carries a negative charge in the transition state. The mechanism was therefore written as annexed. This mechanism also explains the high sensitivity of the reaction to acid-catalysis. Nucleophilic attack on bromine was also favoured as the initial step in the reactions of triphenylphosphine with bromo ketones in view of the inertness of the corresponding chloro ketones.[130]

The kinetics of the reactions of phenacyl bromides with iodide ions and pyridine[131] and with thiobenzamides,[132] and of substituted α-halogenodeoxybenzoins with sodium methoxide,[133] have also been investigated.

The reaction of 3-chloro-1-butene with PhNDMe is slower than that with PhNHMe. The rate law is: Rate $= k_2[RCl][R_2NH] + k_3[RCl][R_2NH]^2$, and the isotope effect is shown only in the third-order term, $k_3^H/k_3^D = 1.38$ at $80.2°$.[134] It was suggested that the part of the reaction which follows third-order kinetics involves electrophilic catalysis by an amine molecule. The kinetics of the reaction between 1,3-dichloropropene and amines have also been investigated.[135]

The solvolyses of [1-^{14}C]ethyl toluene-p-sulphonate in aqueous dioxan, acetic acid, and formic acid proceed without any rearrangement of the label.[136]

There have been several investigations on the stereochemistry of quaternization of amines,[137] the elimination–addition mechanism,[138] the nucleo-

[129] F. Hampson and S. Trippett, *J. Chem. Soc.*, **1965**, 5129.

[130] P. A. Chopard, R. F. Hudson, and G. Klopman, *J. Chem. Soc.*, **1965**, 1379; see also R. F. Hudson in "Reaction Mechanisms", *Chem. Soc. Special Publ. No.*. 19, 1965, p. 93; R. D. Partos and A. J. Speziale, *J. Am. Chem. Soc.*, **87**, 5068 (1965).

[131] A. J. Sisti and W. Memeger, *J. Org. Chem.*, **30**, 2102 (1965).

[132] J. Okamiya, *Nippon Kagaku Zasshi*, **86**, 315 (1965); *Chem. Abs.*, **63**, 4123 (1965).

[133] T. I. Temnikova and V. S. Karavan, *Zh. Organ. Khim.*, **1**, 609 (1965); *Chem. Abs.*, **63**, 1680 (1965).

[134] D. C. Dittmer and A. F. Marcantonio, *J. Am. Chem. Soc.*, **86**, 5621 (1964).

[135] G. Geiseler, P. Herrmann, and W. Siemann, *J. Prakt. Chem.*, **29**, 113 (1965).

[136] C. C. Lee and M. K. Frost, *Can. J. Chem.*, **43**, 526 (1965).

[137] A. T. Bottini, B. F. Dowden, and R. L. VanEtten, *J. Am. Chem. Soc.*, **87**, 3250 (1965); J. McKenna, J. M. McKenna, A. Tulley, and J. White, *J. Chem. Soc.*, **1965**, 1711; J. Mc Kenna, B. G. Hutley, and J. White, *ibid.*, **1965**, 1729; J. McKenna, J. M. McKenna, and J. White, *ibid.*, **1965**, 1733; J. McKenna, J. M. McKenna, and A. Tulley, *ibid.*, **1965**, 5439; A. T. Bottini and R. L. VanEtten, *J. Org. Chem.*, **30**, 575 (1965).

[138] L. K. Montgomery, F. Scardiglia, and J. D. Roberts, *J. Am. Chem. Soc.*, **87**, 1917 (1965); C. J. M. Stirling, *J. Chem. Soc.*, **1964**, 5856, 5863, 5875; D. S. Campbell and C. J. M. Stirling, *ibid.*, **1964**, 5869; G. D. Jones, D. C. MacWilliams, and N. A. Braxtor, *J. Org. Chem.*, **30**, 1994 (1965).

philicity of phenoxide ions,[139] and electrophilic catalysis in nucleophilic substitutions.[140]

A few displacement reactions at unsaturated carbon have been studied. *cis-β*-Bromostyrene reacts with lithium diphenylphosphide in tetrahydrofuran to give *cis-β*-styryldiphenylphosphine uncontaminated with the *trans*-isomer; the same applies to the *trans*-bromide, *mutatis mutandis*; the mechanistic reasons for this stereospecificity are uncertain.[141] We note that *cis*-addition of lithium diphenylphosphide followed by *trans*-elimination of hydrogen bromide explains the facts. It is claimed that the exchange of acetate groups between vinyl acetate and trideuterioacetic acid, catalysed by mercuric acetate, proceeds by direct displacement at the unsaturated carbon rather than by addition-elimination.[142] Nucleophilic attack by ethoxide ions on a series of 3- and 4-substituted 1,2-dichlorocyclobutenes gave mixtures of products formed from the two possible carbanions, such that the apparent stabilization of the carbanion by the substituents on the β-carbon was in the order: dichloro > ethoxy, chloro > fluorochloro > diethoxy ≥ difluoro.[143] The reactions of various arylsulphonylhalogeno-ethylenes, $ArSO_2CH{=}CRHal$, with primary and secondary amines in methanol, ethanol, and propan-2-ol have been studied, and some of the complexities unravelled.[144a] Sodium toluene-*p*-thiolate reacts with tetrachloroethylene in boiling ethanol, to give *trans*-1,2-dichloro-1,2-di-(*p*-tolylthio)ethylene by two addition-elimination processes.[144b]

Other reactions which have received attention include: reactions of epoxides with nucleophiles;[145] hydrolysis of benzylidene di-chloride;[146]

[139] A. Berge and J. Ugelstad, *Acta Chem. Scand*, 1965, **19**, 742; A. Fischer, S. V. Sheat, and J. Vaughan, *J. Chem. Soc.*, **1965**, 1892; J. Ugelstad, A. Berge, and H. Listou, *Acta Chem. Scand.*, **19**, 208 (1965); H. E. Zaugg and A. D. Schaefer, *J. Am. Chem. Soc.*, **87**, 1857 (1965).

[140] R. Anataraman and K. Saramma, *Can. J. Chem.*, **43**, 1770 (1965); *Tetrahedron*, **21**, 535 (1965); R. S. Satchell, *J. Chem. Soc.*, **1964**, 5464; **1965**, 797; R. J. Berni, R. R. Benerito, H. M. Ziifle, *J. Phys. Chem.*, **69**, 1882 (1965); E. Meléndez and C. Prévost, *Bull. Soc. Chim. France*, **1965**, 2103; M. R. V. Sahyun, *Nature*, **206**, 788 (1965); G. D. Parfitt, A. L. Smith, and A. G. Walton, *J. Phys. Chem.*, **69**, 661 (1965); Y. Pocker and D. N. Kevill, *J. Am. Chem. Soc.*, **87**, 4760, 4771, 4778, 5060 (1965).

[141] A. M. Aguiar and D. Daigle, *J. Org. Chem.*, **30**, 3527 (1965).

[142] I. P. Samchenko and A. F. Rekasheva, *Zh. Fiz. Khim.*, **39**, 859 (1965); *Chem. Abs.*, **63**, 2864 (1965).

[143] J. D. Park, J. R. Dick, and J. H. Adams, *J. Org. Chem.*, **30**, 400 (1965).

[144a] S. Ghersetti, G. Lugli, G. Melloni, G. Modena, P. E. Todesco, and P. Vivarelli, *J. Chem. Soc.*, **1965**, 2227.

[144b] W. E. Truce, M. G. Rossmann, F. M. Perry, R. M. Burnett, and D. J. Abraham, *Tetrahedron*, **21**, 2899 (1965).

[145] E. A. S. Cavell, R. E. Parker, and A. W. Scaplehorn, *J. Chem. Soc.*, **1965**, 4780; R. M. Laird and R. E. Parker, *ibid.*, **1965**, 4784; R. E. Parker and B. W. Rockett; *ibid.*, **1965**, 2569; J. K. Addy and R. E. Parker, *ibid.*, **1965**, 644; M. P. Hartshorn and D. N. Kirk, *Tetrahedron*, **21**, 1547 (1965); D. E. Bissing and A. J. Speziale, *J. Am. Chem. Soc.*, **87**, 2683 (1965); G. Berti, B. Macchia, and F. Macchia, *Tetrahedron Letters*, **1965**, 3421; N. N. Lebedev and M. M. Smirnova, *Kinetika i Kataliz*, **6**, 457 (1965); *Chem. Abs.*, **63**,

solvolytic and elimination reactions of cyclooctane and cyclodecane derivatives;[147] solvolyses of toluene-*p*-sulphonates of neopentyl-type spiran alcohols;[148] displacement reactions of the dimethyl-α-methylbenzylsulphonium ion;[149] hydrolysis of 6-trichloromethylpurine;[150] reaction of 2-hydroxy-5-nitrobenzyl bromide;[151] acetolysis of the four 1-decalyl toluene-*p*-sulphonates;[152] hydrolysis of 3,4-dibromo- and 3,4-dichloro-2,2-dimethylchroman;[153] bimolecular nucleophilic substitution of alkyl halides;[154] solvolysis of arylalkyl halides;[155] reaction of methyl bromide and sodium cyanide in the presence of sodium dodecyl sulphate;[156] exchange of hexyl halides and halide ions,[157] of benzyl chloride and hydrogen chloride,[158] of ethyl iodide and iodine,[159] and of methyl 4,6-dideoxy-2,3-di-*O*-benzyl-4-iodo-α-D-gluco- and -galacto-pyranosides with iodide ion;[160] reactions of halogeno carboxylic acids with nucleophiles;[161] acid-catalysed solvolyses of α-diazo sulphones and 9-diazofluorenes;[162] nucleophilic displacement reactions of monochloroaziridines;[163] and reaction of 1,4:3,6-dianhydro-2,6-dideoxy-2,5-diiodoiditol with silver nitrate.[164]

8150 (1965); J. A. Franks, B. Tolbert, R. Steyn, and H. Z. Sable, *J. Am. Chem. Soc.*, 87, 1440; (1965); D. E. Bissing and A. J. Speziale, *J. Am. Chem. Soc.*, 87, 1405 (1965); A. Padwa, *Tetrahedron Letters*, 1965, 1049; J. Meinwald, S. S. Labana, L. L. Labana, and G. H. Wahl, *Tetrahedron Letters*, 1965, 1789; T. B. Zalucky, S. Marathe, L. Malspeis, and G. Hite, *J. Org. Chem.*, 30, 1324 (1965); M. P. Hartshorn and D. N. Kirk, *Tetrahedron*, 21, 1547 (1965); J. Bornstein, M. A. Joseph, and J. E. Shields, *J. Org. Chem.*, 30, 801 (1965); J. C. Leffingwell and E. E. Royals, *Tetrahedron Letters*, 1965, 3829.

[146] T. Ido and K. Tanabe, *J. Res. Inst. Catalysis, Hokkaido Univ.*, 12, 212 (1965).

[147] A. C. Cope, M. Brown, and G. L. Woo, *J. Am. Chem. Soc.*, 87, 3107 (1965).

[148] A. P. Krapcho, J. E. McCullough, and K. V. Nahabedian, *J. Org. Chem.*, 30, 139 (1965).

[149] H. M. R. Hoffmann, *J. Chem. Soc.*, 1965, 823.

[150] S. Cohen and N. Dinar, *J. Am. Chem. Soc.*, 87, 3195 (1965).

[151] H. R. Horton and D. E. Koshland, *J. Am. Chem. Soc.*, 87, 1126 (1965).

[152] C. A. Grob and S. W. Tam, *Helv. Chim. Acta*, 48, 1317 (1965).

[153] R. Binns, W. D. Cotterill, and R. Livingstone, *J. Chem. Soc.*, 1965, 5049.

[154] J.-J. Delpuech, *Tetrahedron Letters*, 1965, 2111; A. Kirrmann and J. J. Delpuech, *Compt. Rend.*, 260, 6600 (1965); J. J. Delpeuch and M. P. Pascal, *ibid.*, 260, 6600 (1965); J. J. Delpuech, *ibid.*, 260, 6355 (1965); J. J. Delpuech, *ibid.*, 259, 4259 (1964); G. C. Lalor and E. A. Moelwyn-Hughes, *J. Chem. Soc.*, 1965, 2201.

[155] E. Berliner, D. M. Falcione, and J. L. Riemenschneider, *J. Org. Chem.*, 30, 1812 (1965); F. F. Guzik and A. K. Colter, *Can. J. Chem.*, 43, 1441 (1965); F. L. Scott and J. B. Aylward, *Tetrahedron Letters*, 1965, 841; M. M. Tessler and C. A. VanderWerf, *J. Org. Chem.*, 30, 405 (1965); R. Bolton, *J. Chem. Soc.*, 1965, 1542.

[156] L. J. Winters and E. Grunwald, *J. Am. Chem. Soc.*, 87, 4608 (1965).

[157] H. Elias, O. Christ, and E. Rosenbaum, *Chem. Ber.*, 98, 2725 (1965).

[158] W. Bruce, M. Kahn, and J. A. Leary, *J. Am. Chem. Soc.*, 87, 2800 (1965).

[159] E. D. Cohen and C. W. Trumbore, *J. Am. Chem. Soc.*, 87, 964 (1965).

[160] C. L. Stevens, K. G. Taylor, and J. A. Valicenti, *J. Am. Chem. Soc.*, 87, 4579 (1965).

[161] J. Leska and M. Capla, *Chem. Zvesti*, 19, 339 (1965); *Chem. Abs.*, 63, 5467 (1965); J. F. Hinton and F. J. Johnson, *J. Phys. Chem.*, 69, 854 (1965).

[162] K. D. Warren and J. R. Yandle, *J. Chem. Soc.*, 1965, 4221; B. Zwanenburg and J. B. F. N. Engberts, *Rec. Trav. Chim.*, 84, 165 (1965).

[163] J. A. Deyrup and R. B. Greenwald, *J. Am. Chem. Soc.*, 87, 4538 (1965).

[164] L. O. Hayward, M. Jackson, and L. G. Csizmadia, *Can. J. Chem.*, 43, 1656 (1965).

CHAPTER 3

Electrophilic Aliphatic Substitution

A monograph on carbanion chemistry has appeared.[1]

Cram and his colleagues have continued their work on the measurement of the rate constants for the base-catalysed hydrogen isotope exchange, k_e, and racemization, k_α, of organic substrates. The ratio of these constants, k_e/k_α, is used as an indication of the steric course of the exchange reaction as follows:

k_e/k_α	Steric course
∞	100% retention
1	100% racemization
0.5	100% inversion
0	100% racemization without exchange (isoracemization)

Thus, Cram and Gosser[2] have examined the factors which control the steric

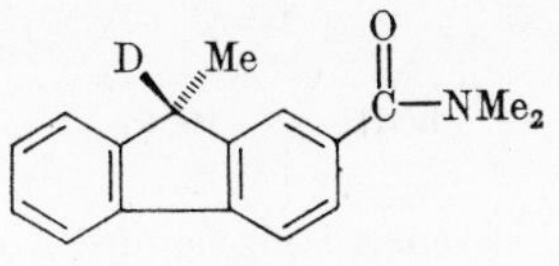

(1)

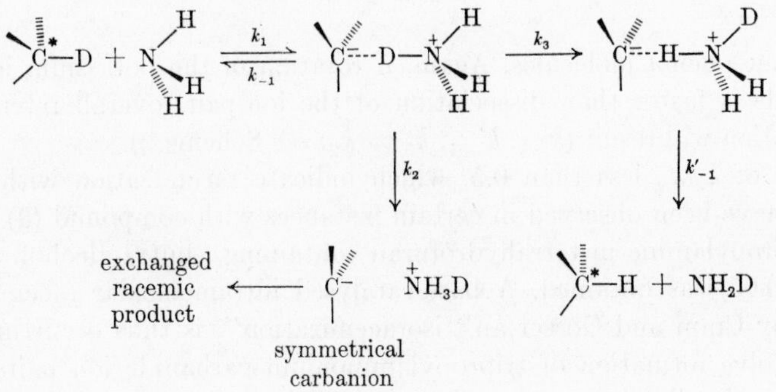

Scheme 1

[1] D. J. Cram "Fundamentals of Carbanion Chemistry", Academic Press, New York, N.Y., 1965.

[2] D. J. Cram and L. Gosser, *J. Am. Chem. Soc.*, **86**, 5445 (1964).

course of deuterium exchange at position 9 of 2-(NN-dimethylcarbamoyl)-9-methylfluorene and its 9-deuterated analogue (1) and obtained values of k_e/k_α of between 150 and 0.2 depending on the conditions. With ammonia or propylamine as base in the poorly dissociating solvent, tetrahydrofuran, a high degree of retention of configuration is observed. This is attributed to the formation of an ion pair which dissociates to free ions more slowly than the ammonium ion rotates and collapses (k_{-1}, k'_{-1}, $k_3 \gg k_2$; see Scheme 1). Addition of a salt or the use of a better ionizing solvent (e.g. dimethyl sulphoxide) causes racemization. Retention of configuration is also observed with potassium phenoxide when 10% phenol in benzene is used as solvent. This reaction was thought to involve the formation, first, of an asymmetrically solvated potassium phenoxide ion pair (2) in which the potassium ion is solvated by one molecule of deuterated phenol just formed and one or

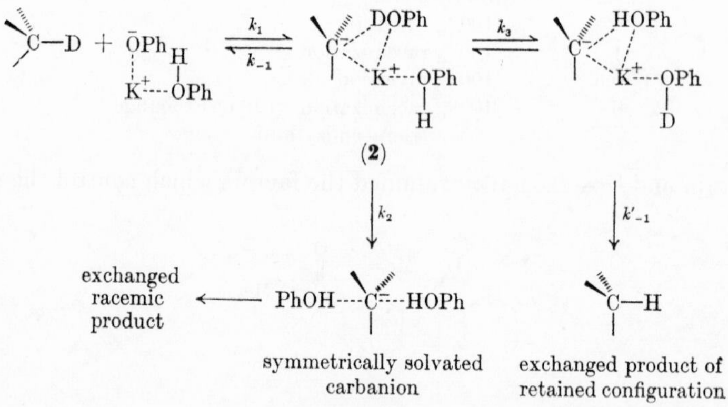

(2)

exchanged
racemic   ◄————   PhOH----C⁝---HOPh
product

symmetrically solvated
carbanion

exchanged product of
retained configuration

Scheme 2

more other phenol molecules. Again, if rotation of the potassium ion with its ligands is faster than dissociation of the ion-pair, overall retention of configuration will result (k_{-1}, k'_{-1}, $k_3 \gg k_2$; see Scheme 2).

Values of k_e/k_α less than 0.5, which indicate racemization without exchange, have been observed in certain instances with compound (3).[3] Thus, with tripropylamine in tetrahydrofuran containing t-butyl alcohol, a value $k_e/k_\alpha = 0.05$ was obtained. A base-catalysed intramolecular racemization, termed by Cram and Gosser an "isoracemization", is thus occurring. This must involve formation of tripropylammonium–carbanide ion pairs which racemize and collapse more rapidly than they undergo deuterium exchange. It was suggested that this involved rotation of the carbanion as shown in Scheme 3.

[3] D. J. Cram and L. Gosser, *J. Am. Chem. Soc.*, **86**, 5457 (1964).

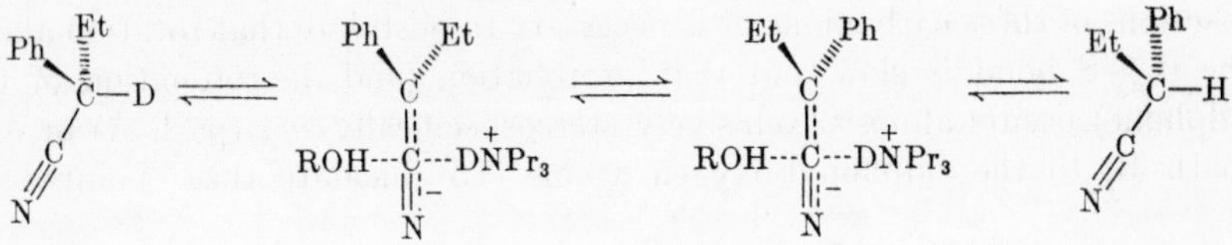

$$\underset{\underset{N}{\overset{C}{\parallel}}}{\overset{Ph}{\underset{|}{\overset{Et}{\underset{|}{C}}}}}\text{—D} \rightleftharpoons \underset{\underset{N}{\overset{|||^{-}}{\text{ROH---}\overset{+}{C}\text{---DNPr}_3}}}{\overset{Ph}{\overset{|}{\underset{||}{C}}}}\overset{Et}{} \rightleftharpoons \underset{\underset{N}{\overset{|||^{-}}{\text{ROH---}\overset{+}{C}\text{---DNPr}_3}}}{\overset{Et}{\overset{|}{\underset{||}{C}}}}\overset{Ph}{} \rightleftharpoons \underset{\underset{N}{\overset{C}{\parallel}}}{\overset{Et}{\underset{|}{\overset{Ph}{\underset{|}{C}}}}}\text{—H}$$

(3)

Scheme 3

The potassium *t*-butoxide-catalysed hydrogen exchange in *t*-butyl[2H]-alcohol of the vinylic hydrogen of *cis*-stilbene occurs about 2500 times faster than isomerization of *trans*-stilbene. It was, therefore, concluded that the vinylic anion of *cis*-stilbene has considerable geometric stability.[4] The *t*-butoxide-catalysed rearrangement of 1,3-diphenylpropene into *cis*- and *trans*-α-methylstilbene occurs with 55% and 36% intramolecularity.[4] Other intramolecular base-catalysed 1,3- and 1,5-proton shifts have also been reported.[5,6,7]

The trifluoromethyl group has been shown to be considerably less able to stabilize the configuration of an adjacent carbanion centre than is a cyano or phenyl group.[8] This was attributed to the ability of the three fluorine atoms to act as ligands to the potassium ion and cause the ion to pass into the plane of symmetry of the carbanion which thus becomes symmetrically solvated.

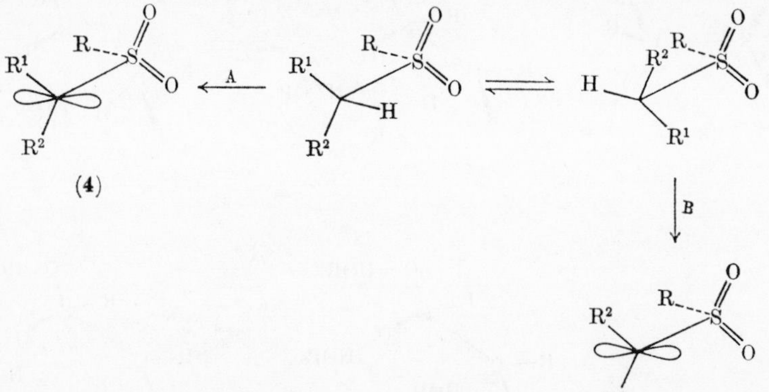

(4)

The question whether the carbanion-carbon of α-sulphonyl carbanions generated in hydrogen–deuterium exchange reactions of sulphones is planar and sp^2 hybridized, as (4), or pyramidal, as (8), has been discussed by Corey and Lowry.[9] In the former case, to explain the high stereospecificity of the

[4] D. H. Hunter and D. J. Cram, *J. Am. Chem. Soc.*, **86**, 5478 (1964).
[5] D. J. Cram and R. T. Uyeda, *J. Am. Chem. Soc.*, **86**, 5466 (1964).
[6] D. J. Cram, F. Willey, H. P. Fischer, and D. A. Scott, *J. Am. Chem. Soc.*, **86**, 5370 (1964).
[7] D. J. Cram and R. D. Guthrie, *J. Am. Chem. Soc.*, **87**, 397 (1965).
[8] D. J. Cram and A. S. Wingrove, *J. Am. Chem. Soc.*, **86**, 5490 (1964).
[9] E. J. Corey and T. H. Lowry, *Tetrahedron Letters*, **1965**, 793.

reactions of these carbanions, it is necessary to postulate that rotation about the $C_{(\alpha)}$–S bond is slow and that protonation (and deprotonation of the sulphone) occurs almost exclusively stereospecifically *syn* (path A) or *anti* (path B) to the sulphonyl-oxygen atoms. To elucidate these points, the

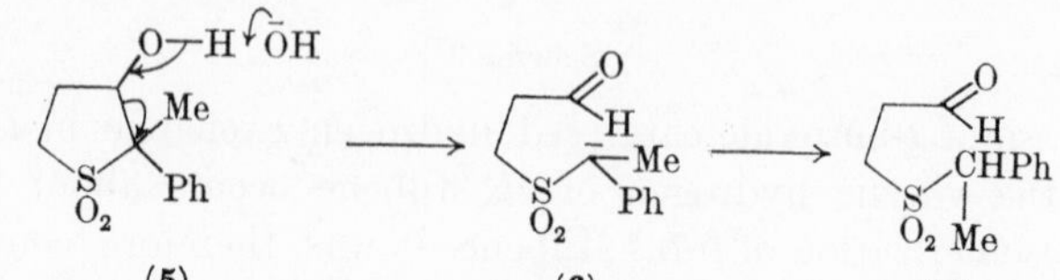

carbanion (6) was generated from compound (5) of known absolute configuration, and the configuration of the resulting sulphone (7) was determined. It was shown to have been formed with inversion of configuration; hence protonation of the carbanion had occurred *syn* to the sulphone oxygens, i.e. as path A. It was considered that this was not due to shielding by the carbon leaving group. It was also thought that this result could not be reconciled with a pyramidal asymmetric carbon atom which is constrained

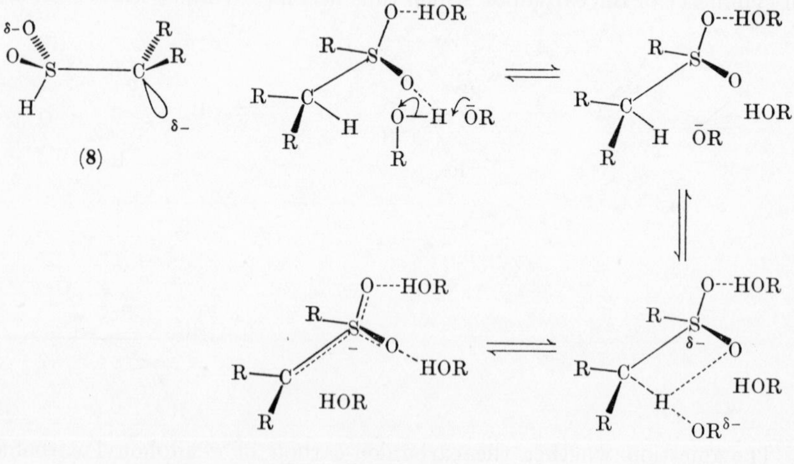

Scheme 4

to hold conformation (8) because of "electrostatic inhibition of inversion", since this would predict retention of configuration in the reaction studied. It was thought that possibly the formation and protonation of the α-sulphonyl carbanions by path A might be due to hydrogen bonding between the sulphonyl oxygen atoms and solvent molecules as shown in Scheme 4. Evidence has also been presented that the decarboxylation of (9) to give

$$
\underset{\substack{\text{Ph} \\ (-) \\ (9)}}{\text{Me}-\underset{\substack{| \\ }}{\overset{\substack{\text{CO}_2\text{H} \\ |}}{\text{C}}}-\text{SO}_2\text{Ph}} \longrightarrow \underset{\substack{\text{Ph} \\ (+) \\ (10)}}{\text{Me}-\underset{\substack{| \\ }}{\overset{\substack{\text{H} \\ |}}{\text{C}}}-\text{SO}_2\text{Ph}}
$$

(10) proceeds with retention of configuration.[10] It was thence concluded that the carbon dioxide must be departing *syn* to the sulphonyl-oxygen.

Changes in the electronic spectrum of fluorenylsodium in tetrahydrofuran with temperature have been interpreted in terms of varying amounts of solvent-separated and intimate ion pairs.[11] The NMR spectra of complexed fluorenyl-lithium in different solvents have also been reported.[12]

The rate of the *t*-butoxide-catalysed deuterium exchange of compound **(11)** in dimethyl sulphoxide occurs 3×10^4 times faster than that of **(12)**, which has been ascribed to intervention of the non-classical carbanion **(13)**.[13]

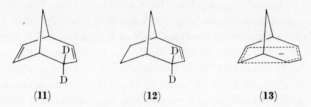

The equilibrium acidities of a number of carbon acids in cyclohexylamine have been measured.[14] 9,10-Dihydro-9,9-dimethyl-10-phenylanthracene is a stronger acid than triphenylmethane by 3.5 pK units. This was attributed to planarity of two of the phenyl rings of the former. On this basis it was estimated that, of the 14 pK unit difference in acidity between triphenyl-methane and 9-phenylfluorene, about one-third comes from the increased planarity of the rings in the latter carbanion and two-thirds from the distinctive anion-stabilizing ability of the five-membered ring.

The base-catalysed equilibration *exo*- and *endo*-2-norbornenyl cyanide and of *exo*- and *endo*-norbornyl cyanide have been studied.[15]

The equilibrium:

$$\text{RLi} + \text{R'I} \rightleftharpoons \text{RI} + \text{R'Li}$$

[10] E. J. Corey and T. H. Lowry, *Tetrahedron Letters*, **1965**, 803.
[11] T. E. Hogen-Esch and J. Smid, *J. Am. Chem. Soc.*, **87**, 669 (1965).
[12] J. A. Dixon, P. A. Gwinner, and D. C. Lini, *J. Am. Chem. Soc.*, **87**, 1379 (1965).
[13] J. M. Brown and J. L. Occolowitz, *Chem. Comm.*, **1965**, 376; see also H. Prinzbach, W. Eberbach, and G. von Veh, *Angew. Chem. Intern. Ed. Engl.*, **4**, 436 (1965).
[14] A. Streitwieser, J. I. Brauman, J. H. Hammons, and A. H. Pudjaatmaka, *J. Am. Chem. Soc.*, **87**, 384 (1965).
[15] P. Wilder and D. B. Knight, *J. Org. Chem.*, **30**, 3078 (1965).

in ether–light petroleum at $-70°$ has been studied by quenching the equilibrium mixture with an excess of benzaldehyde and determining the composition of the resulting alcohols.[16] It was shown in this way that the order of stability of the lithium alkyls was 1-norbornyl > s-butyl > bicyclo[2.2.2]-octan-1-yl > t-butyl. It was therefore concluded that increased bridgehead strain results in increased carbanion stability. 1-Adamantyl-lithium has been reported.[17]

On attempted carboxylation of the organolithium compound, $Ph_2C:CClLi$ in ether, more than 98% of phenylacetylene was formed, but in tetrahydrofuran the expected acid was obtained in 94% yield.[18] It was suggested that the lithium ion is more strongly solvated in tetrahydrofuran and so is less able to act as an electrophilic catalyst for the α-elimination.

The rates of ionization of a number of carbon acids have been measured by trapping the anion with molecular oxygen which is reported to react rapidly, yielding the corresponding alcohol.[19]

A number of investigations of the structure, configurational stability, and reactions of Grignard reagents have been reported. Ashby[20] has crystallized the ethylmagnesium bromide prepared from ethyl bromide and magnesium in triethylamine into seven fractions each of which had a Mg:Br:N ratio of 1.0:1.0:1.0. On the grounds that $MgBr_2$ is highly insoluble and Et_2Mg highly soluble in triethylamine it was concluded that the product must be a single species $EtMgBr.Et_3N$. When a diethyl ether solution of the Grignard reagent from ethyl bromide and magnesium was poured into triethylamine, the complex $ErMgBr.Et_3N$ was again obtained, and no $MgBr_2·Et_3N$ was isolated although it is highly insoluble in triethylamine. It was, therefore, concluded that in diethyl ether solution the Grignard reagent consists of RMgX (as monomer or dimer).[20] Ebullioscopic measurements also support a monomeric structure $EtMgBr$.[21]

It has now been shown that when magnesium-25 bromide is added to a solution of a Grignard reagent the label becomes statistically distributed between the different forms of magnesium and hence the equilibrium:

$$2EtMgBr \rightleftharpoons Et_2Mg + MgBr_2$$

or some equivalent process must be set up.[22]

Several products obtained from the evaporation of ethereal solutions of

[16] P. T. Lansbury and J. D. Sidler, *Tetrahedron Letters*, **1965**, 691.

[17] P. T. Lansbury and J. D. Sidler, *Chem. Comm.*, **1965**, 373.

[18] G. Köbrich, H. R. Merkle, and H. Trapp, *Tetrahedron Letters*, **1965**, 969.

[19] G. A. Russell and A. G. Bemis, *Chem. Ind. (London)*, **1965**, 1262.

[20] E. C. Ashby, *J. Am. Chem. Soc.*, **87**, 2509 (1965).

[21] M. B. Smith and W. E. Becker, *Tetrahedron Letters*, **1965**, 3843.

[22] D. O. Cowan, J. Hsu, and J. D. Roberts, *J. Org. Chem.*, **29**, 3688 (1964); see also R. E. Dessy, S. E. I. Green, and R. M. Salinger, *Tetrahedron Letters*, **1964**, 1369; C. Blomberg, A. D. Vreugdenhil, B. Van Zanten, and P. Vink, *Rec. Trav. Chim.*, **84**, 828 (1965).

Grignard reagents have been found by X-ray crystallography to have the composition, $R_2Mg + MgX_2$.[23]

It has been shown by NMR spectroscopy that inversion of the $-CH_2Mg$ centres of 3,3-dimethylbutylmagnesium chloride and bis-(3,3-dimethylbutyl) magnesium occur rapidly at room temperature.[24] The mechanisms of the reactions of Grignard reagents with alkynes,[25, 26, 27] epoxides,[28] and benzonitrile[29] have been investigated.

Other reactions involving carbanions which have been investigated include: *t*-butoxide-catalysed deuterium exchange of trideuteriotoluene in dimethyl sulphoxide;[30] base-catalysed olefin isomerizations;[31] base-catalysed reactions of 1*H*-undecafluorobicyclo[2.2.1]heptane;[32] deuteroxide-catalysed exchange of dimethyl sulphoxide[33] and methylenecyclopropanecarboxylic acids;[34] tautomerism of unsaturated ethers;[35] dimerization of propene in the presence of potassium;[36] and deuterium exchange of thioanisole.[37]

Stable carbanions which have been studied include the cyclononatetraenide anion,[38, 39, 40] the cyclooctatetraene dianion,[41] the heptaphenylcycloheptatrienyl anion,[42] the dibenzocyclobutadiene dianion,[43] the naphthalene dianion,[44] and the malonaldehyde anion.[45]

It has been shown that the reaction of R-$(-)$-dibutyl α-methylbenzylboronate (**14**) with mercuric chloride in water–glycerol–acetone containing sodium chloride and sodium acetate yields $R(+)$-α-methylbenzylmercuric chloride (**15**).[46] *exo*-Bicyclo[2.2.2]oct-5-en-2-ylboronic acid (**16**) reacts much

[23] E. Weiss, *Chem. Ber.*, **98**, 2805 (1965).
[24] G. M. Whitesides, M. Witanowski, and J. D. Roberts, *J. Am. Chem. Soc.*, **87**, 2854 (1965).
[25] T. L. Jacobs and R. A. Meyers, *J. Am. Chem. Soc.*, **86**, 5244 (1964).
[26] H. Hashimoto, T. Nakano, and H. Okada, *J. Org. Chem.*, **30**, 1234 (1965).
[27] J. H. Wotiz and G. L. Proffitt, *J. Org. Chem.*, **30**, 1240 (1965).
[28] H. Felkin and G. Roussi, *Tetrahedron Letters*, **1965**, 4153.
[29] A. A. Scala and E. I. Becker, *J. Org. Chem.*, **30**, 3491 (1965).
[30] J. E. Hofman, A. Schriesheim, and R. E. Nickols, *Tetrahedron Letters*, **1965**, 1745.
[31] S. Bank, A. Schriesheim, and C. A. Rowe, *J. Am. Chem. Soc.*, **87**, 3244 (1965); S. Bank, *ibid.*, **87**, 3245 (1965); R. Kuhn and D. Rewicki, *Tetrahedron Letters*, **1965**, 3513.
[32] S. F. Campbell, R. Stephens, and J. C. Tatlow, *Chem. Comm.*, **1965**, 134.
[33] E. Buncel, E. A. Symons, and A. W. Zabel, *Chem. Comm.*, **1965**, 173.
[34] A. T. Bottini and A. J. Davidson, *J. Org. Chem.*, **30**, 3302 (1965).
[35] C. D. Broaddus, *J. Am. Chem. Soc.*, **87**, 3706 (1965).
[36] A. W. Shaw, C. W. Bittner, W. V. Bush, and G. Holzman, *J. Org. Chem.*, **30**, 3286 (1965); W. V. Bush, G. Holzman, and A. W. Shaw, *ibid.*, **30**, 3290 (1965).
[37] A. I. Shatenshtein, E. A. Rabinovich, and V. A. Pavlov, *Zh. Obshch. Khim.*, **34**, 3991 (1964).
[38] T. J. Katz and P. J. Garrett, *J. Am. Chem. Soc.*, **86**, 5194 (1964).
[39] H. E. Simmons, D. B. Chesnut, and E. A. LaLancette, *J. Am. Chem. Soc.*, **87**, 982 (1965).
[40] E. A. LaLancette and R. E. Benson, *J. Am. Chem. Soc.*, **87**, 1941 (1965).
[41] T. S. Cantrell and H. Shechter, *J. Am. Chem. Soc.*, **87**, 136 (1965).
[42] R. Breslow and H. W. Chang, *J. Am. Chem. Soc.*, **87**, 2200 (1965).
[43] N. L. Bauld and D. Banks, *J. Am. Chem. Soc.*, **87**, 128 (1965).
[44] J. Smid, *J. Am. Chem. Soc.*, **87**, 655 (1965).
[45] N. Bacon, W. O. George, and B. H. Stringer, *Chem. Ind. (London)*, **1965**, 1377.
[46] D. S. Matteson and R. A. Bowie, *J. Am. Chem. Soc.*, **87**, 2587 (1965).

$$\underset{\textbf{(14)}}{H\!-\!\overset{\displaystyle Ph}{\underset{\displaystyle Me}{C}}\!-\!B(OBu)_2} \quad \xrightarrow{\;HgCl_2\;} \quad \underset{\textbf{(15)}}{H\!-\!\overset{\displaystyle Ph}{\underset{\displaystyle Me}{C}}\!-\!HgCl}$$

faster than its *endo*-isomer with mercuric chloride and yields tricyclo-[2.2.2.0^{2,6}]octan-3-ylmercuric chloride **(17)**.[47] This is then another[48] example of an intramolecular electrophilic substitution proceeding with inversion of configuration.

(16)   (17)

Other reactions involving organometallic compounds which have been investigated include mercuration of acetic acid by mercuric acetate,[49] reaction between mercuric chloride and diazodiphenylmethane,[50] protolysis of dibenzylchromium,[51a] of dibenzylmercury,[51b] and of *cis*- and *trans*-2-chlorovinylmercuric chloride,[52] and the reactions of these chlorides with iodine.[53]

Mechanism[54] and reactivity[55] in aliphatic electrophilic substitution have been discussed and redistribution reactions reviewed.[56]

The ring opening of bicyclo[2.1.0]pentane **(18)** with chlorine yields cyclopentyl chloride (5%), *trans*-1,2-dichlorocyclopentane (62%), and *trans*- (8%) and *cis*-1,3-dichlorocyclopentane (5%).[57] The reaction was envisaged as proceeding via the 1,3-chloronium ion **(19)** which then underwent ring opening or 1,2 hydride migration to a 1,2-chloronium ion **(20)**.

[47] D. S. Matteson and M. L. Talbot, *Chem. Ind. (London)*, **1965**, 1378.
[48] For an earlier example see D. S. Matteson and J. O. Waldbillig, *J. Am. Chem. Soc.*, **86**, 3778 (1964).
[49] W. Kitching and P. R. Wells, *Australian J. Chem.*, **18**, 305 (1965).
[50] A. Ledwith and L. Phillips, *J. Chem. Soc.*, **1965**, 5969.
[51a] J. K. Kochi and D. Buchanan, *J. Am. Chem. Soc.*, **87**, 853 (1965).
[51b] I. P. Beletskaia, L. A. Fedorov, and O. A. Reutov, *Dokl. Akad. Nauk SSSR*, **163**, 1381 (1965).
[52] I. P. Beletskaya, V. I. Karpov, V. A. Moskalenko, and O. A. Reutov, *Dokl. Akad. Nauk SSSR*, **162**, 88 (1965); *Chem. Abs.*, **63**, 4111 (1965).
[53] I. P. Beletskaya, V. I. Karpov, and O. A. Reutov, *Dokl. Akad. Nauk SSSR*, **161**, 586 (1965); *Chem. Abs.*, **63**, 1676 (1965).
[54] N. A. Nesmeyanov and O. A. Reutov, *Tetrahedron*, **20**, 2803 (1964).
[55] M. H. Abraham and J. A. Hill, *Chem. Ind. (London)*, **1965**, 561.
[56] R. E. Dessy, T. Psarras, and S. Green, *Ann. N.Y. Acad. Sci.*, **125**, 43 (1965).
[57] R. T. LaLonde, *J. Am. Chem. Soc.*, **87**, 4217 (1965).

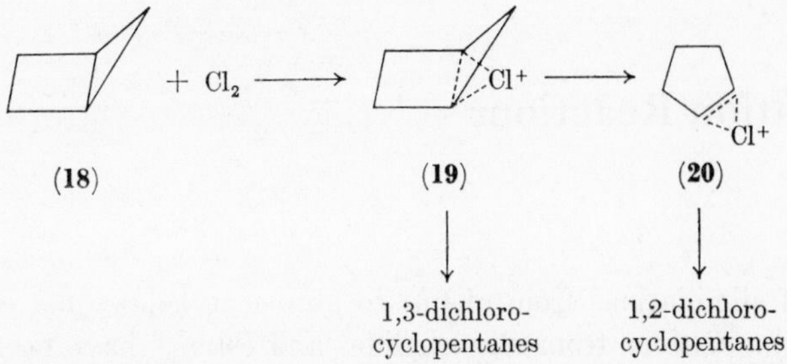

(18)   (19)   (20)

1,3-dichloro-   1,2-dichloro-
cyclopentanes   cyclopentanes

Further evidence has been presented that the side-chain chlorination of hexamethylbenzene proceeds with initial nuclear attack *via* the benzenonium ion (21).[58] The relative rates of bromination of some 4-substituted cyclo-

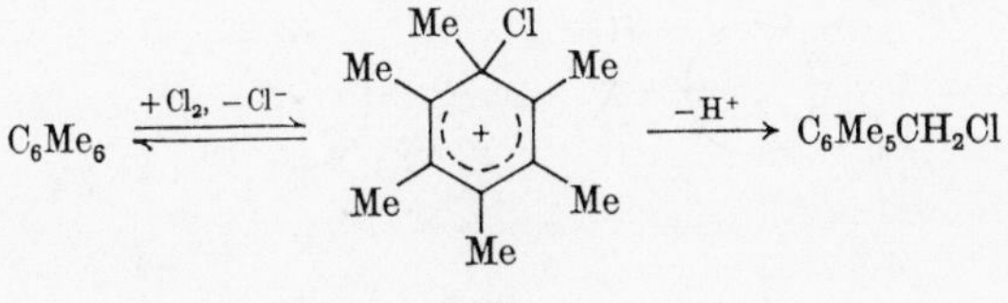

(21)

hexanecarboxylic acids and 4-substituted 2-phenylpropionic acids in thionyl chloride have been measured by a competition method and the mechanism has been discussed.[59]

[58] E. Bacioicchi, A. Ciana, G. Illuminati, and C. Pasini, *J. Am. Chem. Soc.*, **87**, 3953 (1965).
[59] H. Kwart and F. V. Scalzi, *J. Am. Chem. Soc.*, **86**, 5496 (1964).

CHAPTER 4

Elimination Reactions[1]

Reviews of eliminations from olefins to give acetylenes,[2] and of gas-phase pyrolytic eliminations from alkyl halides and esters,[3] have been published this year.

Interest continues in the possibility that concerted eliminations in which the groups being eliminated are *cis* and coplanar, as in (1), may occur almost

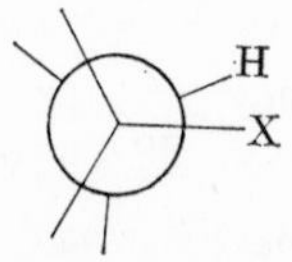

(1) 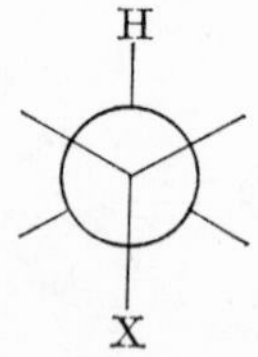　　　　　(2)

as readily as when they are antiperiplanar, as in (2).[4] DePuy and his co-workers have now published full details of their work on *cis*-elimination from *trans*-2-phenylcyclopentyl toluene-*p*-sulphonate (4) with potassium *t*-butoxide in *t*-butyl alcohol.[5] Their results (Table 1) were that in the 2-

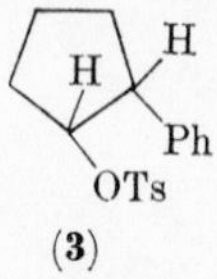

(3) 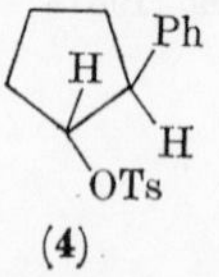　　　　(4) 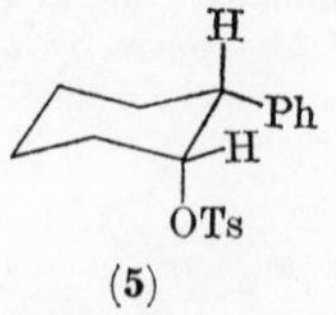　　　　(5) 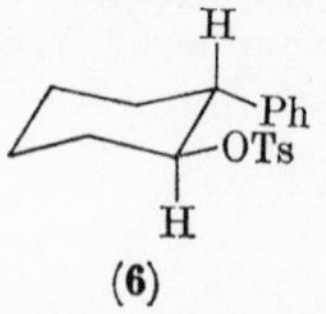　　　　(6)

phenylcyclopentyl system *trans*-elimination of toluene-*p*-sulphonic acid is only about 9 times as fast as *cis*-elimination, whereas with the 2-phenyl-

---

[1] For reviews see: D. V. Banthorpe, "Elimination Reactions", Elsevier, London, 1963; J. F. Bunnett, *Angew. Chem. Intern. Ed. Engl.*, 1, 225 (1962); C. K. Ingold, *Proc. Chem. Soc.*, 1962, 265.

[2] G. Kobrich, *Angew. Chem. Intern. Ed. Engl.*, 4, 49 (1965).

[3] A. Maccoll, *Adv. Phys. Org. Chem.*, 3, 91 (1965).

[4] For recent work previous to 1965, see: M. M. Kreevoy, J. W. Gilje, L. T. Ditsch, W. Batorewicz, and M. A. Turner, *J. Org. Chem.*, 27, 726 (1962); L. C. Schaleger, M. A. Turner, T. C. Chamberlin, and M. M. Kreevoy, *ibid.*, 27, 3421 (1962); N. A. LeBel, P. D. Beirne, E. R. Karger, J. C. Powers, and P. M. Subramanian, *J. Am. Chem. Soc.*, 85, 3199 (1963); N. A. LeBel, P. D. Beirne, and P. D. Subramanian, *ibid.*, 86, 4144 (1964); H. Kwart, T. Takeshita, and J. L. Nyce, *ibid.*, 86, 2606 (1964).

[5] C. H. DePuy, G. F. Morris, J. S. Smith, and R. J. Smat, *J. Am. Chem. Soc.*, 87, 2421 (1965).

cyclohexyl compounds *trans*-elimination occurs at least 10^4 times faster. This difference is attributed to the fact that eliminating groups can take up a *cis*-planar arrangement easily in the cyclopentyl but not in the cyclohexyl system, since in the latter conversion into a boat conformation would be

Table 1. Second-order rate constants for the elimination reactions with potassium *t*-butoxide in *t*-butyl alcohol at 50°.

Toluene-*p*-sulphonate		Type of elimination	$10^4 k_{E2}$ (l mole^{-1} sec^{-1})	k_{trans}/k_{cis}
cis-2-Phenylcyclopentyl	**(3)**	*trans*	26.4⎫	9.1
trans-2-Phenylcyclopentyl	**(4)**	*cis*	2.9⎭	
cis-2-Phenylcyclohexyl	**(5)**	*trans*	1.92 ⎫	$> 10^4$
trans-2-Phenylcyclohexyl	**(6)**	*cis*	No reaction⎭	

necessary. Evidence that this *cis*-elimination is concerted and does not involve an intermediate carbanion comes from the ρ-value obtained on introducing substituents into the phenyl ring and from isotope effects. Thus *cis*-elimination of toluene-*p*-sulphonic acid from *trans*-2-arylcyclopentyl toluene-*p*-sulphonates with potassium *t*-butoxide in *t*-butyl alcohol has a ρ-value of 2.8, intermediate between that for the *trans*-eliminations of *cis*-2-arylcyclopentyl toluene-*p*-sulphonates ($\rho = 1.5$) and 2-arylethyl toluene-*p*-sulphonates ($\rho = 3.4$). Then, on the thesis that the smaller the ρ-value the closer the mechanism is to being $E1$, while the larger the ρ-value the closer it is to being $E1cB$, the transition state for the *cis*-elimination is intermediate in carbanionic character between the two *trans*-eliminations. It was thought[5] that a reaction with a pure $E1cB$ mechanism would show a ρ-value of 4—5. The kinetic isotope effect k_H/k_D is 5.6 at 50°, close to the theoretical maximum; but if a carbanion were involved, a smaller value would be expected[6] since then the proton would be almost completely transferred in the transition state.

A *cis*-elimination mechanism has also been suggested by Závada and Sicher[7] for some reactions of the trimethyl-5-nonylammonium and dimethyl-5-nonylsulphonium ions. The ratios of *cis*- to *trans*-4-nonene obtained from the reactions between the trimethyl-5-nonylammonium ion and *t*-BuO⁻ in *t*-BuOH, EtO⁻ in EtOH, and MeO⁻ in MeOH, are 26:74, 76:24, and 81:19, respectively, and a similar trend was observed with the dimethyl-5-nonyl-sulphonium ion. It was suggested that on going from *t*-butyl alcohol to methanol the reaction becomes more $E1cB$-like owing to the increasing

[6] F. H. Westheimer, *Chem. Rev.*, **61**, 265 (1961).
[7] J. Závada and J. Sicher, *Collection Czech. Chem. Commun.*, **30**, 438 (1965).

ionizing power of the solvent. The higher proportion of *cis-* than of *trans-*4-nonene in the products of the reactions with EtO^- and MeO^- was taken to indicate that these cannot involve antiperiplanar transition states, such as (7) and (8), since then a higher proportion of *trans-*product should always be

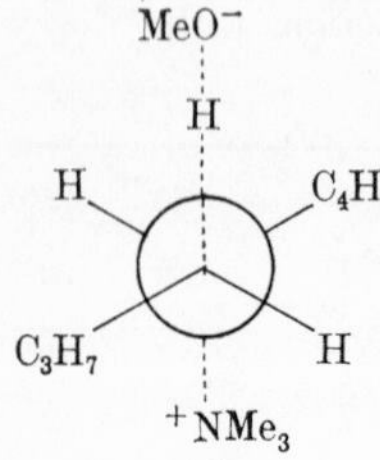

formed. Instead, it was suggested that, in accord with an earlier prediction by Ingold,[8] these *E1cB-*like reactions involve synclinal transition states such as (9) and (10), which would favour the formation of the *cis-*cycloalkene if double-bond character were not well developed. The basis of Ingold's prediction was that an elimination is to be regarded as an S_N2 process at the α-carbon atom, coupled to an S_N2 process at the β-carbon, and that the steric course of the latter should change from retention to inversion as the transition state becomes more ionic, so that *cis-*elimination should then be observed.

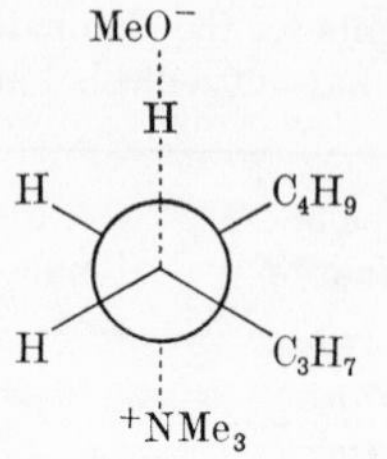

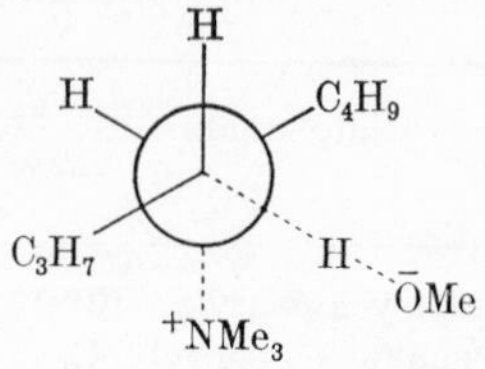

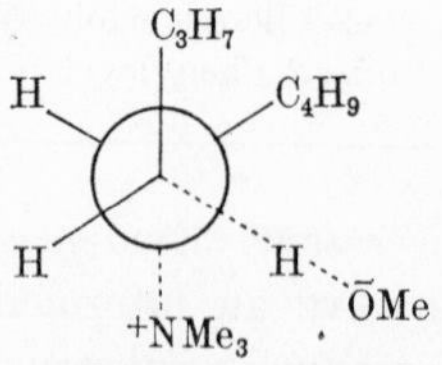

Saunders and his coworkers,[9] in a careful investigation, have shown that the proportion of 1-pentene obtained from the ethoxide-catalysed elimination from 2-halogenopentanes increases in the order $I < Br < Cl < F$, and that the proportion of 2-methyl-2-butene from 2-halogeno-2-methylbutanes increases in the order $Br < Cl < F$. It was hoped that these experiments might provide an answer to the question whether the formation of the least highly substituted olefin (Hofmann rule) is a result of steric[10] or electronic[8] effects, and at first sight the results obtained appear to support Ingold's view[8] that it is an electronic effect. Thus the proportion of most highly

[8] C. K. Ingold, *Proc. Chem. Soc.*, **1962**, 265.
[9] W. H. Saunders, S. R. Fahrenholtz, E. A. Caress, J. P. Lowe, and M. Schreiber, *J. Am. Chem. Soc.*, **87**, 3401 (1965).
[10] H. C. Brown and I. Moritani, *J. Am. Chem. Soc.*, **78**, 2203 (1956).

substituted double bond increases with increasing steric requirements of the leaving group (if this is assumed to be I > Br > Cl > F), which is contrary to what would be expected if steric effects were dominant. However, it has been shown[11] that the proportion of axial conformer in the halogenocyclo-hexanes in CS_2 solution at $-82°$ increases in the order Cl < Br < I < F, and if this indicates the steric requirements of the halogens in the compounds whose eliminations have been studied, under the experimental conditions, only the high proportion of 1-pentene and 2-methyl-1-butene obtained from the fluoride is inconsistent with Brown's view[10] of steric control. Also, there must be some uncertainty whether conclusions drawn from experiments with compounds where the leaving group is halide may be extrapolated to compounds containing leaving groups with larger steric requirements, e.g. SMe_2 and NMe_3. However, the results are entirely consistent with Ingold's view[8] that the formation of the least highly substituted double bond is controlled by the carbanionic character of the transition state.

Eliminations in dimethyl sulphoxide (DMSO) and mixtures of dimethyl sulphoxide with other solvents have been investigated by several groups of workers,[12,13] and some of these results also provide evidence on the origin of the Saytzeff and the Hofmann rule. Froemsdorf and McCain[14] have shown that the proportion of terminal olefin (i.e. the Hofmann product) from the potassium ethoxide-catalysed eliminations of s-butyl and 1-methylbutyl toluene-p-sulphonate increases when the solvent is changed from ethanol to dimethyl sulphoxide, and that even higher proportions of the terminal olefin are obtained when the base is potassium t-butoxide (see Table 2). The ratio of trans- to cis-2-alkene is also increased when the solvent is dimethyl sulphoxide. It was suggested[14] that these reactions could not involve a cis-elimination[15,16] since the cycloalkene from trans-2-methylcyclohexyl toluene-p-sulphonate consisted entirely of 3-methylcyclohexene although that from the cis-isomer contained 59% of 1- and 41% of 3-methylcyclohexene. Instead, the results were explained by assuming that, on change to dimethyl sulphoxide as solvent, the reaction becomes more E1cB-like and that the product ratio is

---

[11] A. J. Berlin and F. R. Jensen, *Chem. Ind.* (*London*), **1960**, 998; the ΔG values for chloro-, bromo-, and iodo-cyclo-hexane are, respectively, 513, 480, and 341 cal mole^{-1}, i.e. the differences between them are very small.

[12] For work previous to 1965 see: C. H. Snyder, *Chem. Ind.* (*London*), **1963**, 121; J. E. Hofmann, T. J. Wallace, P. A. Argabright, and A. Schriesheim, *ibid.*, **1963**, 1243; T. J. Wallace, J. E. Hofmann and A. Schriesheim, *J. Am. Chem. Soc.*, **85**, 2739 (1963); J. E. Hofmann, T. J. Wallace, and A. Schriesheim, *ibid.*, **86**, 1561 (1964).

[13] For a review of the solvent properties of dimethyl sulphoxide see C. Agami, *Bull. Soc. Chim. France*, **1965**, 1021.

[14] D. H. Froemsdorf and M. E. McCain, *J. Am. Chem. Soc.*, **87**, 3983 (1965).

[15] C. H. Snyder and A. R. Soto, *Tetrahedron Letters*, **1965**, 3261.

[16] C. H. Snyder and A. R. Soto, *J. Org. Chem.*, **30**, 673 (1965).

Table 2. Products from the elimination of s-butyl and 1-methylbutyl
toluene-p-sulphonate at 55°.

Alkyl group	Solvent	Base	1-ene (%)	2-ene: *trans/cis*
s-Butyl	EtOH	KOEt	35	1.95
s-Butyl	DMSO	KOEt	54	2.34
s-Butyl	DMSO	KOBu-t	61	2.53
1-Methylbutyl	EtOH	KOEt	42	1.90
1-Methylbutyl	DMSO	KOEt	66	3.07
1-Methylbutyl	DMSO	KOBu-t	72	3.26

controlled by the relative acidities of the terminal and non-terminal protons.
Similar proportions of 1-octene and of *cis*- and *trans*-2-octene were obtained
by Snyder and Soto in reactions of a series of 1-methylheptyl arenesulphonates
and sodium methoxide in dimethyl sulphoxide,[15,16] but these workers
preferred a *cis*-elimination mechanism and explained their results in terms
of steric effects.

The effects of solvent and base on the products of elimination from s-butyl
bromide have also been investigated by Froemsdorf and his co-workers.[17]
The results (Table 3) were discussed in terms of the ratio of C–H to C–Br
bond stretching in the transition state. It was suggested that when this is

Table 3. Products from the elimination of s-butyl bromide at 55°.

Solvent	Base	1-ene (%)	2-ene: *trans/cis*
EtOH	KOEt	19	3.35
t-BuOH	KOEt	38	2.19
DMSO	KOEt	27	3.57
t-BuOH	KOBu-t	53	1.64
DMSO	KOBu-t	31	3.65

high (i.e. the reaction is *E1cB*-like) a high proportion of 1-butene is formed,
and when it is low and C–H and C–Br stretching are highly concerted the
ratio of *trans*- to *cis*-2-butene is increased because then the eclipsing effects
become more important in the transition state. Thus, in *t*-butyl alcohol the
ratio of C–H to C–Br stretching is high because this solvent enhances the
basic strength of alkoxide ions but is poorly solvating for bromide ions, and
this results in a high proportion of 1-butene and a low ratio of *trans*- to *cis*-2-
butene. These results, therefore, fit well Ingold's[8] view "that the Hofmann
rule is a manifestation of a transition state in which C–H stretching is much

[17] D. H. Froemsdorf, M. E. McCain, and W. W. Wilkison, *J. Am. Chem. Soc.*, **87**, 3984 (1965).

greater than C–X stretching, while the Saytzelf rule is the result of a transition state in which the ratio of C–H to C–X stretching approaches one or less."[17]

The observation[18] that the secondary α-deuterium isotope effect in ethoxide- and hydroxide-catalysed elimination from the trimethylphenethylammonium ion is only 1—2% per deuterium compared with the 9% per deuterium observed in the corresponding reactions of phenethyl bromide also supports this view, since it suggests that the transition states for the former reaction have much greater carbanionic character.

There has been another report that, whereas primary alkyl halides give predominantly olefins on reaction with potassium t-butoxide in dimethyl sulphoxide, the corresponding toluene-p-sulphonates give mainly the t-butyl ether.[19]

Another example has been found[20] where the proportion of olefin formed in the solvolysis of a tertiary alkyl halide, RX, depends on X. t-Butyl iodide, bromide, and chloride, on solvolysis in dimethyl sulphoxide containing 2% of water, yield 48%, 60%, and 98% of olefin, respectively. Presumably an ion pair is involved in which the counter-ion assists in the removal of a proton from the t-butyl cation, and Cl^-, Br^-, and I^- do this at different rates. It was also shown that the proportion of olefin formed in the solvolysis of t-butyl chloride in mixtures of dimethyl sulphoxide and water increases linearly with the mole fraction of dimethyl sulphoxide, and that dimethyl sulphoxide is more effective in promoting elimination than are several other organic solvents. Thus, the percentage of olefin from mixed solvents in which the mole fraction of water is 0.717 is 25.0, 13.0, 9.7, and 6.8 when the organic component of the mixture is Me_2SO, Me_2CO, EtOH, and MeOH, respectively.

Colter and McKelvey[21] have studied t-butoxide-catalysed elimination from a series of 1-ethyl-2-methylpropylarenesulphonates, $CH_3 \cdot CH_2 \cdot CHX \cdot CH(CH_3)_2$, in t-butyl alcohol–dioxan and –dimethyl sulphoxide mixtures. Of particular interest is the observation that in 25% t-butyl alcohol–dimethyl sulphoxide these compounds yield a ratio of *trans*- to *cis*-4-methyl-2-pentene in the rate 21:1 to 35:1 although the *trans*:*cis* equilibrium mixture of these olefins in pure dimethyl sulphoxide is only 6:1. Thus, except in the unlikely event that this ratio is changed markedly on going to a 25% t-butyl alcohol–dimethyl sulphoxide solution, the elimination yields more *trans*-olefin than is present in the equilibrium mixture. This result therefore lays open to question

[18] S. Asperger, L. Klasinc, and D. Pavlovic, *Croat. Chem. Acta*, **36**, 159 (1964); *Chem. Abs.*, **63**, 2863 (1965).
[19] N. F. Wood and F. C. Chang, *J. Org. Chem.*, **30**, 2054 (1965); see also P. Veeravagu, R. T. Arnold, and E. W. Eigenmann, *J. Am. Chem. Soc.*, **86**, 3072, 5711 (1964).
[20] K. Heinonen and E. Tommila, *Suomen Kemistilehti*, Sect. B, **38**, 9 (1965); see also M. Cocivera and S. Winstein, *J. Am. Chem. Soc.*, **85**, 1702 (1963).
[21] A. K. Colter and D. R. McKelvey, *Can. J. Chem.*, **43**, 1282 (1965).

the simple view that the factors determining the proportion of *trans-* to *cis-* olefin at equilibrium also determine this proportion in the product of elimination and that the latter proportion is a function of the amount of double-bond character in the transition state.

A systematic investigation of the effect of pressure on elimination reactions has been carried out by Brower and Chen[22] who showed that the proportions of olefin formed in unimolecular solvolyses of *t*-pentyl chloride and the dimethyl-*t*-pentyl sulphonium ion are independent of pressure, indicating that the volume of activation for elimination and substitution are the same. The solvolysis of the chloride shows a negative volume of activation (see Table 4), attributed to the attraction of solvent by the largely ionic transition state, while that for the sulphonium ion is positive, which is thought to

Table 4. Volume of activation for some elimination reactions.

Reaction	Solvent	Temp.	$\Delta V^{\ddagger}$ (ml mole^{-1})
t-Pentyl chloride, solvolysis	EtOH (80%)	34.2°	-18
	MeOH (80%)	29.8	-16
Dimethyl-*t*-pentylsulphonium iodide, solvolysis	EtOH (80%)	53.8	$+14$
s-Butyl bromide and NaOEt	EtOH	47.8	-10
Trimethyl-*t*-pentylammonium iodide and KOH	EtOH	84.8	$+15$
Triethylsulphonium bromide and NaOH	H$_2$O	109.0	$+11^a$, $+8^b$
Triethylsulphonium bromide and NaOMe	MeOH	50.9	$+15^a$
Tetraethylammonium bromide and NaOMe	MeOH	104.8	$+20$

a Elimination.   *b* Substitution.

result from the expansion of the ion along the axis of the breaking carbon–sulphur bond. The proportion of olefin on bimolecular elimination from *s*-butyl bromide is also pressure-invariant, and the volume of activation (-10 ml mole^{-1}) is of the order of magnitude normally found for bimolecular reactions in which the total number of ionic species remains constant.[23] The proportion of olefin from the reaction of the triethylsulphonium ion and sodium hydroxide in water varies with pressure, and the substitution and elimination must, therefore, have different volumes of activation. However, the values of these, as well as those of the bimolecular eliminations of the other ionic substrates studied, are positive; it was considered that the increase in volume caused by release of molecules of solvation on formation of the

[22] K. R. Brower and J. S. Chen, *J. Am. Chem. Soc.*, **87**, 3396 (1965).
[23] E. Whalley, *Adv. Phys. Org. Chem.*, **2**, 93 (1964).

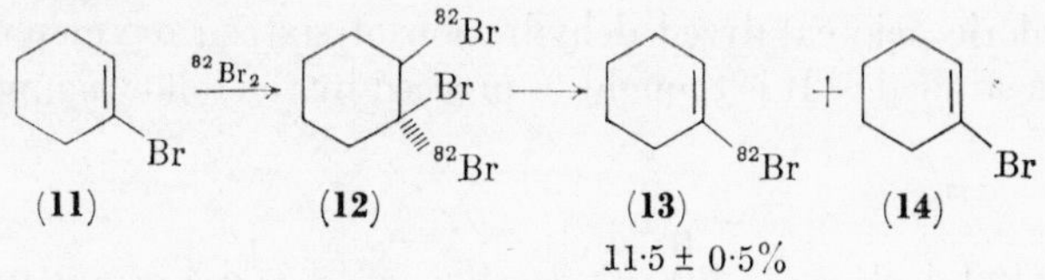

(11) (12) (13) (14)

$11.5 \pm 0.5\%$

transition state was greater than any squeezing out of space between base and substrate on formation of a covalent bond.

Stevens and Valicenti[24] have investigated the steric course of the zinc-promoted debromination of dibromides by studying the addition–elimination sequence, (11) → (14), of 1-bromocyclohexene with labelled bromine. It was first shown that the addition of labelled bromine to 1-bromocyclohexene followed by the iodide-promoted elimination of bromine resulted in retention of effectively none (0.3 ± 0.5%) of the label, which is consistent with both processes occurring in a stereospecific, presumably *trans*, manner. However, in the zinc-promoted debromination 11.5 ± 0.5% of the label was retained in the final 1-bromocyclohexene. It is not at present clear whether this non-specificity results from a *cis*-elimination or from a zinc-catalysed epimerization of the tribromide (12). Stevens and Valicenti even consider that the ionic nature of these zinc-promoted eliminations is by no means certain and that they may perhaps involve radicals.

The deoxymercuration of a series of oxymercurials of structure (15) by hydrochloric acid has been investigated.[25] The reaction follows a rate law:

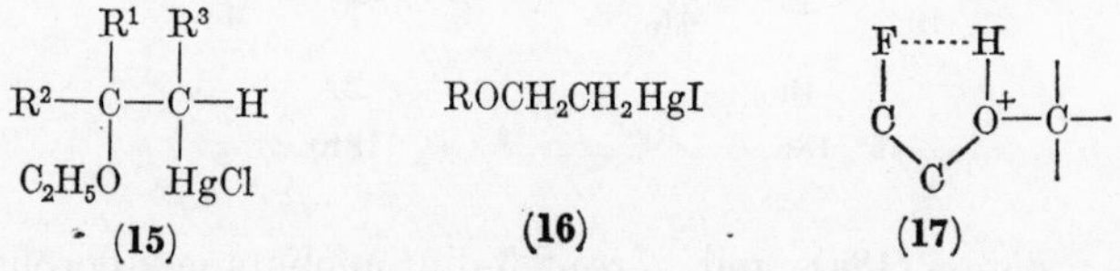

(15) (16) (17)

rate = $k[S][H^+][Cl^-]$ and changing R^1, R^2, or R^3 from hydrogen to methyl produces a 5—15-fold rate enhancement. The mechanism illustrated was suggested.

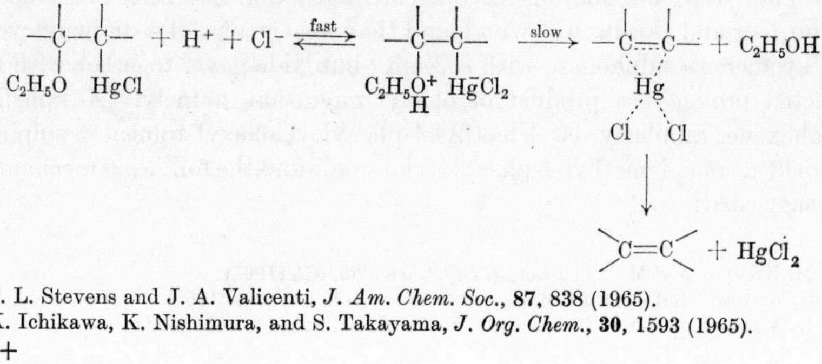

[24] C. L. Stevens and J. A. Valicenti, *J. Am. Chem. Soc.*, **87**, 838 (1965).
[25] K. Ichikawa, K. Nishimura, and S. Takayama, *J. Org. Chem.*, **30**, 1593 (1965).

4+

The perchloric acid-catalysed dehydration of sixteen oxymercurials (16) has also been studied.[26] It is thought to proceed by the following mechanism:

$$\text{RO}-\overset{|}{\underset{|}{\text{C}}}-\overset{|}{\underset{|}{\text{C}}}-\text{HgI} + \text{H}^+ \underset{\text{}}{\overset{\text{fast}}{\rightleftharpoons}} \underset{+}{\text{RO}}-\overset{\text{H}}{\underset{|}{\text{C}}}-\overset{|}{\underset{|}{\text{C}}}-\text{HgI} \overset{\text{slow}}{\longrightarrow} -\underset{+}{\text{C}}\overset{\cdots}{=}\overset{\text{HgI}}{\text{C}}- + \text{ROH}$$

$$-\underset{\underset{\text{HgI}}{+}}{\text{C}}\overset{\cdots}{=}\text{C}- + \text{RO}-\overset{|}{\underset{|}{\text{C}}}-\overset{|}{\underset{|}{\text{C}}}-\text{HgI} \overset{\text{series of fast steps}}{\longrightarrow \longrightarrow \longrightarrow} \overset{}{\underset{}{\diagup}}\text{C}{=}\text{C}\overset{}{\underset{}{\diagdown}} + \text{HgI}_2 + \text{RO}-\overset{|}{\underset{|}{\text{C}}}-\overset{|}{\underset{|}{\text{C}}}-\text{Hg}^+$$

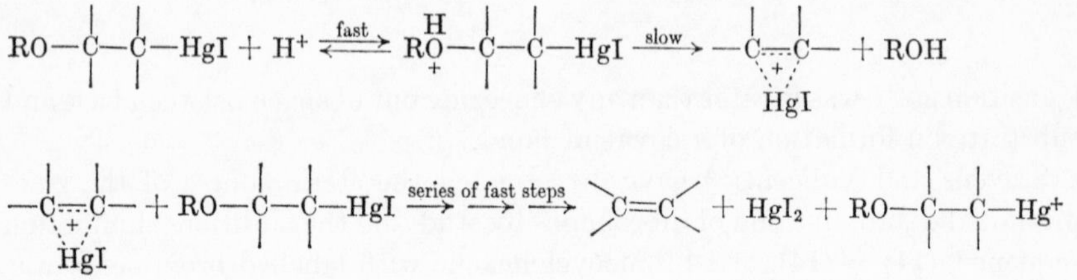

However, there is only a limited correlation with the Taft σ^* constant to give a ρ^* value of approximately -1.4, and it was suggested that this failure of the $\rho^*\sigma^*$ relationship to correlate all the rates was due partly to steric effects and partly, when R was F_3CCH_2, $NCCH_2CH_2$, $CH_3COCH_2CH_2$, or $HOCH_2CH_2$, to intramolecular hydrogen bonding (e.g. as in 17) which stabilized the congugate acid.

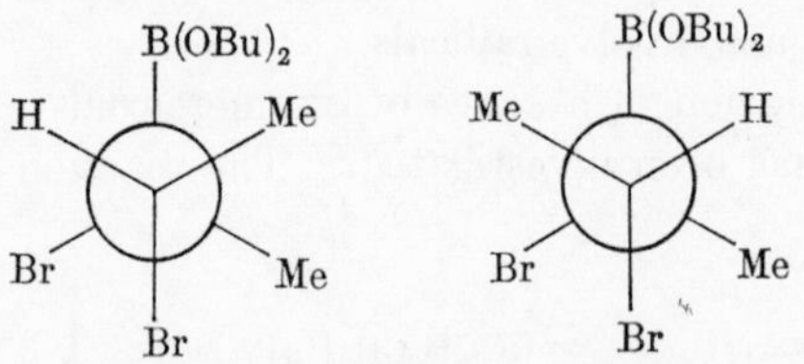

$$\begin{array}{cc}
\text{B(OBu)}_2 & \text{B(OBu)}_2 \\
\end{array}$$

(18a)   (18b)

Dibutyl *erythro*- (18a) and *threo*-2,3-dibromobutane-2-boronate (18b) yield exclusively *cis*- and *trans*-2-bromobut-2-ene, respectively, on treatment with water or base, showing that these deboronations involve *trans*-elimination.[27]

An interesting elimination-rearrangement reaction has been observed by Abdun-Nur and Bordwell,[28] who found that treatment of 4,4-diphenylcyclohexyl toluene-*p*-sulphonate with sodium *t*-butoxide gave, together with the expected products, a product of phenyl migration, namely, 1,4-diphenylcyclohexene. Similarly, *cis*-4-methyl-4-phenylcyclohexyl toluene-*p*-sulphonate yielded some 1-methyl-4-phenylcyclohexene and the following mechanism was suggested:

[26] M. M. Kreevoy and M. A. Turner, *J. Org. Chem.*, **30**, 373 (1965).
[27] D. S. Matteson and J. D. Liedtke, *J. Am. Chem. Soc.*, **87**, 1526 (1965).
[28] A. R. Abdun-Nur and F. G. Bordwell, *J. Am. Chem. Soc.*, **86**, 5695 (1964).

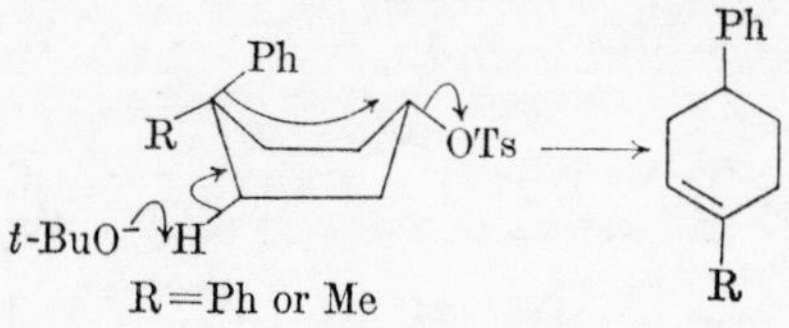

R=Ph or Me

On Hofmann elimination compound **(19)** yielded the α,β-unsaturated amide **(20)**.[29] The deuterated analogue **(21)** yielded **(22)**, in which deuterium had been lost from the position α to the carbonyl group but in which there had been no incorporation of deuterium into the dimethylamino-group. This result excludes an ylid mechanism and led the authors to suggest an $E1cB$ mechanism.[30]

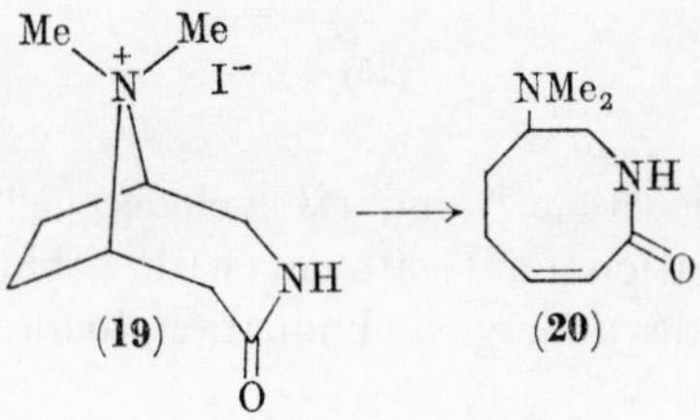

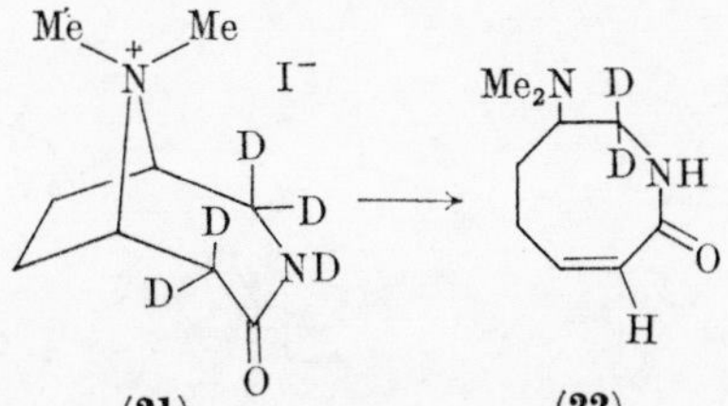

The transient existence of a bridgehead (anti-Bredt) olefin **(24)** has been postulated to occur in the reaction of undecafluorobicyclo[2.2.1.]heptyl-lithium **(23)** that affords 1-iodononafluorobicyclo[2.2.1.]hept-2-ene **(25)**,[31] and this is supported by the observation that, in the presence of furan, **(23)** yields a compound which is probably the adduct **(26)**. An attempt to generate the anti-Bredt olefin **(28)** as a vicinal diradical by pyrolysis of 1-vinylnortri-cyclene **(27)** failed.[32]

[29] L. A. Paquette and L. D. Wise, *J. Org. Chem.*, **30**, 228 (1965).

[30] For a penetrating criticism of the validity of this kind of evidence for an $E1cB$ mechanism see R. Breslow, *Tetrahedron Letters*, **1964**, 399.

[31] S. F. Campbell, R. Stephens, and J. C. Tatlow, *Tetrahedron*, **21**, 2997 (1965).

[32] J. A. Berson and M. R. Willcott, *J. Org. Chem.*, **30**, 3569 (1965).

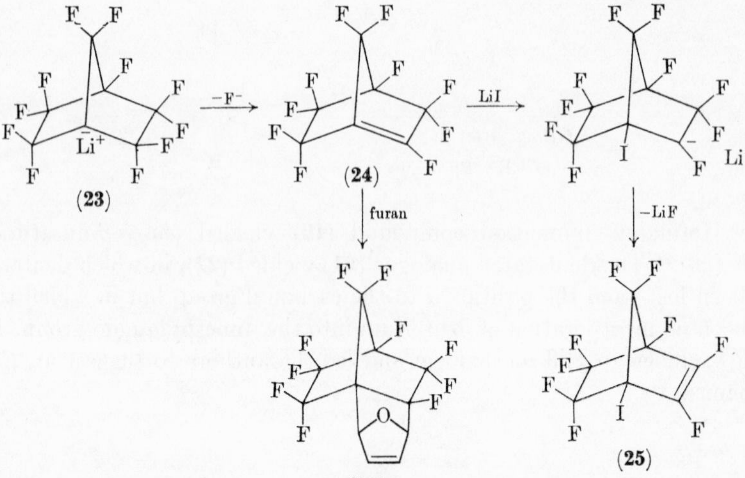

$-F^-$ → (24) $\xrightarrow{\text{LiI}}$

(23)

furan

$-\text{LiF}$

(26)

(25)

Unlike the *cis*-cyclooctene[33] and *cis*-cyclodecene[34] oxides which yield carbenes by α-elimination on treatment with lithium diethylamide, *cis*- and *trans*-4-octene oxides undergo β-elimination, to form *trans*-5-octen-4-ol.[35]

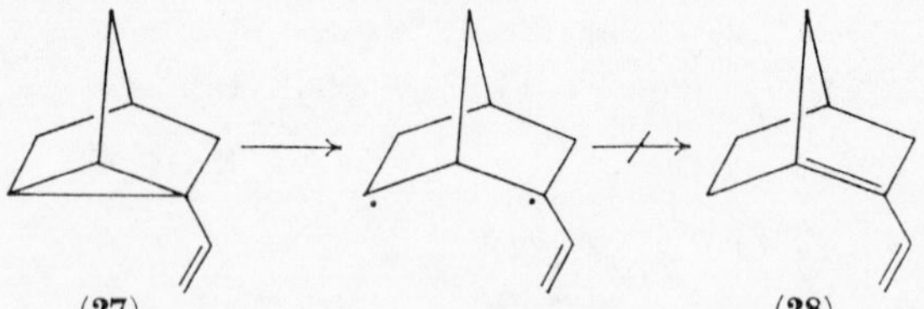

(27) (28)

That this is a true β-elimination and does not involve a carbene intermediate was shown by carrying out the reaction with the deuterated compound (29), which reacted without loss of deuterium.[35]

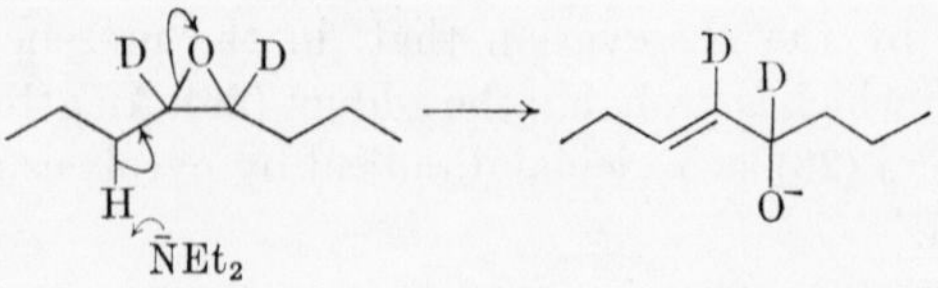

(29)

[33] A. C. Cope, H.-H. Lee, and H. E. Petree, *J. Am. Chem. Soc.*, **80**, 2849 (1958).
[34] A. C. Cope, M. Brown, and H. H. Lee, *J. Am. Chem. Soc.*, **80**, 2855 (1958).
[35] A. C. Cope and J. K. Heeren, *J. Am. Chem. Soc.*, **87**, 3125 (1965).

An α-elimination mechanism has, however, been found[36] for the gas-phase pyrolysis of 1,1,2,2-tetrafluoroethylsilanes (e.g. $CHF_2CF_2 \cdot SiF_3$, $CHF_2CF_2 \cdot SiMe_3$) which undergo a homogeneous reaction according to the equation:

$$CHF_2CF_2 \cdot SiX_3 \longrightarrow CHF:CF_2 + SiX_3F$$

However, in the presence of an olefin (e.g. propene, 2-butene), a large amount of the product is diverted to a cyclopropane without any change in the rate. The reaction is therefore formulated as involving an intermediate carbene:

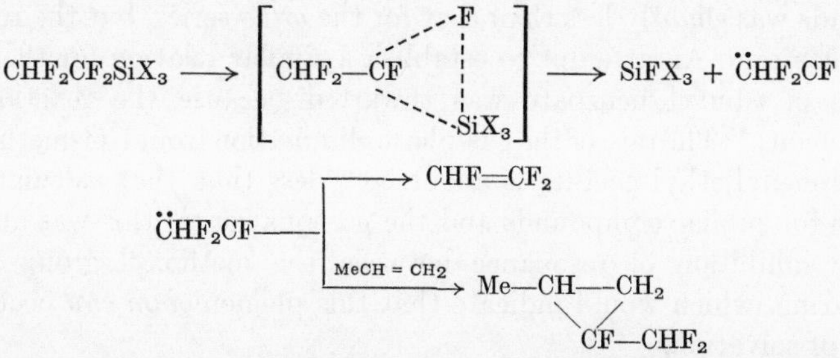

The gas-phase elimination from 2-chloroethyldiethylchlorosilane is 160 times faster than that of 2-chloroethyltrichlorosilane, a result which was interpreted as indicating that there is incipient siliconium ion formation in the transition state.[37]

Evidence for the proposal that gas-phase pyrolytic eliminations from alkyl halides require an $E1$ mechanism involving unimolecular heterolysis of the carbon–halogen bond has been reviewed.[3] The main evidence is that the rates of the gas-phase reactions closely parallel those for the unimolecular solvolyses; however, Herndon and his co-workers[38] have observed that, although the solvolysis in 80% aqueous ethanol at 85° of α-methylbenzyl chloride is almost as fast as that of *t*-butyl chloride, the rate of gas-phase pyrolytic elimination at 400° is only a thirteenth of this. They therefore questioned the validity of the $E1$-like mechanism for the gas-phase reaction. This criticism has been countered by Hoffmann and Maccoll[39] who point out that the solvolytic reactions of α-methylbenzyl chloride frequently have a bimolecular component,[40] and that when a reaction in solution which is purely unimolecular (e.g. elimination from bromides in acetonitrile) is used

[36] G. Fishwick, R. N. Haszeldine, C. Parkinson, P. J. Robinson, and R. F. Simmons, *Chem. Comm.*, **1965**, 382.

[37] I. M. T. Davidson and M. R. Jones, *J. Chem. Soc.*, **1965**, 5481.

[38] W. C. Herndon, J. M. Sullivan, M. B. Henley, and J. M. Manion, *J. Am. Chem. Soc.*, **86**, 5691 (1964).

[39] H. M. R. Hoffmann and A. Maccoll, *J. Am. Chem. Soc.*, **87**, 3774 (1965).

[40] It should be noted, however, that according to E. D. Hughes, C. K. Ingold, and A. D. Scott (*J. Chem. Soc.*, **1940**, 1203), the mechanism of the solvolysis of this compound in 80% aqueous ethanol is unimolecular and that a bimolecular component is only observed when the solvent is anhydrous methanol and ethanol.

a good correlation between the rates of the gas- and solution-phase reactions is obtained.

The rates of the gas-phase pyrolytic eliminations from a series of *ortho*-substituted isopropyl benzoates have been measured and used to define a series of *ortho*-substituent constants.[41] For pyrolysis of a series of *ortho*-, *meta*-, and *para*-substituted α-methylbenzyl methyl carbonates there was a linear free-energy relation between their rates and those of the corresponding 1-arylethyl acetates;[42] the slope of the line for the *meta*- and *para*-substituted compounds was slightly less than that for the *ortho*-series, but the reason for this is uncertain. An attempt to establish a similar relation for the rates of pyrolysis of *t*-butyl benzoate was thwarted because the reactions were heterogeneous.[43] The rate of the gas-phase elimination from 1-(4-methoxy-3,5-dimethylphenyl)ethyl acetate is about 50% less than that calculated from the rates for similar compounds and the $\rho\sigma$ equation;[44] this was attributed to steric inhibition of resonance between the methoxyl group and the benzene ring, which would indicate that this phenomenon can occur in the absence of solvation.

The isomerization of 2,2-dimethyl-1-*p*-nitrobenzoylaziridine (**30**) in diglyme at 72—101° to *N*-(2-methylallyl)-*p*-nitrobenzamide (**31**) has been classed as a pyrolytic *cis*-elimination on account of the close correspondence between its activation parameters and those for such reactions.[45]

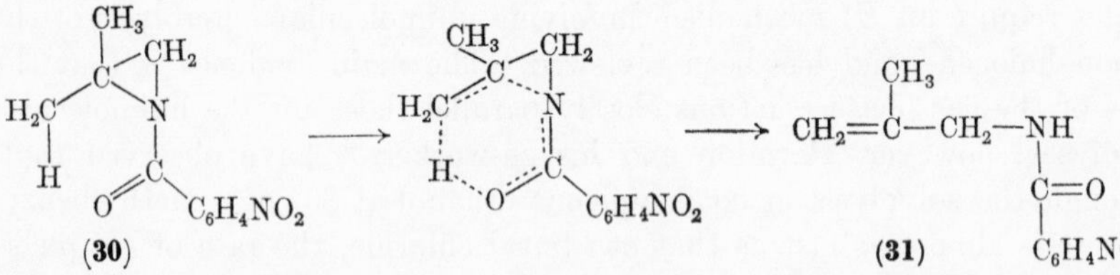

The following reactions have also been investigated: base-catalysed dehydrobromination of tetra-*O*-acetyl-α-D-glucopyranosyl,[46] alkyl, and cyclo-alkylmethyl bromides;[47] liquid-phase pyrolysis of some tertiary alkyl hydrogen phthalates;[48] gas-phase pyrolysis of pentyl chloride;[49] homo-

[41] D. A. D. Jones and G. G. Smith, *J. Org. Chem.*, **29**, 3531 (1964).
[42] G. G. Smith and B. L. Yates, *J. Org. Chem.*, **30**, 434 (1965).
[43] G. G. Smith and B. L. Yates, *Can. J. Chem.*, **43**, 702 (1965).
[44] G. G. Smith and D. V. White, *J. Org. Chem.*, **29**, 3533 (1964).
[45] P. E. Fanta and M. K. Kathan, *J. Heterocyclic Chem.*, **1**, 293 (1964); see also D. V. Kashelikar and P. E. Fanta, *J. Am. Chem. Soc.*, **82**, 4930 (1960); T. Taguchi, Y. Kawazoe, K. Yozhihira, H. Kanayama, M. Mori, K. Tabata, and K. Harano, *Tetrahedron Letters*, **1965**, 2717.
[46] R. U. Lemieux and D. R. Lineback, *Can. J. Chem.*, **43**, 94 (1965).
[47] G. Le Ny, G. Roussi, and H. Felkin, *Bull. Soc. Chim. France*, **1965**, 1920.
[48] K. G. Rutherford and D. P. C. Fung, *Can. J. Chem.*, **42**, 2657 (1964).
[49] R. C. S. Grant and E. S. Swinbourne, *J. Chem. Soc.*, **1965**, 4423.

geneous and heterogeneous pyrolytic elimination from *s*-butyl chloride;[50] thermal decomposition of dialkyl oxalates;[51] dehydration of alcohols on an alumina surface;[52] formation of allyl alcohol from glycerol;[53] dehydration of the diastereoisomeric 3-hydroxy-2,3-diphenylpropionic acids in acetic anhydride;[54] dehydration 3-hydroxy-3-phenylbutyric acid in 50% sulphuric acid;[55] dehydrogenation of 2,4,5-triphenyl-2-imidazoline;[56] elimination in the hydrogenolysis of halogen compounds;[57] vinylogous dehydration of 5,10-dihydroacepleiadene-5,10-diol and 5-(α-hydroxybenzyl)acenaphthene;[58] and base-catalysed elimination from carbomenthyl toluene-*p*-sulphonate.[59]

[50] K. A. Holbrook and J. J. Rooney, *J. Chem. Soc.*, **1965**, 247.
[51] G. J. Karabatsos, J. M. Corbett, and K. L. Krumel, *J. Org. Chem.*, **30**, 689 (1965).
[52] J. R. Jain and C. N. Pillai, *Tetrahedron Letters*, **1965**, 675.
[53] O. Cervinka and O. Kriz. *Collection. Czech. Chem. Commun.*, **30**, 1334 (1965).
[54] C. G. Kratchanov and B. J. Kurtev, *Tetrahedron Letters*, **1965**, 3085.
[55] D. S. Noyce and R. A. Heller, *J. Am. Chem. Soc.*, **87**, 4325 (1965).
[56] T. Hayashi, M. Kuyama, E. Takizawa, and M. Hata, *Bull. Chem. Soc. Japan*, **37**, 1702 (1964).
[57] D. A. Denton, F. J. McQuillin, and P. L. Simpson, *J. Chem. Soc.*, **1964**, 5535.
[58] M. P. Cava, K. E. Merkel, and R. H. Schlessinger, *Tetrahedron*, **21**, 3059 (1965).
[59] Z. Chabudzinski and J. Kuduk, *Roczniki Chem.*, **39**, 1037 (1965).

CHAPTER 5

Addition Reactions

Electrophilic Additions

Addition of halogens and nitrosyl halides. An attempt has been made to obtain evidence, other than from stereochemistry, about the structure of the intermediate cation (**1**, **2**, or **3**) formed in the ionic addition of bromine to olefins.[1] The rates of addition of bromine to styrenes containing electron-withdrawing substituents, in anhydrous acetic acid, are given by

$$\nu = k_2[\text{Styrene}][\text{Br}_2] + k_3[\text{Styrene}][\text{Br}_2]^2$$

The separated second-order rate coefficients give a linear Hammett plot with a ρ value of -2.23, which is much smaller than that (-4.01) for the analogous addition of chlorine to cinnamic acids and those (-4.0 to -4.7) typical of solvolyses involving intermediate carbonium ions similar to (**1**). The ρ value for addition of chlorine is considered to indicate a classical chlorocarbonium ion of type (**1**; Cl in place of Br) where charge delocalization into

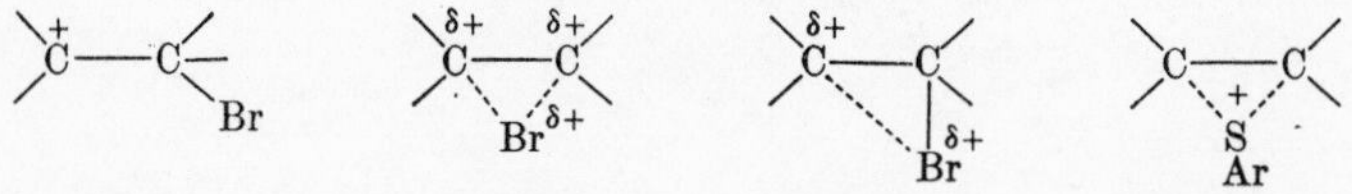

(**1**)   (**2**)   (**3**)   (**4**)

the adjacent phenyl ring is important, partly because chlorine is a less effective neighbouring group than bromine. The ρ value for addition of bromine is more consistent with a brominium ion, (**2**) or (**3**), where charge delocalization in the side chain is more important. A very similar ρ value (-2.20) had been reported[2] for addition of 2,4-dinitrobenzenesulphenyl chloride to styrenes in acetic acid, consistent with a cyclic intermediate (**4**), and the entropies of activation (ca. -27 e.u.) are almost the same for the two reactions, suggesting comparable charge separation and solvent restriction in the two transition states.[1] Support for an episulphonium ion intermediate in the reactions of 4-substituted 2-nitrobenzenesulphenyl chloride to cyclohexene has been provided.[3] Further support has been provided for an open carbonium ion intermediate in the addition of chlorine to

[1] K. Yates and W. V. Wright, *Tetrahedron Letters*, **1965**, 1927.
[2] W. L. Orr and N. Kharasch, *J. Am. Chem. Soc.*, **78**, 1201 (1956).
[3] C. Brown and D. R. Hogg, *Chem. Comm.*, **1965**, 357.

cis- and *trans*-1-phenylpropene; however, the bridged chloronium ion is indicated in the addition to *cis-* and *trans*-2-butene which proceeds exclusively *trans* in acetic acid, to give 2,3-dichlorobutane and 2-chloro-1-methylpropyl acetate.[4]

The reaction between chlorine and alkenes, as neat liquids in the presence of oxygen to inhibit radical reactions, has been carefully reinvestigated.[5] Relative reactivities support an electrophilic mechanism. 1-Substituted and 1,2-disubstituted ethylenes with no chain branching at the double bond give predominantly stereospecific *trans*-addition, with minor amounts of allylic substitution products; the tendency for branched-chain compounds to give substitution products has been confirmed. The relative reactivities of the olefins towards chlorine could be correlated with the sum of Taft's σ^* constants for the alkyl substituents regardless of their orientation on the double bond. A transition state involving partial bonding of a chlorine molecule to both ends of the double bond with little development of positive charge on carbon (5) is favoured for these non-polar media; the intermediate

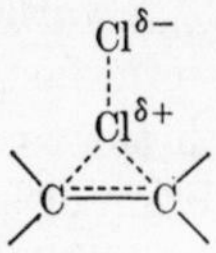

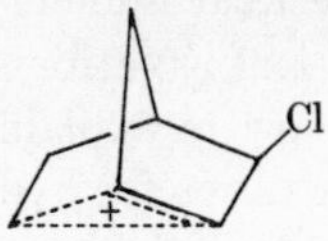

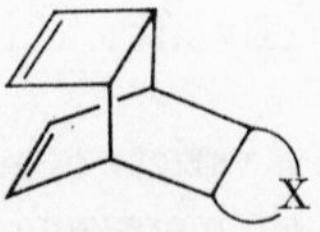

(5) may proceed to a cyclic chloronium ion, which undergoes *trans* ring-opening by chloride ion, or to an open carbonium ion (if this is sufficiently stabilized by alkyl substitution), which suffers preferential proton loss.[5] The reaction of norbornene with chlorine, in carbon tetrachloride in the presence of oxygen, has also been reinvestigated and the pattern of products explained by invoking two intermediates (6) and (7) with the latter (non-classical or rapidly equilibrating pair of classical) ions predominating. This is in agreement with other electrophilic additions to norbornene, which are discussed.[6]

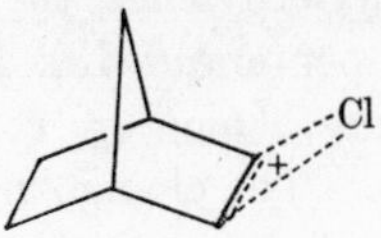

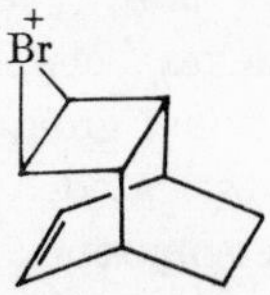

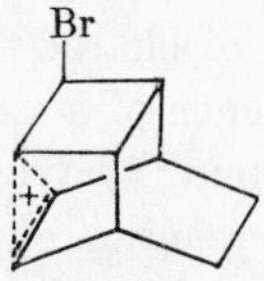

In the addition of bromine to the cyclobutene double bond of compounds of

[4] R. C. Fahey and C. Schubert, *J. Am. Chem. Soc.*, **87**, 5172 (1965).
[5] M. L. Poutsma, *J. Am. Chem. Soc.*, **87**, 4285 (1965).
[6] M. L. Poutsma, *J. Am. Chem. Soc.*, **87**, 4293 (1965).

4*

structure (8) participation by the other double bond decreases, as it is made less nucleophilic by electron-withdrawal by the group X. For example, as X is varied from —CH_2OCH_2— to —CH_2OCO— to —$COOCO$— the reaction is progressively slower and the cis/trans ratio decreases since more of the reaction is forced through a cation of type (9), leading to trans-addition, rather than (10).[7]

The greatly increased tendency to halogenonium ion formation in the order $Cl^+ < Br^+ < I^+$ is shown by the influence of solvent polarity in additions to cis- and trans-stilbenes, provided that polar solvents are assumed to stabilize open carbonium ions more than halogenonium ions.[8]

Dubois and his co-workers have published several more communications on the addition of bromine to olefins. The reactivity of ethylene and its polymethyl derivatives towards bromine and other electrophilic reagents has been measured in various solvents; and its variation with structure, and the sensitivity of the different additions towards changes in structure, have been discussed. Bromine in methanol is the most selective reagent.[9] In the addition of bromine, in methanol containing sodium bromide, to pairs of geometrical isomers of 1,2-dialkylethylenes, the ratio k_{cis}/k_{trans} varied little (1.0—1.8) save for 4-methyl-2-pentene (8.18). Polar and steric factors are necessary to explain these values, which are compared with the somewhat higher ratios for hydroboration.[10] The change in rate of addition of bromine to olefins with change of solvent from water to methanol, through mixtures of the two, was very large but was more or less independent of the olefin.[11] The role of charge-transfer complexes between olefin and bromine,[12] and the influence of a variety of functional groups not directly attached to the double bond,[13] have also been investigated.

In the reaction of bromine with isobutene alone or in a variety of solvents the proportion of substitution products decreased with increasing dipole moment of the solvent.[14] Addition of bromine to olefins in methanol often gives mixtures of dibromide and methoxy-bromide; with phenanthrene the reaction is more complex, giving a variety of products considered to arise from the 9,10-bromonium ion and reaction of this with free bromide ions or the solvent; only the trans-isomer of 9,10-dibromo-9,10-dihydrophenanthrene

[7] D. G. Farnum and J. P. Snyder, Tetrahedron Letters, 1965, 3861.

[8] G. Heublein, Angew. Chem. Intern. Ed. Engl., 4, 881 (1965).

[9] J. E. Dubois and G. Mouvier, J. Chim. Phys., 62, 696 (1965).

[10] J. E. Dubois and G. Mouvier, Tetrahedron Letters, 1965, 1629; Compt. Rend., 259, 2101 (1964).

[11] J. E. Dubois and G. Barbier, Tetrahedron Letters, 1965, 1217; J. E. Dubois, F. Garnier, and H. Viellard, ibid., 1965, 1227.

[12] J. E. Dubois and F. Garnier, Tetrahedron Letters, 1965, 3961.

[13] J. E. Dubois and E. Goetz, Tetrahedron Letters, 1965, 303.

[14] I. V. Bodrikov, Z. S. Smolyan, and G. A. Korchagina, Zh. Obschch. Khim., 35, 933 (1965); Chem. Abs., 63, 8146 (1965).

was detected.[15] Treatment of *trans*-cinnamic acid with chlorine in acetic acid gave the *erythro*- and *threo*-dichloro- and -acetoxychloro-products; in the presence of lithium acetate two additional products are formed, that are considered most likely to arise by loss of carbon dioxide from a zwitterion:[16]

$$\text{Ph}\overset{+}{\text{C}}\text{HCHClCO}_2{}^- \longrightarrow \text{PhCH=CHCl} \longrightarrow \underset{\overset{|}{\text{OAc}}}{\text{PhCHCHCl}_2} + \underset{\overset{|}{\text{Cl}}}{\text{PhCHCHCl}_2}$$

The mechanism and stereochemistry of the addition of bromine to cyclo-octatetraene,[17] and to stilbenes and tolans,[18] and of the halogenation and halogenomethoxylation of D-glucal triacetate, D-galactal triacetate, and 3,4-dihydropyran,[19] have been studied. Synthetic use has been made of the treatment of alkenes with chlorine in the presence of a third, nucleophilic, component to give chlorinated ethers and esters.[20]

Benson has extended his semi-ion pair model of the transition state for four-centre reactions[21] to include the addition of HX and X_2 (X = halogen) to a carbon–carbon bond. Activation energies calculated for a large number of these reactions on the basis of a simple electrostatic model of point dipoles are within 1.3 kcal mole^{-1} of the observed values.[22] Molecular-orbital calculations based on the "frontier electron" concept, where attention is focused on interaction of the highest occupied MO of one component and the lowest vacant MO of the other in the transition state, have been made for the problem of stereoselectivity in additions to unsaturated centres. It is deduced that with a conjugated diene 1,2-addition will occur *trans* and 1,4-addition will occur *cis*.[23] From a perturbation treatment of the system formed when one of the carbon atoms of ethylene is approached by a proton from above the molecular plane it is considered that twisting of ethylene is induced, and it is suggested that electrophilic addition to the double bond goes through a twisted configuration; however, in view of the stereospecificity of these reactions, the twisting cannot be very far developed before the second step occurs.[24]

In the addition of nitrosyl chloride to thirty olefins in chloroform at 0° the

[15] J. van der Linde and E. Havinga, *Rec. Trav. Chim.*, **84**, 1047 (1965).
[16] M. C. Cabaliero and M. D. Johnson, *Chem. Comm.*, **1965**, 454.
[17] R. Huisgen and G. Boche, *Tetrahedron Letters*, **1965**, 1769.
[18] H. Sinn, S. Hopperdietzel, and D. Sauermann, *Monatsh. Chem.*, **96**, 1036 (1965).
[19] R. U. Lemieux and B. Fraser-Reid, *Can. J. Chem.*, **43**, 1460 (1965).
[20] G. Sumrell, R. G. Howell, B. M. Wyman, and M. C. Harvey, *J. Org. Chem.*, **30**, 84 (1965).
[21] S. W. Benson and A. N. Bose, *J. Chem. Phys.*, **39**, 3463 (1963).
[22] S. W. Benson and G. R. Haugen, *J. Am. Chem. Soc.*, **87**, 4036 (1965).
[23] K. Fukui, *Tetrahedron Letters*, **1965**, 2427.
[24] L. Burnelle, *Tetrahedron*, **21**, 49 (1965).

rates were primarily influenced by electronic effects, and by steric effects in highly branched olefins; they were strongly solvent-dependent but not altered by Lewis acids.[25] The stereochemical course of nitrosyl chloride addition (to cyclohexene) was also solvent-dependent, giving the *trans*-dimeric adduct in liquid sulphur dioxide and the *cis*-isomer in methylene chloride, chloroform, or trichloroethylene.[26] Nitrosyl chloride added to *cis*, *trans,trans*-1,5,9-cyclododecatriene entirely at a *trans*-double bond, as a result, it was suggested, of less eclipsing in the four-centre transition state for *cis*-addition at the *trans*-centre.[27]

Hydration and related additions. The acid-catalysed high-pressure hydrations of propene and isobutene recently reported[28] have now been extended to ethylene. In dilute aqueous perchloric acid at 180° the energy, entropy, and volume of activation are approximately 33.3 kcal mole^{-1}, -5.7 cal deg^{-1} mole^{-1}, and -15.5 cm^3 mole^{-1}, respectively. A loss of volume of this magnitude (almost the volume of a water molecule) shows that the transition state for the hydration of ethylene contains at least one firmly bound molecule of water. As with propene and isobutene this rules out a mechanism in which a π-protonated olefin isomerizes unimolecularly to a carbonium ion, and the entropies of activation in each case support this conclusion. The energies of activation show a steady decrease of about 5 kcal mole^{-1} in the series C_2H_4, C_3H_6, C_4H_8, no doubt reflecting the increased stability of the corresponding incipient carbonium ions.[29]

The kinetics of hydration of 2-arylpropenes closely resemble those of other alkenes, and it is concluded that the transition states are all very similar and that there is no mechanistic difference between simple alkyl- and aryl-ethylenes. Interruption of the reaction of 2-phenylpropene with EtOD–D$_2$SO$_4$ showed very little incorporation of deuterium in the unchanged alkene, and interruption of the same reaction of *cis*-2-phenyl-2-butene showed none. Thus the alkene is not in reversible equilibrium with the carbonium ion before ether formation. $(+)$-2-Phenyl-2-butanol in 8.5% sulphuric acid in ethanol racemizes 5 times faster than it adds ethanol. Thus the carbonium ion is an intermediate, about 80% undergoing internal return and 20% reacting with ethanol to give racemic ether. By analogy it is inferred that the carbonium ion is also an intermediate between alcohol and transition state in the dehydration of alcohols in aqueous acid.[30] A ratio for the comparative rates of anti-Markownikow and Markownikow hydration of propane in D$_2$O–D$_2$SO$_4$, $k_{antiM}/k_M < 4 \times 10^{-4}$ at 60° in 55% D$_2$SO$_4$, has been directly

[25] T. Beier, H. G. Hauthal, and W. Pritzkow, *J. Prakt. Chem.*, **26**, 304 (1964).
[26] M. Ohno, M. Okamoto, and K. Nakoda, *Tetrahedron Letters*, **1965**, 4047.
[27] M. Ohno, M. Okamoto, and N. Naruse, *Tetrahedron Letters*, **1965**, 1971.
[28] B. T. Baliga and E. Whalley, *Can. J. Chem.*, **42**, 1019 (1964).
[29] B. T. Baliga and E. Whalley, *Can. J. Chem.*, **43**, 2453 (1965).
[30] N. C. Deno, F. A. Kish, and H. J. Peterson, *J. Am. Chem. Soc.*, **87**, 2157 (1965).

estimated from an NMR study of the rate of disappearance of the tertiary propan-2-ol proton.[31]

Acid-catalysed addition of methanol and acetic acid to bicyclo[3.1.0]hex-2-ene (11) under conditions of thermodynamic control gave 4-methoxy- and 4-acetoxy-cyclohexene, respectively. Under conditions of kinetic control addition of methanol gave mainly *cis*- and *trans*-2-methoxybicyclo[3.1.0]hexane. Acid-catalysed addition of methan[2H]ol was stereospecifically *cis*, to give *trans*-3-deuterio-*trans*-2-methoxybicyclo[3.1.0]hexane (12), thus re-

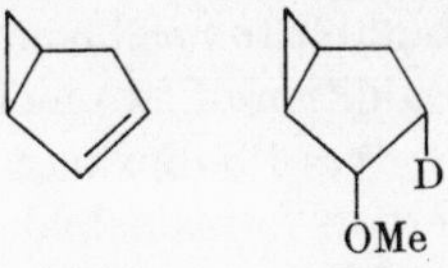

(11)   (12)

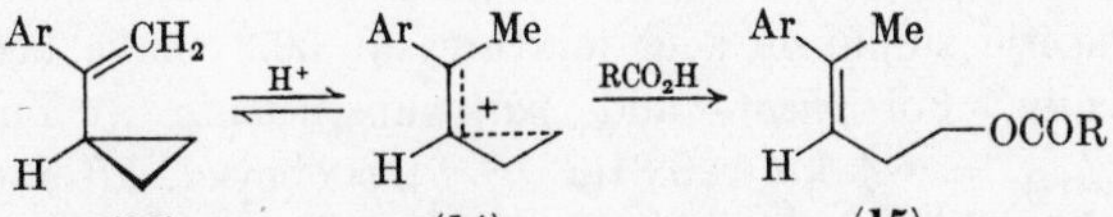

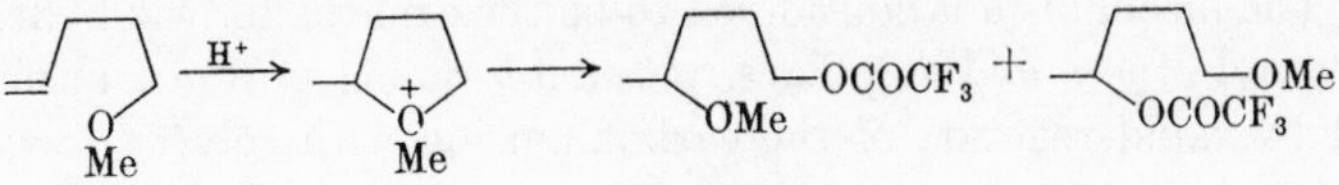

Scheme 1

$$CH_2O + H^+ \rightleftharpoons CH_2\overset{+}{O}H \xrightarrow{CH_2O} \overset{+}{C}H_2{-}O{-}CH_2OH$$

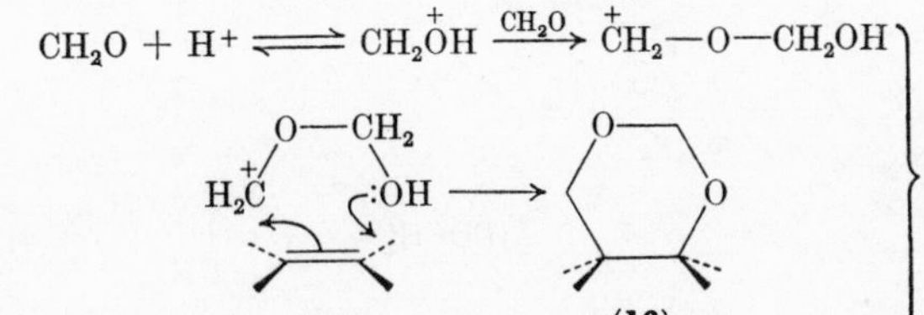

(16)

Scheme 2

presenting another example of stereospecific *cis* polar addition to alkenes.[32] Vinylcyclopropanes (13) react with acetic acid at the boiling point or with trifluoroacetic acid at 0° to give the rearranged *cis*-adducts (15). This probably involves conjugative 1,5-addition with opening of the cyclopropane ring via the homoallylic carbonium ion (14). The reaction is stereospecific since the ion (14) giving the *cis*-product is more stable than that where the aryl group and the emerging acyloxyethyl group are eclipsed. Electron-release from the aryl group facilitates the reaction and vice versa.[33] Peterson

[31] S. Ehrenson, S. Seltzer, and R. Diffenbach, *J. Am. Chem. Soc.*, **87**, 563 (1965).

[32] P. K. Freeman, M. F. Grostic, and F. A. Raymond, *J. Org. Chem.*, **30**, 771 (1965).

[33] S. Sarel and R. Ben-Shoshan, *Tetrahedron Letters*, **1965**, 1053.

et al. have published full details of their interesting work on the rates of addition of trifluoroacetic acid to 1-alkenes having chains of 4—11 carbon atoms. Large rate decreases are caused by certain substituents such as acetoxy, trifluoroacetoxy, and cyano-groups, even when remote from the double bond, and this is attributed partly to the enhanced inductive effect of substituents which form hydrogen bonds with trifluoroacetic acid. Inductive rate depressions have been measured for substituents up to ten carbon atoms from the double bond, and the attenuation per methylene is constant (0.66-fold) and surprisingly small. 5-Methoxy- and 5-halogeno-1-pentenes shown enhanced rates attributable to neighbouring group participation, e.g. Scheme 1.[34] Acid-catalysed addition of two molecules of formaldehyde to a double bond (Prins reaction) gives 1,3-dioxanes (16); NMR analysis of these indicates *cis*-addition of the two formaldehyde units, and the mechanism shown in Scheme 2 is proposed.[35]

The relative rates of hydration of a range of substituted phenylacetylenes by aqueous acetic–sulphuric acid mixtures at 50.2° have been measured spectrometrically. For *meta*- and *para*-substituents a Yukawa–Tsuno equation, $\log k_{rel.} = -4.3[\sigma + 0.81(\sigma^+ - \sigma)]$, is obeyed. After 50% reaction with 2-phenyl-1-tritioacetylene, unchanged material had lost none of its tritium. The mechanism is considered to be the same as for the hydration of alkyl alkynyl ethers and sulphides, rate-determining proton transfer being followed by rapid reaction of the carbonium ion with solvent (Scheme 3).

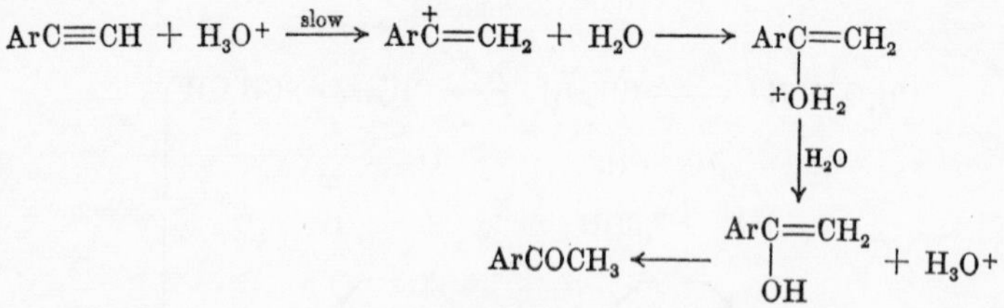

$$ArC{\equiv}CH + H_3O^+ \xrightarrow{\text{slow}} Ar\overset{+}{C}{=}CH_2 + H_2O \longrightarrow ArC{=}CH_2$$
$$\underset{+OH_2}{|}$$
$$\downarrow H_2O$$
$$ArCOCH_3 \longleftarrow \quad \underset{OH}{\overset{ArC=CH_2}{|}} \quad + H_3O^+$$

Scheme 3

The deactivating effect of *ortho*-substituents, surprisingly large for *o*-methoxyl, is attributed to steric hindrance to solvation of the initial carbonium ion.[36] The addition of trichloroacetic acid to phenylacetylene in benzene to give 1-phenylvinyl trichloroacetate is of the first order in the acetylene and of the second order in the acid. The acid is dimeric in benzene and presumably proton transfer from the dimer to the triple bond is involved.[37]

[34] P. E. Peterson, C. Casey, E. V. P. Tao, A. Agtarap, and G. Thompson, *J. Am. Chem. Soc.*, **87**, 5163 (1965).
[35] E. E. Smissman, R. A. Schnettler, and P. S. Portoghese, *J. Org. Chem.*, **30**, 797 (1965).
[36] R. W. Bott, C. Eaborn, and D. R. M. Walton, *J. Chem. Soc.*, **1965**, 384.
[37] A. G. Evans, E. D. Owen, and B. D. Phillips, *J. Chem. Soc.*, **1964**, 5021.

Diphenylacetylene-2,2′-dicarboxylic acid isomerizes very readily to the lactone (18). Measurements in aqueous buffers show that the rate is proportional to the monoanion concentration and this, together with the very much slower reaction in the absence of the second *ortho*-carboxylic acid group,

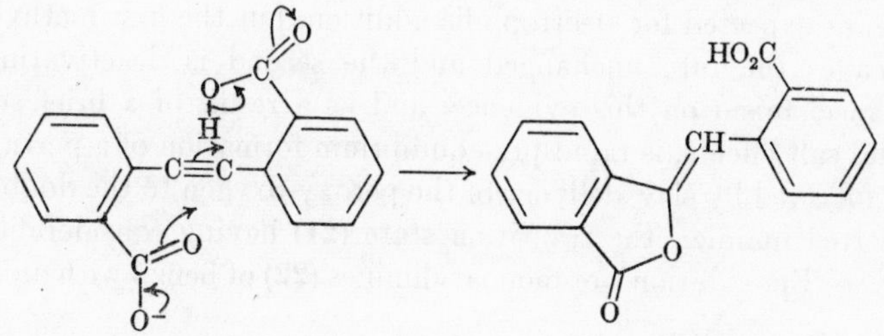

(17)   (18)

leads to the suggestion of intramolecular catalysis by the carboxylic acid group of attack on the triple bond by the carboxylate anion, as shown in (17).[38]

The hydration of furanose glycals has also been studied,[39] and qualitative[40] and quantitative[41] aspects of the hydration of C=N bonds in nitrogen heteroaromatic compounds have been comprehensively reviewed.

Epoxidations. The rates of epoxidation with *m*-chloroperbenzoic acid in ether at 25° of a number of 3- and 4-alkyl cyclohexanes suggest that the effect of these alkyl groups on the reactivity is steric rather than electronic in origin. Thus all the alkyl groups are deactivating, and 3,3- and *cis*-4,5-dimethylcyclohexene both react at approximately one-half the rate of cyclohexene, suggesting that an axial methyl group effectively prevents attack on its side of the ring; in the latter case a preponderance of *trans*-epoxide is formed, as expected.[42] The rates of epoxidation of some bicycloheptenes with *m*-chloroperbenzoic acid in chloroform are found to increase in the order: bicyclo[2.2.1]hept-5-ene-*endo*-2,3-dicarboxylic anhydride (19; X = O) < the *exo*-isomer of (19; X = O) < the *N*-phenethylimide (19; X = NCH₂CH₂Ph) < the *exo*-isomer of (19; X = NCH₂CH₂Ph). All are oxidized more slowly than *cis*-1,2,3,6-tetrahydrophthalic anhydride. This sequence is explained in terms of ease of approach of the peracid and a field effect favouring electrophilic attack more in the nitrogen than the oxygen heterocyclic compounds.[43] Epoxidation of bicyclo[3.2.1]oct-2-ene with buffered

[38] R. L. Letsinger, E. N. Oftedall, and J. R. Nazy, *J. Am. Chem. Soc.*, **87**, 742 (1965).
[39] M. Haga and R. K. Ness, *J. Org. Chem.*, **30**, 158 (1965).
[40] A. Albert and W. L. F. Armarego, *Adv. Heterocyclic Chem.*, **4**, 1 (1965).
[41] D. D. Perrin, *Adv. Heterocyclic Chem.*, **4**, 43 (1965).
[42] B. Rickborn and S.-Y. Lwo, *J. Org. Chem.*, **30**, 2212 (1965).
[43] A. P. Gray and D. E. Heitmeier, *J. Org. Chem.*, **30**, 1226 (1965).

peracetic acid gave the *exo*-epoxide predominantly, as does norbornene. Bromine adds to the same olefin without rearrangement.[44]

Insight into the mechanism of epoxidation of allyl alcohol by hydrogen peroxide and tungstic acid has been sought by measuring the reaction rate for methyl-substituted allyl alcohols. Methyl groups on the double bond are activating, as expected for electrophilic addition, but the first methyl group on $C_{(1)}$ leaves the rate unchanged and the second is deactivating. The favoured mechanism on this evidence, and as a result of a brief study of solvent and salt effects, is rapid pre-equilibrium formation of a pertungstate ester (20) followed by slow delivery of the peroxy-oxygen to the double bond in a concerted manner, the transition state (21) having considerable polar character.[45] "Epoxidation" of monoarylimines (22) of benzil with anhydrous

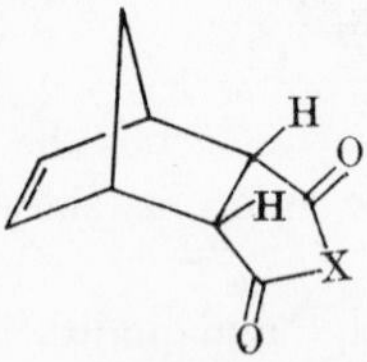

(19) 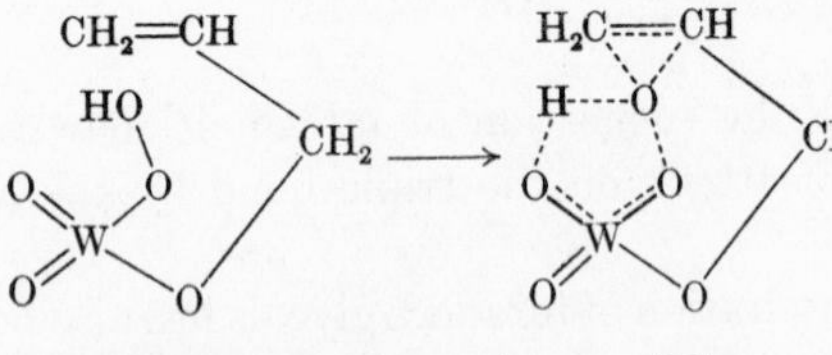   (20)   (21)

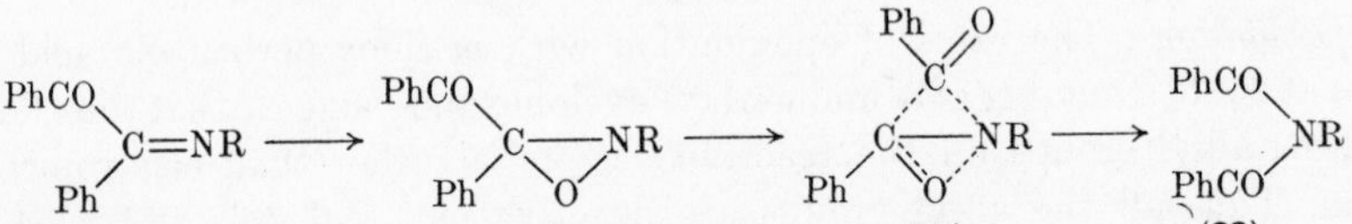

(22)   (24)   (25)   (23)

peracetic acid or *m*-chloroperbenzoic acid gives the *N,N*-dibenzoylanilines (23) in high yield, via the corresponding 3-benzoyloxazirine (24). 3-Benzoyl-2-cyclohexyl-3-phenyloxazirine (24; $R = C_6H_{11}$) is stable enough to be isolated, and its thermal rearrangement to (23; $R = C_6H_{11}$) in acetonitrile followed first-order kinetics. The rearrangement probably involves nucleophilic attack by the nitrogen lone-pair on the neighbouring benzoyl group with electron reorganization as shown (25).[46] Neighbouring tertiary nitrogen has been found to participate, and to cause rearrangement, in the hydroxylation of olefins with silver iodide dibenzoate (Prévost reaction);[47] the mechanism of hydroxylation of elaidic acid and cyclohexene with Caro's acid in ^{18}O-water has also been studied.[48]

[44] R. R. Sauers, H. M. How, and H. Feilich, *Tetrahedron*, **21**, 983 (1965).
[45] H. C. Stevens and A. J. Kaman, *J. Am. Chem. Soc.*, **87**, 734 (1965).
[46] A. Padwa, *J. Am. Chem. Soc.*, **87**, 4365 (1965).
[47] A. Ferretti and G. Tesi, *J. Chem. Soc.*, **1965**, 5203.
[48] L. V. Sulima, *Zh. Org. Khim.*, **1**, 71 (1965); *Chem. Abs.*, **62**, 16074 (1965).

Nucleophilic Additions

Factors which govern the rates of nucleophilic addition to acrylonitrile of thiols, amino-thiols, and related peptide model compounds have been investigated. The rates were measured as a function of pH, and a quantitative estimate of steric and polar factors was obtained from a Hammett–Taft free-energy relationship. The reactive nucleophiles are the mercaptide ions and the free amines, as expected; their response to polar and steric changes are similar and their relative reactivities towards other vinyl compounds remain essentially constant. Sulphur anions are about 280 times more reactive than amino groups in a similar environment.[49] Addition of secondary amines, and the non-catalysed addition of alcohols, to dimethyl acetylenedicarboxylate gives *cis*-products, whilst the tertiary amine-catalysed addition of alcohols gives the usual *trans*-products; formation of the *cis*-isomer is explained as occurring through a cyclic proton transfer in the initially formed ammonium (26) or oxonium ion.[50] The kinetics of nucleophilic addition of halide and thiocyanate ions to dimethyl acetylenedicarboxylate have also been measured.[51]

What is thought to be the first authenticated case of addition of a Grignard reagent to an unconjugated carbon–carbon double bond is provided by the reaction of 1,1-diphenyl-3-butanol with allylmagnesium bromide to give 1,1-diphenyl-6-hepten-1-ol.[52] The addition of vinyllithium to 1,1-diphenylethylene in tetrahydrofuran is slow enough to be followed spectroscopically,

(26)   (27)   (28)

and from the kinetic form of the reaction it is concluded that vinyllithium is associated, possibly to trimers, in this solution.[53] Pines and his co-workers have reported several studies of alkali-metal-catalysed reactions which involve inter- and intra-molecular addition of carbanions to carbon–carbon double bonds;[54-56] in the intramolecular reactions, e.g. (27) to (28), the size

[49] M. Friedman, J. F. Cavins, and J. S. Wall, *J. Am. Chem. Soc.*, **87**, 3672 (1965).

[50] E. Winterfeldt and H. Preuss, *Angew. Chem. Intern. Ed. Engl.*, **4**, 689 (1965).

[51] G. F. Dvorko and D. F. Mironova, *Ukr. Khim. Zh.*, **31**, 195 (1965); *Chem. Abs.*, **63**, 2865 (1965).

[52] J. J. Eisch and G. R. Husk, *J. Am. Chem. Soc.*, **87**, 4194 (1965).

[53] R. Waack and P. E. Stevenson, *J. Am. Chem. Soc.*, **87**, 1183 (1965); see also R. Waack, M. A. Doran, and P. E. Stevenson, *J. Organomet. Chem.*, **3**, 481 (1965).

[54] H. Pines and N. C. Sih, *J. Org. Chem.*, **30**, 280 (1965).

[55] N. C. Sih and H. Pines, *J. Org. Chem.*, **30**, 1462 (1965).

[56] H. Pines, N. C. Sih, and E. Lewicki, *J. Org. Chem.*, **30**, 1457 (1965).

of ring formed appears to be determined by the stability of the intermediate carbanion. The addition of organotin hydrides (hydrostannation) to strongly electrophilic acetylenes has been shown to involve nucleophilic attack by the hydride hydrogen in the rate-determining step.[57]

Radical Additions

Free-radical addition to alkenes has been reviewed.[58]

Tedder and Walton have investigated the addition of the trichloromethyl radical to ethylene and to a series of fluorinated ethylenes.[59] The rate differences depend almost entirely on the activation energy of the process, and the pre-exponential factors are sensibly constant both for different compounds in the series and for addition at each end of unsymmetrical compounds. Furthermore, the activation energies depend largely on the substituents on the carbon atom which is attacked. They are much less dependent on the substituents on β-carbon atom. These results lead to the conclusion that orientation in addition reactions is primarily determined, not by the stability of the adduct radicals, but by the relative strengths of the new bonds being formed (the intermediacy of bridged radicals in this reaction is discounted). The key to the above conclusion was the use of an addition reaction in which products of initial radical attack at both ends of an unsymmetrical double bond could be isolated. In the majority of radical additions only one product is observed, and use of the new concept in predicting orientation of addition will normally lead to the same conclusions as those based on relative stability of adduct radicals.

Arrhenius parameters for the addition of ethyl radicals to a series of olefins have been studied by James and his co-workers.[60,61] For a series of vinyl derivatives, $CH_2 = CHX$ (where attack is at the β-carbon), there is a wide variation of activation energy with X, but the pre-exponential factor is again sensibly constant.[60] The lowest activation energies were found for acrylonitrile (X = CN) and styrene (X = Ph), both of which are known readily to undergo radical polymerization. The possibility that this activation energy term might be used as a measure of the intrinsic reactivity of a monomer was discussed.

The reactions of ethyl radicals with methacrylonitrile and *cis*- and *trans*-crotononitrile differ from the above series, in showing similar activation energies while having different pre-exponential factors.[61] The latter effect is attributed to steric factors in the reactions of the crotononitriles.

[57] A. J. Leusink and J. W. Marsman, *Rec. Trav. Chim.*, **84**, 1123 (1965).
[58] J. I. G. Cadogan and M. J. Perkins in "The Chemistry of Alkenes", ed. S. Patai, Interscience Publishers, London, New York, and Sydney, 1964.
[59] J. M. Tedder and J. C. Walton, *Proc. Chem. Soc.*, **1964**, 420.
[60] D. G. L. James and D. MacCallum, *Can. J. Chem.*, **43**, 633 (1965).
[61] D. G. L. James and T. Ogawa, *Can. J. Chem.*, **43**, 640 (1965).

Abell has investigated competitive additions of hydrogen bromide to a series of olefins under radical conditions in the gas-phase.[62] The negative temperature coefficient of the rates of addition is consistent with initial reversible addition of Br·. The resulting bromoalkyl radical then abstracts hydrogen from HBr in a reaction of moderate activation energy. The overall relative reactivities are determined by the magnitude of this activation energy, together with the electronegativity of the olefin with respect to addition of Br·.

The stereochemistry of thiol addition to *t*-butylcyclohexenes has been studied.[63,64] Methanethiol adds to 4-*t*-butylcyclohexene to give the thermo-dynamically less stable, axially substituted sulphides (**29** and **30**) as predominant products (together *ca.* 90% of the 1:1 adducts).[63] In the reaction of thiolacetic acid with 4-*t*-butyl-1-methylcyclohexene, only two products are formed, each by thiyl radical addition at the position 2. The predominant

(**29**)    (**30**) 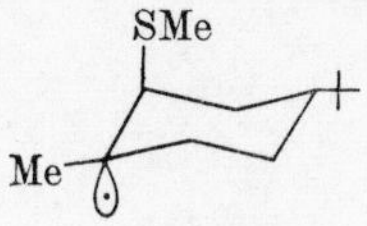   (**31**)

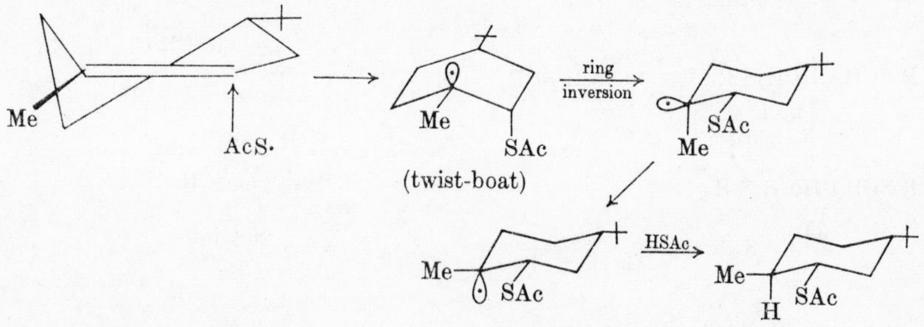

$$(1)$$

isomer is again formed by axial attack, and chain transfer leads to overall *trans*-addition which, it is suggested, involves the pyramidal intermediate (**31**). The minor product arises from overall *cis*-addition believed to occur by the sequence of reaction (1). It is suggested that factors affecting product distribution include steric interactions with pseudo-axial hydrogen atoms, and the relative stabilities of chair and twist-boat intermediates. The results of additions to substituted methylenecyclohexanes support the tetrahedral-radical formulation as an explanation of axial hydrogen transfer in the final step.[64]

[62] P. I. Abell, *Trans. Farad. Soc.*, **60**, 2214 (1964).
[63] E. S. Huyser and J. R. Jeffrey, *Tetrahedron*, **21**, 3083 (1965).
[64] F. G. Bordwell, P. S. Landis, and G. S. Whitney, *J. Org. Chem.*, **30**, 3764 (1965).

Isomerization of *cis,trans,trans*-1,5,9-cyclododecatriene to the all-*trans*-isomer by thioglycollic acid or *N*-bromosuccinimide is surprising in view of this compound's usual reaction at a *trans*-double bond.[65] *cis,cis*-Cycloocta-1,5-diene is unusual in its reaction with thiols, in that a normal 1:1 adduct is formed instead of a derivative of the bicyclo[3.3.0]octane system commonly encountered in radical reactions of this diene.[66]

Several radical additions to allenes have been studied. The majority of radicals add to allene itself predominantly or exclusively at the terminal carbon. The addition of hydrogen bromide is exceptional in that Br· appears to attack only the central carbon atom. Abell and Anderson explained this by initial addition of Br· at terminal carbon, followed by rearrangement of the initial adduct to the more stable 2-bromoallyl radical.[67] However, the results do not appear to exclude direct addition at the central carbon atom, and the

$$CH_2=C=CH_2 + RSH$$

$$\downarrow$$

$$CH_2=CH-CH_2SR$$

$$R'S\cdot \;\Big\vert\!\uparrow\; -R'S\cdot$$

$$\longrightarrow RSCH_2\overset{\centerdot}{C}HCH_2SR$$

$$R'SCH_2CH_2CH_2SR \xleftarrow{\;R'SH\;}{(or\;RSH)} \quad R'SCH_2\overset{\centerdot}{C}HCH_2SR \qquad\qquad \Big\downarrow \begin{array}{l} R'SH \\ (or\;RSH) \end{array}$$

$$RS\cdot \;\Big\vert\!\uparrow\; -RS\cdot \qquad\qquad RSCH_2CH_2CH_2SR$$

$$R'SCH_2\overset{\centerdot}{C}HCH_2S'R \underset{-R'S\cdot}{\overset{R'S\cdot}{\rightleftarrows}} \quad R'SCH_2-CH=CH_2 \qquad\qquad (32)$$

$$\Big\downarrow \begin{array}{l} R'SH \\ (or\;RSH) \end{array}$$

$$R'SCH_2CH_2CH_2S'R$$

$$(33)$$

$$\begin{array}{c} R \quad R \\ R-\overset{\centerdot\centerdot}{\underset{\;}{\cdots}}\!-R \\ Sn(Me)_3 \\ (34) \end{array}$$

possibility of bridging in the initial adduct was not considered. Unlike the addition to an isolated double bond, the Br· addition to allene is effectively irreversible.

The reversible addition of thiyl radicals to isolated double bonds leads to an apparent substitution (to give **32** and **33**) when allene is subjected to successive addition of one mol. each of two different thiols.[68] Hydrogen sulphide adds to allene at the terminal carbon atom.[69]

In an extensive study of the addition of trimethyltin hydride, Kuivila,

[65] E. W. Duck and J. M. Locke, *Chem. Ind. (London)*, **1965**, 507.

[66] J. M. Locke and E. W. Duck, *Chem. Comm.*, **1965**, 151.

[67] P. I. Abell and R. S. Anderson, *Tetrahedron Letters*, **1964**, 3727.

[68] D. N. Hall, A. A. Oswald, and K. Griesbaum, *J. Org. Chem.*, **30**, 3829 (1965).

[69] K. Griesbaum, A. A. Oswald, E. R. Quiram, and P. E. Butler, *J. Org. Chem.*, **30**, 261 (1965).

Rahman, and Fish have shown that for a series of alkyl-substituted allenes the predominant attack of the nucleophilic $Me_3Sn\cdot$ radical is at the central carbon atom.[70] This situation is reversed for allene itself. The results are discussed in terms of steric effects on the direction of approach of $Me_3Sn\cdot$, and hyperconjugative stabilization of the intermediate allyl radical (**34**).

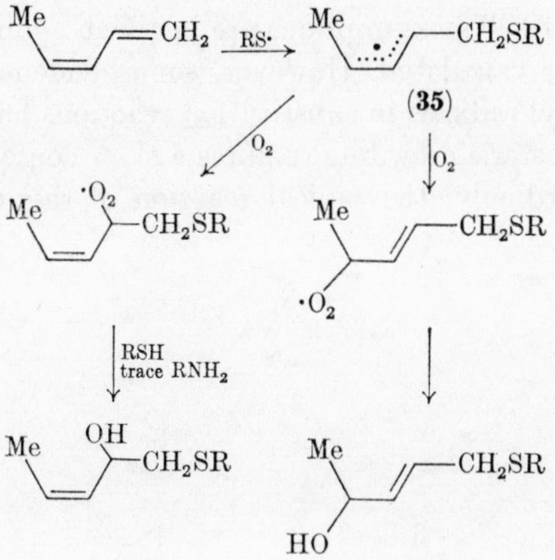

New evidence concerning the conformational stability of allyl radicals has been obtained from a study of thiol addition to *cis*- and *trans*-1,3-pentadiene.[71] If the intermediate allyl radical (e.g. **35**) is rapidly intercepted, as in the presence of oxygen, the 1,2-adducts retain the stereochemistry of the 3,4-double-bond. This indicates that the stereochemistry of the 1,4-adducts should reveal the stereochemistry (*cisoid* or *transoid*) of the 2,3-bond in the reacting diene. Both isomers were found to react in a predominantly *transoid* conformation (e.g. giving product **36**).

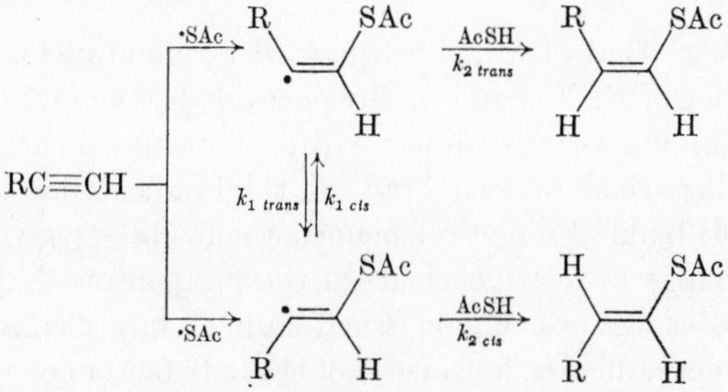

[70] H. G. Kuivila, W. Rahman, and R. H. Fish, *J. Am. Chem. Soc.*, **87**, 2835 (1965).
[71] W. A. Thaler, A. A. Oswald, and B. E. Hudson, *J. Am. Chem. Soc.*, **87**, 311 (1965).

A study of the stereochemistry of addition of thiolacetic acid to 1-hexyne shows predominantly *trans*-addition.[72] Whether this preference is dictated by the stereochemistry of the initial addition step ($k_2 \gg k_1$ in Scheme 4) or by the relative magnitude of the rate constants $k_{2(trans)}$ and $k_{2(cis)}$ ($k_1 \gg k_2$) was not determined. There is a substantial error in Skell's value of > 18 kcal mole^{-1} for the activation energy for geometrical isomerization of $CH_3\dot{C}{=}CHBr$.[72] If Skell's assumptions are justified, a minimum value of 10 kcal mole^{-1} may be calculated. However, some evidence for retention of configuration of vinyl radicals in substitution reactions has been obtained.[73]

In the diene (37), steric crowding requires a skew conformation about the central bond. Accordingly, the radical reaction of this diene with $BrCCl_3$

(37)

(38)   (39)

is similar in rate to that of a non-conjugated terminal olefin.[74] The predominant 1:1 adduct of $BrCCl_3$ and methylenecyclodecane (38) is (39), formed by a process involving a transannular hydrogen atom transfer.[75]

Unlike the bulky trichloromethyl radical, thiyl radicals add to the strained endocyclic double bond of 5-methylenenorbornene (40), to give the products shown.[76] The change in relative yields of the products with dilution of the reactants mitigates against non-classical radical intermediates. Relief of strain in the adduct radical is, however, not the only factor governing addition of thiyl radicals to this system. Reversibility of thiyl radical addition must also be taken into account.[77] In a competitive study of addition of thiols to norbornene and methylenenorbornane, it has been shown that reversibility is more important for the exocyclic double bond, but that the effect of this is reduced by lowering the reaction temperature: at lower temperatures the exocyclic double bond appears to be relatively more reactive.

[72] J. A. Kampmeier and G. Chen, *J. Am. Chem. Soc.*, **87**, 2608 (1965).
[73] I. P. Beletskaya, V. I. Karpov, and O. A. Reutov, *Dokl. Akad. Nauk SSSR*, **161**, 586 (1965); *Chem. Abs.*, **63**, 1676 (1965).
[74] E. S. Huyser, F. W. Siegert, and H. Wynberg, *Tetrahedron Letters*, **1965**, 2569.
[75] M. Fisch and G. Ourisson, *Chem. Comm.*, **1965**, 407.
[76] S. J. Cristol, T. W. Russell, and D. I. Davies, *J. Org. Chem.*, **30**, 207 (1965).
[77] E. S. Huyser and R. M. Kellogg, *J. Org. Chem.*, **30**, 3003 (1965).

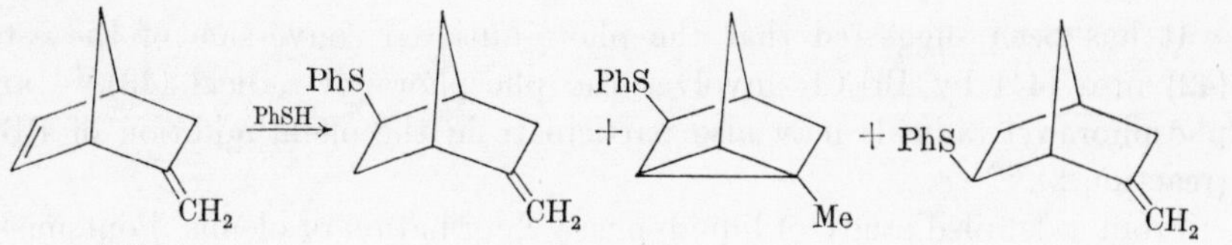

In the addition of thiols to hexachloronorbornadiene (**41**),[78] products of a 1,2-vinyl shift are observed. It is suggested that the driving force for the rearrangement is the stabilization of the rearranged radical by the α-chlorine substituent.

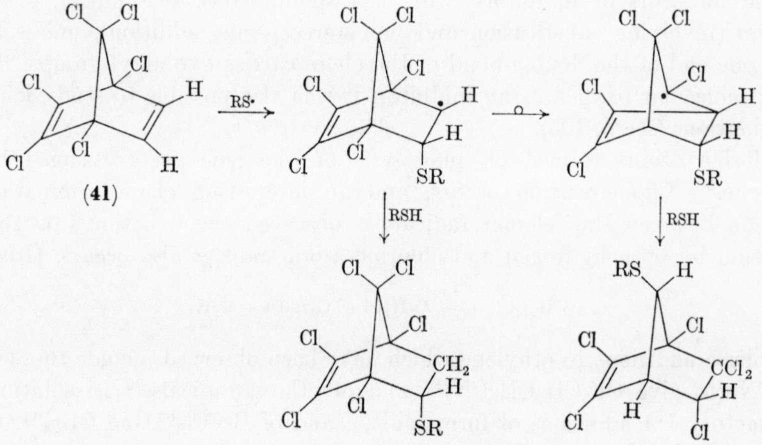

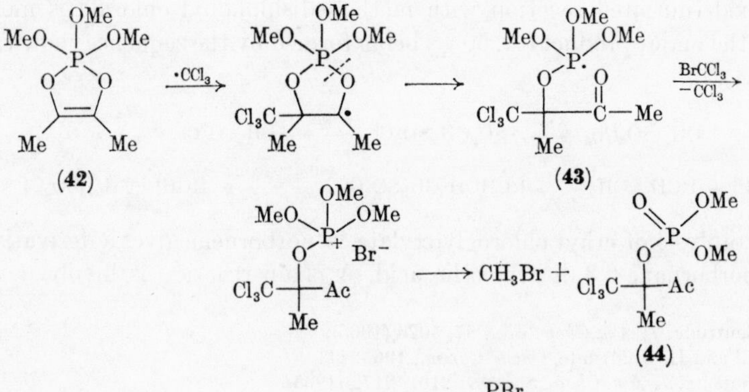

$$RCH{=}CH_2 + Br\cdot \longrightarrow R\dot{C}HCH_2Br \xrightarrow{PBr_3} \overset{\cdot PBr_3}{\underset{}{R\dot{C}HCH_2Br}} \tag{2}$$

$$\xrightarrow{RCH=CH_2} RCH(PBr_2)CH_2Br + R\dot{C}HCH_2Br$$

[78] J. A. Claisse and D. I. Davies, *Chem. Comm.*, **1965**, 209.

It has been suggested that the photo-initiated conversion of the ester
(42) into (44) by $BrCCl_3$ involves the phosphoranyl radical (43),[79a] and
phosphoranyl radicals may also participate in the olefin addition of PBr_3
(reaction 2).[79b]

From a detailed study of liquid–phase chlorination of olefins, Poutsma[79c]
has advanced the following rule: chlorination in the dark, in the absence of
radical initiators or inhibitors, will be a spontaneous, free-radical, chain
process (involving substitution and non-stereospecific addition), unless at
least one end of the double bond of the olefin carries two alkyl groups. By
using molecular oxygen as an inhibitor, it was also possible to study ionic
chlorinations (see p. 105).

Chlorine atoms formed on photolysis of phosgene are scavenged by
ethylene.[80] Telomerization occurs, and an interesting chain-termination
reaction between the telomer radicals is observed, e.g. reaction (3). Termination by both hydrogen and chlorine atom transfer also occurs. Other

$$2 \cdot (CH_2)_4Cl \longrightarrow C_4H_7Cl + C_2H_5Cl + C_2H_4 \tag{3}$$

gas–phase additions to ethylene which have been observed include those of
CF_3CN (to give $CF_3CH_2CH_2CN$)[81] and of ethoxy-radicals.[82] In solution,
satisfactory 1:1 additions of formamide,[83] and of $BrCCl_3$[84] and CBr_4[84], to
non-terminal olefins have been achieved.

Methanesulphonyl chloride adds to terminal olefins under the influence of
ultra-violet light, to give β-chlorosulphones

$$(RCH{=}CH_2 \longrightarrow RCHClCH_2SO_2CH_3).^{85a}$$

The peroxide-initiated reaction with methanedisulphonyl chloride is more
complex, the major product (*ca.* 50%) being formed by the sequence shown in
reaction (4).[85a]

$$CH_2(SO_2Cl)_2 \xrightarrow{X\cdot} \cdot SO_2CH_2SO_2Cl \xrightarrow{-SO_2} \cdot CH_2SO_2Cl$$

$$\cdot CH_2SO_2Cl + RCH{=}CH_2 \longrightarrow R\overset{\cdot}{C}HCH_2CH_2SO_2Cl \xrightarrow{CH_2(SO_2Cl)_2} RCHCl(CH_2)_2SO_2Cl \tag{4}$$

The photolysis of ethyl chloroglyoxylate in norbornene gives a derivative
of *trans*-norbornane-2,3-dicarboxylic acid. A chain reaction is involved, in

[79a] W. G. Bentrude, *J. Am. Chem. Soc.*, **87**, 4026 (1965).
[79b] B. Fontal and H. Goldwhite, *Chem. Comm.*, **1965**, 111.
[79c] M. L. Poutsma, *J. Am. Chem. Soc.*, **87**, 2161, 2172 (1965).
[80] J. Heicklen, *J. Am. Chem. Soc.*, **87**, 445 (1965).
[81] N. A. Gac and G. J. Janz, *J. Am. Chem. Soc.*, **86**, 5059 (1964).
[82] J. C. J. Thynne, *J. Chem. Soc.*, **1964**, 5882.
[83] D. Elad and J. Rokach, *J. Org. Chem.*, **30**, 3361 (1965).
[84] D. Lefort and D. Blanchet, *Bull. Soc. Chim. France*, **1965**, 2353.
[85a] H. Goldwhite, M. S. Gibson, and C. Harris, *Tetrahedron*, **21**, 2743 (1965).

which the adduct of norbornene and an ethoxycarbonyl radical abstracts a chloroformyl group from ethyl chloroglyoxylate.[85b]

Radical amination (by $R_2N\cdot$ or $R_2NH\cdot^+$) of olefins,[86,87] allene,[86] and aromatic compounds[87] has been reported, and the ESR spectra of the initial adducts of $H_2N\cdot$ and a series of olefins have been determined.[88]

Cadogan and Hey have presented the details of their work on addition of acetic acid derivatives to terminal olefins.[89] By proper control of the reaction conditions (high addendum to olefin ratio), the yield of 1:1 adduct $(C_9H_{19}CO_2H)$ of 1-octene to acetic acid was 69% (based on olefin). Other yields included Ac_2O 73%, MeCN 18%, $MeCONH_2$ 29%, and MeCOCl 43%. Simple ketones and esters were also added. For low-boiling addenda, iso-propyl peroxydicarbonate $(i\text{-}PrOCO_2)_2$ was a useful initiator, and isolation of $C_6H_{13}CH_2CH_2OCO_2Pr\text{-}i$ after reaction with acetyl chloride defined the initiation process as that shown in (5)—(7).

$$(i\text{-}PrOCO_2)_2 \longrightarrow 2\ i\text{-}PrOCO_2\cdot \tag{5}$$

$$i\text{-}PrOCO_2\cdot \xrightarrow{C_6H_{13}CH=CH_2} C_6H_{13}\overset{\cdot}{C}HCH_2OCO_2Pr\text{-}i \tag{6}$$

$$C_6H_{13}\overset{\cdot}{C}HCH_2OCO_2Pr\text{-}i \xrightarrow{CH_3COCl} C_6H_{13}CH_2CH_2OCO_2Pr\text{-}i + \cdot CH_2COCl \tag{7}$$

These radical additions have been extended to cyclization reactions, e.g. (8), which are, however, non-chain processes.[90] Julia and Le Goffic have

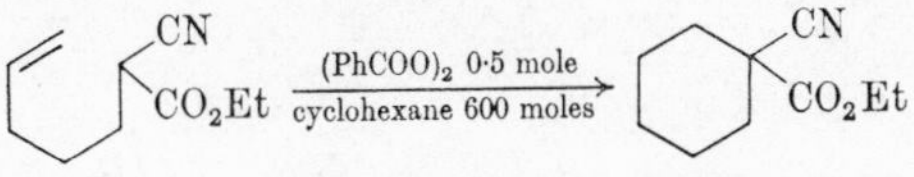

$$(70\%) \tag{8}$$

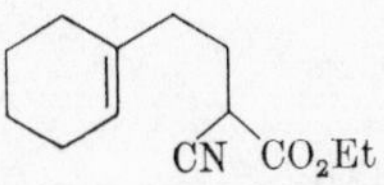

(45) 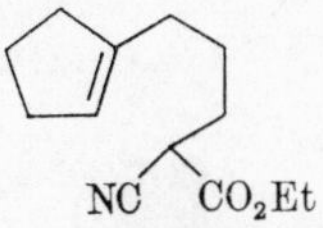 (46) 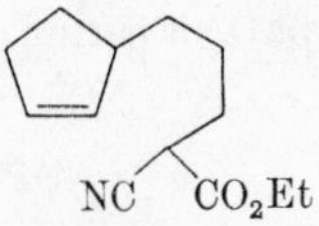 (47)

prepared five-membered rings in this way, as well as hydroindane derivatives.[91] Precursors (45) and (46) lead to mixtures of products with cis- and trans-ring fusion, isomerization in the intermediate adduct radical being possible. However, exclusively cis-fusion is obtained from (47), where the

[85b] C. Pac and S. Tsutsumi, *Tetrahedron Letters*, **1965**, 2341.
[86] R. S. Neale, *J. Am. Chem. Soc.*, **86**, 5340 (1964).
[87] F. Minisci and R. Galli, *Tetrahedron Letters*, **1965**, 1679.
[88] J. Dewing, G. F. Longster, J. Myatt, and P. F. Todd, *Chem. Comm.*, **1965**, 391.
[89] J. C. Allen, and J. I. G. Cadogan, D. H. Hey, *J. Chem. Soc.*, **1965**, 1918.
[90] J. I. G. Cadogan, D. H. Hey, and S. H. Ong, *J. Chem. Soc.*, **1965**, 1932.
[91] M. Julia and F. Le Goffic, *Bull. Soc. Chim. France*, **1965**, 1550, 1555.

intermediate radical site is not at the ring-junction, and stereochemistry is dictated by the approach of the initial radical to the double bond.

In an attempt to extend these cyclizations to allyl esters, it was found instead that products of addition of the solvent (cyclohexane) were formed. This reaction was, in turn, extended to addition to other terminal olefins, and a satisfactory route to, for instance, n-octylcyclohexane disclosed.[92]

In an attempt to functionalize C_{18} of an unsaturated steroid by the Barton reaction, the epoxide of partial structure (49) was isolated (as the oxime).[93] Before the intramolecular addition whereby this is formed, the intermediate alkoxy radical (48) is considered to become inverted, in order to give the stereochemistry assigned to the product (49). The evidence for

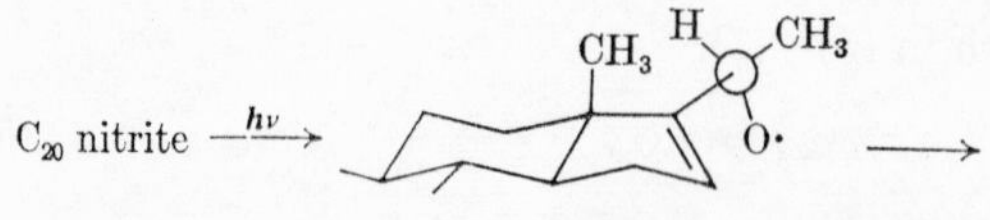

(48)

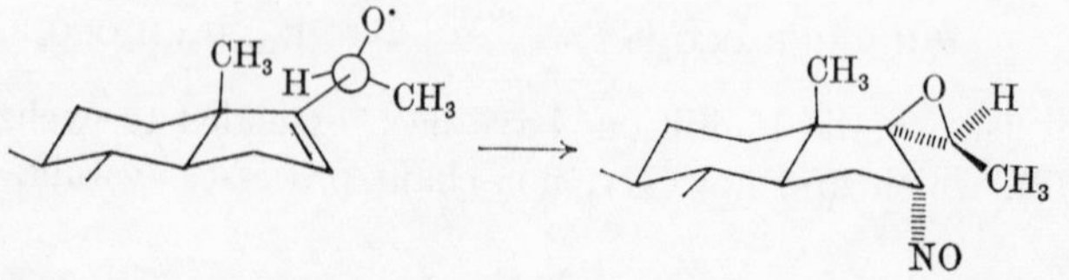

(49)

this seems rather tenuous; for example, the quoted precedents for this inversion appear to require that the alkoxy-radical carbon atom should form part of a ring.

$$RCO \cdot + \underset{\underset{COR'}{|}}{\overset{\overset{COR'}{|}}{C}}=O \rightleftarrows RCO-\underset{\underset{COR'}{|}}{\overset{\overset{COR'}{|}}{C}}-O \cdot \rightleftarrows RCO-\overset{\overset{COR'}{|}}{C}=O + R'CO \cdot \qquad (10)$$

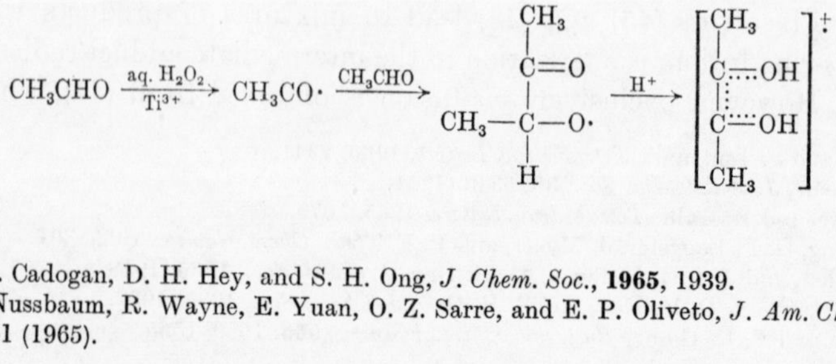

$$CH_3CHO \xrightarrow[Ti^{3+}]{aq.\ H_2O_2} CH_3CO \cdot \xrightarrow{CH_3CHO} \qquad (11)$$

[92] J. I. G. Cadogan, D. H. Hey, and S. H. Ong, *J. Chem. Soc.*, **1965**, 1939.

[93] A. L. Nussbaum, R. Wayne, E. Yuan, O. Z. Sarre, and E. P. Oliveto, *J. Am. Chem. Soc.*, **87**, 2451 (1965).

Cyclization of the 2-allyloxyethyl radical in toluene to tetrahydro-3-methylfuran,[94] rather than to tetrahydropyran, provides a further example of preferential formation of a 5- rather than a 6-membered ring in a radical cyclization.[95]

Methyl-radical addition to nitrogen in azotrifluoromethane is probably involved in the displacement reaction (9).[96] Examples of acyl-radical addition

$$CH_3 \cdot + CF_3N{=}NCF_3 \longrightarrow CH_3N{=}NCF_3 + \cdot CF_3 \qquad (9)$$

to the carbonyl-carbon atom of a vicinal triketone is shown in (10)[97] and to acetaldehyde in (11).[98]

Diels–Alder Reactions

The mechanistic subtleties of the Diels–Alder reaction continue to challenge and inspire organic chemists. The major problem is still that of the exact timing of bond-making in the forward reaction, and of bond-breaking in its retrogression, i.e. whether the cyclization is truly concerted or whether the two new bonds are formed sequentially with a second, lower, energy barrier giving rise to a reaction intermediate. It is not, of course, to be assumed that the same detailed mechanism will be favoured for the vast range of Diels–Alder reactions and good evidence for both one-step and two-step processes has been obtained this year. Lambert and Roberts[99] presented convincing evidence for a concerted mechanism by finding none for an intermediate in a system deliberately designed to favour the diradical likely to be involved in a two-step mechanism. Their choice was based upon the known formation of a diradical, sufficiently long-lived to undergo internal rotation with concomitant loss of initial stereochemistry, in 1,4-cycloaddition of 2,4-hexadiene and 1,1-dichloro-2,2-difluoroethylene.[100] A much more stabilized diradical intermediate (50) could be involved in the Diels–Alder addition of hexachlorocyclopentadiene to α-methylstyrene, one radical being tertiary and benzylic and the other allylic and stabilized by the α-chlorine. The stereochemical course of this reaction was followed by NMR location in the product of deuterium initially on the styrene vinyl-carbon atom. The *exo/endo* distribution of deuterium in the product was identical, within experimental error, with its *cis/trans* (relative to phenyl) distribution in the

[94] R. C. Lamb, J. C. Pacifici, and P. W. Ayers, *J. Org. Chem.*, **30**, 3099 (1965).
[95] B. Capon and C. W. Rees, *Ann. Reports Chem. Soc.* (*London*), **61**, 261 (1964).
[96] L. Batt and J. M. Pearson, *Chem. Comm.*, **1965**, 575.
[97] W. H. Urry, M. H. Pai, and C. Y. Chen, *J. Am. Chem. Soc.*, **86**, 5342 (1964).
[98] J. R. Steven and J. C. Ward, *Chem. Comm.*, **1965**, 273.
[99] J. B. Lambert and J. D. Roberts, *Tetrahedron Letters*, **1965**, 1457.
[100] P. D. Bartlett, L. K. Montgomery, and B. Seidel, *J. Am. Chem. Soc.*, **86**, 616 (1964); L. K. Montgomery, K. Schueller, and P. D. Bartlett, *ibid.*, **86**, 622 (1964); P. D. Bartlett and L. K. Montgomery, *ibid.*, **86**, 628 (1964).

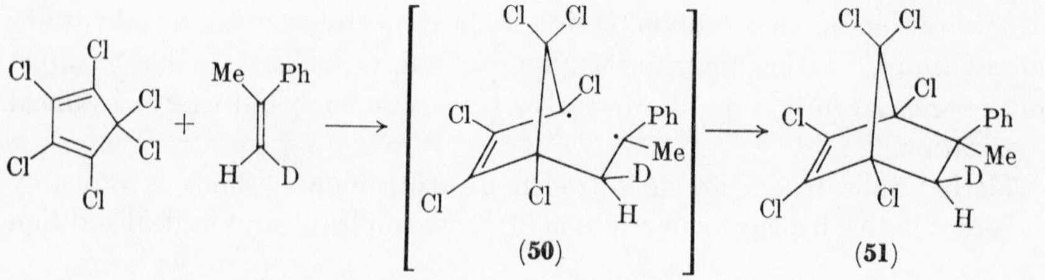

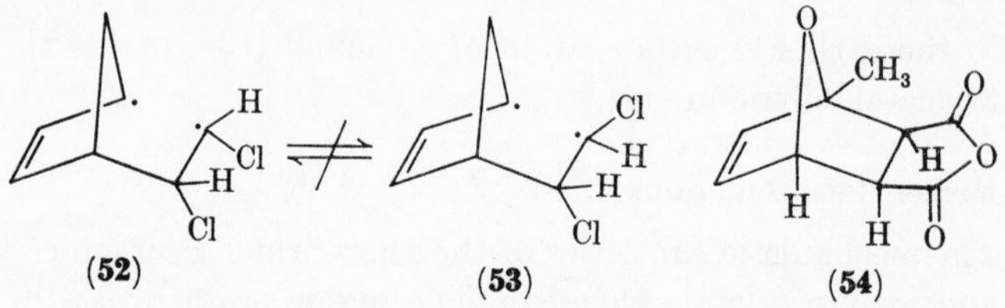

(50)   (51)

(52)   (53)   (54)

styrene. Hence, even in this case, heavily weighted in favour of diradical formation, the stereochemical relation between the groups on the double bond is maintained precisely, with no sign of free rotation. The phenyl group is exclusively *exo* in the product (51) of this reaction and it was conceivable that the reaction did go through the diradical (50) with the phenyl groups always finally in its more stable *exo*-position. This thermodynamic control would explain the observed results. To overcome this objection the separate addition of *cis*- and *trans*-1,2-dichloroethylene to cyclopentadiene, where both products were available, was studied by gas chromatography. In each case the addition was stereospecific, and if diradicals (52) and (53) were formed they showed no tendency for interconversion, though steric strain would favour (52) → (53). These results strongly support a one-step, concerted mechanism.[99] Seltzer[101] has now published full details of his elegant demonstration of simultaneous bond rupture in the dissociation of the 2-methylfuran–maleic anhydride adduct (54). The reaction rates for five deuterated isomers, with one or more of the hydrogen shown variously replaced by deuterium, were measured in isooctane at about 50°. The magnitude of the secondary α- and β-deuterium isotope effects are compatible only with simultaneous rupture of both bonds in a concerted process, and if this is true for this unsymmetrical adduct it is presumably true, at least, for similar symmetrical adducts such as those from furan and 2,5-dimethylfuran. By the principle of microscopic reversibility the forward Diels–Alder reactions must also involve simultaneous formation of both bonds, equally advanced in the transition state.

The opposite conclusion is drawn by Goldstein and Thayer from the results of their even deeper probe into the timing of bond-breaking, in an admittedly

[101] S. Seltzer, *J. Am. Chem. Soc.*, **87**, 1534 (1965).

somewhat atypical retro-Diels–Alder reaction, by measuring heavy-atom primary isotope effects at both reaction sites. Decomposition of the 2-pyrone-maleic anhydride adduct (55) into *cis*-1,2-dihydrophthalic anhydride (57) and carbon dioxide in dimethyl phthalate at 130–160° is shown to have the necessary characteristics of a Diels–Alder retrogression.[102] Carbon (k_{12}/k_{13} = 1.030) and oxygen (k_{16}/k_{18} = 1.014) kinetic isotope effects were determined, by consecutive mass-spectral analyses of the carbon dioxide produced, and shown to require a transition state with the C–C bond shown in (56) effectively broken and the C–O bond virtually intact.[103]

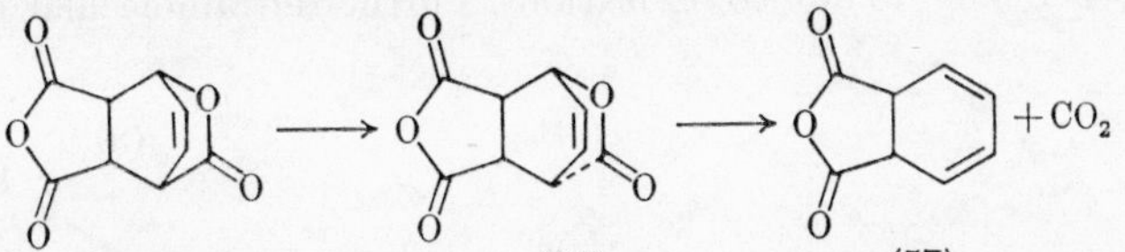

Good evidence for the two-step mechanism involving a true intermediate would be provided by the observation of a diene–dienophile reaction, involving a common transition state, that led simultaneously to a cyclohexene and a vinylcyclobutane. A reaction which possibly fulfils these requirements has been described.[104] Butadiene adds to 1-cyanovinyl acetate in benzene at 150° to give products (60) and (61) in ratio 7 : 1. The overall rate was relatively insensitive to changes in solvent, the relative rates did not vary appreciably with temperature, and the isomer ratios were insensitive to changes in solvent and temperature. This could, of course, result from two separate reactions, with different transition states, whose activation parameters responded very similarly to changes in solvent. The author (Little) suggests, however, that the reactions proceed through a single transition state (58) to an intermediate (59)[105] which then continues through either of two lower-energy paths to (60) and (61).

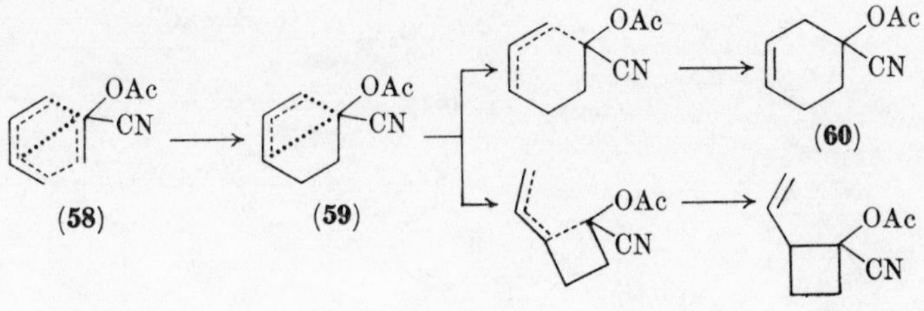

[102] M. J. Goldstein and G. L. Thayer, *J. Am. Chem. Soc.*, 87, 1925 (1965).
[103] M. J. Goldstein and G. L. Thayer, *J. Am. Chem. Soc.*, 87, 1933 (1965).
[104] J. C. Little, *J. Am. Chem. Soc.*, 87, 4020 (1965).
[105] Dashes symbolise partial bonds, and dotted lines symbolise "secondary attractive forces"; cf. R. B. Woodward and T. J. Katz, *Tetrahedron*, 5, 70 (1959).

Because of its bearing on the Diels–Alder mechanism the isomerization of various cyclopentadiene adducts have been much studied, and in particular it has been shown in the past that conversion of the *endo*-maleic anhydride adduct (62) into its *exo*-isomer proceeds by both an "external" pathway (retro-Diels–Alder followed by alternative recombination) and an "internal" pathway not involving dissociation into kinetically free fragments. Roberts and his co-workers[106] have now shown that, at least in *t*-pentylbenzene and probably in other solvents, there is no internal mechanism and the rearrangement occurs entirely by dissociation-recombination, as is well established for certain other *endo*- to *exo*-isomerizations. Further intramolecular Diels–Alder

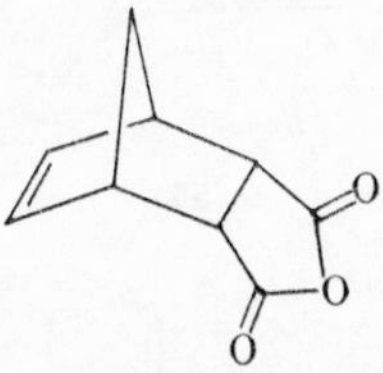

(62) 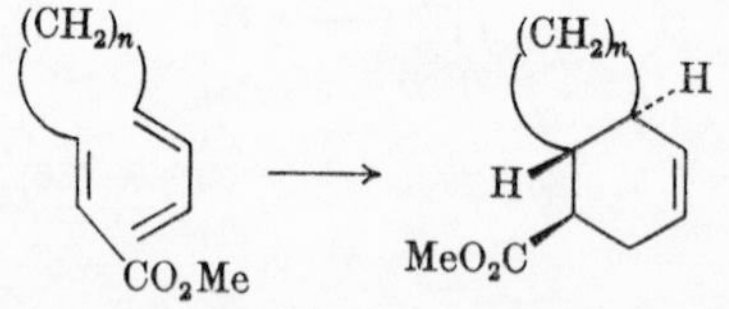   (63)

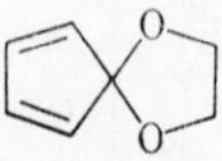

(64) 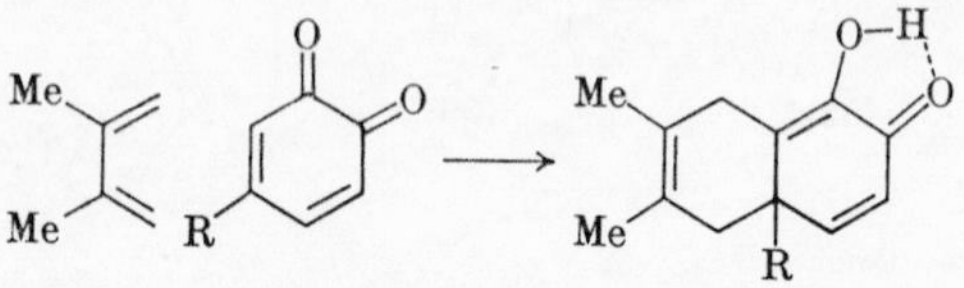   (65)

reactions have been reported: methyl *trans,trans*- (63; n = 3) and *cis,cis*-2,7,9-decatrienoate both cyclized stereospecifically but the corresponding nonatrienoates (n = 2) did not.[107] Aluminium chloride-catalysis has now been observed in the condensation of butadienes with methyl acrylate and methacrylate and acrylonitrile; other Lewis acids were ineffective.[108]

Cyclopentadienone ketals have been synthesized and found to be very reactive dienes; the ethylene ketal (64) dimerises nearly 500,000 times faster than cyclopentadiene.[109] *o*-Benzoquinones with electron-withdrawing substituents show enhanced dienophilic properties and condense readily with 2,3-dimethylbutadiene, e.g. to give (65; R = CO_2Me or CN).[110] Polar and steric effects in the addition of aromatic nitroso compounds to many unsymmetrical butadienes have been discussed further.[111] Various alkenes undergo

[106] C. Ganter, U. Scheidegger, and J. D. Roberts, *J. Am. Chem. Soc.*, 87, 2771 (1965).
[107] H. O. House and T. H. Cronin, *J. Org. Chem.*, 30, 1061 (1965).
[108] T. Inukai and M. Kasai, *J. Org. Chem.*, 30, 3567 (1965).
[109] P. E. Eaton and R. A. Hudson, *J. Am. Chem. Soc.*, 87, 2769 (1965).
[110] M. F. Ansell and A. F. Gosden, *Chem. Comm.*, 1965, 520.
[111] G. Kresze and J. Firl, *Tetrahedron Letters*, 1965, 1163.

Diels–Alder addition to dimethyl 1,2,4,5-tetrazine-3,6-dicarboxylate, followed by loss of nitrogen, to give dihydropyrazines.[112] When heated in the gas phase, norbornene decomposes smoothly to give cyclopentadiene and ethylene in a first-order homogeneous reaction unaffected by radical scavengers.[113]

A monograph on Diels–Alder reactions with particular emphasis on physical chemical aspects,[114] and an excellent review,[115a] have recently appeared.

Other Cycloaddition Reactions [115b]

The formation of a single bond between the ends of a linear conjugated polyunsaturated chain containing π-electrons, symbolized by **(66)** $\rightarrow$ **(67)**, and the converse process have been defined as electrocyclic transformations.[116] Geometrical isomerism in the acyclic system will lead to geometrical isomerism in the cyclic system and there are two possibilities, called disrotatory (**68** $\rightleftharpoons$ **69**) and conrotatory (**70** $\rightleftharpoons$ **71**). All the known thermal and

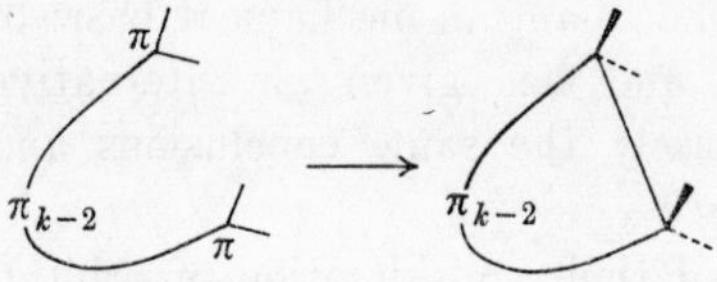

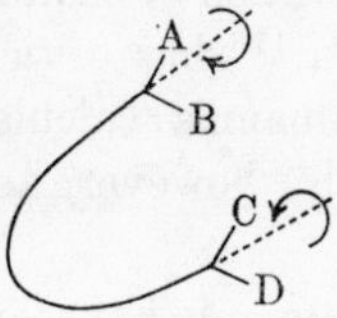

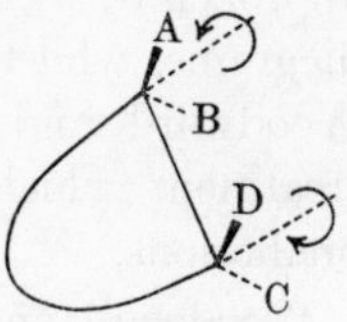

photochemical transformations of this type, such as the conversion of cyclobutene into butadiene and the cyclization of hexatrienes to cyclohexadienes, proceed in a highly stereospecific manner. Woodward and Hoffmann[116] have given a theoretical interpretation of this stereospecificity based on stereochemical control by the symmetry of the highest occupied MO of the acyclic reactant. Extended Hückel calculations support this idea and show that it also applies to the ring-opening process. In an acyclic

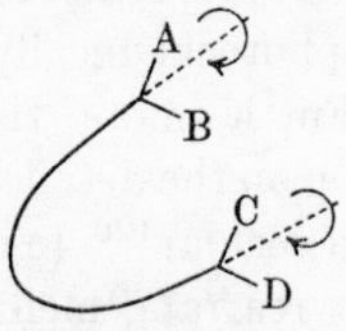

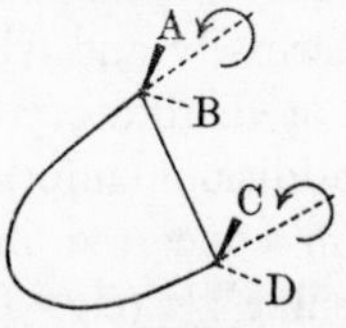

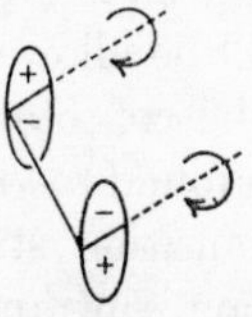

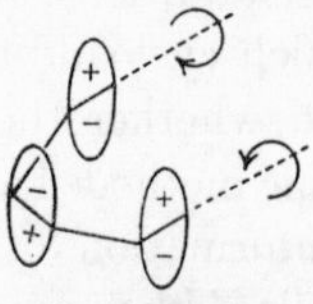

[112] J. Sauer, A. Mielert, D. Lang, and D. Peter, *Chem. Ber.*, **98**, 1435 (1965).

[113] B. C. Roquitte, *J. Phys. Chem.*, **69**, 1351 (1965).

[114] A. Wassermann, "Diels–Alder Reactions", Elsevier, London, 1965.

[115] R. Huisgen, R. Grashey, and J. Sauer in "The Chemistry of the Alkenes", ed. S. Patai, Interscience Publishers, London, 1964, (a) p. 878, (b) p. 739; in (b) are a definition and discussion of cycloaddition reactions and a comprehensive review of the cycloaddition to olefins.

[116] R. B. Woodward and R. Hoffmann, *J. Am. Chem. Soc.*, **87**, 395 (1965).

system containing $4n$ π-electrons the symmetry of the highest occupied ground-state orbital (e.g. **72**) is such that bonding between the termini must involve overlap of orbital lobes on opposite sides of the system and this is achieved by the conrotatory process; if there are $(4n + 2)$ π-electrons the overlap must then be between orbital lobes on the same side of the molecule achieved by the disrotatory process, e.g. (**73**). Promotion of an electron to the first excited state leads to a reversal of terminal orbital symmetry, so that the stereochemistry of an electrocyclic process induced photochemically will be the opposite of that of the thermal process. This hypothesis, therefore, accommodates all the known stereospecifities of thermal and photochemical ring openings and cyclizations of this type; it is also applicable to electro-cyclic processes of ions and radicals and has been used to predict the stereo-chemistry of some of these; one such prediction has already been confirmed (see p. 44). Longuet-Higgins and Abrahamson[117] have supplemented this approach by a consideration of orbital and electron configuration correlation diagrams, whilst Fukui[118] has criticized the quantum-mechanical basis of Woodward and Hoffmann's calculations and has given an alternative treatment which leads, however, to precisely the same conclusions and predictions.

Woodward and Hoffmann have extended their consideration of orbital-symmetry relations to the general problem of concerted intermolecular cycloadditions, and have been able to deduce a set of selection rules predicting which cycloadditions will be allowed thermally and which photochemically, and whether they will be conrotatory or disrotatory.[119] Furthermore, the same methods have been found to yield additional information on the detailed conformation of the transition state in concerted intermolecular[120] (e.g. Diels–Alder reaction) and intramolecular[121] (e.g. Cope rearrangement) cycloadditions.

The scope, orientation, and stereochemistry of the photochemical cyclo-additions of 2-cyclohexenone and other conjugated enones to a variety of, mostly unsymmetrically, substituted ethylenes has been determined. The reactions are facilitated by electron release to the double bond (e.g. 1,1-dimethoxyethylene > methoxyethylene > isobutene $\gg$ acrylonitrile). They are characterized by considerable orientational specificity, e.g. (**74**) is formed in much larger amount than (**75**) from isobutene, and by frequent formation of the less stable *trans*-fused ring systems, e.g. *trans*-(**74**) predominates. The following tentative mechanism is proposed: the ketone, K, is photoexcited to

[117] H. C. Longuet-Higgins and E. W. Abrahamson, *J. Am. Chem. Soc.*, **87**, 2045 (1965).
[118] K. Fukui, *Tetrahedron Letters*, **1965**, 2009.
[119] R. Hoffmann and R. B. Woodward, *J. Am. Chem. Soc.*, **87**, 2046 (1965).
[120] R. Hoffmann and R. B. Woodward, *J. Am. Chem. Soc.*, **87**, 4388 (1965).
[121] R. Hoffmann and R. B. Woodward, *J. Am. Chem. Soc.*, **87**, 4389 (1965).

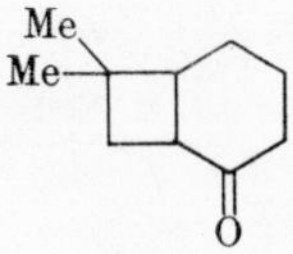

(74) 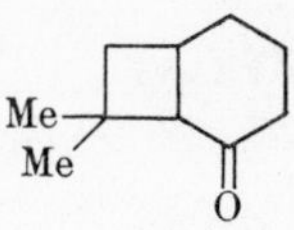   (75)

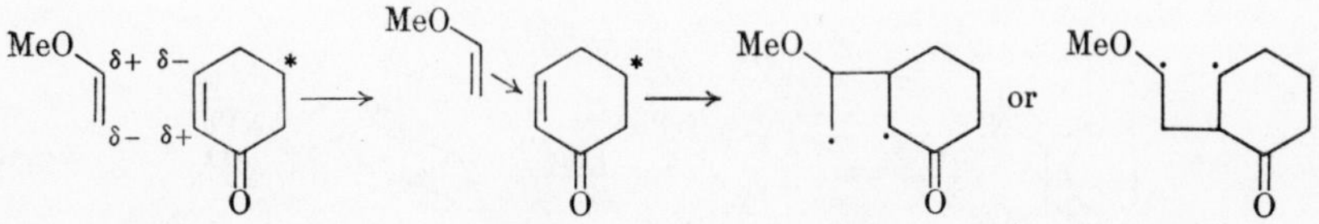

(76)   (77)   (78)   or   (79)

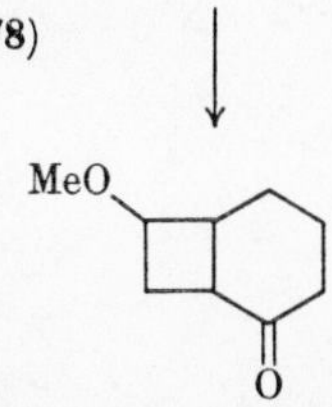

K* (the olefins do not absorb the ultraviolet light used) which forms reversibly an orientated π-complex (77) with the olefin; this complex is in equilibrium with a diradical (78) or (79), that finally closes to the cyclobutane. It is the geometry of the π-complex which is thought to control the orientation of the product. In the n–π^* excited state of planar $\alpha\beta$-unsaturated ketones $C_{(\beta)}$ is considered to be more negative than $C_{(\alpha)}$ (as in 76) and therefore the observed orientation, shown for methoxyethylene, follows.[122]

Heating the phenyl azide–cyclopentene adduct with phenyl isothiocyanate has been shown to give 2-anilinocyclopentenethiocarboxanilide (80), and not the previously proposed symmetrical thiourea expected from trapping the assumed 1,3-dipolar intermediate (81). Similar decomposition of the norbornene–phenyl azide adduct with phenyl isocyanate does not give the symmetrical urea expected from the dipole (82) but, instead, the rearranged product (83), a reasonable mechanism for its formation being as shown. These structural reassignments emphasize the need for caution in interpreting reactions according to unifying mechanistic patterns, and many such "1,3-dipolar" additions may not involve 1,3-dipolar intermediates.[123] The structure of the adduct from tetracyanoethylene and methyl-1-azepinecarboxylate has now been shown to be (84), arising from a 2 + 4 concerted cycloaddition rather than from a 2 + 6 non-concerted addition as previously suggested.[124] The addition of a series of *meta*- and *para*-substituted phenyl azides to norbornene in ethyl acetate to give triazolines, stable in the absence

[122] E. J. Corey, J. D. Bass, R. Le Mahieu, and R. B. Mitra, *J. Am. Chem. Soc.*, **86**, 5570 (1964).
[123] J. E. Baldwin, G. V. Kaiser, and J. A. Romersberger, *J. Am. Chem. Soc.*, **87**, 4114 (1965).
[124] J. E. Baldwin and R. A. Smith, *J. Am. Chem. Soc.*, **87**, 4819 (1965).

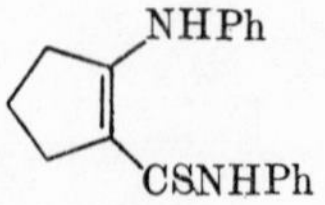

(80) 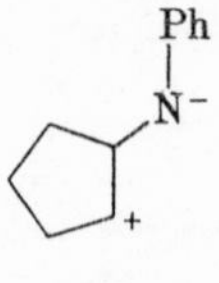   (81) 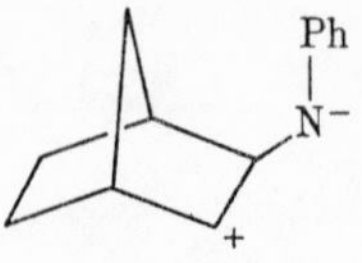   (82)

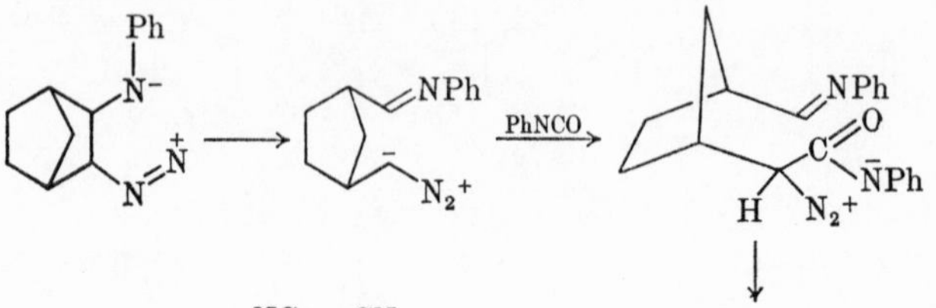

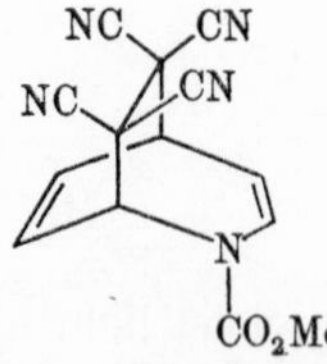

(84) 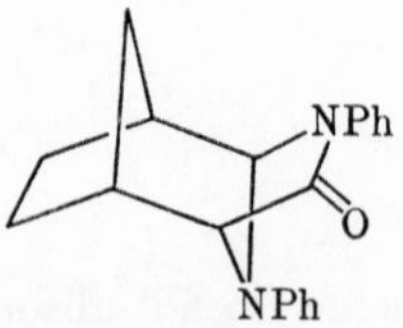   (83)

of light, is of the first order in each reactant, and irreversible. The reaction is facilitated by electron-withdrawing groups and slightly retarded by electron-releasing groups ($\rho = +0.84$ at 25°). The reactions are insensitive to solvent polarity and have large negative entropies of activation consistent with synchronous 1,3-dipolar addition in a highly ordered transition state, but in view of substituent effects it is assumed that the new bonds are not equally developed in the transition state (85), which is charged.[125]

Tetracyanooxirane adds readily to olefins, acetylenes, and some aromatic compounds at 130–150° with cleavage of the carbon–carbon bond of the

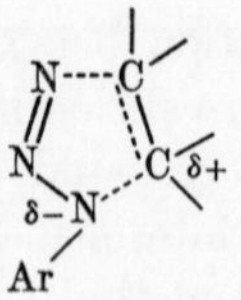

(85) 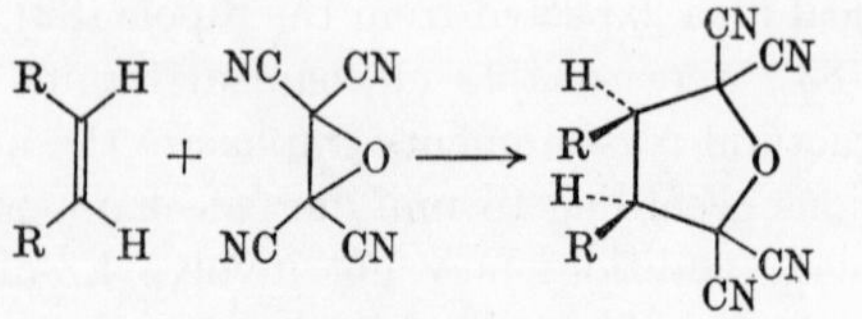   (86)

$$86 \underset{k_{-1}}{\overset{k_1}{\rightleftharpoons}} 86^*$$

$$86^* + \text{olefin} \xrightarrow{k_2} \text{product}$$

Scheme 5

[125] P. Scheiner, J. H. Schomaker, S. Deming, W. J. Libbey, and G. P. Nowack, *J. Am. Chem. Soc.*, **87**, 306 (1965).

epoxide ring, to give tetracyano-tetrahydro- and -dihydro-furans. The addition to olefins is stereospecific since *cis-* and *trans-*2-butene, -stilbene, and -1,2-dichloroethylene each give the corresponding *cis-* and *trans-*adduct.[126] The rate of addition of the oxirane (86) to *trans-*stilbene is given

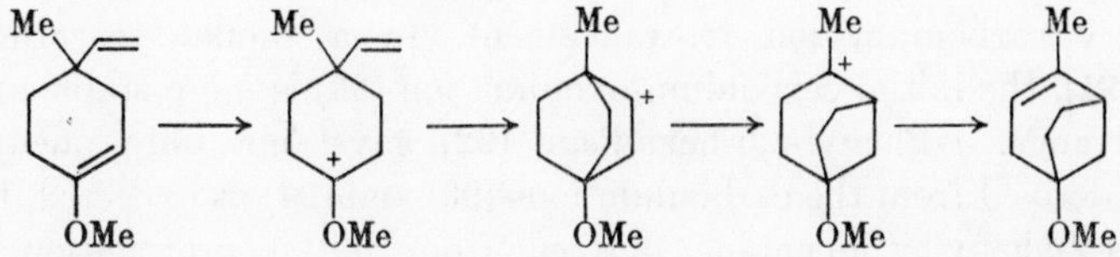

(87)　　　　　　　　　　　　　　　　　　　　(88)

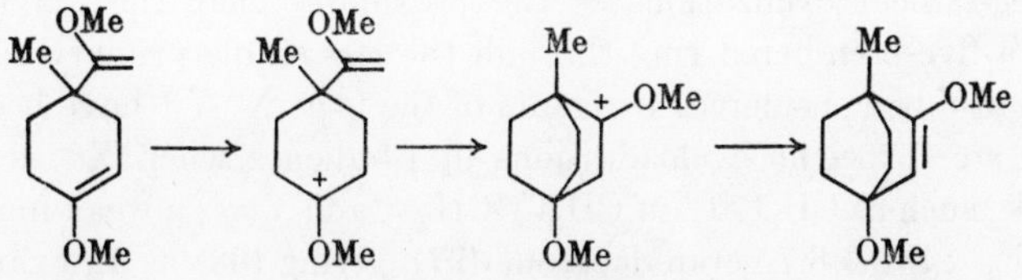

(89)　　　　　　(90)　　　　　(91)

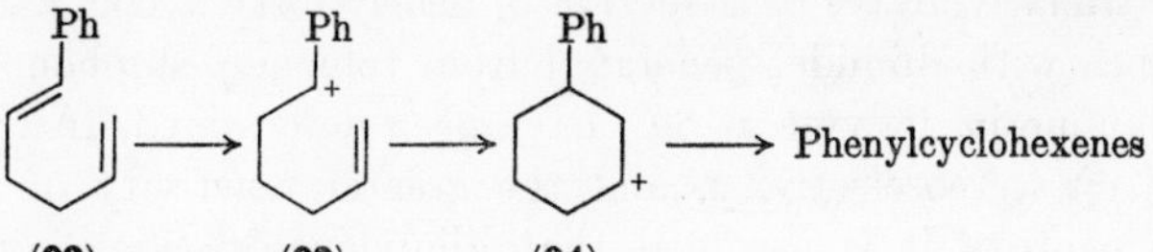

(92)　　　(93)　　　(94)　　　⟶ Phenylcyclohexenes

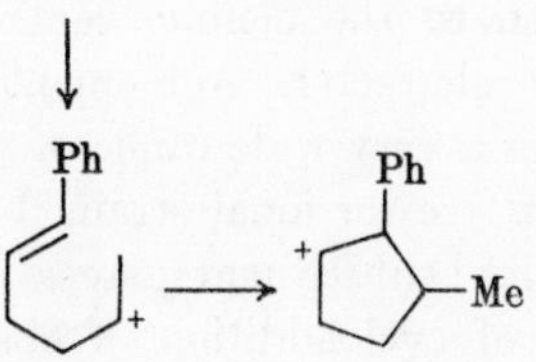

(95)　　　(96)

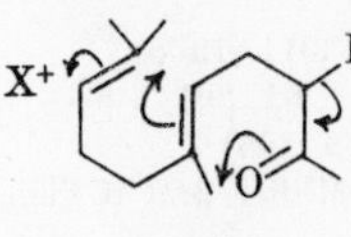

(97) 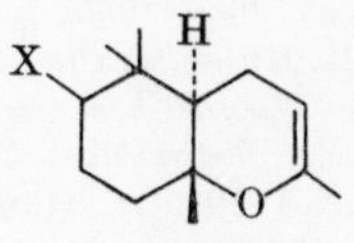　　　　　　　　(98) 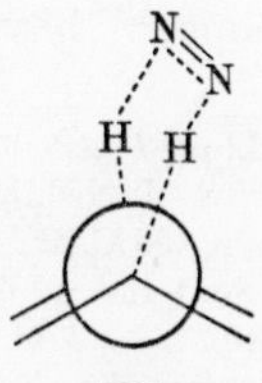　　　　　　　(99)

by rate $= k_1k_2[86][\text{Stilbene}]/(k_{-1} + k_2[\text{Stilbene}])$ in agreement with the mechanism shown in Scheme 5; (86*) is a thermally activated form of (86), best represented as a symmetrical hybrid of the many dipolar forms possible

126 W. J. Linn and R. E. Benson, *J. Am. Chem. Soc.*, **87**, 3657 (1965).

with the carbon–carbon bond broken. The overall reaction is relatively insensitive to environmental and structural changes, and the second step has all the appearance of a concerted multicentre cycloaddition.[127]

Cyclization of the vinylcyclohexene (87) catalyzed by boron trifluoride etherate gave (88), presumably by the route shown, involving the secondary-to-tertiary carbonium ion rearrangement. In a similar rearrangement, (89) → (91), the initial carbonium-oxonium ion (90) is more stable and does not rearrange. 1-Phenyl-1,5-hexadiene (92) gave, not only phenylcyclo-hexenes derived from the carbonium ions (93) and (94) as expected, but also 1-methyl-2-phenylcyclopentene derived from initial protonation of the terminal double bond, via (95) and (96).[128] By analogy with a number of similar free-radical cyclizations[129] the possibility that the ion (93) may cyclize to a five-membered ring through the less stable primary carbonium ion should also be considered. Reagents of the type X^+Y^- have been shown to initiate stereospecific cycloadditions in 1,5-dienes when X^+ is a strong electrophile, such as CH_3CO^+ or $CH_3COCH_2^+$, and Y^- is a weak nucleophile, such as BF_4^-; *trans*-6,7-geranylacetone (97), giving (98), is an example.[130]

Hydrogenations with diimide have been reviewed and reaction mechanisms discussed.[131] Relative rates of reduction of nearly forty cyclic, exocyclic, or acyclic alkenes with diimide, generated from toluene-*p*-sulphonohydrazide and triethylamine in diglyme at 80°, have been determined. In view of the known high *cis*-stereoselectivity, and the greater reactivity of less polar unsaturated centres, it is assumed that diimide reductions involve syn-chronous *cis*-addition of neutral hydrogen to the olefinic centre via the transition state (99) of negligible ionic character. Agreement between observed and calculated relative rates, over a very wide range of reactivity, suggests that the major controlling factors are torsional strain, bond-angle bending strain, and the polar effects of α-alkyl substituents; steric effects are not very important.[132] The mechanisms of cycloaddition of enamines to diphenylcyclopropenone,[133] and of α,β-unsaturated ketones to 3,4-dihydro-isoquinolines[134] have also been studied.

[127] W. J. Linn, *J. Am. Chem. Soc.*, **87**, 3665 (1965).

[128] K. Morita, M. Nishimura, and H. Hirose, *J. Org. Chem.*, **30**, 3011 (1965).

[129] B. Capon and C. W. Rees, *Ann. Reports Chem. Soc. (London)*, **61**, 261 (1964).

[130] W. A. Smit and A. V. Semenovsky, *Tetrahedron Letters*, **1965**, 3651.

[131] C. E. Miller, *J. Chem. Educ.*, **42**, 254 (1965); S. Hünig, H. R. Müller, and W. Thier, *Angew. Chem. Intern. Ed. Engl.*, **4**, 271 (1965).

[132] E. W. Garbisch, S. M. Schildcrout, D. B. Patterson, and C. M. Sprecker, *J. Am. Chem. Soc.*, **87**, 2932 (1965).

[133] J. Ciabattoni and G. A. Berchtold, *J. Am. Chem. Soc.*, **87**, 1404 (1965).

[134] C. Szantay and J. Rohaly, *Chem. Ber.*, **98**, 557 (1965); *Magy. Kem. Folyoirat*, **70**, 478 (1964).

CHAPTER 6

Nucleophilic Aromatic Substitution

The Bimolecular Mechanism

Bunnett and Zahler's classic review in 1951[1] stimulated much of the later work in this area, and the mechanistic advances have been reviewed by Bunnett[2] and by Ross.[3] The most important and vexed question has been whether the bimolecular displacements are synchronous, one-step processes involving one transition state or two-step processes involving a tetrahedral intermediate, on the normal reaction path, the formation or decomposition of which may be rate-determining. Considerable evidence consistent with the latter has been accumulated over the last eight years, mostly by Bunnett and his co-workers,[2] and further notable contributions along broadly similar lines have been made this year by him and by Jencks.

Kirby and Jencks[4] believe that unambiguous kinetic evidence for the existence of a reaction intermediate is provided by their results on the base-catalysed reactions of dimethylamine and piperidine with p-nitrophenyl phosphate to give the corresponding p-nitroanilines. These reactions in aqueous solution are catalysed by hydroxide ions, and the former is also subject to general-base-catalysis by a second mole of dimethylamine. As the catalyst concentration is increased, the initial rapid increase in rate levels off and the second-order rate constants become almost independent of catalyst concentration. Thus there is a change from a rate-determining step at low basicity, which is strongly base-catalysed, to a rate-determining step at high basicity, which shows little or no such catalysis. This demands the existence of an intermediate, and a charge-transfer complex is rejected in favour of a tetrahedral addition compound. Comparison with the reaction of dimethylamine with p-chloronitrobenzene, which is not base-catalysed, suggested that the catalysed step occurs after formation of the intermediate. Mechanisms similar to those proposed by Bunnett were considered in detail and the most probable was considered to be that shown in Scheme 1 where the conjugate acid of the general base catalyst donates a proton to the conjugate base of the intermediate, making the phosphate a better leaving group.

[1] J. F. Bunnett and R. E. Zahler, *Chem. Rev.*, **49**, 273 (1951).

[2] J. F. Bunnett, *Quart. Rev.*, **12**, 1 (1958).

[3] S. D. Ross, "Progress in Physical Organic Chemistry", Vol. I, ed. S. G. Cohen, A. Streitwieser, and R. W. Taft, Interscience Publishers, New York, p. 31 (1963).

[4] A. J. Kirby and W. P. Jencks, *J. Am. Chem. Soc.*, **87**, 3217 (1965).

The slower reaction of p-nitrophenyl phosphate with methylamine is not base-catalysed in aqueous solution and here the initial nucleophilic attack on the aromatic ring appears to be rate-determining.[4] Other such reactions are those between 1-fluoro-2,4-dinitrobenzene and butylamine in methanol, and aniline in methanol, t-butyl alcohol, and aqueous dioxan, described by

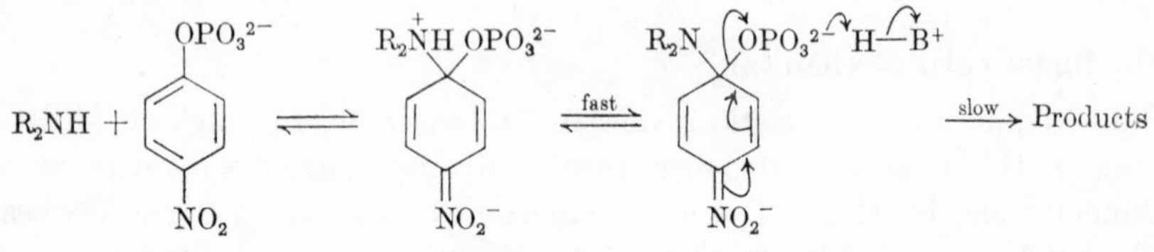

Scheme 1

Bunnett and Garst,[5] who discuss critically the incidence and significance of base-catalysis in nucleophilic aromatic substitution.[5,6] These reactions, like many other such displacements, are only mildly accelerated, if at all, by added weak and strong bases, and they are considered to occur by the intermediate-complex mechanism with the first step rate-limiting. In contrast, the same workers[6] find that the reaction of 2,4-dinitrophenyl phenyl ether with piperidine in 3:2 dioxan–water is strongly base-catalysed. The second-order rate coefficient is linearly dependent on the concentration of piperidine. However, for the sodium hydroxide-catalysed reaction a plot of the second-order rate coefficient against hydroxide concentration is curved. The nature of this non-linear relation is shown to fulfil a particular requirement of the intermediate-complex mechanism, of the form given in Scheme 1 and is considered strong additional evidence for this mechanism. Another

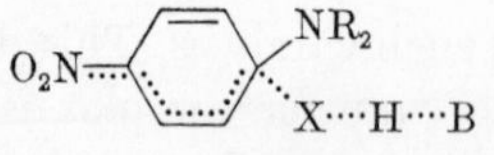

(1) 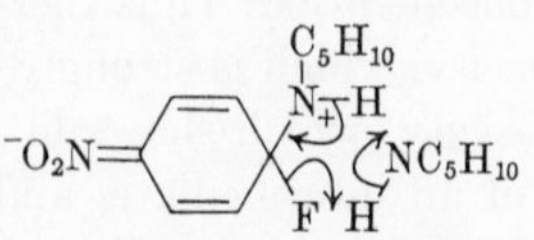   (2)

such requirement fulfilled by the present results is that base-catalysis should be observed in reactions of amines with 1-X-2,4-dinitrobenzenes where X is a relatively poor leaving group, and indeed the phenyl ether was chosen with this in mind. Detailed consideration was given to how the base could catalyse expulsion of a leaving group, and the most probable mechanism[7a] is thought to be reversible transformation of the intermediate complex into its con-

---

[5] J. F. Bunnett and R. H. Garst, *J. Am. Chem. Soc.*, **87**, 3875 (1965).
[6] J. F. Bunnett and R. H. Garst, *J. Am. Chem. Soc.*, **87**, 3879 (1965).
[7a] In aqueous dioxan; an alternative suggestion of the Reviewers[7b] was considered to be attractive for poorly ionizing solvents such as benzene.

jugate base, followed by general-acid-catalysed separation of the leaving group with the transition state (1), in very good agreement with Kirby and Jencks's proposal.[4]

Further convincing evidence for this mechanism has been provided by Bunnett and Bernasconi[8] in the kinetic form of the reactions of piperidine with 2,4-dinitrophenyl ethers, $(O_2N)_2C_6H_3OR$, in 1:9 dioxan–water. The observed decrease in base-catalysis as OR becomes a better leaving group, the curvilinear form of the plots of second-order rate coefficients for the formation of 1-(2,4-dinitrophenyl)piperidine against base-catalyst concentration, the demonstration of general-base-catalysis, and the variation of solvent isotope effect with base concentration, are all shown to be in complete agreement with the intermediate-complex mechanism.[8] The evidence for this mechanism, at least over a fair range of activated aromatic substrates, is thus compelling.

Pietra and Fava had recently shown[9] that in the reaction of 1-fluoro-2,4-dinitrobenzene with piperidine in benzene the order with respect to piperidine changed from 1 to 2 with increasing piperidine concentration, whilst with 1-chloro-2,4-dinitrobenzene the order remained 1 at such high concentrations; neither the order nor the rate was changed significantly by a much higher concentration of the tertiary base, triethylamine. It was suggested that the second molecule of piperidine was not simply involved in rate-determining proton abstraction but in rate-determining separation of the fluoride ion by hydrogen bonding. This was supported by the effect of adding methanol, which increased the rate at low piperidine concentration, decreased it at high piperidine concentration, and reduced the order in piperidine to 1 throughout. [It was further suggested[7b] that the second molecule of piperidine, or the molecule of methanol, might assist in concerted separation of both the hydrogen and the fluorine in a six-membered cyclic hydrogen-bonded transition state (2).] This work has now been extended[10] to 2-methylpiperidine and *cis*-2,6-dimethylpiperidine, the results for which exactly parallel those of piperidine, and the second-order rate coefficients for 1-chloro-2,4-dinitrobenzene have been compared with the second-order component of the 1-fluoro-2,4-dinitrobenzene reaction. These reactivity ratios, ArF/ArCl, are 8, 36, and 7 for piperidine and its mono- and its di-methyl derivative, respectively, and thus do not depend greatly on the steric requirement of the nucleophile. This contrasts with earlier work[11] and casts doubt on the earlier deductions about the relative amounts of bond-breaking in the transition

[7b] B. Capon and C. W. Rees, *Ann. Reports Chem. Soc. (London)*, **60**, 279 (1963).
[8] J. F. Bunnett and C. Bernasconi, *J. Am. Chem. Soc.*, **87**, 5209 (1965).
[9] F. Pietra and A. Fava, *Tetrahedron Letters*, **1963**, 1535.
[10] F. Pietra, *Tetrahedron Letters*, **1965**, 745, 1432.
[11] G. S. Hammond and L. R. Parks, *J. Am. Chem. Soc.*, **77**, 340 (1955).

state. Pietra's results also require a stepwise mechanism for simple rationalization. Bernasconi and Zollinger[12] have also established the same general kinetic pattern (second-order) for reaction of 1-chloro-2,4-dinitrobenzene with piperidine in benzene, whilst reaction of the fluoro analogue is catalysed by piperidine, 1,4-diazabicyclo[2,2,2]octane, methanol, and pyridine with effectiveness in the ratios $250:27:10:1$, respectively. They also consider attractive the possibility of more or less synchronous separation of proton and fluoride, as in (2). However, they regard the catalysis by piperidine to be general-base-catalysis and not electrophilic catalysis, with the non-effectiveness of triethylamine attributed to steric causes (contrast diazabicyclooctane) rather than to its not being a proton donor.

Like Bunnett and Garst, Pietra[13] has found that the reaction of 2,4-dinitrophenyl phenyl ether with piperidine in benzene is of the third order, and the reaction is also catalysed by pyridine (which does not itself react with the ether) and is therefore general-base-catalysed. Furthermore, a primary isotope effect has been observed for the first time in this class of reactions; replacing piperidine by 1-deuteriopiperidine slows the reaction $1 \cdot 27$-fold. The same intermediate complex mechanism is favoured.

A similar kinetic pattern has been found[14] in the reactions of hexachlorocyclotriphosphazatriene with piperidine and with ethylamine in toluene at $0°$. The reactions follow a mixed second- and third-order rate law:

$$\text{Rate} = k_2[\text{N}_3\text{P}_3\text{Cl}_6][\text{R}_2\text{NH}] + k_3[\text{N}_3\text{P}_3\text{Cl}_6][\text{R}_2\text{NH}]^2$$

and the piperidine reaction was shown to be catalysed by tri-n-butylamine. Here the third-order term makes a greater relative contribution than in the

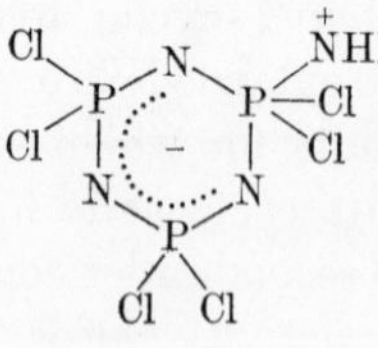

(3) 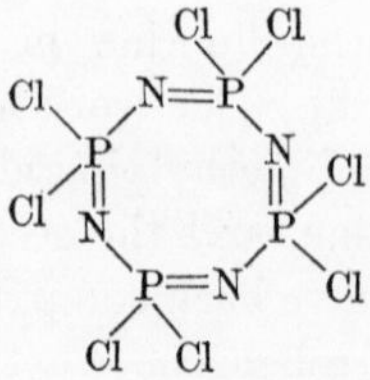   (4)

benzene series; an analogous mechanism involving the quinquecoordinated phosphorus complex (3) is proposed. The stereochemical course of such displacements in the phosphazenes is unknown, but attack perpendicular to the ring, causing inversion of configuration, is favoured. Interestingly, the

[12] C. Bernasconi and H. Zollinger, *Tetrahedron Letters*, **1965**, 1083.
[13] F. Pietra, *Tetrahedron Letters*, **1965**, 2405.

cyclic tetramer, octachlorocyclotetraphosphazatetraene (4) reacts 10^2—10^3 times faster than the trimer with diethylamine.[14]

Meisenheimer and Related Complexes

Interest has recently been renewed in Meisenheimer complexes, such as (5), as models for the tetrahedral intermediates of the last section. Demonstration of their existence does not, of course, prove that such complexes lie on the reaction pathway. A methanolic solution of 2,4,6-trinitro[methyl-^{14}C]anisole with sodium methoxide at room temperature immediately became red, and the radioactivity could not be extracted into toluene; decomposition of the complex with dilute acid gave nearly 50% decrease in activity of the labelled compound, indicating the identity of the methoxyl groups in the complex. Similar treatment of 2,4-dinitroanisole did not give a

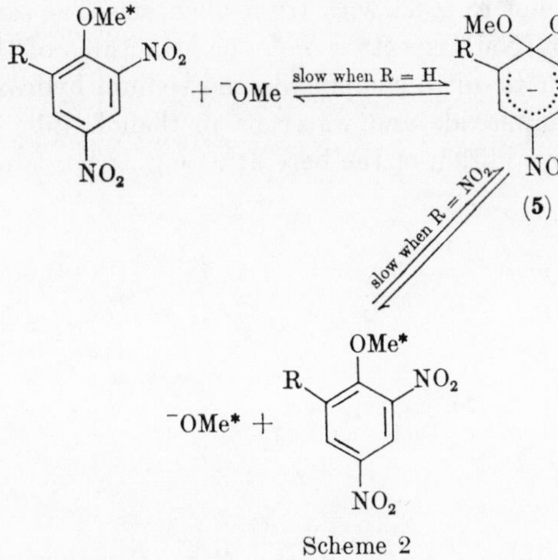

Scheme 2

red solution. Variation of rate coefficient with change in methoxide concentration, the Arrhenius parameters for these symmetrical methoxyl exchanges, and the heat of formation of the trinitroanisole–sodium methoxide complex, indicate that the methoxyl exchanges are two-step reactions, the rate-determining step being unimolecular heterolysis of the intermediate complex for the trinitro compound and formation of the complex for the dinitro compound, as shown in Scheme 2.[15]

Crampton and Gold[16] have investigated the interaction of 1,3,5-trinitrobenzene with aliphatic amines in anhydrous dimethyl sulphoxide by NMR

[14] B. Capon, K. Hills, and R. A. Shaw, *J. Chem. Soc.*, **1965**, 4059.
[15] J. H. Fendler, *Chem. Ind. (London)*, **1965**, 764.
[16] M. R. Crampton and V. Gold, *Chem. Comm.*, **1965**, 549.

spectroscopy, electrical conductivity measurements, and by ultraviolet-visible spectrophotometry. All the results agree with an equilibrium involving one molecule of trinitrobenzene and two of the primary or secondary amine (Scheme 3). The observation of similar absorption spectra in certain other

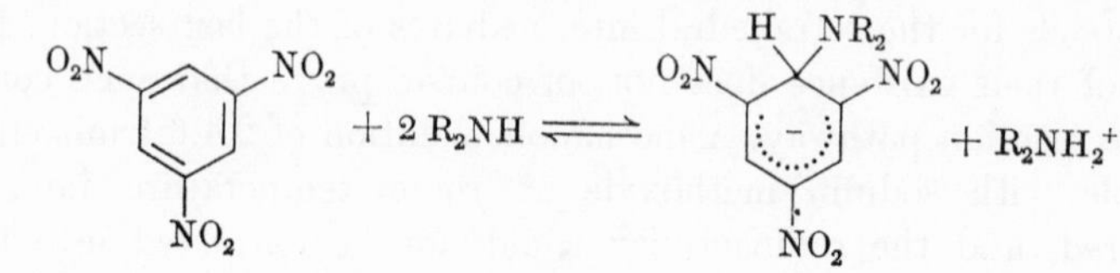

Scheme 3

solvents suggests that this 1:2 interaction may be more general than the formation of 1:1 zwitterionic complexes. Tertiary amines that cannot give a 1:2 complex appear not to react with trinitrobenzene. The same workers[17] have also studied, by NMR spectroscopy, the structures of the complexes formed between aromatic nitro compounds and sodium hydroxide or methoxide in dimethyl sulphoxide and water or methanol. Loss of an active proton, if present, or addition of the base at a ring carbon atom, is usually observed.

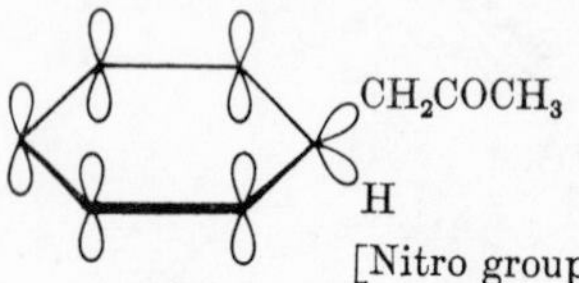

(6)   [Nitro groups and charge omitted]   (7)

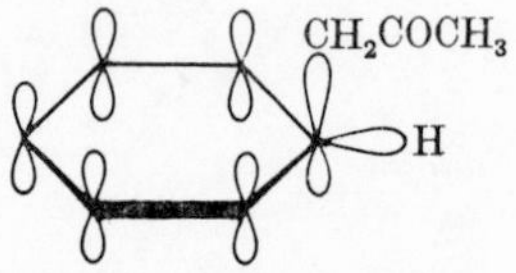

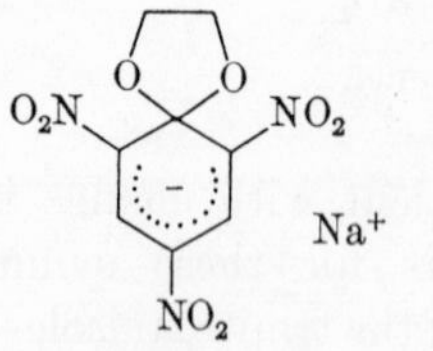

(8) 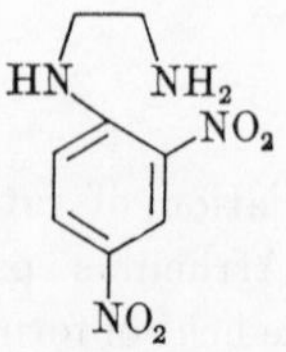   (9)

A detailed spectroscopic investigation[18] of the coloured complexes formed in the Janovsky reaction, for example that between *m*-dinitrobenzene and alkali in acetone shows that these complexes differ in a number of ways from the Meisenheimer complexes (e.g. 5 and 6) to which they are normally considered analogous. It is tentatively proposed, therefore, that the Janovsky

[17] M. R. Crampton and V. Gold, *Chem. Comm.*, **1965**, 256.
[18] R. J. Pollitt and B. C. Saunders, *J. Chem. Soc.*, **1965**, 4615.

complexes are not fully tetrahedral as in (6), but are better represented as in (7), where "the acetonate ion is situated above one of the ring-carbon atoms and is donating the lone pair on the methylene group to the lowest available π^* orbital of the benzene ring to form a dative σ bond."[18] The hydrogen atom is still near its original position and the π-structure is not grossly disturbed. It is not made clear, however, why (7) should represent a stable structure. The spiro Meisenheimer complex (8) has been isolated after treatment of 1-(2-hydroxyethoxy)-2,4,6-trinitrobenzene with sodium in glycol,[19] but attempts to detect a rapid intramolecular nucleophilic substitution in N-(2,4-dinitrophenyl)ethylenediamine (9) in deuterated nitromethane failed —the NMR spectrum, which showed two triplets for the methylene groups, was unchanged at 90°.[20]

The rates of formation and decomposition of the Meisenheimer complex of 2,4,6-trinitroanisole and sodium methoxide in methanol have been measured.[21]

Substitution in Polyfluoro Aromatics

There have been several papers this year, mostly from Burdon and Tatlow and their co-workers, concerned with the nucleophilic displacement of fluorine from tetra- and penta-fluorobenzene derivatives. Pentafluoronitrobenzene reacts with predominant displacement of a *para*-fluorine atom by most nucleophiles, but of an *ortho*-fluorine atom by ammonia, methylamine, or dimethylamine which are capable of hydrogen bonding to the nitro group (cf. ref. 22). The proportion of *ortho*-substitution decreases markedly as the size of the amine increases, and this is ascribed to steric hindrance to hydrogen bonding.[23] However, displacement of *ortho*-fluorine by methoxide ion in ether–methanol mixtures has been observed, and it increases as the methanol content decreases; the $o:p$ ratio is $1:1$ in 3.8% methanol in ether.[24] Further, no comparable steric hindrance to hydrogen bonding was observed in the reactions of 1-(pentafluorophenyl)piperidine oxide (10) (or the corresponding morpholine compound) with dimethylamine or with piperidine (or morpholine). Again, piperidine (or morpholine) rapidly displaces *ortho*-fluorine, this time both atoms, and hydrogen bonding between the N-oxide-oxygen and the secondary amine provides the explanation.[25] Similar factors may be operative in the analogous reactions of pentafluoronitrosobenzene, which

[19] R. Foster, C. A. Fyfe, and J. W. Morris, *Rec. Trav. Chim.*, **84**, 516 (1965).

[20] B. Lamm, *Acta Chem. Scand.*, **19**, 1492 (1965).

[21] T. Abe, T. Kumai, and H. Arai, *Bull. Chem. Soc. Japan*, **38**, 1526 (1965).

[22] B. Capon and C. W. Rees, *Ann. Reports Chem. Soc. (London)*, **60**, 277 (1963).

[23] J. G. Allen, J. Burdon, and J. C. Tatlow, *J. Chem. Soc.*, **1965**, 1045.

[24] J. Burdon, D. Fischer, D. King, and J. C. Tatlow, *Chem. Comm.*, **1965**, 65.

[25] M. Bellas and H. Suschitzky, *Chem. Comm.*, **1965**, 367.

appears to give some *ortho*-displacement with methylamine, but all the nitroso compounds involved were very unstable.[26]

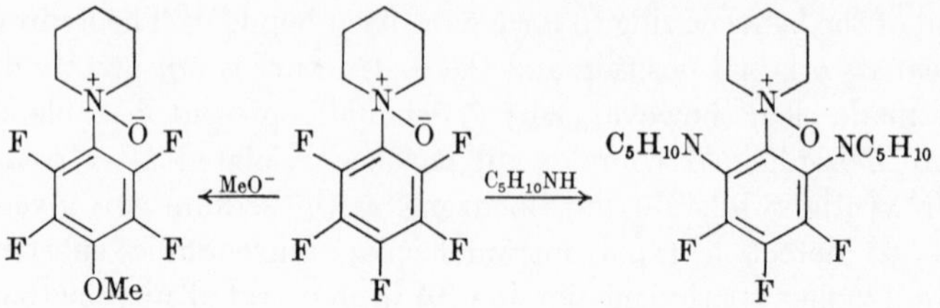

(10)

Nucleophilic attack by sodium methoxide, hydrazine, and lithium aluminium hydride on the two tetrafluorobenzenes **(11)** and **(12)** results in exclusive displacement of the fluorine atoms (circled) which are *para* to hydrogen. On the assumption that *para*-quinonoid contributions to the transition state are greater than *ortho*-quinonoid, these and other results are explained if a negative charge on carbon is stabilized more when the carbon is attached to hydrogen, as in **(13)**, than to fluorine. This requires fluorine to be electron-repelling in π-electron systems (I_π repulsion effect).[27] Various

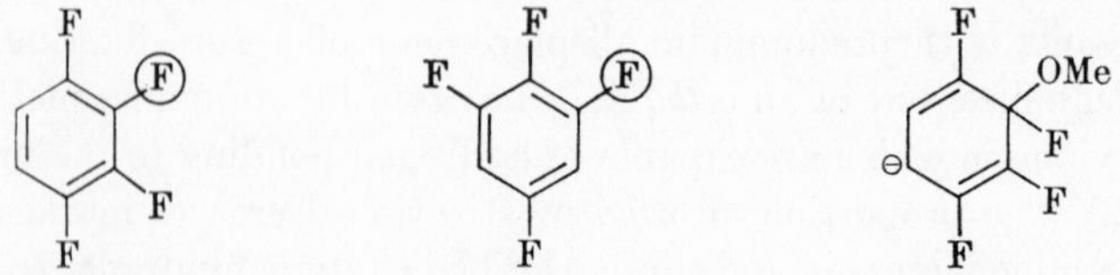

(11)   **(12)**   **(13)**

nucleophiles attack pentafluoroaniline, its *N*-methyl, and *N,N*-dimethyl derivative, replacing *m*- and *p*-fluorine in the approximate ratios, $m:p = 7$, 1, and 0.07, respectively. This is attributed to steric inhibition of the conjugation of the amino group, increasing in this order; hence deactivation of the *para*-position decreases in this order.[28]

Kinetics of the second-order formation of pentafluoroanisole from hexafluorobenzene and sodium methoxide,[29] nucleophilic displacements on pentafluorobenzoic acid,[30] and the *m*-:*p*-fluorine replacement ratios in the reactions of various C_6F_5X with nucleophiles[31] have also been studied.

[26] J. Burdon and D. F. Thomas, *Tetrahedron*, **21**, 2389 (1965).
[27] J. Burdon and W. B. Hollyhead, *J. Chem. Soc.*, **1965**, 6327.
[28] J. G. Allen, J. Burdon, and J. C. Tatlow, *J. Chem. Soc.*, **1965**, 6329.
[29] V. A. Sokolenko, L. V. Orlova, N. A. Gershtein, and G. G. Yakobson, *Kinetika i Kataliz*, **6**, 365 (1965); *Chem. Abs.*, **63**, 4124 (1965).
[30] J. Burdon, W. B. Hollyhead, and J. C. Tatlow, *J. Chem. Soc.*, **1965**, 6336.
[31] J. Burdon, W. B. Hollyhead, and J. C. Tatlow, *J. Chem. Soc.*, **1965**, 5152.

Several reactions of C_6F_5X with sodium methoxide in methanol obeyed second-order kinetics, and relative rates varied with X in the expected manner.[32]

Heterocyclic Systems

Full details of the first quantitative work on the effect of substituents on the orientation of the entering amide group in the sodamide amination of pyridines (Tschitschibabin reaction) have now appeared, and the addition-elimination and the elimination-addition (aryne) mechanism have been discussed critically.[33] The absence of deuterium kinetic isotope effects, and the orientation observed for substituted pyridines (e.g. 3-picoline gives 2-amino-3-picoline), clearly eliminate the intervention of 2,3-dehydropyridine [for which special stability resulting from interaction of the vacant carbon sp^2 orbitals with the nitrogen lone pair, (14), had been suggested][34] and 2,6-dehydropyridine. Quantitative competitive experiments show that the methyl group in 3-picoline is mildly deactivating in the amination, as expected for normal nucleophilic attack, and that this attack predominates at $C_{(2)}$ rather than $C_{(6)}$. The reason for this last feature, which is common to many reactions with 3-substituted pyridines, is not clear.[33] Cleavage of 2-[^{18}O]-

(14)   (15)   (16)

methoxypyridimine in aqueous sulphuric acid gave methanol containing 92% of the label, showing that nucleophilic aromatic substitution via the conjugate acids of (15) represents the major pathway; displacement on the methoxyl-carbon is a minor pathway, and in hydrochloric acid some methyl chloride was formed in addition to methanol.[35]

Benzimidazoles can form cations and anions, and their reactivity towards nucleophiles would be expected to decrease in the order cation > neutral molecule > anion, if bond-formation is rate-determining. In keeping with this, 1-unsubstituted 2-chlorobenzimidazoles form the anion on treatment with sodium alkoxide and are inert, whilst the corresponding 1-substituted compounds react normally with alkoxides unless the substituent is large,

[32] J. Burdon, W. B. Hollyhead, C. R. Patrick, and K. V. Wilson, *J. Chem. Soc.*, **1965**, 6375.
[33] R. A. Abramovitch, F. Helmer, and J. G. Saha, *Can. J. Chem.*, **43**, 725 (1965).
[34] H. L. Jones and D. L. Beveridge, *Tetrahedron Letters*, **1964**, 1577.
[35] R. Daniels, L. T. Grady, and L. Bauer, *J. Am. Chem. Soc.*, **87**, 1531 (1965).

such as isopropyl.[36] The kinetics of the reactions of piperidine in ethanol with 2-, 6-, and 8-chloro-9-methylpurine (see **16**) where N-methylation precludes anion formation, shows the order of reactivity to be $6 > 8 \gg 2$. This pattern of reactivity would probably hold for the chloropurines (neutral molecules) themselves, save that the 8-isomer may be relatively more reactive since it will have been deactivated most by the 9-methyl group.[37] 1-Methoxy-3-methylbenzimidazolium iodide (**17**) reacts with a wide range of nucleophiles in water or alcohols, mostly at room temperature, to give the 2-substituted 1-methylbenzimidazole (**18**) in high-to-quantitative yield. These reactions may proceed by the more usual addition-elimination via

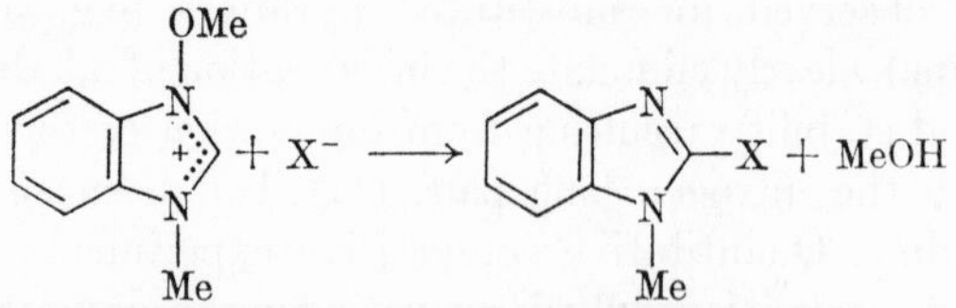

(**17**)   (**18**)

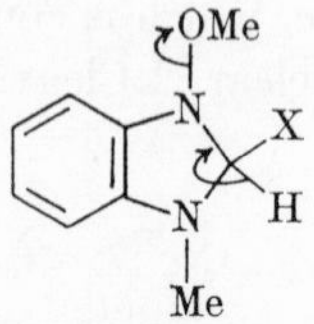

(**19**) 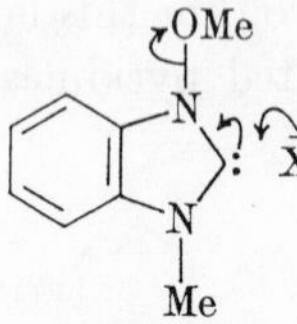   (**20**)

(**19**) or by prior ionization of the 2-hydrogen to give the resonance-stabilized carbene (**20**). The latter is favoured, since $H_{(2)}$ is unusually acidic (it exchanges with deuterium in deuterium oxide at room temperature in the absence of catalysts, and has the very low τ value of -1.00), and since triethylamine in refluxing methyl cyanide converts (**17**) into a dimeric compound.[38]

Kinetics of the reactions of the 2-halogenobenzothiazoles and their 6-nitro derivatives with various nucleophiles have been measured and the relative reactivities of the halogens and of the nucleophiles discussed.[39] The chemical shifts of the proton on $C_{(2)}$ in 4-, 5-, 6-, and 7-substituted benzothiazoles are related to the chemical reactivity of the corresponding 2-chloro

[36] D. Harrison and J. T. Ralph, *J. Chem. Soc.*, **1965**, 236.
[37] G. B. Barlin and N. B. Chapman, *J. Chem. Soc.*, **1965**, 3017.
[38] S. Takahashi and H. Kano, *Tetrahedron Letters*, **1965**, 3789.
[39] A. Ricci, P. E. Todesco, and P. Vivarelli, *Gazz. Chim. Ital.*, **95**, 101, 478, 490 (1965); M. Foa, A. Ricci, P. E. Todesco, and P. Vivarelli, *Boll. Sci. Fac. Chim. Ind. Bologna*, **23**, 65 (1965); A. Ricci, M. Foa, P. E. Todesco, and P. Vivarelli, *Tetrahedron Letters*, **1965**, 1935; P. E. Todesco, P. Vivarelli, and A. Ricci, *Tetrahedron Letters*, **1964**, 3703; M. Foa, A. Ricci, and P. E. Todesco, *Boll. Sci. Fac. Chim. Ind. Bologna*, **23**, 229 (1965).

compounds towards methoxide, both being correlated by a Hammett plot.[40] The reaction of 2-chlorobenzothiazole with the sodium salt of 2-mercapto-benzothiazole in dimethylformamide,[41] of a series of 5-substituted 2-bromo-3-nitrothiophenes with piperidine in ethanol,[42] the orientation of nucleophilic attack in excited molecules of nitrogen heterocyclic quanternary salts,[43] and hydrogen exchange of deuterated thiophenes and selenophenes with *t*-butoxides in dimethyl sulphoxide[44] have been investigated.

A detailed, comprehensive and critical review of nucleophilic hetero-aromatic substitution has appeared; particular attention is paid to six-membered mono- and poly-aza-hydrogen systems, but there is also much valuable mechanistic discussion of the related carbocyclic systems.[45]

Diazonium Decomposition

Thermal decomposition of N,N-dicyclohexylbenzamide-*o*-diazonium (21) hydrogen sulphate and fluoroborate in various solvents gave the product (23) of 1,5-hydride-ion shift in the carbonium ion (22), together with various products of the reaction of this carbonium ion with solvent or other external nucleophile. The variation in the ratio of intra- and inter-molecular reactions with solvent can be explained if it is assumed that the former is a reaction of the free, i.e. unpaired, diazonium ion (21) whilst the latter are reactions of diazonium ion pairs. Thus the reaction with external anions competes with hydride transfer only if the anion is associated with the solvated cation.[46]

(21)   (22)   (23)

The rates of thermal decomposition of this diazonium fluoroborate in methanol and in acetic acid, both 1M in sulphuric acid, were accurately of the first order, of the same order of magnitude in both solvents, and not affected appreciably by the presence of a variety of radical-chain inhibitors. Further

[40] F. Taddei, P. E. Todesco, and P. Vivarelli, *Gazz. Chim. Ital.*, **95**, 499 (1965).

[41] J. J. D'Amico, R. H. Campbell, S. T. Webster, and C. E. Twine, *J. Org. Chem.*, **30**, 3625 (1965).

[42] C. Dell'Erba and D. Spinelli, *Tetrahedron*, **21**, 1061 (1965).

[43] G. G. Dyadyusha and E. A. Ponomareva, *Teor. i Eksperim. Khim., Akad. Nauk Ukr. SSR.*, **1**, 117 (1965); *Chem. Abs.*, **63**, 2886 (1965).

[44] I. O. Shapiro, A. G. Kamrad, Y. I. Ranneva, and A. I. Shatenshtein, *Latvijas PSR Zinatnu Akad. Vestis. Kim. Ser.*, **1964**, 535; *Chem. Abs.*, **62**, 8955 (1965); see also *Chem. Abs.* **62**, 8960 (1965).

[45] R. G. Shepherd and J. L. Fedrick, *Adv. Heterocyclic Chem.*, **4**, 285 (1965).

[46] T. Cohen and J. Lipowitz, *Tetrahedron Letters*, **1964**, 3721.

support is, therefore, given to an ionic mechanism in which differences in product composition in different solvents are determined by the state of ion association of the salt, rather than by change to a radical mechanism.[47] The carbonium ion (22) from the diazonium fluoroborate was also found, somewhat surprisingly, to abstract a chloride ion from chlorinated hydrocarbon solvents; the yield of aryl chloride decreased sharply as the dielectric constant decreased and radicals are thought not to be involved.[46] The reaction between *p*-nitrobenzenediazonium ions and bromide ions in dimethyl sulphoxide is markedly accelerated by light. The dark reaction is approximately of first order in each ion, and both light and dark reactions gave the same yield (70 ± 5%) of *p*-bromonitrobenzene. The order of reactivity of halide ions in this reaction ($I^- > Br^- > Cl^-$) is opposite to that usual for aprotic solvents and appears to exclude straightforward nucleophilic displacement of nitrogen for this bimolecular reaction.[48]

Other Reactions

The intriguing suggestion has been made[49] that the hydrated electron ($e_{aq.}^-$)[50] may be considered as the most elementary nucleophile and by far the most reactive one. The rates of reaction of $e_{aq.}^-$ with many aromatic compounds in homogeneous, deaerated, weakly alkaline aqueous solution were measured by pulse radiolysis. Second-order rate coefficients were calculated from first-order rates of disappearance of $e_{aq.}^-$ (ranging from 1×10^5 to 3×10^6 sec^{-1}) corrected for the spontaneous disappearance of $e_{aq.}^-$ (2.5 to 3.5×10^4 sec^{-1}). The rate for many of the more reactive aromatic substrates approached the diffusion-controlled limit ($\sim 10^{10}$ mole^{-1} sec^{-1}). It was shown that the effects of substituents in these reactions directly paralleled those for normal nucleophilic aromatic substitution, electron-withdrawing substituents being strongly activating, and electron-releasing ones deactivating. Quantitatively, these effects fitted a Hammett plot, but better correlation was obtained with the $e_{aq.}^-$ rate constants and normal σ values derived from electrophilic substitution. It was claimed that this was because the hydrated electron interacts with the orbitals of the ring as does the electrophile in electrophilic aromatic substitution. It is perhaps surprising that Hammett free-energy relationships appear to hold up to the limit of diffusion-controlled processes.[49]

The first few examples of nucleophilic aromatic substitution catalyzed or

[47] J. Lipowitz and T. Cohen, *J. Org. Chem.*, **30**, 3891 (1965).

[48] M. D. Johnson, *J. Chem. Soc.*, **1965**, 805.

[49] M. Anbar and E. J. Hart, *J. Am. Chem. Soc.*, **86**, 5633 (1964).

[50] J. W. Boag and E. J. Hart, *Nature*, **197**, 45 (1963); J. P. Keene, *ibid.*, **197**, 47 (1963); E. J. Hart and J. W. Boag, *J. Am. Chem. Soc.*, **84**, 4090 (1962).

induced by light were reported in 1964.[51] Letsinger *et al.*[52] have now given full details of their photo-induced nucleophilic displacements of nitro groups from substituted nitrobenzenes, together with some new interesting results. When irradiated with light of wavelength greater than 289 mμ, disodium *p*-nitrophenyl phosphate (24) and *p*-nitroanisole react rapidly with pyridine in cold dilute aqueous solution, to give 1-arylpyridinium nitrites. Nitrobenzene, *m*-nitroanisole, and *p*-dinitrobenzene did not react similarly and the *p*-phosphoryl and *p*-methoxyl groups must be activating substituents for displacement of the nitro group from the aromatic substrate in an excited

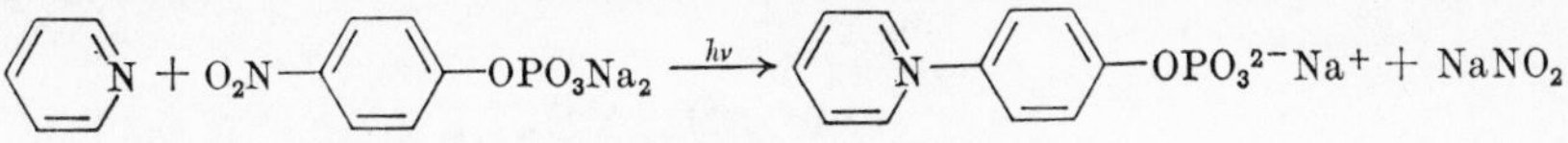

(24)

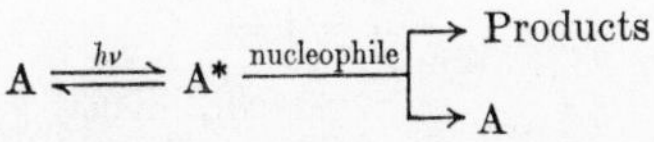

$$A \underset{}{\overset{h\nu}{\rightleftharpoons}} A^* \xrightarrow{\text{nucleophile}} \begin{array}{l} \text{Products} \\ A \end{array}$$

state, in contrast to their effect on ground-state molecules. Rate data for the *p*-nitroanisole reaction with pyridine or 4-picoline and with hydroxide ions are consistent with a mechanism in which the nitro compound, A, is the species activated in the initial photochemical step, giving an excited state, A*, which reacts competitively with the nucleophile to give products, or reverts to the ground state by two quenching processes, one dependent upon, and the other independent of, the nucleophile, as shown.[52] Ultraviolet irradiation of 2-bromo-4-nitroanisole in aqueous alkaline tetrahydrofuran led to the displacement of bromine by hydroxyl; this substitution is rather specific since it was not observed with *m*-nitrobromobenzene, 3,5-dinitrobromobenzene, or 2- or 4-bromoanisole.[53]

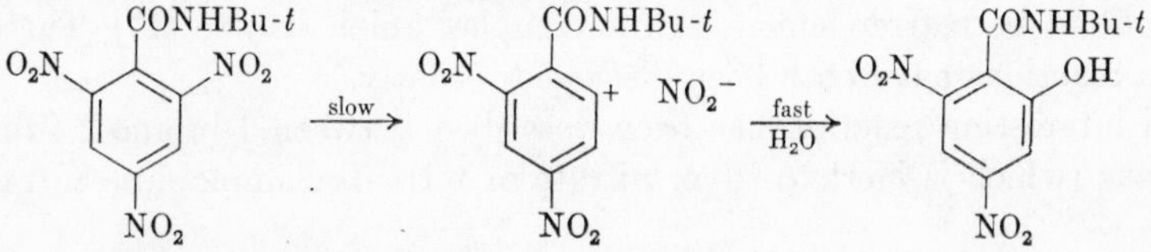

(25)

N-*t*-Butyl-2,4,6-trinitrobenzamide (25) reacts with dilute aqueous-methanolic sodium hydroxide with the liberation of nitrite ions. The amide group is stable under the reaction conditions, and it is assumed that an *o*-nitro group is displaced by hydroxide ion since with methanolic sodium methoxide the 2-methoxyl derivative was isolated—its structure was

[51] B. Capon and C. W. Rees, *Ann. Reports Chem. Soc. (London)*, **61**, 278 (1964).
[52] R. L. Letsinger, O. B. Ramsay, and J. H. McCain, *J. Am. Chem. Soc.*, **81**, 2945 (1965).
[53] D. F. Nijhoff and E. Havinga, *Tetrahedron Letters*, **1965**, 4199.

deduced for the NMR spectrum. After allowance for the equilibrium formation of a red, anionic, presumably 1:1, complex between the reactants, the displacement reaction appears to be of zero order in hydroxide ion. On the basis of these preliminary results the surprising claim is made that the rate-controlling step is unimolecular heterolysis of the C–N bond facilitated by steric inhibition of resonance between the nitro and the bulky amide group.[54]

A new reaction, described as an aromatic nucleophilic substitution, was observed when fluoro-, chloro-, and bromo-benzene and anisole were treated with trichloramine and aluminium chloride. The proposed mechanism, as illustrated below, must be considered very tentative. The initial step is

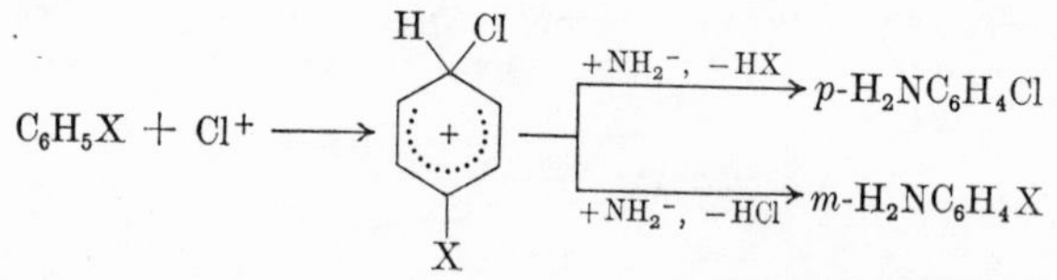

actually electrophilic attack; the nature of the nucleophile in the second steps, written as NH_2^-, is unknown.[55]

Fanta has extended his earlier review[56a] of the Ullmann synthesis of biaryls to cover the period 1945–1963 and has included a further discussion of the reaction mechanism. The circumstantial evidence so far available suggests that the most likely mechanism is nucleophilic attack on the aryl halide by copper, to form an activated complex at the metal surface, followed by reaction of this with a second molecule of aryl halide.[56b] Further work by Weingarten[57] on the kinetics of the reaction of bromobenzene and potassium phenoxide in diglyme, catalysed by copper salts (the Ullmann condensation), has led to a more detailed formulation of a mechanism whose key step is thought to be rearrangement of the complex anion (26) to (27). These Ullmann reactions have both been discussed recently.[58]

An interesting reaction has been described between 1-bromo-2,4-dinitro-benzene (which is inert to silver nitrate or tetraalkylammonium nitrates in

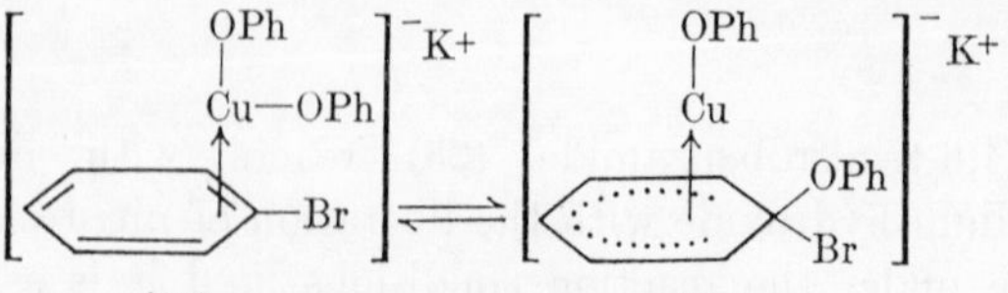

(26)   (27)

[54] P. J. Hutchison and R. J. L. Martin, *Austral. J. Chem.*, **18**, 699 (1965).
[55] P. Kovacic, J. J. Hiller, J. F. Gormish, and J. A. Levisky, *Chem. Comm.*, **1965**, 580.
[56] P. E. Fanta, *Chem. Rev.*, (*a*) **38**, 139 (1946); (*b*) **64**, 613 (1964).
[57] H. Weingarten, *J. Org. Chem.*, **29**, 3624 (1964).
[58] R. G. R. Bacon and H. A. O. Hill, *Quart. Rev.*, **19**, 95 (1965).

refluxing acetonitrile) and molten tetraalkylammonium nitrates at 75–125°: it gives 2-bromo-4,6-dinitrophenol and picric acid. It is autocatalytic and catalyzed by nitrite ions, and a mechanistic scheme involving nucleophilic aromatic substitution by nitrite ions was proposed. Nucleophilic anions display greatly enhanced reactivity in these melts, where ion association and anion solvation are unimportant.[59]

The following reactions have also been studied: 2,4- and 2,6-dinitrophenyl benzenesulphonates with aromatic amines;[60] 1-halogeno-2,4-dinitrobenzene and picryl chloride with alkali halides in acetone;[61] isotopic exchange between 1-iodo-2,4-dinitronaphthalene and potassium iodide in methanol;[62] 1-halo-geno-2,4-dinitrobenzene and -2,4-dinitronaphthalene with MOH (M = Li, Na, K, and Bu_4N);[63] acidic and alkaline hydrolyses of p-nitrophenyl benzene-sulphonate in ^{18}O-enriched water;[64] p-fluoro- and p-iodo-nitrobenzene and 1-fluoro- and 1-iodo-2,4-dinitronaphthalene with thiomethoxide ions in methanol;[65] alkaline and neutral hydrolyses of fluoronitrobenzenes in aqueous dimethyl sulphoxide mixtures;[66] 1-chloro-2,4-dinitrobenzene with tri-t-butylamine in ethanol at pressures up to 1000 kg cm^{-2};[67] a wide variety of p-nitrophenyl derivatives with piperidine in dimethyl sulphoxide[68] and in a range of other solvents;[69] 1-fluoro-2,4-dinitrobenzene with sodium methoxide in methanol;[70] and nine 6-substituted 1-chloro-2-nitrobenzenes with piperi-dine in benzene.[71] In the last reaction hydrogen bonding in the transition state is again apparent; in spite of their being in the *ortho*-position, the effects of the substituents fit a Hammett plot well, save for the methyl and bromo derivatives; it is suggested that steric effects are negligible with substituents having a van der Waals radius of less than 1.9 Å.[71]

Benzyne and Related Intermediates

Flash-initiated decomposition of benzenediazonium-2-carboxylate gave N_2, CO_2, and benzyne as the primary products, as shown by the agreement of ultraviolet-absorption and mass-spectral measurements as functions of time.

[59] J. E. Gordon, *J. Am. Chem. Soc.*, **87**, 1499 (1965).

[60] R. V. Vizgert, S. M. Kononenko, and I. M. Ozdrovskaya, *Zh. Organ. Kkim.*, **1**, 264 (1965); *Chem. Abs.*, **62**, 16,006 (1965).

[61] C. W. L. Bevan, J. Hirst, and E. C. Okafor, *J. Chem. Soc.*, **1964**, 6248.

[62] C. A. Marcopoulos, *J. Chem. Soc.*, **1965**, 4613.

[63] J. D. Reinheimer and W. Hostetler, *Ohio J. Sci.*, **64**, 275 (1964); *Chem. Abs.*, **63**, 5470 (1965).

[64] S. Oae and R. Kiritani, *Bull. Chem. Soc. Japan*, **38**, 765 (1965).

[65] J. Miller and K. W. Wong, *J. Chem. Soc.*, **1965**, 5454.

[66] J. Murto and A. M. Hiro, *Suomen Kemistilehti*, **37**, Pt. B, 177 (1964).

[67] N. I. Prokhorova and M. G. Gonikberg, *Izv. Akad. Nauk SSSR, Otd. Khim. Nauk*, **1965**, 1188.

[68] H. Suhr, *Chem. Ber.*, **97**, 3268 (1964).

[69] H. Suhr, *Chem. Ber.*, **97**, 3277 (1964).

[70] J. Miller and K. W. Wong, *Austral. J. Chem.*, **18**, 117 (1965).

[71] N. E. Sbarbati, *J. Org. Chem.*, **30**, 3365 (1965).

The time scale for the disappearance of mass 76 (benzyne) and appearance of mass 152 (biphenylene) was the same as for the disappearance of the continuum due to benzyne and the appearance of the biphenylene spectrum in the ultraviolet region. A second-order rate constant of not less than 7×10^8 mole^{-1} sec^{-1} is estimated for the dimerization of gaseous benzyne, implying that the chance of reaction was about 1 in 150 collisions under the experimental conditions.[72] The photolytic decomposition of the two isomeric diazonium carboxylates was studied in the same way. The species of mass 76 from the *para*-isomer was much more stable than benzyne, persisting for about 2 minutes; it is thought most likely to be 1,4-dehydrobenzene, of

(28)   (29)   (30)   (31)   (32)

structure (28) or (29), or possibly even (30). In contrast to the *ortho*-decomposition, masses 152 and 228 were absent.[73] Decomposition of the *meta*-isomer gave a transient species of mass 76, different from either of its isomers, much less stable than 1,4-dehydrobenzene, but with a lifetime in the gas phase comparable with that of benzyne. Comparison of the fragmentation patterns suggests 1,3-dehydrobenzene as the most likely structure, possibly in the form (31) or (32). There is indication in this case of a dimer peak at 152, but this is only transient.[74] A different approach to the generation of a "*meta*-benzyne," 1,8-dehydronaphthalene (34), has been reported in the oxidation of the aminotriazine (33) with lead tetraacetate in cold benzene, which gives, not the dimer perylene, but fluoranthene (35) which presumably results from 1,2-addition of (34) to benzene, followed by oxidation of the dihydro derivative.[75] This failure of species (34) to dimerize and the reaction with benzene show the naphthalyne to be much more reactive than benzyne generated in an analogous way.[76] Addition of (34) to *cis*- and *trans*-dichloroethylene is almost completely stereospecific.[75] Photolysis of compound (36) gave a 25% yield of the sulphone (37), which on pyrolysis at 240° gave 0.4% of perylene, possibly via 1,8-dehydronaphthalene.[77]

A careful analysis of substituent effects on the relative rates of proton capture by, and chloride ion loss from, *o*-chlorophenyl anions has been made by interrupting the reactions between substituted *o*-deuteriochlorobenzenes

[72] M. E. Schafer and R. S. Berry, *J. Am. Chem. Soc.*, **87**, 4497 (1965).
[73] R. S. Berry, J. Clardy, and M. E. Schafer, *Tetrahedron Letters*, **1965**, 1003.
[74] R. S. Berry, J. Clardy and M. E. Schafer, *Tetrahedron Letters*, **1965**, 1011.
[75] C. W. Rees and R. C. Storr, *Chem. Comm.*, **1965**, 193.
[76] C. D. Campbell and C. W. Rees, *Proc. Chem. Soc.*, **1964**, 296.
[77] R. W. Hoffmann and W. Sieber, *Angew. Chem. Intern. Ed. Engl.*, **4**, 786 (1965).

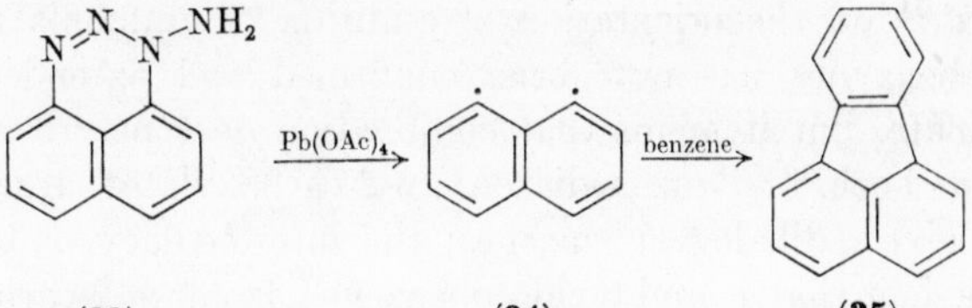

(33)   (34)   (35)

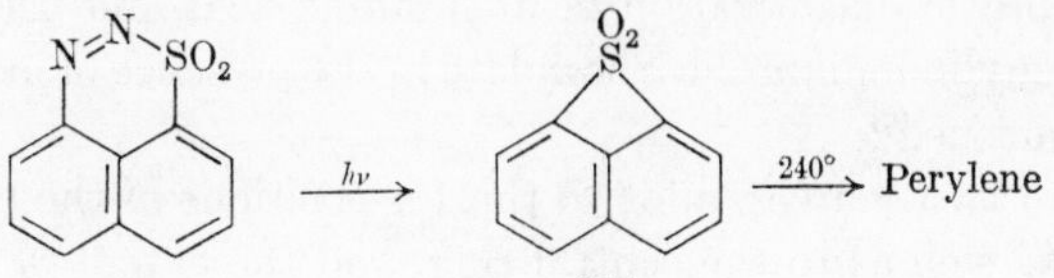

(36)   (37)

(mixed with the corresponding undeuterated compound) and potassamide in 6:4 liquid ammonia—ether. The chloride liberated and the deuterium content of the recovered chlorobenzene were measured, and the proton-capture to chloride-ion-loss rate-ratios were calculated. All substituents (Cl, Me, CF$_3$, and MeO) increased this ratio. The ratios could be explained in terms of reasonable substituent effects on an anion-like transition state (38) for proton capture and an aryne-like transition state (39) for chloride loss. Reaction of *m*-dichlorobenzene with potassamide occurred to the extent of 93% via 3-chlorobenzyne and 3% via 4-chlorobenzyne.[78] The relative rates of proton

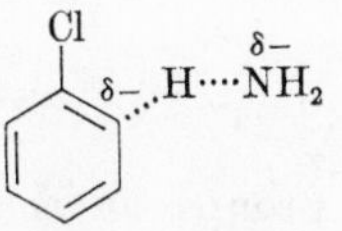

(38) 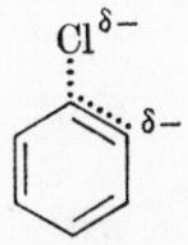   (39)

capture by, and bromide ion loss from, the *o*-bromophenyl anion in ethanolic sodium ethoxide were independent of the method of generating the anion, whether by fragmentation of an arylazocarbonyl compound, alkaline cleavage of the aryltosylhydrazone, or oxidation of the arylhydrazine. Proton capture was about 10 times faster than bromide loss.[79]

The ratio of isomers formed in the addition of a variety of nucleophiles to unsymmetrical benzynes is not constant but depends upon the nucleophilicity of the reagent and the solvating properties of the solvent.[80] Earlier work by

[78] J. A. Zoltewicz and J. F. Bunnett, *J. Am. Chem. Soc.*, **87**, 2640 (1965).
[79] R. W. Hoffmann, *Chem. Ber.*, **98**, 222 (1965).
[80] R. W. Hoffmann, G. E. Vargas-Núñez, G. Guhm, and W. Sieber, *Chem. Ber.*, **98**, 2074 (1965).

Roberts *et al.*[81] on the orientation of addition of ammonia to 3- and 4-substituted benzynes has now been confirmed and extended. Inductive effects dominate, but it seems that conjugation does play a minor role in additions to both.[82] Benzenediazonium-2-carboxylates react with the halogens to give *o*-dihalogenobenzenes; the intermediacy of benzynes was shown by the fact that 4- and 5-chlorobenzenediazonium-2-carboxylate with iodine monochloride gave the same mixture of 2,4- and 2,5-dichloro-1-iodobenzene. Aprotic diazotization of anthranilic acid and thermal decomposition of diphenyliodonium-2-carboxylate in the presence of iodine similarly gave *o*-diiodobenzene.[83]

Benzyne, from two sources, adds to the 1,4-positions of the terminal rings of anthracene, as well as to the central ring, and the extent of this addition

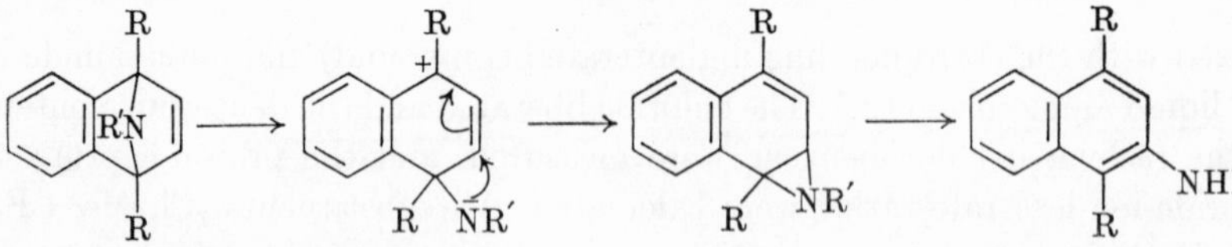

(40)   (41)

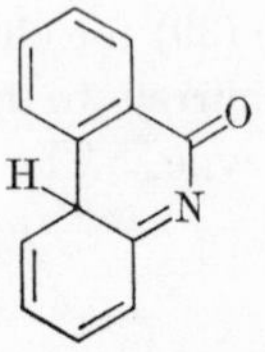

(42) 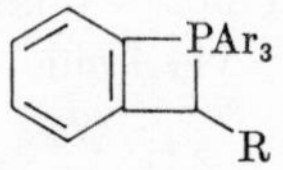   (43)

increases with electron-withdrawing groups at the 9,10-positions. It is suggested that the ratio of terminal- to central-ring addition to 9-cyano- or 9,10-dicyano-anthracene is diagnostic for free benzyne.[84] Addition of benzyne to 2,5-dimethyl-1-phenyl-, 1-methyl-2,5-diphenyl-, and 1-benzyl-2,5-dimethyl-pyrrole gave 1,4-dimethyl-*N*-phenyl-, *N*-methyl-1,4-diphenyl, and *N*-benzyl-1,4-dimethyl-2-naphthylamine (41), respectively, presumably via the Diels–Alder adducts (40), though these could not be isolated.[85] Aliphatic and aromatic azides[86] and aromatic carbonyl and sulphonyl azides[87] act as

[81] J. D. Roberts, C. W. Vaughan, L. A. Carlsmith, and D. A. Semenow, *J. Am. Chem. Soc.*, 78, 611 (1956).

[82] G. B. R. de Graaff, H. J. den Hertog, and W. Ch. Melger, *Tetrahedron Letters*, 1965, 963.

[83] L. Friedman and F. M. Logullo, *Angew. Chem. Intern. Ed. Engl.*, 4, 239 (1965).

[84] B. H. Klanderman, *J. Am. Chem. Soc.*, 87, 4649 (1965).

[85] E. Wolthuis, D. Vander Jagt, S. Mels, and A. De Boer, *J. Org. Chem.*, 30, 190 (1965).

[86] G. A. Reynolds, *J. Org. Chem.*, 29, 3733 (1964).

[87] W. Ried and M. Schön, *Chem. Ber.*, 98, 3142 (1965).

good traps for benzyne, giving the 1-substituted benzotriazole. Benzyne also reacts with phenyl isocyanate, giving phenanthridone and 9-phenoxyphenanthridine in low yields, presumably via the adduct (42).[88] Cycloaddition of benzyne to alkenes, to give benzocyclobutenes, is rare but does occur with the nucleophilic systems, ethyl vinyl ether and vinyl acetate, which afford benzocyclobutenyl ethyl ether (40%) and benzocyclobutenyl acetate (45%).[89] Addition of benzyne to alkylidenephosphoranes, $Ar_3P{=}CHR$, to give *ortho*-disubstituted benzenes is considered to involve the analogous intermediate (43).[90]

The rate of decomposition of 1,2,3-benzothiadiazole 1,1-dioxide into benzyne, sulphur dioxide, and nitrogen is insensitive to changes in solvent (a factor of 4 from tetrahydrofuran to dimethyl sulphoxide), to added lithium perchlorate, and to substitution in the benzene ring. The most likely mechanism is thus concerted fragmentation (44).[91] Pyrolysis of phthalic anhydride in benzene at 690° under nitrogen appears to duplicate its decomposition on electron impact (mass spectrometer) in giving consecutively carbon dioxide, carbon monoxide, and benzyne. Of the key pyrolysis products, acetylene, biphenylene, and naphthalene, only the last is found, in traces, in the absence of phthalic anhydride. Naphthalene and acetylene probably arise from pyrolysis of benzocyclooctatetraene (45), and benzobicyclo[2,2,2]-octatriene (46) which are, respectively, the 1,2- and 1,4-addition products of

(44)   (45)   (46)

benzyne to benzene.[92] Biphenylene could not be detected amongst the photolysis products of 2-iodobiphenyl and *o*-diiodobenzene in benzene.[93] Benzyne is probably an intermediate in the pyrolysis of indanetrione, either alone at 600°/0.7 mm or in benzene at 500°; biphenylene is isolated from the former and biphenyl and naphthalene from the latter reaction.[94] *o*-Chlorobenzeneboronic acid with potassium *t*-butoxide in tetrahydrofuran or dimethoxyethane, or with butyllithium in ether in the presence of furan, gave benzyne,

[88] J. C. Sheehan and G. D. Daves, *J. Org. Chem.*, **30**, 3247 (1965).
[89] H. H. Wasserman and J. Solodar, *J. Am. Chem. Soc.*, **87**, 4002 (1965).
[90] E. Zbiral, *Tetrahedron Letters*, **1964**, 3963.
[91] R. W. Hoffmann, W. Sieber, and G. Guhn, *Angew. Chem. Intern. Ed. Engl.*, **4**, 704 (1965); *Chem. Ber.*, **98**, 3470 (1965).
[92] E. K. Fields and S. Meyerson, *Chem. Comm.*, **1965**, 474.
[93] N. Kharasch and T. G. Alston, *Chem. Ind.* (*London*), **1965**, 1463.
[94] R. F. C. Brown and R. K. Solly, *Chem. Ind.* (*London*), **1965**, 181, 1462.

as shown by isolation of 1-naphthol after hydrolysis. Carbonation of the butyllithium reaction mixture at $-78°$ gave o-chlorobenzoic acid, suggesting the intermediate formation of o-chlorophenyllithium.[95]

Extensive reviews of heterocyclic arynes (hetarynes) with five-, six-, and seven-membered rings are mainly concerned with these species as intermediates in nucleophilic heteroaromatic substitution.[96] 4-Substituted 5-bromopyrimidines (47) react with potassamide in liquid ammonia to give the 6- rather than the 5-amino isomer, exclusively; this is attributed to significant interaction between the nitrogen lone pair and the triple bond (cf. 48) that deactivates the 5-position. Only with the 4-t-butyl compound is a little 5-amino isomer formed.[97] 3-Bromo-4-ethoxypyridine with lithium piperidide

in piperidine gave a little 2,4-dipiperidinopyridine (1–2%), presumably via 4-piperidino-2,3-pyridyne.[98] 3-Chloropyridine 1-oxide and 3-bromoquinoline 1-oxide react slowly with piperidine to give the 4-piperidino derivatives, as well as the 3-isomers, but none of the 2-isomers, suggesting an aryne mechanism: the 2- and 4-chloro isomers react almost quantitatively with piperidine to give the normal non-rearranged products.[99] 5- and 6-Chloro- and -bromo-quinoline react with lithium piperidide and piperidine via 5,6-quinolyne, and 7-chloro- and 7-bromo-quinoline react similarly via 7,8-quinolyne. In contrast, 8-chloro-, 8-bromo-, and 8-iodo-quinoline react by the normal mechanism to give 8-piperidinoquinoline, exclusively.[100] All these elimination addition reactions involve halide as the leaving group and may be contrasted with the Tschitschibabin reaction (cf. ref. 33) which does not.

Tetraphenylbenzyne adds normally to furan, giving the corresponding naphthalene 1,4-epoxide.[101] Cyclopentyne and cyclohexyne have been generated by heating bromomethylene-cyclobutane and -cyclopentane,

[95] G. Cainelli, G. Zubiani, and S. Morrocchi, *Chim. Ind. (Milan)*, **46**, 1489 (1964); *Chem. Abs.*, **62**, 16154 (1965).
[96] T. Kauffmann, *Angew. Chem. Intern. Ed. Engl.*, **4**, 543 (1965); H. J. den Hertog, and H. C. van der Plas, *Adv. Heterocyclic Chem.*, **4**, 121 (1965).
[97] H. C. van der Plas, *Tetrahedron Letters*, **1965**, 555.
[98] H. C. van der Plas, T. Hijwegen, and H. J. den Hertog, *Rec. Trav. Chim.*, **84**, 53 (1965).
[99] T. Kauffmann and R. Wirthwein, *Angew. Chem. Intern. Ed. Engl.*, **3**, 806 (1964).
[100] T. Kauffmann, J. Hansen, and R. Wirthwein, *Ann. Chem.*, **680**, 31 (1964).
[101] D. Seyferth and H. H. A. Menzel, *J. Org. Chem.*, **30**, 649 (1965).

respectively, with potassium *t*-butoxide in toluene, and were trapped with 1,3-diphenylisobenzofuran; alkylidenecarbenes, $R_2C=C$: may be intermediates.[102]

Benzyne chemistry has been reviewed.[103]

[102] K. L. Erickson and J. Wolinsky, *J. Am. Chem. Soc.*, **87**, 1142 (1965).

[103] W. Czuba, *Wiadomosci Chem.*, **19**, 157 (1965); G. Wittig, *Angew. Chem. Intern. Ed. Engl.*, **4**, 731 (1965).

CHAPTER 7

Radical and Electrophilic Aromatic Substitution

Radical Substitution

The outstanding mechanistic study of homolytic aromatic substitution in recent years must be Rüchardt's elucidation[1] of the mechanism of homolytic arylation in the Gomberg reaction and in the decomposition of N-acyl-N-nitrosoarylamines (Hey reaction). Neither of these reactions produces the hydroaromatic by-products of the types shown in (1), which are normally

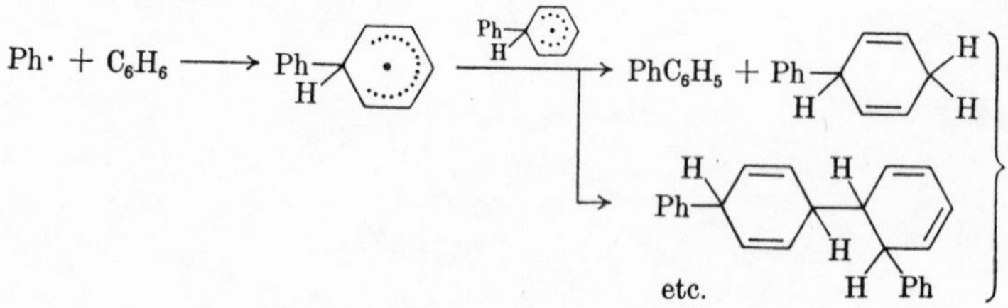

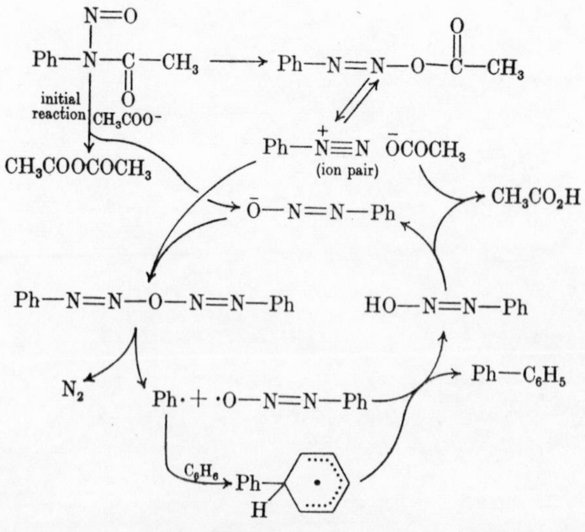

Scheme 1

[1] C. Rüchardt and E. Merz, *Tetrahedron Letters*, **1964**, 2431; C. Rüchardt and B. Freudenberg, *ibid.*, **1964**, 3623; C. Rüchardt, B. Freudenberg, and E. Merz, *Chem. Soc. Special Publication No. 19*, **1965**, p. 168; C. Rüchardt, *Angew. Chem. Intern. Ed. Engl.*, **4**, 964 (1965).

observed in arylation by diaroyl peroxides.[2] Furthermore, in the Hey reaction, the acyl group appears, in almost quantitative yield, as the corresponding carboxylic acid. Early reaction schemes required that this acid be formed from the appropriate acyloxy radical, which is known to be extremely unstable with respect to decarboxylation. Rüchardt's scheme, as outlined for the reaction of nitrosoacetanilide with benzene (Scheme 1), readily rationalizes these two anomalies. In the first place, reactions between two phenylcyclohexadienyl radicals are eliminated because the concentration of the relatively stable species $PhN=N-O\cdot$ builds up to a level which is sufficient to scavenge phenylcyclohexadienyl radicals before they can dimerize or disproportionate according to (1). Secondly, it can be seen that the carboxylic acid, formed as a by-product, is obtained in an ionic reaction, so that no exceptional stabilization of acyloxy radicals need be invoked.

Previously unreported products of the thermal decomposition of phenylazotriphenylmethane in benzene are the *cis*- and *trans*-isomers of the cyclohexadiene (1). Examination of this class of product from a cross-over experiment with two different azo-compounds provides a test for the participation of *free* radicals in the reaction. The positive result of such a test is consistent

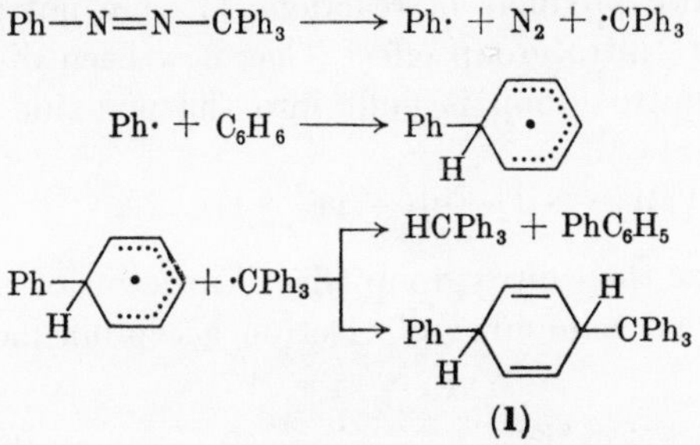

Scheme 2

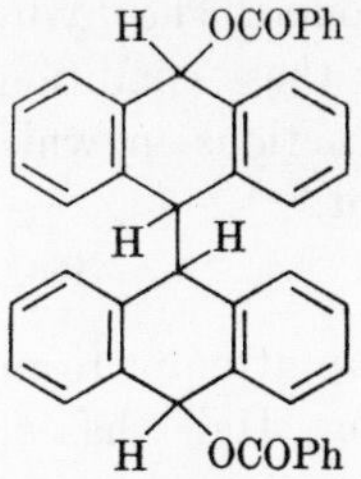

with the reaction scheme shown. Again, products of reaction (1) are absent, and here a high stationary concentration of the relatively stable $Ph_3C\cdot$ takes the place of $PhN=N-O\cdot$ in scavenging phenylcyclohexadienyl radicals.[3] A similar dominance of reactions involving triphenylmethyl radicals is apparent in the products of decomposition of phenylazotriphenylmethane in *p*-xylene.[4]

Naphthalene is known to be more reactive towards free radicals than is benzene. Thus, when it is employed as a solvent for the decomposition of benzoyl peroxide, significant yields of naphthyl benzoates are formed by

[2] E. L. Eliel, J. G. Saha, and S. Meyerson, *J. Org. Chem.*, **30**, 2451 (1965).
[3] D. H. Hey, M. J. Perkins, and G. H. Williams, *J. Chem. Soc.*, **1965**, 110.
[4] M. J. Perkins, *J. Chem. Soc.*, **1964**, 5932.

benzoyloxylation, along with the expected phenylnaphthalenes. DeTar and Long[5] have suggested that the binaphthyls which are formed in the reaction are also derived from benzoyloxylation, as, e.g. in reaction (2); some support

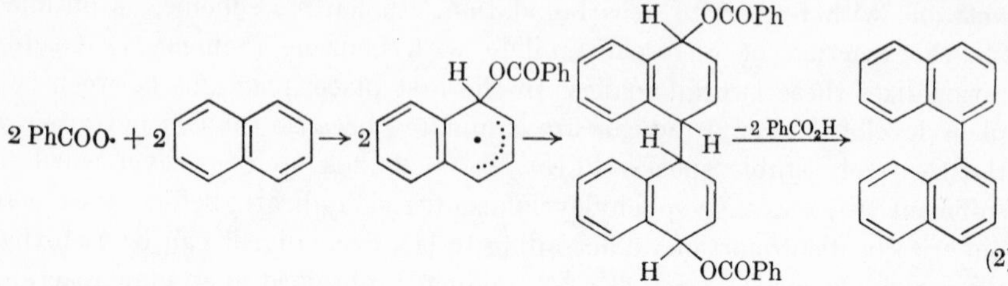

$$2 \; PhCOO\cdot + 2 \quad \xrightarrow{\quad} \quad 2 \quad \xrightarrow{\quad} \quad \xrightarrow{-2 \; PhCO_2H} \quad \tag{2}$$

was obtained for this when the tetrahydrobianthryl (2) was isolated from the decomposition products of benzoyl peroxide in the presence of anthracene.[6]

It was recently demonstrated[7] that in the decomposition of benzoyl peroxide in benzene, a "clean" reaction (3) could be produced by effecting the decomposition in the presence of a small quantity of an aromatic nitro-compound. The hydroaromatic products of reaction (1) were not formed under these conditions. This "nitro group effect" has now been discussed for reactions in which the nitro group is built into the peroxide or the solvent.[6]

$$(PhCOO)_2 + PhH \longrightarrow PhCO_2H + Ph_2 + CO_2 \tag{3}$$

In an attempt to rationalize the "nitro group effect" in terms of electron transfer, Hall[8] has shown that a number of electron acceptors including

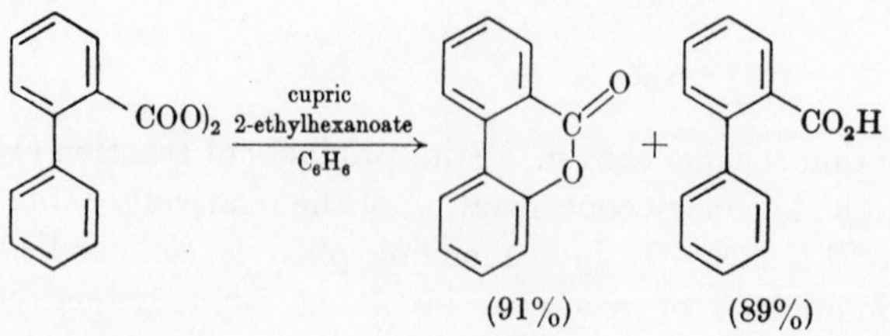

$$\xrightarrow[\text{C}_6\text{H}_6]{\substack{\text{cupric} \\ \text{2-ethylhexanoate}}}$$

(91%)   (89%)

(3)   (4)

quinones, tetracyanoethylene, and tetranitromethane modify the reaction in a similar fashion. However, he was unable to relate the magnitude of the

[5] D. F. DeTar and R. A. J. Long, *J. Am. Chem. Soc.*, **80**, 4742 (1958).
[6] M. Mingin and K. H. Pausacker, *Austral. J. Chem.*, **18**, 821, 831 (1965).
[7] D. H. Hey, M. J. Perkins, and G. H. Williams, *Chem. Ind. (London)*, **1963**, 83.
[8] C. D. Hall, *Chem. Ind. (London)*, **1965**, 384.

effect to the electron affinity of the additives. In the same way, Kochi and Gilliom[9] have found that peroxide decomposition leading to intramolecular substitution may be modified by the presence of copper salts as additives. Thus decomposition of the peroxide (3) in the presence of a catalytic quantity of a copper salt gives almost molar yields of the corresponding acid and of benzocoumarin (4).

In a detailed kinetic and product study of the uncatalysed decomposition of benzoyl peroxide in benzene, Gill and Williams[10] arrive at the rate equation:

$$-d[P]/dt = k_1[P] + k_1[P] + k_{3/2}[P]^{3/2}$$

Two terms ($k_1[P]$ and $k_{3/2}[P]^{3/2}$) relate to induced decomposition and arise from different chain-termination reactions. The first-order induced term ($k_1[P]$) is small.

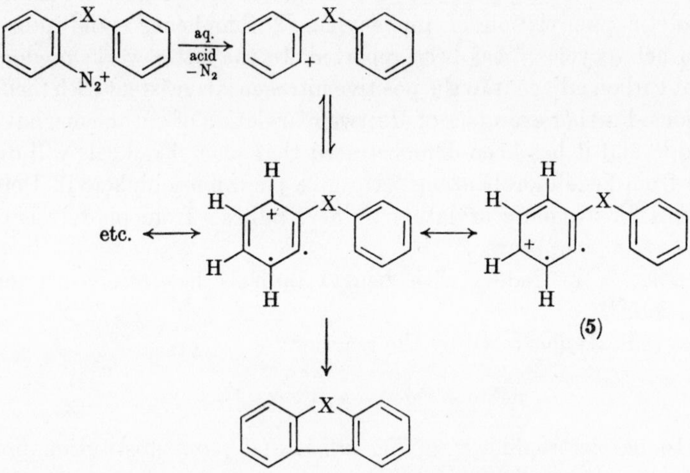

(5)

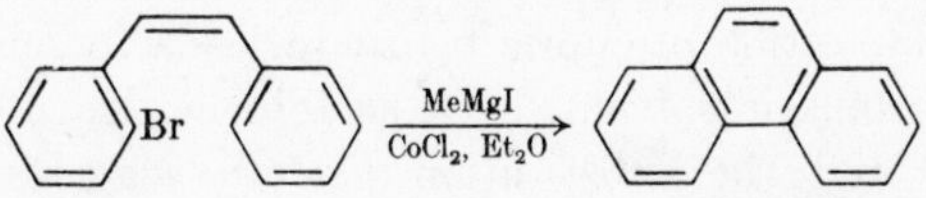

(4)

(75%)

Further evidence for a triplet cation intermediate (5) in the Pschorr reaction has been obtained.[11] The intermediate (the structural type was first proposed by Taft[12]) exhibits properties expected for a highly electrophilic

[9] J. K. Kochi and R. D. Gilliom, *J. Am. Chem. Soc.*, **86**, 5251 (1964).
[10] G. B. Gill and G. H. Williams, *J. Chem. Soc.*, **1965**, 995.
[11] R. A. Abramovitch and G. Tertzakian, *Can. J. Chem.*, **43**, 940 (1965).
[12] R. W. Taft, *J. Am. Chem. Soc.*, **83**, 3350 (1961).

free radical. A novel procedure (reaction 4) for effecting internuclear radical cyclization has also appeared.[13]

Examination of the reactivity of different aromatic substrates towards radical phenylation continues to attract attention. Hey's group have completed a series of papers with a study of the reactions of *meta*-substituted phenyl radicals.[14] The spectrum of results obtained from the reactions of diversely substituted aryl radicals with various aromatic substrates is discussed in terms of polarization in the radicals. Migita, Simamura, and their colleagues have summarized the results of similar work.[15] They find that the relative rates of *meta*-attack on monosubstituted benzenes, show good Hammett σ correlations. For attack by $p\text{-}O_2NC_6H_4\cdot$, $\rho = -0.81$; by $p\text{-}ClC_6H_4\cdot$, $\rho = -0.27$; by $C_6H_5\cdot$, $\rho = 0.05$; by $p\text{-}CH_3C_6H_4\cdot$, $\rho = 0.03$; and by $p\text{-}CH_3OC_6H_4\cdot$, $\rho = 0.09$. Clearly, *para*-nitrophenyl and *para*-chlorophenyl radicals are electrophilic species. For attack at positions *ortho* and *para* to the substituent, pronounced conjugative effects were revealed.

Homolytic phenylation of biphenylene,[16] thiophene,[17] and protonated nitrogen heterocycles[18] has been reported. In the latter work, predominant attack at carbon adjacent to the positive nitrogen is consistent with theoretical predictions. Further examples of Meerwein arylation of anthracene have been reported,[19] and it has been demonstrated that phenyl radicals will displace fluorine from hexafluorobenzene[20] (to give pentafluorobiphenyl). Details of procedures for aromatic arylation by aryl radicals from photolysis of aryl iodides have now appeared.[21a]

The reaction of indole with benzyl radicals has received a cursory examination.[21b]

Amino radicals generated by the reaction:

$$R_2NCl + Fe^{2+} \longrightarrow R_2N\cdot + FeCl^{2+}$$

appear to be electrophilic,[22] giving *ortho*- and *para*-substitution products with anisole, and entering at position 4 of 1,3-dimethoxybenzene.

The thermal decomposition of cupric benzoate leads to salicylic acid; in the presence of an aromatic solvent, such as toluene, benzoyloxylation of the solvent occurs, and the substitution pattern suggests a homolytic

[13] M. Tiecco, *Chem. Comm.*, **1965**, 555.
[14] D. H. Hey, S. Orman, and G. H. Williams, *J. Chem. Soc.*, **1965**, 101.
[15] R. Itô, T. Migita, N. Morikawa, and O. Simamura, *Tetrahedron*, **21**, 955 (1965).
[16] S. C. Dickerman, N. Milstein, and J. F. W. McOmie, *J. Am. Chem. Soc.*, **87**, 5521 (1965).
[17] C. E. Griffin and K. R. Martin, *Chem. Comm.*, **1965**, 154.
[18] H. J. M. Dou and B. M. Lynch, *Tetrahedron Letters*, **1965**, 897.
[19] S. C. Dickerman, M. Klein, and G. B. Vermont, *J. Org. Chem.*, **29**, 3695 (1964).
[20] P. A. Claret, J. Coulson, and G. H. Williams, *Chem. Ind.* (*London*), **1965**, 228.
[21a] W. Wolf and N. Kharasch, *J. Org. Chem.*, **30**, 2493 (1965).
[21b] J. Hutton and W. A. Waters, *J. Chem. Soc.*, **1965**, 4253.
[22] F. Minisci and R. Galli, *Tetrahedron Letters*, **1965**, 433.

mechanism must be involved. Free benzoyloxy radicals rapidly lose carbon dioxide in toluene, and a concerted mechanism is therefore proposed, as shown in the chart for the formation of salicylic acid.[23]

$$
\begin{array}{ccc}
\underset{\text{Ph}-\text{C}=\text{O}}{\underset{\overset{|}{\text{O}}-\text{Cu}^{II}}{\underset{\overset{|}{\text{H}}}{\underset{\text{(benzoyloxy lactone, H, O—Cu}^{II})}{}}}} & \longrightarrow & \underset{\text{OCOPh}}{\underset{\bullet}{\text{(radical, H, OCu}^{I})}}
\end{array}
$$

$$\downarrow \text{Cu}^{II}(\text{OCOPh})_2$$

$$
\underset{\text{work-up}}{\longleftarrow}
$$

CO$_2$H / OH (salicylic acid)   ←work-up—   (o-C$_6$H$_4$, C=O, OCu^I, OCOPh)

$$+ \ \text{Cu}^I\text{OCOPh}$$
$$+ \ \text{PhCO}_2\text{H}$$

Electrophilic Substitution

There has been a recent renewal of interest in aspects of electrophilic substitution. Recent reviews of some of these topics have been written by Olah[24] (intermediate complexes), Perkampus and Baumgarten[25] (proton addition complexes), and Roberts[26] (Friedel–Crafts reaction). A useful monograph on electrophilic aromatic substitution has also been published.[27a]

Olah's work on electrophilic substitution is well known. One aspect of this has been the demonstration that certain highly reactive substituting agents, typically nitronium tetrafluoroborate, participate in competitive reactions with unusually low substrate selectivity but with normal positional selectivity in any individual substrate. This had been attributed to rate-determining π-complex formation, followed by collapse of the π-complex to isomeric σ-complexes in the normal way. It appeared to have been clearly demonstrated that, in spite of the exceptional reactivity of such reagents as nitronium salts, no criticisms could be upheld against the competitive method employed. However, after a first hint that these reactions might not be

[23] W. W. Kaeding and G. R. Collins, *J. Org. Chem.*, **30**, 3750 (1965); W. W. Kaeding, H. O. Kerlinger, and G. R. Collins, *ibid.*, **30**, 3754 (1965).

[24] G. A. Olah, "Organic Reaction Mechanisms", *Chem. Soc. Special Publ. No.* 19, **1965**, p. 21.

[25] H.-H. Perkampus and E. Baumgarten, *Angew. Chem. Intern. Ed. Engl.*, **3**, 776 (1964).

[26] R. M. Roberts, *Chem. and Eng. News*, Jan, 25th 1965, p. 96.

[27a] R. O. C. Norman and R. Taylor, "Electrophilic Substitution in Benzenoid Compounds", Elsevier, Amsterdam, 1965.

kinetically "well-behaved,"[27b] one of Olah's former co-workers has adduced compelling evidence that the unusually low selectivity (e.g. Olah reports that toluene is only 1.7 times as reactive as benzene in nitration by $NO_2^+ BF_4^-$ in sulpholane) is due to unsatisfactory mixing of the reagent with the competing solvents.[28] With so reactive a reagent, at the point of mixing the solution is depleted of molecules of the more reactive substrate, whence reaction with the less reactive one may occur to a very considerable extent. Consistent with this view was the observation that appreciable quantities of dinitrotoluenes were obtained from a competitive reaction between benzene and toluene which was taken only to a few per cent conversion. More conclusive, however, was the observation that by use of less concentrated nitronium salt solutions, and addition of these to substrate mixtures subjected to efficient vibrational stirring, relative reactivities for toluene approaching 25–30 could be obtained, these values being close to those reported for the nitration of toluene by classical procedures. Kreienbühl and Zollinger[29] have, moreover, reported that pentamethylbenzene and nitronium tetrafluoroborate form, instantaneously, a complex which only slowly decomposes to pentamethylnitrobenzene and unidentified by-products. This reaction occurs exclusively, even when benzene is present as a competing substrate. Nitrobenzene is formed in the competition experiment only if the molar proportion of reagent is greater than that of pentamethylbenzene. Kreienbühl and Zollinger relate these results to Olah's work and question the applicability of the competition method reactions of nitronium fluoroborate; however, the nature of these criticisms appear to require revision in the light of the results of the efficient mixing experiments reported above.

Rate-determining π-complex formation in certain Friedel–Crafts alkylations has also come under attack. Spectroscopic identification of 1:1:1 complexes of boron trifluoride, alkyl halide, and alkylbenzene has been made at low temperatures (130–170°K).[30] These complexes are different from Olah's σ-complexes,[31] as it can be shown, spectroscopically and by measurement of thermodynamic boron isotope effects for the equilibrium between the solutions of the complex and gaseous boron trifluoride, that they do not contain the fluoroborate anion. Furthermore, reversible decomposition of some of the complexes occurs readily, instead of exclusive decomposition to substitution products. It is thought that these complexes are best interpreted as oriented π-complexes such as (6); the mechanism for boron trifluoride-catalysed alkylation may then be written as shown. The reversible

[27b] C. D. Ritchie and H. Win, *J. Org. Chem.*, **29**, 3093 (1964).
[28] W. S. Tolgyesi, *Can. J. Chem.*, **43**, 343 (1965).
[29] P. Kreienbühl and H. Zollinger, *Tetrahedron Letters*, **1965**, 1739.
[30] R. Nakane, A. Natsubori, and O. Kurihara, *J. Am. Chem. Soc.*, **87**, 3597 (1965).
[31] G. A. Olah and S. J. Kuhn, *J. Am. Chem. Soc.*, **80**, 6541 (1958).

low-temperature formation of the π-complex, and the kinetic unimportance of proton loss, leave conversion of the π-complex into a σ-complex as the only reasonable rate-determining step, at least in alkylations catalysed by boron trifluoride.

$$RX + BF_3 \rightleftharpoons \overset{\delta+}{R}-\overset{\delta-}{X}\cdots BF_3$$

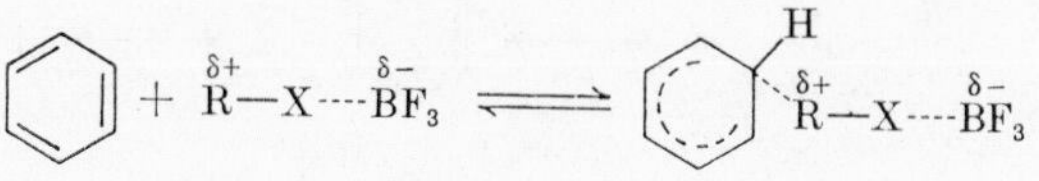

(6)

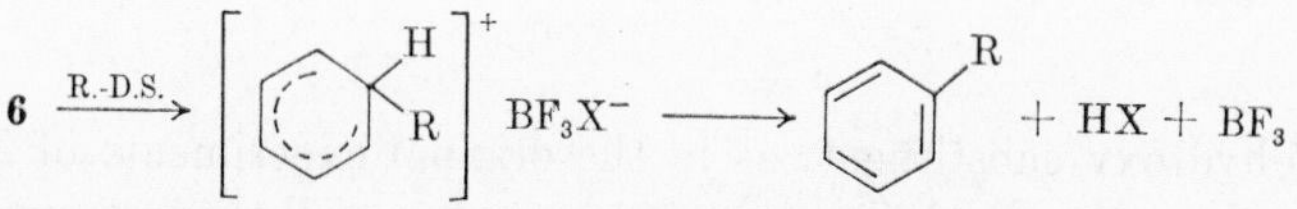

In the crystal, it has been found that the 1:1 complex of aluminium chloride and benzoyl chloride is coordinated through oxygen, and not through chlorine.[32] This may, however, have no bearing on the nature of the reactive species present in solution.

There have been several publications dealing with electrophilic substitution into an aromatic ring fused to a strained alicyclic system. The Mills–Nixon effect, originally attributed to partial bond-fixation in the aromatic portion of such a system, has since received several alternative rationalizations. However, the NMR spectrum of the anion (7) shows that $C_{(6)}$ carries a higher density of negative charge than does $C_{(4)}$,[33] which is consistent with the Mills–Nixon hypothesis that it is best represented in the canonical form shown.

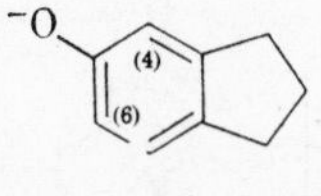

(7)

Vaughan and his co-workers[34] have examined the reactions of unsubstituted indane and tetralin, and find that in several reactions the former gives a higher proportion of β-substitution product than the latter. This is considered in terms, not of bond-fixation in the ground-state hydrocarbons,

[32] S. E. Rasmussen and N. C. Broch, *Chem. Comm.*, **1965**, 289.
[33] W. Koch and H. Zollinger, *Helv. Chim. Acta*, **48**, 1791 (1965).
[34] J. Vaughan, G. J. Welch, and G. J. Wright, *Tetrahedron*, **21**, 1665 (1965).

6+

but of double-bond character in the transition state, if we assume that a ring-junction double bond in the indane system provides a situation of relatively high energy, then the intermediate (and the transition state) for β-substitution will be seen (by inspection of the location of double bonds in the canonical forms of the intermediates) to be more favoured than that for α-substitution. Extension of this argument to substitution in compounds

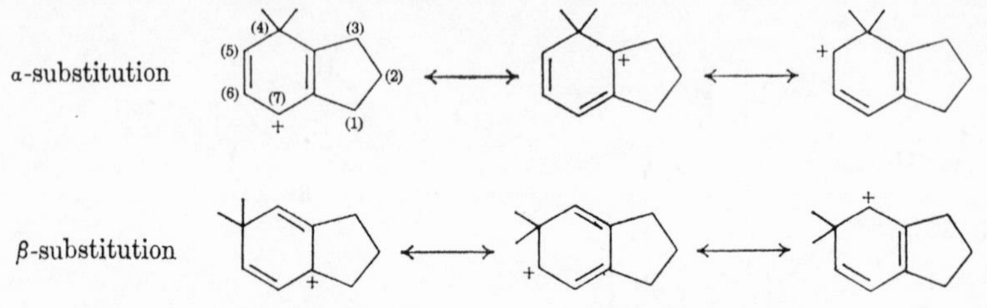

with a 5-hydroxy substituent, as in the original experiments of Mills and Nixon, leads to the conclusion that 6-substitution will there be preferred to 4-substitution by an even greater factor.

Substitution in benzocyclobutene has received further attention.[35,36] Here substitution is also at the β-position. Products of α-substitution could not be detected. However, a second reaction path involves opening of the cyclobutene ring by electrophilic dealkylation. The extent of the latter process is relatively small during halogenation (ca. 15%), but when the electrophile is better able to delocalize the positive charge on the leaving carbonium ion, ring opening becomes the major reaction path, as e.g. in reaction (5).[35]

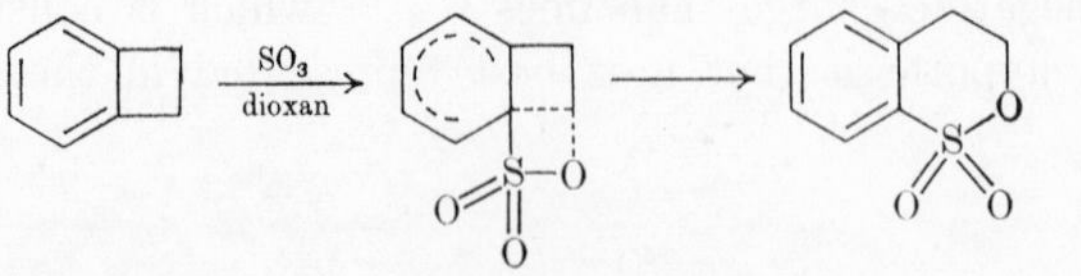

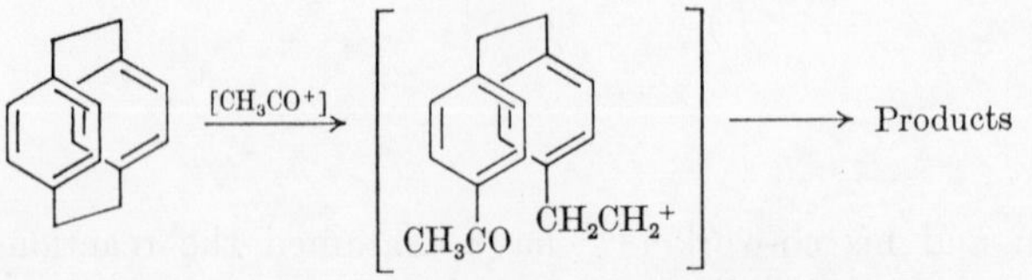

Electrophilic dealkylation in a strained ring system has also been observed during Friedel–Crafts acetylation of [2.2]paracyclophane [reaction (6)].[37]

[35] J. B. F. Lloyd and P. A. Ongley, *Tetrahedron*, **21**, 245 (1965).
[36] L. Horner, P. V. Subramanian, and K. Eiben, *Tetrahedron Letters*, **1965**, 247.
[37] D. J. Cram and H. P. Fischer, *J. Org. Chem.*, **30**, 1815 (1965).

The results of an investigation of electrophilic attack on benzene fused to bridged-ring systems have now been set out in full.[38] Some relative rates for nitration are shown on structures (8) to (11). It has been suggested that the unusually high reactivity of the β-position of benzonorbornene (10) in nitration may be due to strain effects, or perhaps to σ-bond participation of the type shown. The particularly low reactivity of the α-positions of benzo-

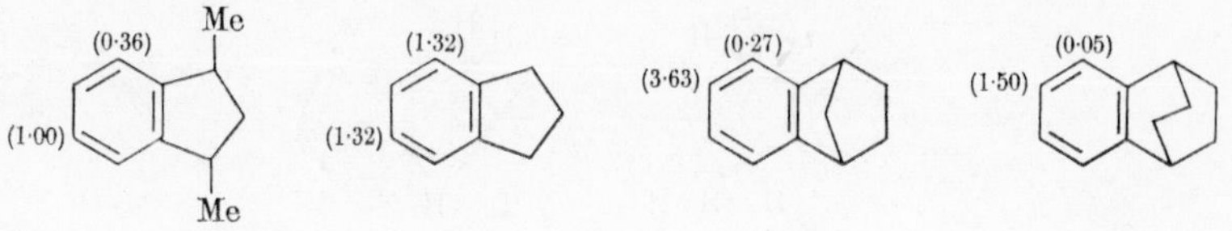

(8)   (9)   (10)   (11)

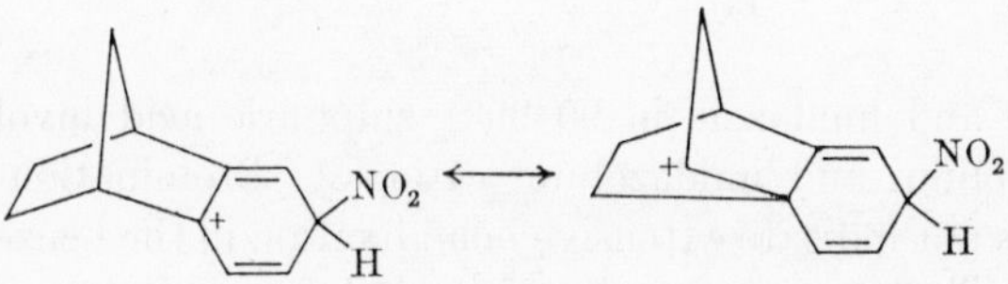

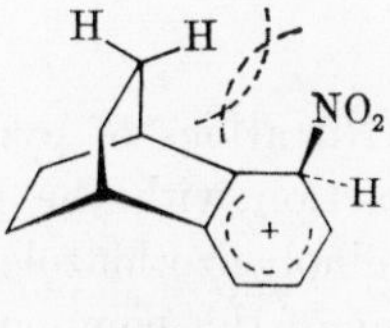

(12)

norbornene, and especially of benzobicyclo[2.2.2]octene (11), was attributed to a "fused *ortho*-effect" (12). Implied in this is the inability of the bridged-ring system to rotate, or flex, out of the path of the approaching electrophile.

The chlorination of biphenyl, fluorene, and other 2,2′-bridged biphenyls has been examined.[39] Reaction is predominantly in the 4-position, even for compounds which show appreciable deviation from co-planarity. The effect of strain on rate of bromination of acenaphthene has also been discussed,[40] and the Freidel–Crafts *t*-butylation of this compound has been studied.[41] With ferric chloride catalysts the 1- and 3-derivatives are formed, but aluminium chloride gives the thermodynamically more stable 2-isomer.

Kovacic and Hiller[42a] have introduced the concept of "linear coordination" to explain unusually high $o:p$ ratios observed in mercuration and alkylation of chlorobenzene and anisole. The concept requires coordination of the reagent at the Lewis base substituent, followed by transfer to the *ortho*-position via an unspecified bridging mechanism supposed to involve the

[38] H. Tanida and R. Muneyuki, *J. Am. Chem. Soc.*, **87**, 4794 (1965).
[39] P. B. D. de la Mare, E. A. Johnson, and J. S. Lomas, *J. Chem. Soc.*, **1964**, 5317; P. B. D. de la Mare and J. S. Lomas, *ibid.*, **1965**, 5739.
[40] E. Berliner, D. M. Falcione, and J. L. Riemenschneider, *J. Org. Chem.*, **30**, 1812 (1965).
[41] A. T. Peters, *Nature*, **205**, 170 (1965).
[42] P. Kovacic and J. J. Hiller, *J. Org. Chem.*, **30**, (a) 1581, (b) 2871 (1965).

π-electrons of the ring. It is thus distinguished from *ortho*-substitution directed by chelation to the substituent already present. Linear coordination is also suggested to explain the high *ortho*-nitration of anisole by nitronium fluoroborate.[42b]

In further studies of proton exchange of azulene[43] it was shown[44] that treatment with aqueous acid gave salts of the dication **(13)**.

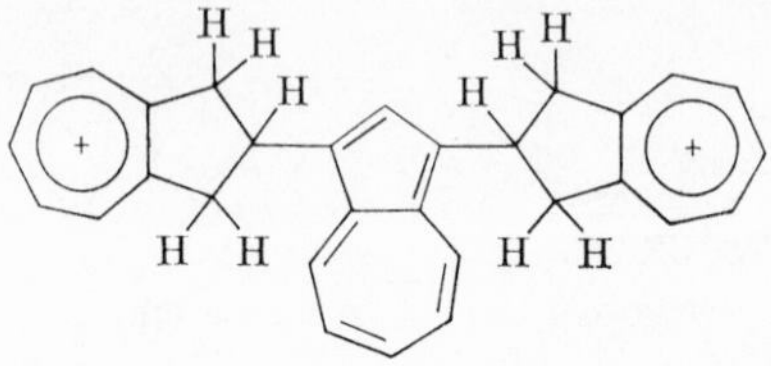

(13)

Nitrations of pyrazole and imidazole in 90–99% sulphuric acid involve reaction with the pyrazolium and imidazolium cations.[45] Bromination of aminobenzothiazoles does not indicate extensive bond fixation in the benzene ring of this compound.[46] The reactions of nitronium and nitrosonium tetra-fluoroborate with pyridine in protic solvents lead to *N*-nitro- and *N*-nitroso-derivatives;[47] the reagent must be kept in excess, as traces of base cleave the pyridine ring.

Orientation of electrophilic substitution in 1,3,5-[48] and 1,2,3-triphenyl-benzene[49] has been examined.

The 2,4-dinitrobenzenediazonium ion will couple *para* to OMe in a benzene ring, but *para* to SMe only if a second activating group is present.[50] It does not react *para* to SeMe. These results are consistent with known mesomeric effects of Group VI elements.

It is now known that nitration of the anilinium ion gives both *meta*- and *para*-substitution (the *para*-position being deactivated least by a direct inductive effect). Electrophilic substitution in this species has been treated theoretically by Chandra and Coulson.[51]

In a study of protodesilylation of a series of phosphorus-substituted aromatic compounds, it was found that p-PMe$_3{}^+$ is more deactivating than

[43] B. C. Challis and F. A. Long, *J. Am. Chem. Soc.*, **87**, 1196 (1965).
[44] P. C. Myhre and R. D. Andersen, *Tetrahedron Letters*, **1965**, 1497.
[45] M. W. Austin, J. R. Blackborow, J. H. Ridd, and B. V. Smith, *J. Chem. Soc.*, **1965**, 1051.
[46] E. R. Ward and C. H. Williams, *J. Chem. Soc.*, **1965**, 2248.
[47] G. A. Olah, J. A. Olah, and N. A. Overchuk, *J. Org. Chem.*, **30**, 3373 (1965).
[48] G. E. Lewis, *J. Org. Chem.*, **30**, 2798 (1965).
[49] D. Buza and W. Polaczkowa, *Roczniki Chem.*, **39**, 557 (1965).
[50] N. Marziano and R. Passerini, *Gazz. Chim. Ital.*, **94**, 1137 (1964).
[51] A. K. Chandra and C. A. Coulson, *J. Chem. Soc.*, **1965**, 2210.

p-NMe$_3$$^+$.[52] This was attributed to the possibility of mesomeric withdrawal of π-electrons into the vacant d orbitals on phosphorus.

Nitration of the dimethylphenylsulphonium ion has been re-examined and shown to give 6% of *para-* and 4% of *ortho-*substitution (90% *m-*). Here the situation is still more complex.[53] The p_π–d_π conjugative withdrawal is offset by the possibility of conjugation of the lone pair on sulphur into the ring.

Other molecular-orbital calculations pertinent to electrophilic substitution relate to pyridine-like heterocycles[54] and condensed polycyclic hydrocarbons.[55] A simple correction to Hückel localization energy calculations is available.[56]

An unexpectedly large ^{14}C isotope effect has been found in aromatic decarboxylation,[57] and the decarboxylation of some pyridine heterocycles has been discussed.[58]

Two interesting mercurations have been studied. In one, a tracer technique is used to follow the mercury-atom exchange between liquid mercury and a solution of diphenylmercury.[59] A concerted mechanism at the metal surface seems probable. The other involves "transmercuration" [reaction (7)].[60] A study of this process may shed light on the anomalous substitution patterns

$$\text{PhHgX} + \text{C}_6\text{H}_5\text{Z} \rightleftarrows \text{PhH} + \text{XHgC}_6\text{H}_4\text{Z} \tag{7}$$

observed in mercuration. Evidently reaction (7) is capable of equilibrating isomers of XHgC$_6$H$_4$Z, and indeed when X = OAc and Z = F short heating periods (150°, 50 hr) give isomer proportions (*o-*; 20%; *m-*, 4%; *p-*, 76%) characteristic of electrophilic substitution. The figures change drastically on longer heating (430 hr: *o-*, 86%; *m-*, 9%; *p-*, 5%).

Nitrations of methyl- and methoxy-naphthalenes,[61] of toluic acids,[62] and (at the *meso-*positions) of a porphyrin[63] have been studied. Nitration of 2-substituted p-dimethoxybenzenes, where the substituent is a *meta-*directing group, leads predominantly to the 3-nitro-derivative. In the compounds (**14a**) and (**14b**) this directive effect is most pronounced when X = Cl. As the electronegative chlorine would tend to reduce the carbonyl dipole, a

[52] R. W. Bott, B. F. Dowden, and C. Eaborn, *J. Chem. Soc.*, **1965**, 6306.
[53] H. M. Gilow and G. L. Walker, *Tetrahedron Letters*, **1965**, 4295.
[54] R. Zahradník and C. Párkány, *Collection Czech. Chem. Commun.*, **30**, 355 (1965).
[55] M. J. S. Dewar and C. C. Thompson, *J. Am. Chem. Soc.*, **87**, 4414 (1965).
[56] B. G. Ramsey, *J. Am. Chem. Soc.*, **87**, 2502 (1965).
[57] M. Roesseler and H. Koch, *Radiochim. Acta*, **4**, 12 (1965).
[58] P. Haake and J. Mantecón, *J. Am. Chem. Soc.*, **86**, 5230 (1964).
[59] D. R. Pollard and J. V. Westwood, *J. Am. Chem. Soc.*, **87**, 2809 (1965).
[60] E. C. Kooyman, J. Wolters, J. Spierenburg, and J. Reedijk, *J. Organometal. Chem.*, **3**, 487 (1965).
[61] P. G. E. Alcorn and P. R. Wells, *Austral. J. Chem.*, **18**, 1377, 1391 (1965).
[62] A. A. Spryskov and G. A. Polyakova, *Zh. Organ. Khim.* **1**, 24 (1965).
[63] R. Bonnett and G. F. Stephenson, *J. Org. Chem.*, **30**, 2791 (1965).

dipolar attraction between the electrophile and the carbonyl group appears to be excluded.[64]

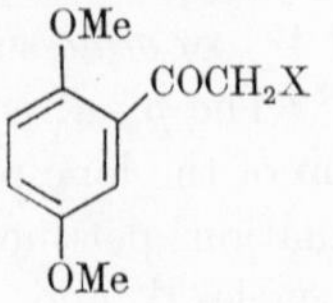

(14)

a X = H

b X = Cl

Sulphonation of toluenesulphonic acids has been studied,[65] and "sulphodesilylation" [reaction (8)] provides a convenient procedure for making isomerically pure sulphonic acids.[66]

$$ArSiMe_3 \xrightarrow[CCl_4]{SO_3} ArSO_2OSiMe_3 \xrightarrow[hydrol.]{Mild} ArSO_3H \ (80\%) \qquad (8)$$

The stereochemistry of rupture of a silicon–aryl bond by bromine has been examined.[67] Optically active (p-anisyl)methyl-1-naphthylphenylsilane, on treatment with bromine in carbon tetrachloride, gives p-bromoanisole, together with the bromosilane in which inversion of configuration at silicon has taken place. This is regarded as disproving a four-centre mechanism and as being most simply explained by the mechanism (9).

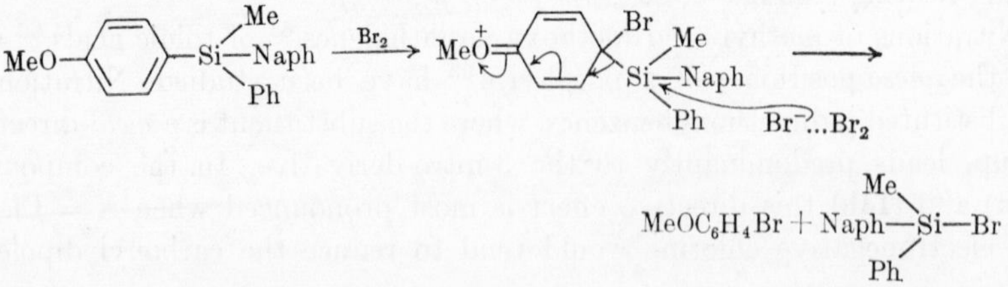

Halogenation of thiophene has been re-examined quantitatively.[68]

From the chlorination of phenanthrene, in acetic acid, a 9-acetoxy-9-chloro-9,10-dihydrophenanthrene has been isolated.[69]

[64] C. A. Howe, A. Howe, C. R. Hamel, H. W. Gibson, and R. R. Flynn, *J. Chem. Soc.*, **1965**, 795.

[65] H. Cerfontain, *Rec. Trav. Chim.*, **84**, 551 (1965).

[66] R. W. Bott, C. Eaborn, and T. Hashimoto, *J. Organometal. Chem.*, **3**, 442 (1965).

[67] C. Eaborn and O. W. Steward, *J. Chem. Soc.*, **1965**, 521.

[68] G. Marino, *Tetrahedron*, **21**, 843 (1965).

[69] P. B. D. de la Mare and R. Koenigsberger, *J. Chem. Soc.*, **1964**, 5327.

Chlorination of mesitylene in the dark gives 10% substitution in a methyl group; this is considered to be produced in an ionic reaction paralleling nuclear substitution.[70] In related work, bromination of 1,3,5-tri-t-butyl-2-deuteriobenzene in the presence of silver perchlorate[71] showed an isotope effect $k_H/k_D \gg 3.6$. This, the largest deuterium isotope effect detected for aromatic bromination, may be due to a change of timing of bond-making and -breaking brought about by steric factors.

An isotope effect (k_H/k_D ca 2) for *ortho*-bromination of N,N-dialkylarylamines is absent in *para*-bromination, indicating reactions of different kinetic form at the two positions.[72]

The kinetics of bromination of phenol and a series of alkylphenols support the hypothesis that conjugative electron-release by the hydroxyl group may be augmented by O–H no-bond resonance.[73] Chlorination of o-cresol gives a mixture of the 4- and the 6-derivatives; the proportions of these reveal steric blocking of the 6-position by hydrogen-bonding to the solvent, should this be possible;[74] thus in carbon tetrachloride the 6-chloro-derivative accounts for 50% of the product: in ether it accounts for 15%.

The kinetics of iodination of tyrosine derivatives have been examined.[75]

Iodination of aromatic compounds by peracetic acid and iodine probably involves acetyl hypoiodite;[76] acetyl hypobromite is involved in bromination of alkylbenzenes by bromine, acetic acid, and mercuric acetate;[77] and chlorination of biphenyl by acetyl hypochlorite has been studied.[78]

A long-known but little studied reaction is the brominative cleavage of arylcarbinols, particularly benzhydrols [reaction (10)]. This has now been examined by Arnett and Klingensmith, who found excellent second-order

$$\text{ArCH(OH)Ar}' \xrightarrow[\text{Aq.AcOH}]{\text{Br}_2} \text{ArCHO} + \text{Ar}'\text{Br} + \text{HBr} \qquad (10)$$

rate constants for the reaction.[79] The rate shows pronounced dependence on substituents in the attacked ring, strong activation being required for

[70] G. Illuminati and F. Stegel, *Ric. Sci., Rend., Sez. A*, **7**, 458 (1964); *Chem. Abs.*, **63**, 5465 (1965).

[71] G. Illuminati and F. Stegel, *Ric. Sci., Rend., Sez.* A, **7**, 460 (1965); *Chem. Abs.*, **63**, 5466 (1965).

[72] J. E. Dubois and R. Uzan, *Tetrahedron Letters*, **1965**, 309.

[73] P. B. D. de la Mare, O. M. H. el Dusouqui, J. G. Tillett, and M. Zeltner, *J. Chem. Soc.*, **1964**, 5306.

[74] A. Campbell and D. J. Shields, *Tetrahedron*, **21**, 211 (1965).

[75] W. E. Mayberry, J. E. Rall, and O. Bertoli, *J. Am. Chem. Soc.*, **86**, 5302 (1964); W. E. Mayberry and D. A. Bertoli, *J. Org. Chem.*, **30**, 2029 (1965); W. E. Mayberry, J. E. Rall, M. Berman, and D. Bertoli, *Biochemistry*, **4**, 1965 (1965).

[76] Y. Ogata and K. Nakajima, *Tetrahedron*, **20**, 2751 (1964).

[77] Y. Hatanaka, R. M. Keefer, and L. J. Andrews, *J. Am. Chem. Soc.*, **87**, 4280 (1965).

[78] M. Hassan and S. A. Osman, *J. Chem. Soc.*, **1965**, 2194.

[79] E. M. Arnett and G. B. Klingensmith, *J. Am. Chem. Soc.*, **87**, 1023, 1032, 1038 (1965).

satisfactory competition with side-reactions. Substituents in the second ring had less effect on the rate, and a factor of only twenty was observed in changing from *p*-nitro to *p*-methoxyl. These rates gave a good σ-correlation with ρ = −1.24. This appears to be the first example of a study of the electronic effects of the leaving group in electrophilic substitution. The kinetics require reversible formation of an intermediate (presumably a σ-complex), and the effect of the leaving group is divided between the formation of this complex (15) and its collapse to products. The relatively small effect of the leaving group in the product-forming step may be due to the overriding

(15)

importance of the hydroxyl oxygen in assisting this process [see structure (15)].

The protio- and deuterio-deiodination of 1-iodo-2,4,6-trimethoxybenzene in the presence of chloride ions show complex kinetic behaviour.[80] However, it proved possible to analyse the results in terms of three concurrent reactions, two of which were catalysed by chloride ion. It was suggested that the catalysed reactions differed in the timing of protonation and of chloride-complexing at iodine (to give ICl). The unexpectedly large uncatalysed rate, compared with that of hydrogen exchange of 1,3,5-trimethoxybenzene, was ascribed to a steric factor.

From the large deuterium isotope effect in the protolysis of ArMR$_3$ (where M = Si, Ge, Sn, or Pb) by hydrogen chloride in aqueous dioxane, Eaborn has concluded that the rate-determining step is proton transfer.[81] The effect is greatest for the lead compound studied, for which the value of k_H/k_D (= 3.05 at 50°) approaches the theoretical maximum and indicates that the extents of O–H bond-breaking and C–H bond-formation in the transition state are comparable. The smaller effects for the other compounds were interpreted in terms of greater C–H bond-formation at the transition state.

Protodesilylation has been used to examine the additivity of substituent effects[82] and to study inductive electron release from silicon.[83] It was found that the *meta*- and *para*-isomers of Me$_3$XCH$_2$C$_6$H$_4$SiMe$_3$, where X = Si,

[80] B. D. Batts and V. Gold, *J. Chem. Soc.*, **1964**, 5753.
[81] R. W. Bott, C. Eaborn, and P. M. Greasley, *J. Chem. Soc.*, **1964**, 4804.
[82] C. Eaborn and D. R. M. Walton, *J. Organometal. Chem.*, **3**, 169 (1965).
[83] D. R. M. Walton, *J. Organometal. Chem.*, **3**, 438 (1965).

were cleaved by acid more rapidly than were the corresponding carbon compounds. This was taken to indicate a greater facility for inductive release of electrons to saturated carbon from silicon than from carbon.

Protolytic demercuration of ArHgCl by aqueous-alcoholic hydrochloric acid is a clean second-order reaction. There is a good σ^+ correlation, with $\rho = -2.44$. Extension of the study to heterocyclic compounds revealed that the relatively high reactivity of the 2-position of furan was attributable to an entropy factor.[84]

Protodeboronation[85] and iodinolysis[86] of thiopheneboronic acids have been examined.

Hydrogen-exchange reactions of fluorene and diphenylmethane have been compared,[87] and detritiation of monosubstituted 1-tritionaphthalenes has been examined.[88] The latter study revealed a conjugative effect from substituents in the 5-position.

Proton exchange at the 2-position of deuterated imidazole in neutral or basic solution involves the imidazolium cation [reaction (11)].[89] Exchange at

$$ (11) $$

positions 3 and 5 of deuterated 4-pyridone in concentrated sulphuric acid involves the free base.[90] In each case these conclusions may be drawn from the near constancy of the rate of exchange with a wide variation in pH (or H_0).

Electrophilic introduction of amino or hydroxy groups has been reported by several groups. Aluminium chloride-catalysed amination of toluene by trichloramine is, however, not strictly in this category.[91] The main products are *m*-toluidine and *o*- and *p*-chloro-toluenes. The *m*-toluidine is considered to arise by "σ-substitution" which involves nucleophilic attack of NCl_3 on a proton–toluene σ-complex (formed from "adventitious water") [reaction (12)].

Aminobiphenyls and diphenylamines are produced in modest yield from

[84] R. D. Brown, A. S. Buchanan, and A. M. Humffray, *Australian J. Chem.*, **18**, 1507, 1513 (1965).
[85] R. D. Brown, A. S. Buchanan, and A. A. Humffray, *Australian J. Chem.*, **18**, 1521 (1965).
[86] R. D. Brown, A. S. Buchanan, and A. A. Humffray, *Australian J. Chem.*, **18**, 1527 (1965).
[87] K. C. C. Bancroft, R. W. Bott, and C. Eaborn, *J. Chem. Soc.*, **1964**, 4806.
[88] R. W. Bott, R. W. Spillett, and C. Eaborn, *Chem. Comm.*, **1965**, 147.
[89] T. M. Harris and J. C. Randall, *Chem. Ind. (London)*, **1965**, 1728.
[90] P. Bellingham, C. D. Johnson, and A. R. Katritzky, *Chem. Ind (London)*, **1965**, 1384.
[91] P. Kovacic, C. T. Goralski, J. J. Hiller, J. A. Levisky, and R. M. Lange, *J. Am. Chem. Soc.*, **87**, 1262 (1965).

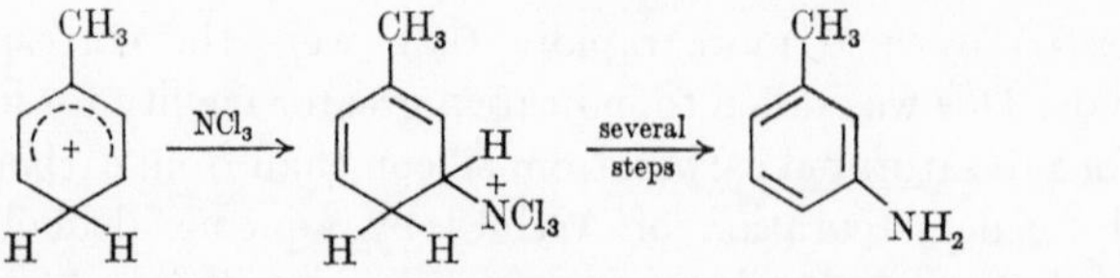

$$\text{(12)}$$

the reaction of HBF_4 with phenylhydroxylamine in sulpholane in the presence of an aromatic compound [reaction (13)].[92] A neighbouring-group effect is postulated for *ortho*-substitution in anisole [reaction (14)].

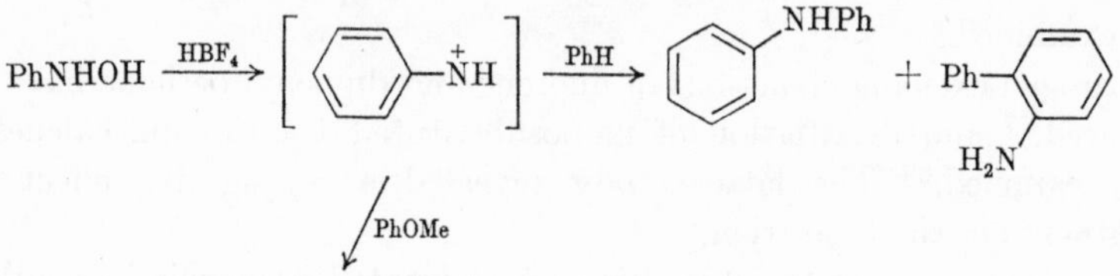

$$\text{PhNHOH} \xrightarrow{\text{HBF}_4} \left[\text{ } \overset{+}{\text{NH}} \right] \xrightarrow{\text{PhH}} \text{ } + \text{Ph} \text{ } \quad \text{(13)}$$

$$\downarrow \text{PhOMe}$$

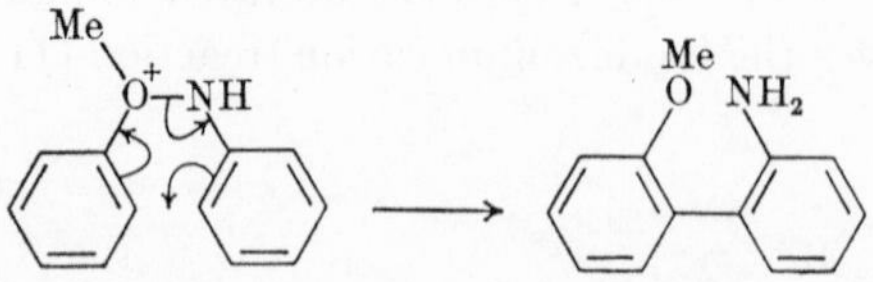

$$\text{(14)}$$

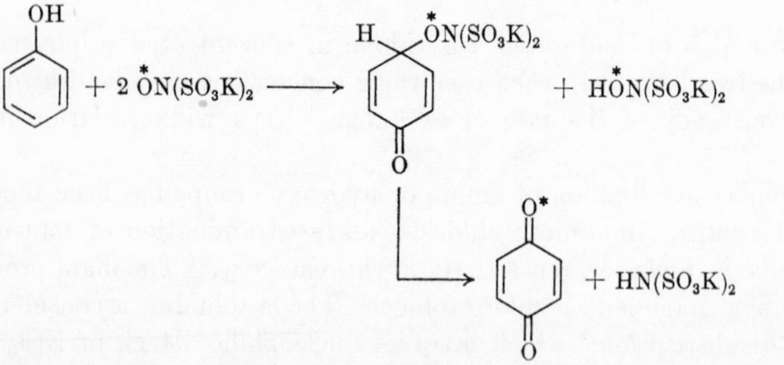

$$\text{(15)}$$

The oxidation of phenols to *p*-quinones by dipotassium nitrosobisulphonate has been demonstrated, by labelling with ^{18}O, to involve transfer of the nitroso-oxygen to the phenol, as in reaction (15).[93] Oxygenation of aromatic compounds by isopropyl peroxydicarbonate catalysed by aluminium chloride gives moderate yields of products (15–70%), showing a characteristic electrophilic substitution pattern.[94] A possible mechanistic scheme is shown in Scheme 1.

[92] J. H. Parish and M. C. Whiting, *J. Chem. Soc.*, **1964**, 4713.
[93] H. J. Teuber and K. H. Dietz, *Angew. Chem. Intern. Ed. Engl.*, **4**, 871 (1965).
[94] P. Kovacic and S. T. Morneweck, *J. Am. Chem. Soc.*, **87**, 1566 (1965), P. Kovacic and M. E. Kurz, *ibid.*, 4811.

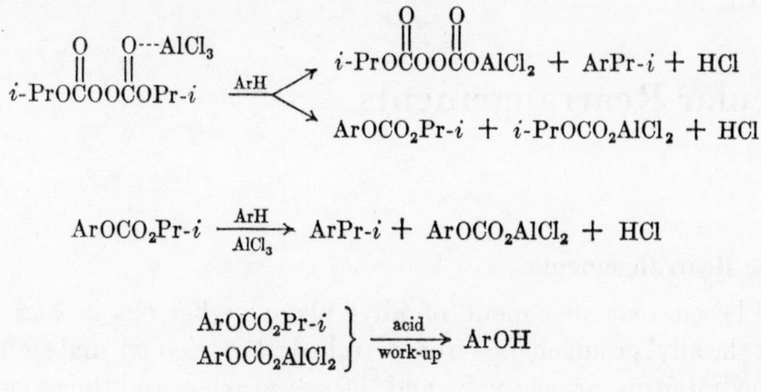

Scheme 1

Relative reactivities and isomer distribution indicate that hydroxylation of aromatic compounds by trifluoroperacetic acid in methylene chloride is also an electrophilic process.[95] This reagent gives hydroxytetralins on reaction with 4-aryl-1-butenes. The reaction was shown not to proceed by acid-catalysed rearrangement of an intermediate epoxide; a concerted mechanism (16) involving aryl participation was advanced to explain the formation of a tetralin rather than an indane derivative.

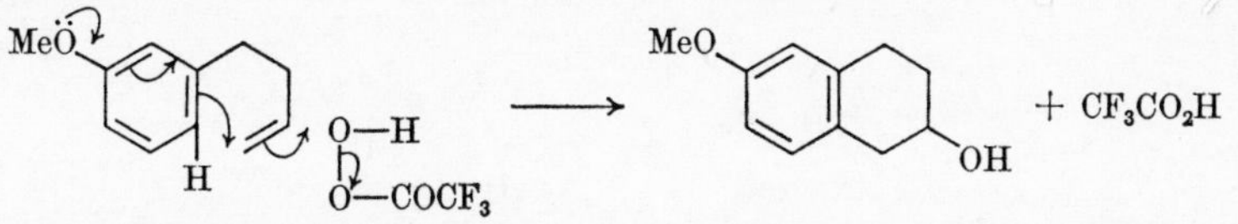

Electrophilic nitration[96] and proton transfer[97] in photoexcited states of aromatic molecules show different reactivity patterns from those observed with the ground states.

[95] A. J. Davidson and R. O. C. Norman, *J. Chem. Soc.*, **1964**, 5404.
[96] C. S. Foote, P. Engel, and T. W. Del Pesco, *Tetrahedron Letters*, **1965**.
[97] D. A. de Bie and E. Havinga, *Tetrahedron*, **21**, 2359 (1965).

CHAPTER 8

Molecular Rearrangements

Aromatic Rearrangements

In the Claisen rearrange ment of allyl phenyl ether the α- and γ-carbon atoms of the allyl group change from tetrahedral to trigonal, and from trigonal to tetrahedral form, respectively, and the *ortho*-carbon and the oxygen atoms undergo similar changes. Assuming that severance and attachment of the allyl group occur on the same side of the aromatic system, the six-membered ring of the cyclic transition state, made up of the three allyl carbon atoms in one plane and $O-C_{(1)}-C_{(2)}$ of the ring in another, roughly parallel, plane will be chair-like (1) or boat-like (2). Also there will be a fixed relation between the configuration of the tetrahedral centres and the geometry about the double bonds and the stereochemistry of the rearrangement could be followed by using optical or geometrical isomerism. Both techniques have been applied this year, the former by Goering and Kimoto,[1] the latter by Marvell, Stephenson, and Ong,[2] and their results are in good agreement.

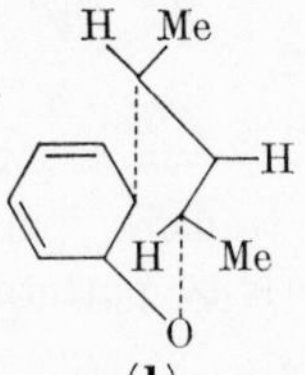

(1) 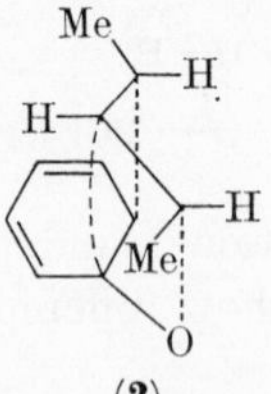   (2)

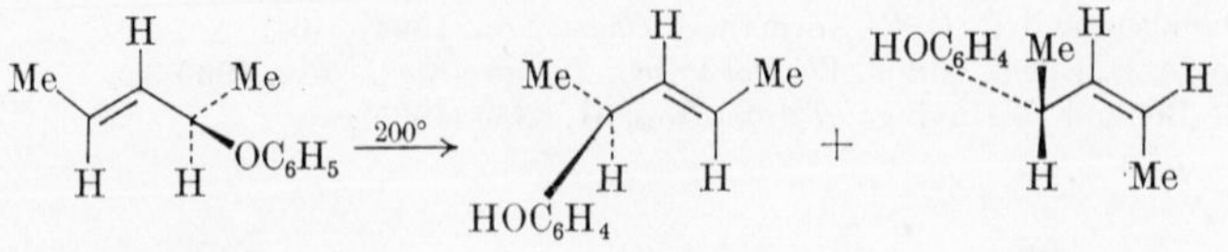

(3) $\xrightarrow{200°}$ (4)   +   (5)

(82%)   (18%)

The relative and absolute configuration of (+)-*trans*-1-methyl-2-butenyl phenyl ether (3) and its rearrangement products *trans*- (4) and *cis-o*-(1-

[1] H. L. Goering and W. I. Kimoto, *J. Am. Chem. Soc.*, **87**, 1748 (1965).
[2] E. N. Marvell, J. L. Stephenson, and J. Ong, *J. Am. Chem. Soc.*, **87**, 1267 (1965).

methyl-2-butenyl)phenol (**5**), formed in the proportions shown, have been determined. Thus, contrary to an earlier suggestion, the optical configuration of the product is the opposite to that of the starting material, and the stereochemistry of the process corresponds to that most reasonably expected from consideration of transition-state stabilities and observed for other intramolecular allylic rearrangements of *trans*-1-methyl-2-butenyl compounds.[1] When *cis*- and *trans*-1-methyl-2-butenyl phenyl ether were heated at 165° in mesitylene the *cis*-ether gave 98 ± 4% of *trans*-product, with no more than a trace of its *cis*-isomer, whilst the *trans*-ether gave 90 ± 4% of *trans*- and 10 ± 5% of *cis*-product. The overall rate constants for dilute solutions in n-octane at the same temperature were 0.95×10^{-5} sec^{-1} for the *cis*- and 1.52×10^{-5} sec^{-1} for the *trans*-ether. The *cis*- and the *trans*-ether and the products were all shown not to isomerize at the double bond under the rearrangement conditions. All the possible stereochemical relationships between factors and products were considered and the above results were explained by assuming a chair-like transition state (**1**), but with a greater distance between the carbon atoms forming the new bond than between the carbon and oxygen of the breaking bond.[2] This stereochemistry also permits rationalization of the observation that the size of an *ortho*-alkyl group does not markedly alter the rate of the *para*-Claisen rearrangement. This was found to be so in the rearrangement of the series of allyl *ortho*-substituted phenyl ethers, where the substituent was Me, Et, *i*-Pr, or *t*-Bu, when the relative amounts of 6-allyl (major) and 4-allyl (minor) products were determined.[3] From rate measurements with these and related 2,6-dialkyl ethers the interesting observation was made that, whilst a single *ortho*-alkyl group depressed the rate of migration of the allyl group to the *para*-position, the presence of alkyl groups in both *ortho*-positions increased the rate 20–40 fold, a very large factor compared with those normally observed in Claisen rearrangements. This was ascribed mainly to steric compression of the buttressed groups which increased the ground-state energy of the ether, and also to higher ratio of the rates of forward to backward reaction of the intermediate dienone, as the substituent became bigger.[4] Similar stereochemical conclusions about the *ortho*-Claisen rearrangement have been drawn by Schmid and his co-workers[5] from a study of *cis*- and *trans*-2-butenyl ethers; the subsequent *para*-rearrangement is, however, said not necessarily to be stereospecific.

The "abnormal" Claisen rearrangement, e.g. (**6**) → (**9**), is known to result from two consecutive processes: normal *ortho*-rearrangement, followed by

[3] E. N. Marvell, B. Richardson, R. Anderson, J. L. Stephenson, and T. Crandall, *J. Org. Chem.*, **30**, 1032 (1965).

[4] E. N. Marvell, B. J. Burreson, and T. Crandall, *J. Org. Chem.*, **30**, 1030 (1965).

[5] H. Schmid, A. Habich, and G. Frater, *Angew. Chem. Intern. Ed. Engl.*, **4**, 443 (1965).

isomerization of the side chain. It was proposed[6] that the latter reaction involved a cyclopropyl–dienone intermediate, e.g. (8). The generality of this type of intramolecular rearrangement has now been demonstrated and the

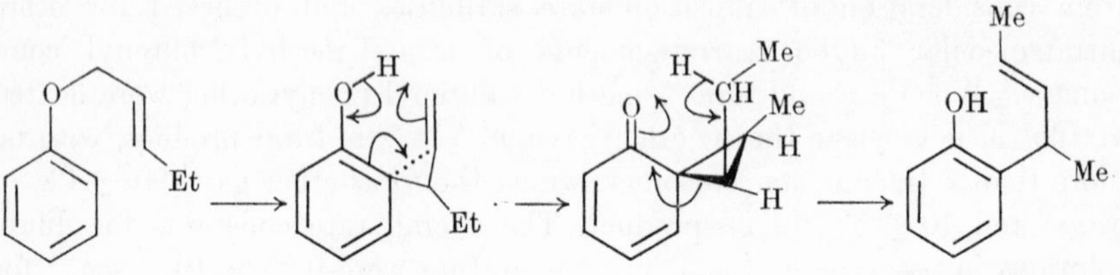

(6)   (7)   (8)   (9)

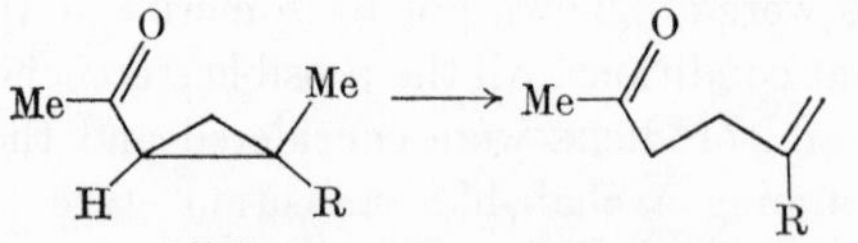

(10)

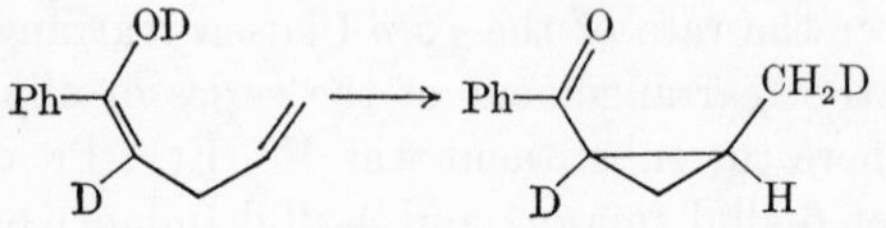

(11) 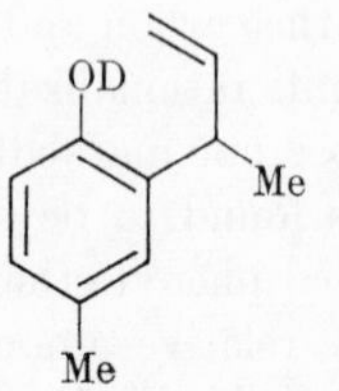   (12)

above proposal thereby strongly supported. 2-Acetyl-1,1-dimethyl- and 1-acetyl-*cis*-2-methyl-cyclopropane (10; R = Me or H) rearranged smoothly to 5-methyl-5-hexen-2-one and 5-hexen-2-one, respectively, whilst the *trans*-isomer was inert.[7] The formation of a cyclopropyl ketone from an aliphatic enol with concerted hydrogen transfer was also demonstrated, by deuterium labelling; when heated at 200°, $PhCOCD_2CH_2CH{=}CH_2$ was shown by NMR analysis to give $PhCOCHDCH_2CH{=}CHD$. This and other results require the intermediate formation of the cyclopropane (11).[8] The above mechanism for the abnormal Claisen rearrangement has also been directly confirmed by the demonstration that the deuterium in (12) is exchanged in an intramolecular process with the terminal methylene and the α-methyl group, but not with hydrogen at the α- or β-allyl position. If the allyl group

[6] E. N. Marvell, D. R. Anderson, and J. Ong, *J. Org. Chem.*, **27**, 1109 (1962); A. Habich, R. Barner, R. M. Roberts, and H. Schmid, *Helv. Chim. Acta*, **45**, 1943 (1962).

[7] R. M. Roberts and R. G. Landolt, *J. Am. Chem. Soc.*, **87**, 2281 (1965).

[8] R. M. Roberts, R. N. Greene, R. G. Landolt, and E. W. Heyer, *J. Am. Chem. Soc.*, **87**, 2282 (1965).

is unsubstituted, deuterium is incorporated only at the terminal methylene group.[9]

Thermal rearrangement of allyl 3-substituted 4-quinolyl ethers gave the corresponding 1-allylquinolones, quantitatively; that this did involve stepwise Claisen rearrangements, via a dienone, was shown by the equivalence of the migrating group with an allyl group on $C_{(3)}$.[10] When there is an (active) methyl group on $C_{(2)}$ in the quinoline the second step terminates on it, giving an "out-of-ring" Claisen rearrangement as the main reaction. Thus, heating allyl 2,3-dimethyl-4-quinolyl ether gave 2-(3-butenyl)-4-hydroxy-3-methylquinoline, presumably via the intermediate (13).[11]

Miller and Margulies[12] have observed some interesting 1,3-allyl migrations in the acid-catalysed rearrangement of 2,6-di-t-butylcyclohexadienones, accompanied by the loss of a t-butyl group. On treatment with sulphuric acid in acetic acid, the dienone (14) gave 2-allyl-6-t-butyl- and 5-allyl-2-t-butyl-4-methylphenol (15) and (16), respectively. The dienone (17) gave a quantitative yield of 2-t-butyl-5-ethyl-4-methylphenol and its acetate, and the dienone

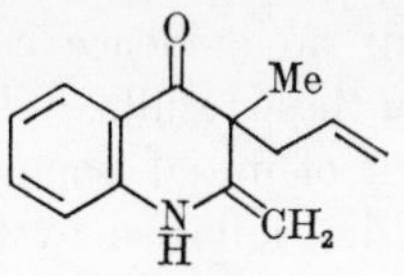

(13)

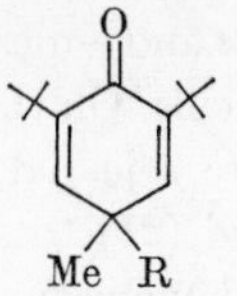

14, R = allyl
17, R = ethyl
18, R = 2-butenyl

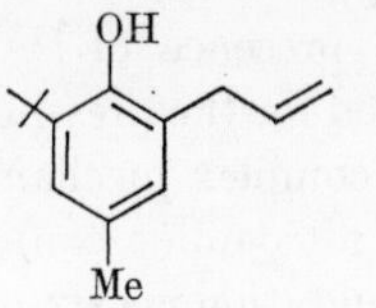

(15)

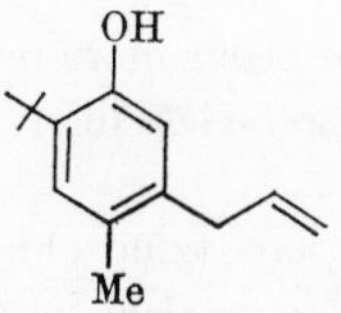

(16) 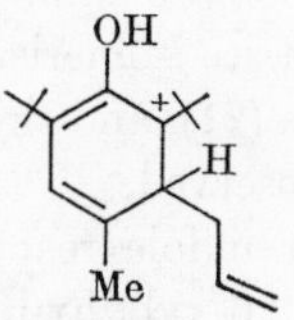   (19) 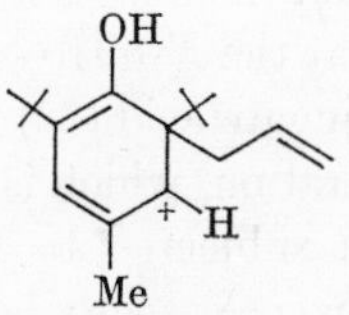   (20)

(18) gave 2,6-di-t-butyl-4-methylphenol and 2-(2-butenyl)-6-t-butyl-4-methylphenol (i.e. not the product of allyl inversion expected from a Cope rearrangement). Appropriate trapping experiments showed, however, that the rearrangements were intramolecular, and the rate of 1,3-migration appeared to depend on the ability of the migrating group to stabilize a positive charge since the proportion of 1,3-migration increased as the substituent R changed

[9] A. Habich, R. Barner, W. von Philipsborn, and H. Schmid, *Helv. Chim. Acta*, **48**, 1297 (1965).

[10] Y. Makisumi, *J. Org. Chem.*, **30**, 1986 (1965).

[11] Y. Makisumi, *J. Org. Chem.*, **30**, 1989 (1965).

[12] B. Miller and H. Margulies, *Tetrahedron Letters*, **1965**, 1727.

from ethyl to allyl to butenyl. Two mechanisms were considered: formation of a π-complex between the migrating group and the phenol ring followed by collapse to carbonium ion (19) or (20); and the formation of (19) followed by its rearrangement to (20). The latter was favoured by the demonstration of a large effect on the ratio of the two products, deuterated (15) and (16), formed from (14) with ring-protium replaced by deuterium. The two consecutive allyl shifts may occur overall without inversion by two migrations with inversion or two without inversion. The observation that (18) rearranges faster than (14) suggests that both occur without inversion since attack by the methyl-substituted double bond of (18) on an atom adjacent to the *t*-butyl group should be sterically retarded.[13]

The kinetics and products of the perchloric and sulphuric acid-catalysed rearrangement of N-nitroaniline to *o*- (mainly) and *p*- but no *m*-nitroaniline, and the effects of *o*- and *p*-deuteration on them, have been studied over a wide range of acidity. The reaction was of the first order in nitroamine, the plot of $\log k_1$ against $-H_0$ was linear, and the solvent isotope effect, k_{D_2O}/k_{H_2O}, was 3·3. The intramolecularity was confirmed by rearrangement in the presence of [^{15}N]-nitric and -nitrous acid; no evidence could be obtained for the presence of radicals. On the basis of these results, radical-ion and π-complex mechanisms were rejected in favour of initial isomerization of the nitroamine conjugate acid (21) to an N-nitrite, followed by Claisen-type migration via six-membered cyclic transition states, with final isomerization of nitrite esters to C-nitro compounds (Scheme 1). Whilst this mechanism satisfactorily explains the intramolecular nature of the *para*-migration, the initial and the final isomerization are, admittedly, less convincing.[14] It seems to us that the N-nitro $\rightarrow$ N-nitrite isomerization might occur more readily in the conjugate acid (22) than in (21), and this would also be less likely to lead to denitration, which is not observed.

When subjected to Fischer indole reaction conditions (zinc chloride in nitrobenzene), acetophenone 2,6-xylylhydrazone (23) gave the indole (24) and the tetrahydro-ψ-indolone (25). The key intermediate (26), which could undergo methyl migration followed by indole formation with retention of $N_{(1)}$ or ring closure on to the ring to give (25) with loss of $N_{(1)}$, had been proposed earlier,[15] and this is now confirmed by a determination of the fate of ^{15}N when this is $N_{(2)}$ of (23).[16]

p-Hydrazobiphenyl has now been shown, in disagreement with an earlier report, to undergo a benzidine rearrangement, as well as disproportionation, on treatment with acid. The lack of rearrangement had been used as support

[13] B. Miller, *Tetrahedron Letters*, 1965, 1733.
[14] D. V. Banthorpe, E. D. Hughes, and D. L. H. Williams, *J. Chem. Soc.*, 1964, 5349.
[15] R. B. Carlin and D. P. Carlson, *J. Am. Chem. Soc.*, 81, 4673 (1959).
[16] R. B. Carlin, A. J. Magistro and G. J. Mains, *J. Am. Chem. Soc.*, 86, 5300 (1964).

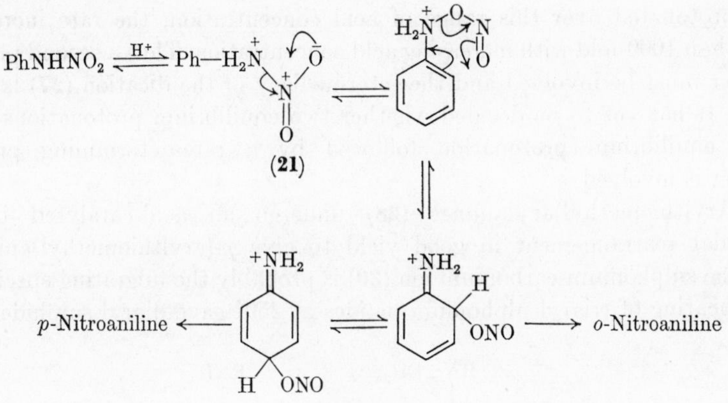

(21)

Scheme 1

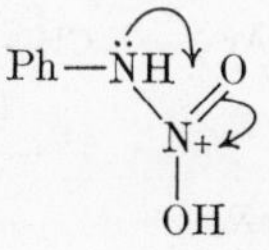

(22) 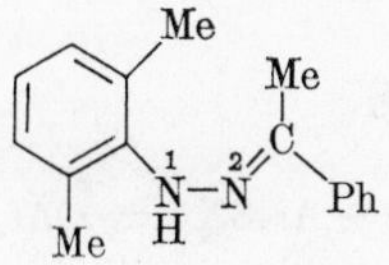   (23) 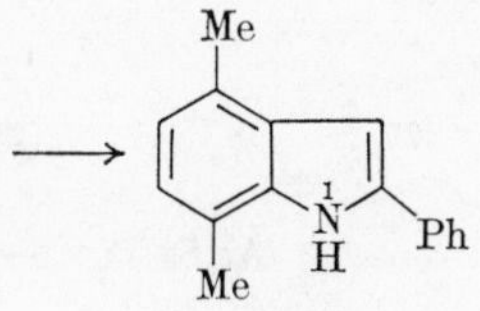   (24)

+

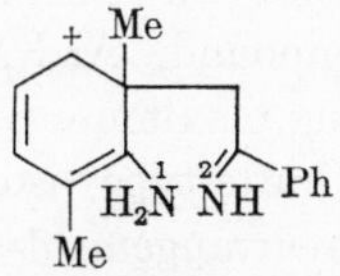

(26) 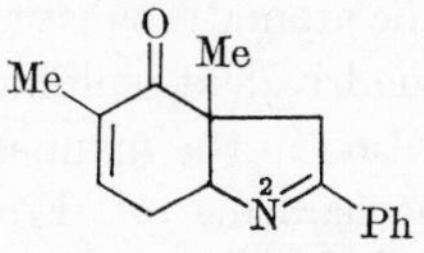   (25)

for the π-complex mechanism for the benzidine rearrangement since the *p*-phenyl groups were held to prevent formation of the π-complex; this is obviously no longer valid.[17] The rates of rearrangement of hydrazobenzene in aqueous ethanol up to 2·5 M in perchloric acid, measured by a potentially valuable new electrochemical technique, are comparable with conventional kinetic data.[18]

Buncel and Lawton[19] have summarized the present evidence on the mechanism of the acid-catalysed rearrangement of azoxybenzene to *p*-hydroxyazobenzene (the Wallach rearrangement) and have measured its rate in 60–95% sulphuric acid. Although azoxybenzene is almost completely

[17] H. J. Shine and J. P. Stanley, *Chem. Comm.*, **1965**, 294.
[18] W. M. Schwarz and I. Shain, *J. Phys. Chem.*, **69**, 30 (1965).
[19] E. Buncel and B. T. Lawton, *Can. J. Chem.*, **43**, 862 (1965).

monoprotonated over this range of acid concentration, the rate increases more than 1000-fold with increasing acid concentration. Thus a second proton transfer must be involved and the intermediacy of the dication (**27**) is proposed. It has yet to be decided whether two equilibrium protonations or a single equilibrium protonation followed by a rate-determining proton transfer is involved.

N-(Arylthiomethyl)arylamines (**28**) undergo an acid-catalysed inter-molecular rearrangement in good yield to give *p*-(arylthiomethyl)anilines (**29**); the sulphonium-carbonium ion (**30**) is probably the migrating species.[20] Brief heating of triarylsulphonium halides at 250° gave diaryl sulphide and

$$Ar_3S^+ \ X^- \longrightarrow Ar_3SX \longrightarrow Ar_2S + ArX \qquad (1)$$

aryl halide, quantitatively [reaction (1)]. The proportions of products formed from mixed sulphonium salts do not accord with uni- or bi-molecular nucleophilic aromatic substitution but can be explained in terms of formation of the quadrivalent sulphur compound, Ar_3SX, and its intramolecular decomposition in the manner giving maximum relief of steric strain.[21]

The mechanisms of Friedel–Crafts-type isomerizations catalysed by aluminium chloride including the rearrangement of methylated and brominated phenols,[22] the isomerization of halogenoethylbenzenes,[23] and of *o*- and *p*-dichlorobenzene,[24] have been further studied, as has methyl migration in the heptamethylbenzenonium ion.[25]

The mechanisms of the cadmium iodide-catalysed rearrangement of potassium phthalate to terephthalate,[26] the rearrangement of 5-bromo-4-methoxy-3-phenyltropone to 6-methoxybiphenyl-3-carboxylic acid,[27] the intramolecular proton and deuterium migration from $C_{(1)}$ to $C_{(3)}$ in indene,[28]

[20] G. F. Grillot and P. T. S. Lau, *J. Org. Chem.*, **30**, 28 (1965).
[21] G. H. Wiegand and W. E. McEwen, *Tetrahedron Letters*, **1965**, 2639.
[22] L. A. Fury and D. E. Pearson, *J. Org. Chem.*, **30**, 2301 (1965).
[23] G. A. Olah, J. C. Lapierre, and C. G. Carlson, *J. Org. Chem.*, **30**, 541 (1965).
[24] Y. G. Erykalov, A. A. Spryskov, and T. A. Rumyantseva, *Zh. Organ. Khim.*, **1**, 21 (1965); *Chem. Abs.*, **62**, 14,451 (1965).
[25] V. A. Koptyug, V. G. Shubin, and D. V. Korchagina, *Tetrahedron Letters*, **1965**, 1535.
[26] Y. Ogata and K. Nakajima, *Tetrahedron*, **21**, 2393 (1965).
[27] Y. Kitahara, I. Murata, and T. Muroi, *Bull. Chem. Soc. Japan.*, **38**, 1195 (1965).
[28] G. Bergson, *Acta Chem. Scand.*, **18**, 2003 (1964).

the *cis–trans* isomerization of 2-hydroxy-5-methylazobenzene in aqueous and ethanolic solution,[29] and the Orton rearrangement[30] have been investigated.

A mechanism has been proposed for the conversion of substituted acetic anhydrides into aldehydes and ketones by pyridine N-oxide according to reaction (2):[31]

$$(RR'CHCO)_2O + 2C_5H_5NO \longrightarrow RR'CO + RR'CHCO_2H + 2C_6H_5N + CO_2 \qquad (2)$$

The reaction of pyridine N-oxide and α-picoline N-oxide with p-nitrobenzene-sulphenyl chloride has also been studied.[32]

Cope and Related Rearrangements; Valence-bond Isomerization

A new thermal rearrangement of cycloheptatrienes (tropilidenes) has been revealed[33] by the pyrolysis (360°) of 3,7,7-trimethylcycloheptatriene (31), which gave the products shown; with recovered (31) the material balance

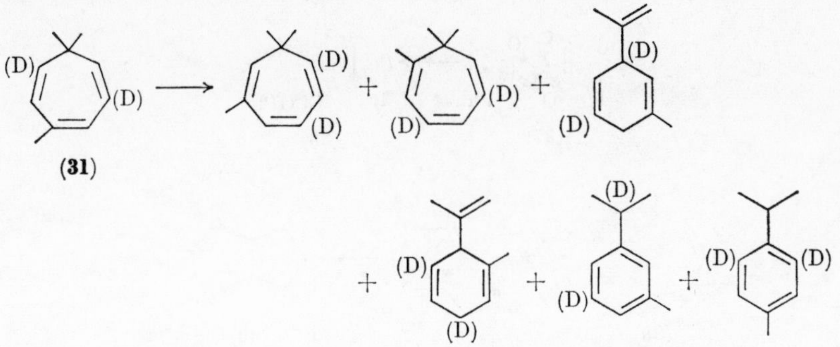

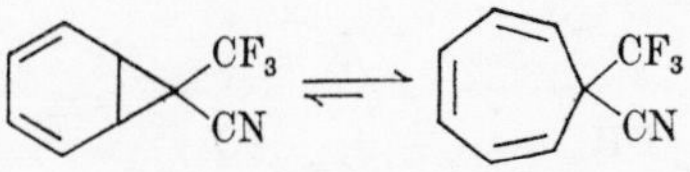

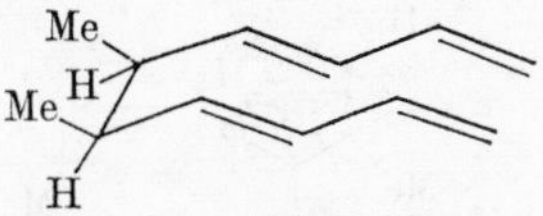

was essentially complete. That a skeletal rearrangement is involved, rather than methyl migration around the ring, in forming the other two tropilidenes, was shown by heating the 1,5-dideuterio-derivative of (31) which gave the same six products, all dideuterated in the positions shown. The rearrangements can be explained by initial Cope rearrangement with formation of a

[29] G. Wettermark, M. E. Langmuir, and D. G. Anderson, *J. Am. Chem. Soc.*, **87**, 476 (1965).

[30] J. M. W. Scott and J. G. Martin, *Can. J. Chem.*, **43**, 732 (1965).

[31] C. Rüchardt, S. Eichler, and O. Krätz, *Tetrahedron Letters*, **1965**, 233; T. Cohen, I. H. Song, and J. H. Fager, *ibid.*, **1965**, 237.

[32] S. Oae and K. Ikura, *Bull. Chem. Soc. Japan*, **38**, 58 (1965).

[33] J. A. Berson and M. R. Willcott, *J. Am. Chem. Soc.*, **87**, 2751, 2752 (1965).

1,6-bond, followed by stepwise migration of the isopropylidene group around the ring. The existence of a rapid norcaradiene–cycloheptatriene equilibrium at room temperature, for which there is much indirect evidence, has been directly observed (**32 ⇌ 33**) in the fluorine and proton NMR spectra of the adduct of cyanotrifluoromethylcarbene and benzene; the cycloheptatriene predominates.[34] The photoisomerization of 7-deuterio- and 7-phenyl-cycloheptatrienes, and transannular 1,5-hydrogen shifts in phenylcyclo-heptatrienes have been reported.[35] In a "vinylogous" Cope rearrangement, racemic and *meso*-5,6-dimethyl-1,3,7,9-decatetraene (**34**) at 375° both gave

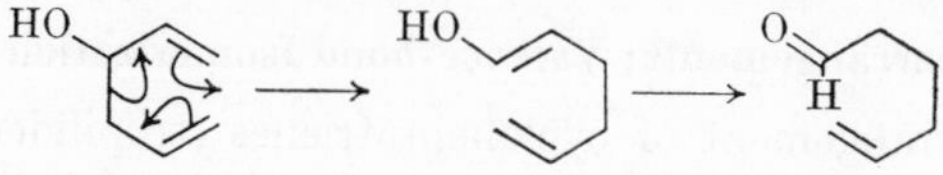

(**35**) (**36**)

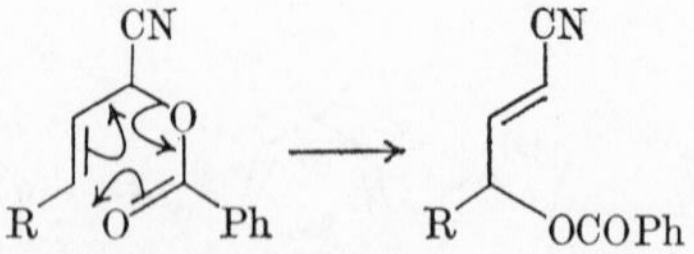

(**37**) (**38**)

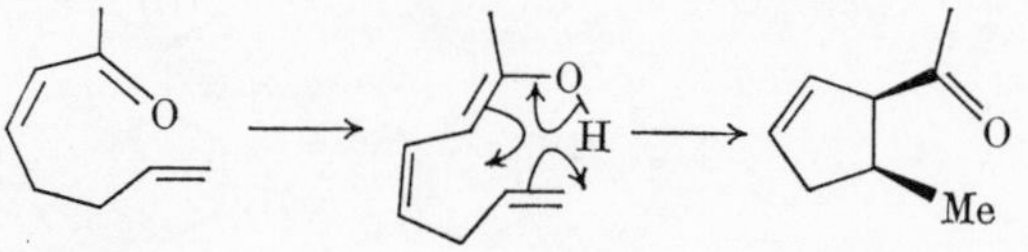

(**39**) (**40**)

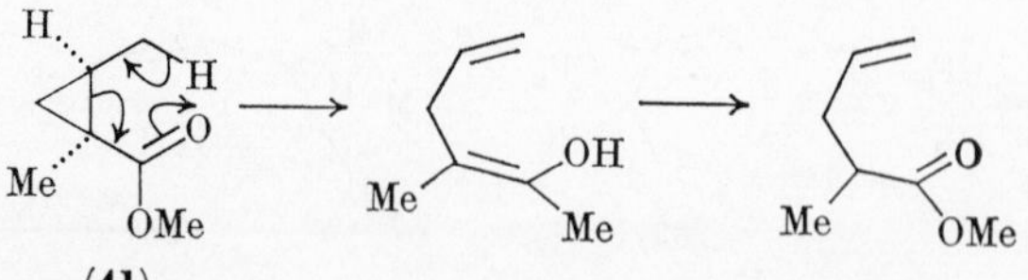

(**41**)

the same complex mixture of products, whose nature suggested initial homolysis into 1-methylpentadienyl radicals followed by recombination in various ways. The formation of C_{12} hydrocarbons possessing one deuterium atom per molecule in the pyrolysis of a mixture of (**34**) and its 1,10-dideuterio derivative supports this mechanism.[36]

The thermal rearrangement of 1,5-hexadien-3-ols to 5,6-unsaturated

[34] E. Ciganek, *J. Am. Chem. Soc.*, **87**, 1149 (1965).
[35] A. P. Ter Borg and H. Kloosterziel, *Rec. Trav. Chim.*, **84**, 241, 245 (1965).
[36] D. H. Gibson and R. Pettit, *J. Am. Chem. Soc.*, **87**, 2620 (1965).

ketones (the "oxy-Cope" rearrangement) was reported last year;[37a] this has now been demonstrated for the simplest case, (35) to (36).[37b] Cyanohydrin benzoates of α,β-unsaturated aldehydes (37) rearrange at high temperature (ca. 500°) to hydroxy-nitrile benzoates (38) as shown.[38] Pyrolysis of α,ϵ-dienones (39) at 250–300° yields the *cis*-cyclopentenes (40), stereospecifically, presumably as shown. At rather higher temperature the corresponding 6-enones react similarly to give the *cis*-cyclopentanes, almost quantitatively.[39] Methyl 1,2-dimethylcyclopropanecarboxylate (41) rearranges completely at 300° when the ring-methyls are *trans*, as shown, but is inert when they are *cis*.[40]

cis,cis,cis-1,3,5-Cyclononatriene (42a) has now been prepared and the rate of its smooth thermal conversion into *cis*-bicyclo[4.3.0]nona-2,4-diene (42b) measured.[41] *trans,cis,cis*-Cyclononatriene is formed from (42b) on photolysis, and heat transforms it into the *trans*-isomer of (42b).[42] The stereochemistry of these two ring closures, and also that of *trans,cis,cis*-2,4,6-octatriene to *cis*-5,6-dimethyl-1,3-cyclohexadiene, are in agreement with the Woodward and Hoffmann predictions[43] (see p. 127). Cope isomerization of the cyclic allene (43) readily gave 2,3-divinylcyclopentene,[44] and the diallene (44) gave 2,3-divinyl-1,3-cyclohexadiene (45).[45] Some novel isomerizations of cyclic and acyclic allenes have been described and rationalized on the basis of the stability of the (assumed) biradical intermediates.[46]

Many other valency-bond isomerizations have been reported, including conversion of the diazanorcaradiene (46) into the diazepine,[47] of the double Schiff's bases of 1,2-diaminocyclopropanes, e.g. (47), into 2,3-dihydro-1,4-diazepines,[48] of *cis*- (48) and *trans*-2-vinylcyclopropyl isocyanate into 3,6-dihydroazepin-2(3H)-one,[49] and of 1-ethyl-2,3-divinylaziridine (49) into 1-ethyl-4,5-dihydroazepine.[50] With these cyclopropanes and aziridines the *cis*-isomers understandably rearrange much faster than the *trans*. In fact, simple *cis*-1,2-divinylcyclopropanes usually rearrange so rapidly that they

[37a] J. A. Berson and M. Jones, *J. Am. Chem. Soc.*, **86**, 5017, 5019 (1964).
[37b] A. Viola and L. A. Levasseur, *J. Am. Chem. Soc.*, **87**, 1150, (1965).
[38] T. Holm, *Acta Chem. Scand.*, **19**, 242 (1965).
[39] J.-M. Conia and P. Le Perchec, *Tetrahedron Letters*, **1965**, 3305; F. Rouessac, P. Beslin, and J.-M. Conia, *ibid.*, **1965**, 3319; F. Rouessac and J.-M. Conia, *ibid.*, **1965**, 3313.
[40] D. E. McGreer, N. W. K. Chiu, and R. S. McDaniel, *Proc. Chem. Soc.*, **1964**, 415.
[41] D. S. Glass, J. W. H. Watthey, and S. Winstein, *Tetrahedron Letters*, **1965**, 377.
[42] E. Vogel, W. Grimme, and E. Dinné, *Tetrahedron Letters*, **1965**, 391.
[43] E. N. Marvell, G. Caple, and B. Schatz, *Tetrahedron Letters*, **1965**, 385.
[44] K. G. Untch and D. J. Martin, *J. Am. Chem. Soc.*, **87**, 4501 (1965).
[45] J. F. Harris, *Tetrahedron Letters*, **1965**, 1359.
[46] L. Skattebøl and S. Solomon, *J. Am. Chem. Soc.*, **87**, 4506 (1965).
[47] G. Maier, *Chem. Ber.*, **98**, 2446 (1965).
[48] H. A. Staab and F. Vogtle, *Chem. Ber.*, **98**, 2701 (1965), *Tetrahedron Letters*, **1965**, 51.
[49] E. Vogel, R. Erb, G. Lenz, and A. A. Bothner-By, *Ann. Chem.*, **682**, 1 (1965).
[50] E. L. Stogryn and S. J. Brois, *J. Org. Chem.*, **30**, 88 (1965).

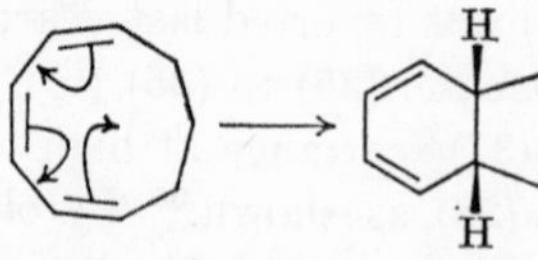

(42a)   (42b) 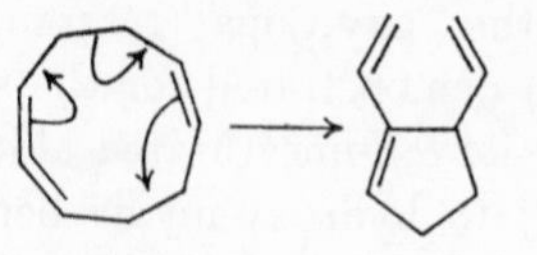   (43)

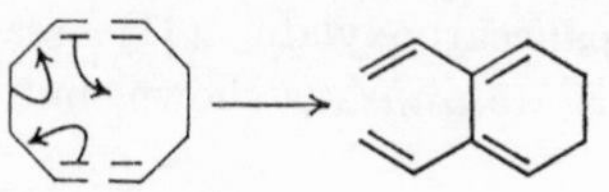

(44)   (45) 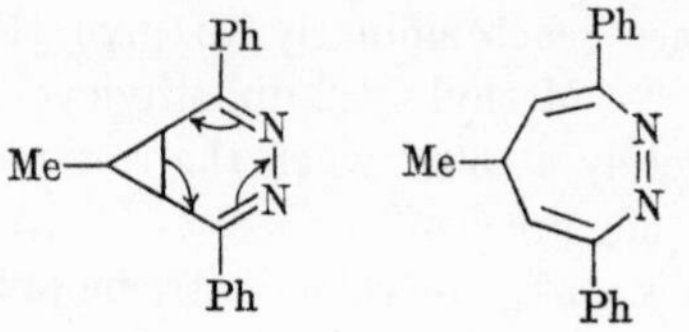   (46)

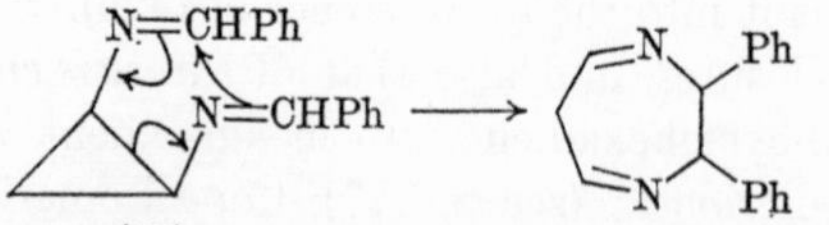

(47)

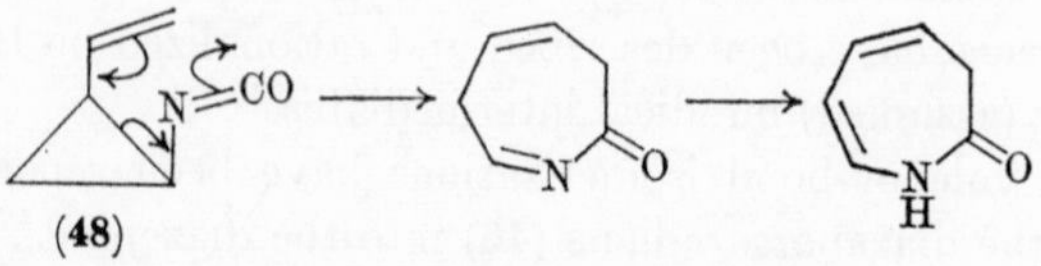

(48)

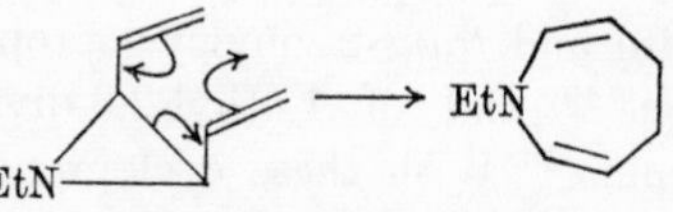

(49)

cannot be isolated; such a compound (50) has now been isolated (half-life of 1 day at 25°) and the rate of its rapid isomerization to (51) measured.[51] A detailed discussion of valence isomerization in cyclobutenes has appeared;[52] and the rapid isomerization of *cis*- and *trans*-1,2,3,4-tetraphenylcyclobutene to *cis,trans*- and *cis,cis*-1,2,3,4-tetraphenylbutadiene, respectively (conrotatory opening), and of the derived cyclobutenyl monoanion has been demonstrated.[53]

A useful review[54] has appeared of molecules which undergo fast, reversible valence-bond isomerization such that the average lifetime of the valence

[51] J. M. Brown, *Chem. Comm.*, **1965**, 226.
[52] R. Criegel, D. Seebach, R. E. Winter, H.-A. Brune, and B. Börretzen, *Chem. Ber.*, **98**, 2339 (1965).
[53] H. H. Freedman, G. A. Doorakian, and V. R. Sandel, *J. Am. Chem. Soc.*, **87**, 3019 (1965).
[54] G. Schröder, J. F. M. Oth, and R. Merényi, *Angew. Chem. Intern. Ed. Engl.*, **4**, 752 (1965).

isomers is less than about 100 seconds at $0°$; these can, at present, be recognized unambiguously only by NMR spectroscopy. The most notable of these molecules, bullvalene, has received more attention: the synthesis and properties of its phenyl[55] and other monosubstituted derivatives[56] have been described, and its degenerate Cope rearrangement, and that of its silver nitrate complex, have been studied more precisely by a spin-echo NMR technique.[57] The kinetics of interconversion of benzene oxide and oxepine (52),[58] and thermal isomerization of *cis*-1,4-hexadiene,[59] have also been reported.

(50)   (51)   (52)

A large number of vapour-phase thermal isomerizations of small-ring compounds have been investigated kinetically, mostly by Frey and his co-workers. These ring-openings and ring-expansions are usually homogeneous and first-order reactions, probably unimolecular; cyclic transition states with concerted bond-making and -breaking are usually indicated. These include the transformation of bicyclobutane into butadiene[60] (though a marked heterogeneous character is also reported for this reaction[61]), of 1,3- and 1,4-dimethylcyclobutene into *trans*-2- and *trans*-3-methyl-1,3-pentadiene, respectively,[62] and of *cis*- and *trans*-3,4-dimethyl-cyclobutene into *cis,trans*- and *trans,trans*-2,4-hexadiene, respectively,[63] and the isomerization of 3-methyl-butene[64] and 1-ethyl-cyclobutene.[65] Many vinylcyclopropane systems have been studied, including the isomerization of *cis*-1-methyl-2-vinylcyclo-propane to *cis*-1,4-hexadiene and of the *trans*-isomer which gave the same product at a higher temperature,[66] of (2-methylpropenyl)cyclopropane,[67]

[55] G. Schröder, *Angew. Chem. Intern. Ed. Engl.*, **4**, 695 (1965).
[56] J. F. M. Oth, R. Merényi, J. Nielsen, and G. Schröder, *Chem. Ber.*, **98**, 3385 (1965).
[57] A. Allerhand and H. S. Gutowsky, *J. Am. Chem. Soc.*, **87**, 4092 (1965).
[58] H. Günther, *Tetrahedron Letters*, **1965**, 4085; E. Vogel, W. A. Böll, and H. Günther, *ibid.*, **1965**, 609.
[59] W. R. Roth and J. König, *Ann. Chem.*, **688**, 28 (1965).
[60] H. M. Frey and I. D. R. Stevens, *Trans. Farad. Soc.*, **61**, 90 (1965).
[61] R. Srinivasan, A. A. Levi, and I. Haller, *J. Phys. Chem.*, **69**, 1775 (1965).
[62] H. M. Frey, D. C. Marshall, and R. F. Skinner, *Trans. Farad. Soc.*, **61**, 861 (1965).
[63] R. E. K. Winter, *Tetrahedron Letters*, **1965**, 1207.
[64] H. M. Frey and D. C. Marshall, *Trans. Farad. Soc.*, **61**, 1715 (1965).
[65] H. M. Frey and R. F. Skinner, *Trans. Farad. Soc.*, **61**, 1918 (1965).
[66] R. J. Ellis and H. M. Frey, *J. Chem. Soc.*, **1964**, 5578.
[67] C. S. Elliot and H. M. Frey, *J. Chem. Soc.*, **1965**, 345.

of 1-isopropenyl-1-methylcyclopropane,[68] of both isomers of 1-(p-methoxy-phenyl)propylcyclopropane,[69] and of 1,1-dicyclopropylethylene (53) which gives (54) and then (55) in consecutive steps.[70] Thermal rearrangement of *para*-substituted α-cyclopropylstyrenes (56), where X = H, Me, MeO, or i-Pr, all proceed at the same rate (though when X = F the rate is halved), showing the insensitivity of the reaction to polar effects.[71] Related kinetic

(53)   (54)   (55)   (56)

(57)   (58)   (59)   (60)

investigations have been made on the thermal isomerization and decomposition of 1-ethyl-[72] and 1,1-diethyl-cyclopropane,[73] the isomerization of trifluoromethyl- and trifluoroethyl-cyclopropane,[74] of cis-2-methyl-1,3-pentadiene,[75] the acid-catalysed isomerization of methylenecyclobutenes with alkyl side chains,[76] and in a further study, the conversion of cyclobutene into butadiene.[77] The reverse of the last reaction has now been observed for the first time in a thermal reaction; cyclization of cis,trans-1,3-cyclooctadiene (57) gave (58), and cis,trans-1,3-cyclononadiene (59) gave (60), both conrotatory processes.[78]

Radical Rearrangements

The most common type of free-radical rearrangement is a 1,2-aryl shift, and several reports involving such rearrangements have appeared during 1965. Thus Rüchardt has studied the decomposition of the β-aryl peresters (61).[79] The rate of thermal decomposition is the same with or without the

[68] C. S. Elliot and H. M. Frey, *J. Chem. Soc.*, **1965**, 4289.
[69] A. J. Berlin, L. P. Fisher, and A. D. Ketley, *Chem. Ind. (London)*, **1965**, 509.
[70] A. D. Ketley and J. L. McClanahan, *J. Org. Chem.*, **30**, 940 (1965).
[71] A. D. Ketley and J. L. McClanahan, *J. Org. Chem.*, **30**, 942 (1965).
[72] M. L. Halberstadt and J. P. Chesick, *J. Phys. Chem.*, **69**, 429 (1965).
[73] H. M. Frey and D. C. Marshall, *J. Chem. Soc.*, **1965**, 191.
[74] D. W. Placzek and B. S. Rabinovitch, *J. Phys. Chem.*, **69**, 2141 (1965).
[75] H. M. Frey and R. J. Ellis, *J. Chem. Soc.*, **1965**, 4770.
[76] R. Criegee, H. Hofmeister, and G. Bolz, *Chem. Ber.*, **98**, 2327 (1965).
[77] R. W. Carr and W. D. Walter, *J. Phys. Chem.*, **69**, 1073 (1965).
[78] K. M. Shumate, P. N. Neuman, and G. J. Fonken, *J. Am. Chem. Soc.*, **87**, 3996 (1965).
[79] C. Rüchardt and R. Hecht, *Chem. Ber.*, **98**, 2460, 2471 (1965).

aryl substituent, from which it may be concluded that there is no aryl participation in the primary peroxide decomposition. The relative yields of products from rearranged (63) and unrearranged (62) arylalkyl radicals on decomposition in different solvents depended on the efficiency of the solvent

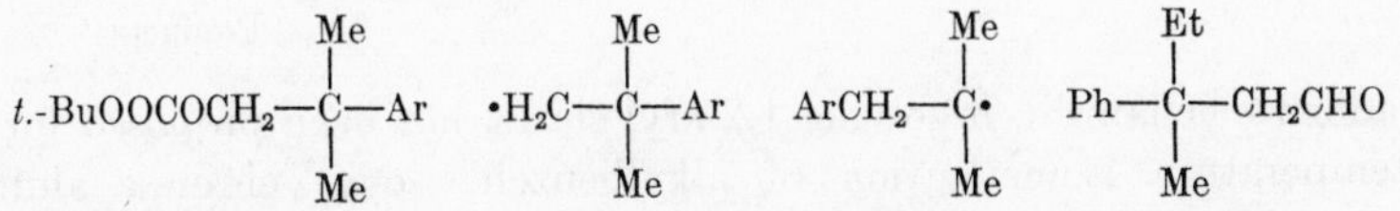

$$t.\text{-BuOOCOCH}_2-\overset{\overset{\text{Me}}{|}}{\underset{\underset{\text{Me}}{|}}{C}}-\text{Ar} \qquad \cdot\text{H}_2\text{C}-\overset{\overset{\text{Me}}{|}}{\underset{\underset{\text{Me}}{|}}{C}}-\text{Ar} \qquad \text{ArCH}_2-\overset{\overset{\text{Me}}{|}}{\underset{\underset{\text{Me}}{|}}{C}}\cdot \qquad \text{Ph}-\overset{\overset{\text{Et}}{|}}{\underset{\underset{\text{Me}}{|}}{C}}-\text{CH}_2\text{CHO}$$

(61)   (62)   (63)   (64)

as a source of hydrogen atoms. The relative rates of aryl migration were: $p\text{-MeOC}_6\text{H}_4$, 0.36; $p\text{-FC}_6\text{H}_4$, 0.40; $p\text{-MeC}_6\text{H}_4$, 0.72; $o\text{-ClC}_6\text{H}_4$, 0.95; C_6H_5, 1.00; $o\text{-MeOC}_6\text{H}_4$, 1.12; $m\text{-ClC}_6\text{H}_4$, 1.38; $m\text{-BrC}_6\text{H}_4$, 1.59; $p\text{-ClC}_6\text{H}_4$, 1.76; $p\text{-BrC}_6\text{H}_4$, 1.91; $p\text{-O}_2\text{NC}_6\text{H}_4$, 31; $p\text{-CNC}_6\text{H}_4$, 35. These are consistent with earlier results for radical rearrangements[80] and require appreciable charge separation in the transition state. Homolytic decarbonylation of ($-$)-3-methyl-3-phenylvaleraldehyde (64) gives a rearranged product which is at least 97.9% racemized, indicating an appreciable lifetime for the rearranged radical.[81] A similar investigation with a p-nitrophenyl- or p-cyanophenyl-substituted aldehyde would be of interest.

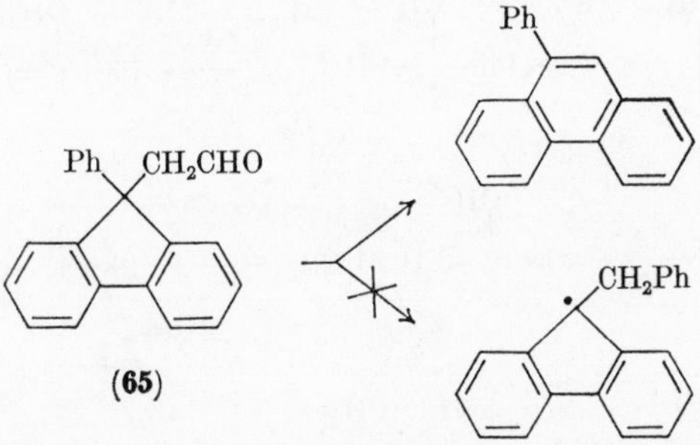

(65)

(66)

Peroxide-initiated decarbonylation of the aldehyde (65) gives the ring-expanded 9-phenylphenanthrene in good yield.[82] No evidence for the alternative phenyl migration (to give radical 66) could be found. Models indicate considerable steric opposition to such a process if it is to have the proper orbital overlap.

[80] C. Rüchardt and S. Eichler, *Chem. Ber.*, **95**, 1921 (1962).
[81] C. Rüchardt and H. Trautwein, *Chem. Ber.*, **98**, 2478 (1965).
[82] B. M. Vittimberga, *Tetrahedron Letters*, **1965**, 2383.

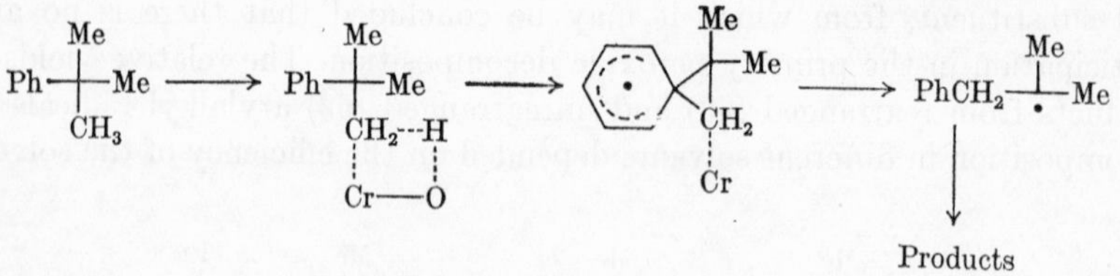

A radical mechanism, involving 1,2-aryl shifts, has been proposed for the high-temperature isomerization of alkylbenzenes over chromia–alumina catalysts, as shown above for $C_6H_5C(CH_3)_3$.[83]

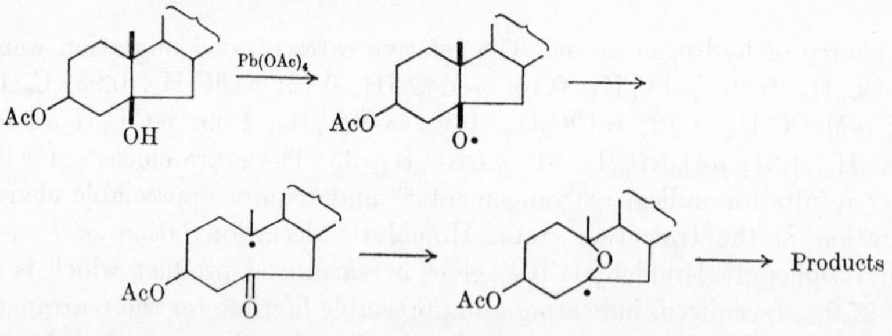

In an attempt to observe aryl migration from silicon to carbon, the peroxide-initiated decarbonylation of triphenylsilylacetaldehyde has been

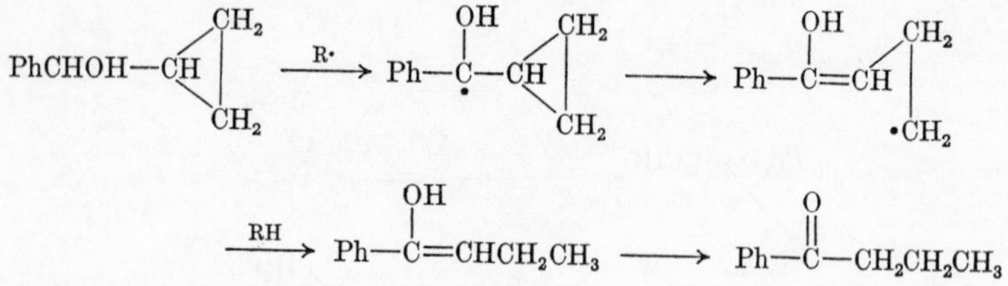

$$PhCHOH{-}CH\overset{CH_2}{\underset{CH_2}{\diagdown}} \xrightarrow{R\cdot} Ph{-}\overset{OH}{\underset{\cdot}{C}}{-}CH\overset{CH_2}{\underset{CH_2}{\diagdown}} \longrightarrow Ph{-}\overset{OH}{C}{=}CH\overset{CH_2}{\underset{\cdot CH_2}{\diagdown}}$$

$$\xrightarrow{RH} Ph{-}\overset{OH}{C}{=}CHCH_2CH_3 \longrightarrow Ph{-}\overset{O}{\overset{\|}{C}}{-}CH_2CH_2CH_3 \quad (2)$$

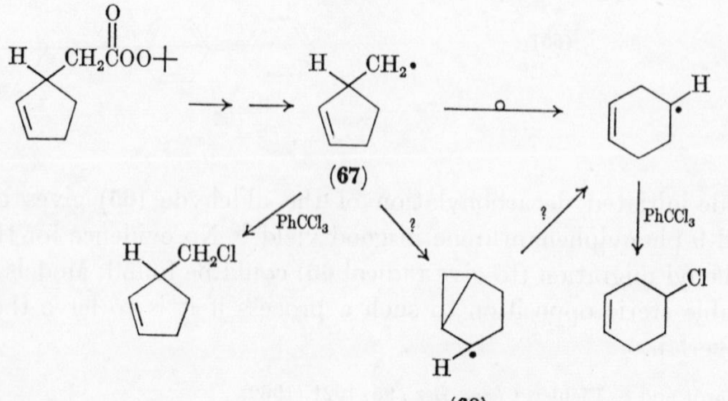

(67)

(68)

[83] H. Pines and C. T. Goetschel, *J. Am. Chem. Soc.*, **87**, 4207 (1965).

investigated.[84] No rearrangement could be detected, methyltriphenylsilane being the major product. The corresponding carbon-to-carbon rearrangement is complete.

1,2-Migration of an alkyl group from carbon to oxygen may be involved in a skeletal rearrangement observed[85] during oxidation of a tertiary alcohol by lead tetraacetate. Alternatively, the reaction could proceed by the dissociation addition mechanism shown opposite, though no ketonic products that might arise from the intermediate carbon radical could be detected.

A novel ring-opening rearrangement of a cyclopropylmethyl radical is shown in reaction (2)[86a]. What is formally a 1,2-vinyl migration in the ring-expansion of the cyclopentenylmethyl radical (67) may also involve a cyclopropylmethyl-type intermediate (68)[86b].

Radical rearrangements and transannular radical reactions are also mentioned in the Section dealing with radical additions (p. 114) and radical displacements (p. 202).

Other Rearrangements

Ring expansion of the dihydropyridine (69) to the dihydroazepine (72) in the presence of potassium cyanide has been investigated.[87] The reaction

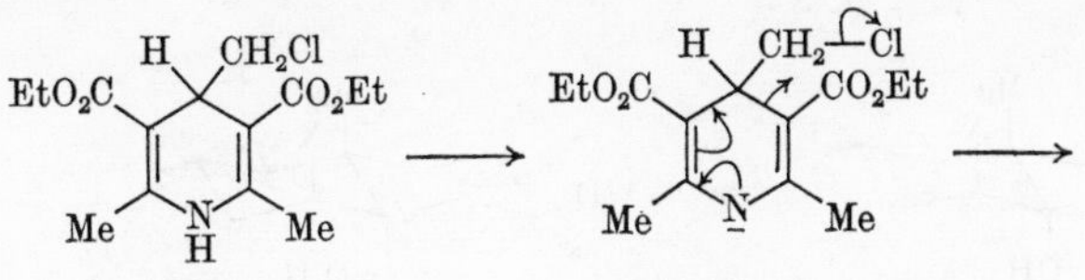

(69) (70)

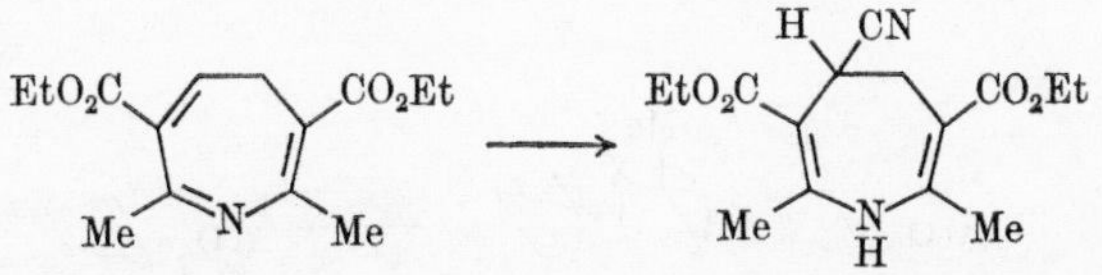

(71) (72)

shows general base catalysis and was thought to involve a rate-determining proton transfer to give the anion (70) which then rearranges to the azepine (71), this being followed by 1,4-addition of HCN; the last step was shown to occur rapidly under the reaction conditions.

[84] J. W. Wilt and O. Kolewe, *J. Am. Chem. Soc.*, **87**, 2071 (1965).

[85] D. Rosenthal, C. F. Lefler, and M. E. Wall, *Tetrahedron Letters*, **1965**, 3203.

[86a] D. C. Neckers, *Tetrahedron Letters*, **1965**, 1889.

[86b] L. H. Slaugh, *J. Am. Chem. Soc.*, **87**, 1522 (1965).

[87] P. J. Brignell, U. Eisner, and H. Williams, *J. Chem. Soc.*, **1965**, 4226.

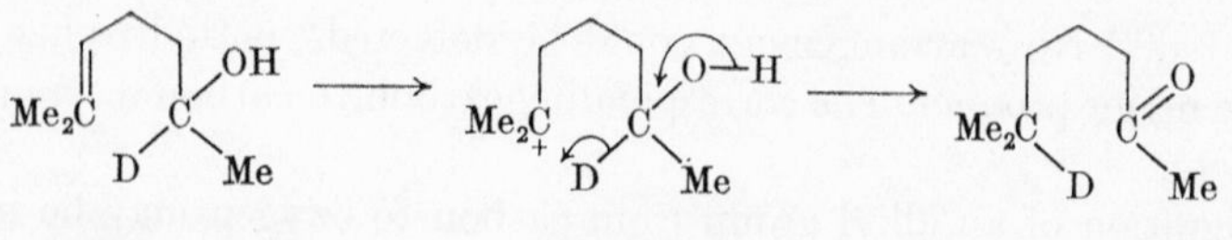

Scheme 3

Polyphosphoric acid-catalysed rearrangement of 6-methylhept-5-en-2-ol to 6-methylheptan-2-one has been shown[88] to occur by an intramolecular 1,5-hydride transfer by demonstration of the specific migration of deuterium as shown in Scheme 3. Similarly 6-phenylhept-5-en-2-ol of known absolute

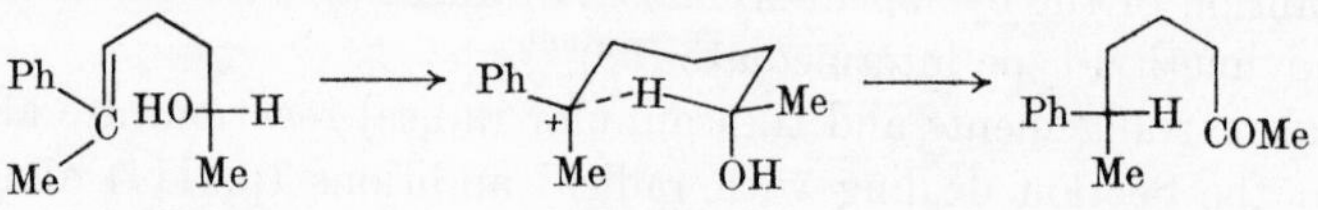

        (73)   **(74)**   **(75)**

configuration (73) gave 6-phenylheptan-2-one of configuration (75), preferentially; this is to be expected if the hydride transfer involves a chair conformation (74) in which the phenyl group and the methyl group occupy equatorial positions preferentially.

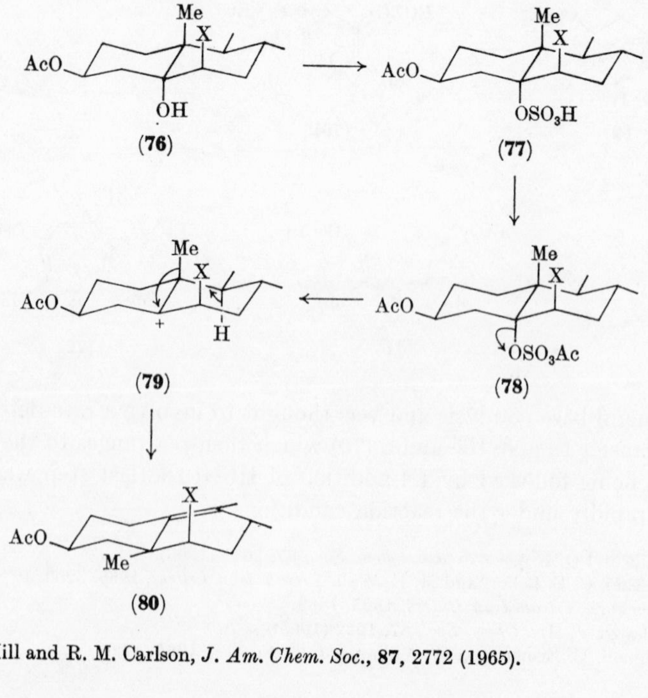

        (76)   **(77)**

        (79)   **(78)**

        (80)

[88] R. K. Hill and R. M. Carlson, *J. Am. Chem. Soc.*, **87**, 2772 (1965).

The Westphalen rearrangement of 3β,6β-diacetoxycholestan-5α-ol (partial structure **76**; X = AcO) with acetic anhydride–acetic acid containing sulphuric acid or potassium hydrogen sulphate gives (**80**; X = OAc).[89] Other acid catalysts (HClO₄, HCl, toluene-*p*-sulphonic acid, etc.) only convert (**76**) into its 5-acetate which does not undergo the rearrangement even in the presence of sulphuric acid and so is not an intermediate. It was also concluded that the reaction does not involve protonation of the hydroxyl group with water as leaving group, since the 5-methyl ether did not undergo the rearrangement. It was thought, therefore, that the hydrogen sulphate (**77**; X = AcO) was the most likely intermediate and it was shown that a similar steroidal alcohol, cholesterol, formed its hydrogen sulphate rapidly under the reaction conditions. Consistent with the rapid formation of the hydrogen sulphate was the demonstration that the rate of rearrangement depended only on the concentration of (**76**) or of sulphuric acid, whichever was in deficiency. The rate also depended on the concentration of acetic anhydride and this was interpreted as indicating the intervention of the acetyl sulphate (**78**; X = AcO) which would have a much better leaving

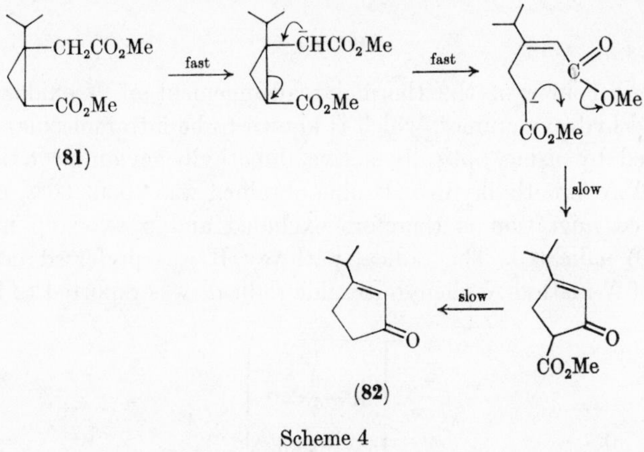

Scheme 4

group than the hydrogen sulphate (**77**). The rates of rearrangement of compound (**76**) with a series of groups X were correlated with the σ* constants to give a ρ* value of −4.8, indicative of a carbonium-ion mechanism. The mechanism was therefore written as shown.[89]

The alkoxide-catalysed conversion of α-thujadicarboxylic ester (**81**) into tanacetophorone (**82**) has been shown to proceed as shown in Scheme 4, by

[89] J. W. Blunt, A. Fischer, M. P. Hartshorn, F. W. Jones, D. N. Kirk, and S. W. Young, *Tetrahedron*, **21**, 1567 (1965).

Crombie and Mitchard, who followed the formation and disappearance of the intermediates by gas chromatography.[90]

Support for the mechanism of Scheme 5 for the base-promoted rearrangement of N,N-dichloro-1,2-diphenylethylamine to desylamine has been provided by Oae and Furukawa[91] who showed that the chloroamine labelled at $C_{(1)}$ with ^{14}C yielded desylamine hydrochloride labelled exclusively at the carbonyl-carbon atom.

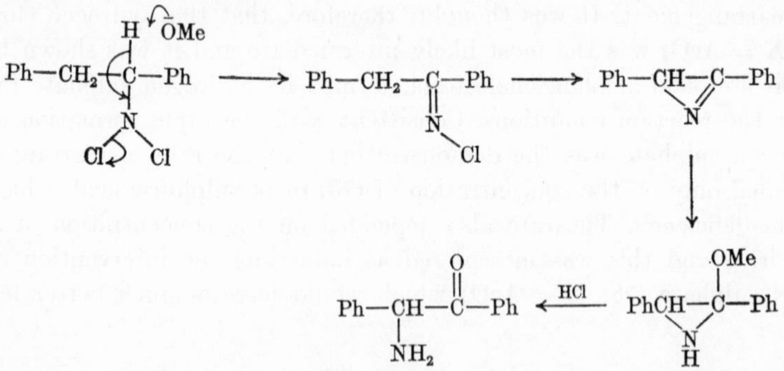

Scheme 5

The steric course of the thermal rearrangement of N-oxides to *ortho*-substituted hydroxylamines, which is known to be intramolecular, has been investigated by using optically active dimethylbenzylamine oxide.[92] The O-benzyl-N,N-dimethylhydroxylamine obtained was about 60% racemized. A concerted migration is therefore excluded and a two-step mechanism (Scheme 6) indicated. The radical pathway B was preferred because the presence of N-methyl-N-phenylnitroxide radicals was reported to have been

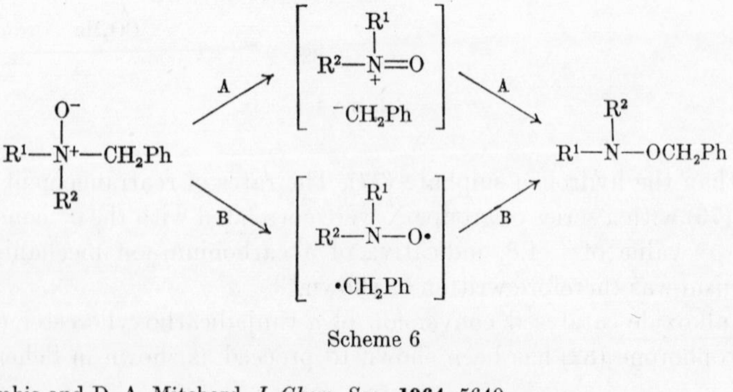

Scheme 6

[90] L. Crombie and D. A. Mitchard, *J. Chem. Soc.*, **1964**, 5640.
[91] S. Oae and N. Furukawa, *Bull. Chem. Soc. Japan*, **38**, 62 (1965).
[92] U. Schöllkopf and H. Schafer, *Ann. Chem.*, **683**, 42 (1965).

demonstrated by ESR in the rearrangement of *N*-benzyl-*N*-methylaniline oxide, and because the influence of substituents in the benzyl residue was less than that expected for an ionic mechanism.

It has been shown that the Beckmann and the Schmidt rearrangement of the ketone (83) proceed with predominant alkyl migration, while the rearrangements of ketone (84) proceed with predominant aryl migration.[93] Since it was considered that the oxime and iminodiazonium ion of ketone

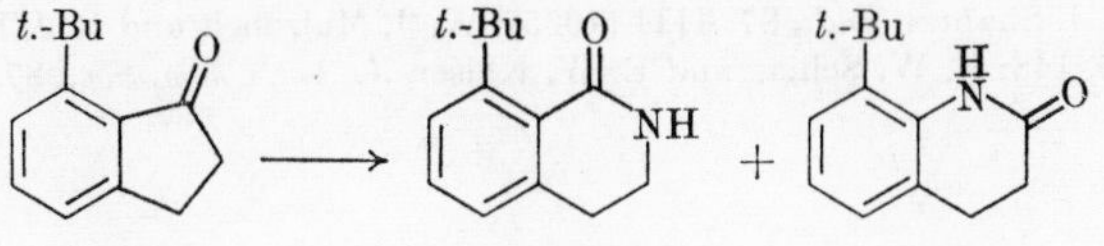

(83) could not exist in the *syn*-aryl forms it was concluded that the predominant reaction must be a *cis*-rearrangement, and this was thought to involve a univalent nitrogen intermediate.

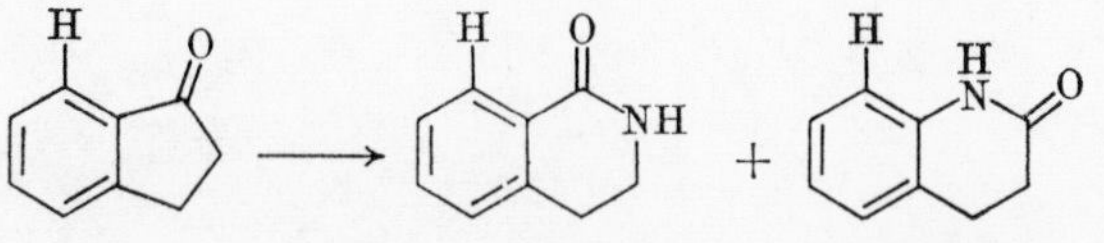

An extensive investigation of solvent effects in the diaxial → diequatorial rearrangement of 2β,3α-dibromocholestane and the analogous bromohydrin toluene-*p*-sulphonates and anisates has been reported.[94]

Other rearrangements that have been studied include those of α-hydroxy-*N*-phenyl-amines,[95] α-amino methyl ketones,[96] 2-halogeno-1-methyl-1-tetralols,[97] bromodecahydroisoquinolines,[98] and aziridines,[99] and the Amadori,[100] Favorski,[101] and Rober rearrangement (of hydrazones to

[93] P. T. Lansbury and N. R. Mancuso, *Tetrahedron Letters*, 1965, 2445.
[94] J. F. King and R. G. Pews, *Can. J. Chem.*, 43, 847 (1965).
[95] C. L. Stevens, A. Thuillier, and F. A. Daniher, *J. Org. Chem.*, 30, 2962 (1965).
[96] C. L. Stevens, I. L. Klundt, M. E. Munk, and M. D. Pillai, *J. Org. Chem.*, 30, 2967 (1965).
[97] J. K. Stille and C. N. Wu, *J. Org. Chem.*, 30, 1222 (1965).
[98] C. A. Grob and R. A. Wohl, *Helv. Chim. Acta*, 48, 1610 (1965).
[99] A. B. Turner, H. W. Heine, J. Irving, and J. B. Bush, *J. Am. Chem. Soc.*, 87, 1050 (1965).
[100] D. Palm and H. Simon, *Z. Naturforsch.*, 20b, 32 (1965).
[101] N. J. Turro and W. B. Hammond, *J. Am. Chem. Soc.*, 87, 3258 (1965); H. O. House and G. A. Frank, *J. Org. Chem.*, 30, 2948 (1965); M. Charpentier-Morize, M. Mayer, and B. Tchoubar, *Bull. Soc. Chim. France*, 1965, 529; C. Rappe, T. Nilson, and G.-B. Carlsson, and K. Andersson, *Arkiv Kemi*, 24, 95, 105, 303, 315 (1965); C. Rappe and R. Adestrom, *Acta Chem. Scand.*, 19, 383 (1965); C. Rappe and R. Adestrom, *ibid.*, 19, 273 (1965); C. Rappe, *ibid.*, 19, 270 (1965); J. Wolinsky and D. Chan, *J. Org. Chem.*, 30, 41 (1965).

amidines).[102] Many investigations of the isomerization of olefins have also been reported.[103]

[102] I. I. Grandberg, Y. A. Naumov, and A. N. Kost, *Zh. Organ. Khim.*, **1**, 805 (1965); *Chem. Abs.*, **63**, 6804 (1965).

[103] A. Maccoll and R. A. Ross, *J. Am. Chem. Soc.*, **87**, 1169, 4997 (1965); K. W. Egger and S. W. Benson, *ibid.*, **87**, 3311, 3314 (1965); G. Geiseler, P. Herrman, and G. Kurzel, *Chem. Ber.*, **98**, 1695 (1965); D. C. Dittmer and A. F. Marcantonio, *J. Org. Chem.*, **29**, 3473 (1964); J. K. Nicholson and B. L. Shaw, *Tetrahedron Letters*, **1965**, 3533; J. Herling, J. Shabtai, and E. Gil-Av, *J. Am. Chem. Soc.*, **87**, 4107 (1965); P. Coppens, E. Gil-Av, J. Herling, and J. Shabtai, *ibid.*, **87**, 4111 (1965); W. J. Muizebelt and R. J. F. Nivard, *Chem. Comm.*, **1965**, 148; E. W. Schlag and E. W. Kaiser, *J. Am. Chem. Soc.*, **87**, 1171 (1965).

CHAPTER 9

Radical Reactions

Topics of current interest in free-radical chemistry,[1a] in particular the concept of bridged free radicals,[1b] were discussed in lectures given at a symposium in Cork, Ireland, in 1964. Also, a multitude of research papers dealing with every aspect of free-radical reactions has appeared during the year. Extensive use has been made of electron spin resonance (ESR) for the study of both stable and short-lived radicals. The present chapter is divided into sections dealing with radical formation, radical reactions, and ESR studies, though it is impossible to avoid extensive overlap. In some instances classification may appear quite arbitrary. Aspects of free-radical chemistry not covered here are dealt with in Chapters 5, 7 and 8.

Radical-forming Reactions

Recent work on the decomposition of azo-compounds includes a study of deuterium and ^{13}C isotope effects in the decomposition of α-methylbenzyl-azomethane (1), which shows that radical formation proceeds by a two-step

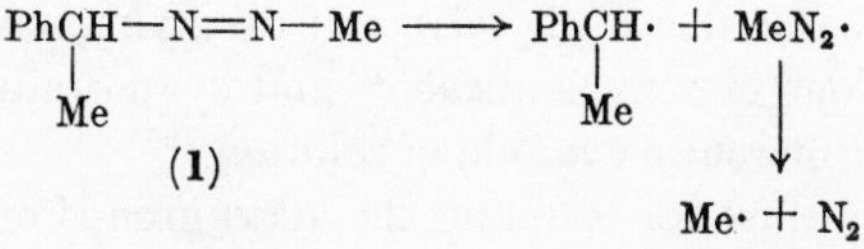

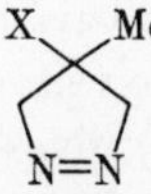

mechanism.[2a] Methyl radicals are formed in the second, kinetically unimportant, step.

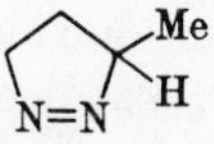

The gas-phase decomposition of 1-pyrazolines was examined by Crawford

---

[1] "Organic Reaction Mechanisms", Chem. Soc. Special Publn. No. 19, 1965, (a) W. A. Waters, p. 71, (b) P. S. Skell, p. 131.
[2a] S. Seltzer and F. T. Dunne, J. Am. Chem. Soc., 87, 2628 (1965).

and his co-workers,[2b] who produced compelling evidence for the formation of an intermediate, presumed to be a 1,3-biradical, in the formation of cyclopropanes in this reaction. Thus a deuterated pyrazoline (**2B**) decomposes at the same first-order rate as its protio analogue (**2A**), but gives different proportions of products (methylcyclopropane and isobutene); this may be explained only in terms of an intermediate whose transformation, but not formation, depends on the deuterium substituent at $C_{(4)}$. The possibility that this intermediate is not a biradical, but a hot (or triplet—see p. 223) cyclopropane appears to be ruled out by the formation of different products (methylcyclopropane and 1- and 2-butene) from the isomeric pyrazoline (**3**). The geometries of the transition state for the decomposition, and of the biradicals, have been discussed.

Waters' School has investigated the possibility of generating benzoyl radicals in a non-chain process.[3] Photolysis of azodibenzoyl, while giving complex products under all conditions studied, appeared to be such a reaction.

Azobisisobutanol   diacetate   [2,2′-azodi-(2-methylpropyl   acetate)], $(=N—CMe_2CH_2OCOMe)_2$, has been proposed as a high-temperature (170–200°) radical initiator.[4]

The kinetics of decomposition of dimethylmercury at 300° have been examined,[5] and Bass has employed the pyrolysis of dibenzylmercury as a source of benzyl radicals.[6] Bis(trimethylsilyl)mercury provides a convenient source of trimethylsilyl radicals on either photolysis or pyrolysis;[7a] for example, pyrolysis in toluene solution under pressure at 195° gives bibenzyl, benzyltrimethylsilane, and trimethylsilane, in addition to mercury and hexamethyldisilane. Gas-phase photolysis of divinylmercury gives products attributed to reactions of vinyl radicals,[7b] and cyano radicals are produced during photolysis of mercuric cyanide in solution.[7c]

An ingenious procedure for coupling the alkyl groups of two molecules of alcohol has been developed,[8a] depending on the variable valency of titanium. Extrusion of titanium dioxide from dibenzyloxytitanium (readily prepared from benzyl alcohol, and decomposed in situ) gives bibenzyl in fair yield. The reaction presumably involves coupling of free benzyl radicals and has been extended to the coupling of other resonance-stabilized radicals.

[2b] R. J. Crawford, R. J. Dummel, and A. Mishra, *J. Am. Chem. Soc.*, **87**, 3023 (1965); R. J. Crawford and A. Mishra, *ibid.*, **87**, 3768 (1965).

[3] D. Mackay, U. F. Marx, and W. A. Waters, *J. Chem. Soc.*, **1964**, 4793.

[4] G. A. Mortimer, *J. Org. Chem.*, **30**, 1632 (1965).

[5] K. B. Yerrick and M. E. Russell, *J. Phys. Chem.*, **68**, 3752 (1964).

[6] K. C. Bass, *J. Organometal. Chem.*, **4**, 1, 92 (1965); K. C. Bass and P. Nababsing, *Chem. Ind.* (*London*), **1965**, 307, 1599; *J. Chem. Soc.*, **1965**, 4396.

[7a] C. Eaborn, R. A. Jackson, and R. W. Walsingham, *Chem. Comm.*, **1965**, 300.

[7b] A. G. Sherwood and H. E. Gunning, *J. Phys. Chem.*, **69**, 2323 (1965).

[7c] K. Yoshida and S. Tsutsumi, *Tetrahedron Letters*, **1965**, 2417.

[8a] E. E. van Tamelen and M. A. Schwartz, *J. Am. Chem. Soc.*, **87**, 3277 (1965).

Products of photolysis of mixed benzoic dithiocarbamic anhydrides (4) in various solvents are best rationalized in terms of radical reactions involving initial cleavage of the sulphur–carbonyl bond.[8b]

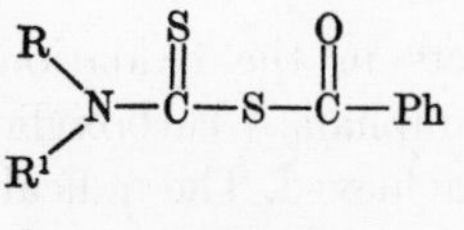

(4) 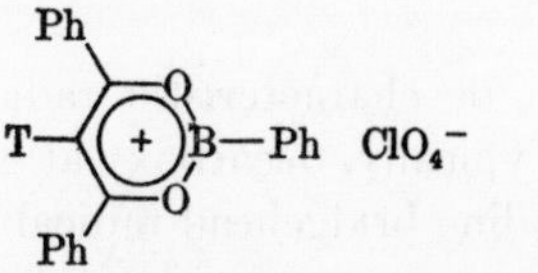   (5)

It has been suggested that formation of tritiobenzene from the perchlorate (5) may involve cleavage of phenyl radicals from boron.[9a] There is good evidence that phenyl radicals are also produced on photolysis of triphenyl-phosphine and tetraphenylphosphonium salts.[9b] The formation of some symmetrical diaryls from arylphosphines is, however, indicative that part of the reaction may not involve *free* radicals.

The reaction of lead tetraacetate with several simple benzene derivatives in acetic acid has been described by Norman, and a diversity of reaction paths defined.[10] With anisole, at 80°, electrophilic attack gives o- and p-acetoxy-anisole, and the lead compound (6), which may be reconverted into anisole by acetic acid at 80°. In benzene at 80° homolysis of the product (6) gives p-methoxyphenyl radicals which produce 4-methoxybiphenyl by substitution. With benzene in refluxing acetic acid, some toluene was formed, and under similar conditions nitrobenzene gave the isomeric nitrotoluenes in proportions characteristic of radical attack. Toluene at 80° gives benzyl acetate as a major product; as there is no evidence for radical formation at this temperature, some type of concerted mechanism, which may be analogous to the allylic acetoxylation of olefins, seems probable.

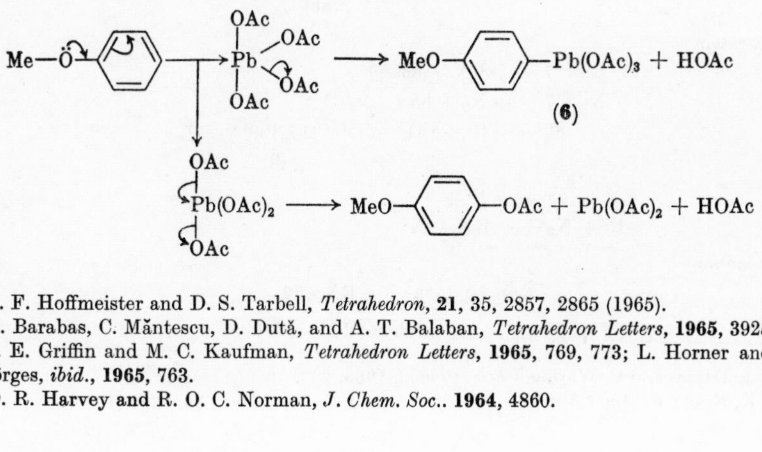

(6)

[8b] E. F. Hoffmeister and D. S. Tarbell, *Tetrahedron*, **21**, 35, 2857, 2865 (1965).
[9a] A. Barabas, C. Măntescu, D. Duță, and A. T. Balaban, *Tetrahedron Letters*, **1965**, 3925.
[9b] C. E. Griffin and M. C. Kaufman, *Tetrahedron Letters*, **1965**, 769, 773; L. Horner and J. Dörges, *ibid.*, **1965**, 763.
[10] D. R. Harvey and R. O. C. Norman, *J. Chem. Soc.*. **1964**, 4860.

Oxidative decarboxylation of carboxylic acids by lead tetraacetate appears to involve radical intermediates [cf. reaction (1)]. This is shown by the

$$RCO_2H \xrightarrow{Pb(OAc)_4} R\cdot + CO_2 \qquad (1)$$

formation of characteristic radical products in the oxidation of several acids.[11] Typically, decarboxylation of apocamphane-1-carboxylic acid to the corresponding bridgehead radical is readily achieved. The radical attacks the solvent, benzene, to give 1-phenylapocamphane in good yield. Decarboxylation would not be expected to be easy if the intermediate were a carbonium ion.

A radical-chain mechanism (Scheme 1) has been advanced for a novel halogenodecarboxylation procedure, in which lead tetraacetate and the carboxylic acid (RCO_2H) in benzene are treated with lithium chloride or bromide.[12] Essentially, this reaction involves coordination of halide to the lead atom, formation of $R\cdot$, and then ligand transfer of halide from lead to $R\cdot$. Halide coordination renders the lead acylate relatively labile, and the reaction is rapid at 80°. Formation of free halogen is probably incidental to the main reaction.

$$RCO_2H + (MeCO_2)_4Pb \rightleftharpoons (MeCO_2)_3PbOCOR + MeCO_2H$$

$$(MeCO_2)_3PbOCOR + n\,X^- \rightleftharpoons [(MeCO_2)_3Pb^{IV}OCORX_n]^{n-}$$

$$(7)$$

$$(n = 1 \text{ or } 2)$$

Initiation

$$7 \longrightarrow R\cdot + CO_2 + [(MeCO_2)_3Pb^{III}X_n]^{n-}$$

$$(8a)$$

$$7 \longrightarrow X\cdot + [(MeCO_2)_3Pb^{III}CO_2RX_{n-1}]^{n-}$$

$$(8b)$$

Propagation

$$R\cdot + 7 \longrightarrow RX + 8b$$

$$X\cdot + 7 \longrightarrow X_2 + 8b$$

$$8b \longrightarrow R\cdot + CO_2 + [(MeCO_2)_3Pb^{II}X_{n-1}]^{n-}$$

$$(9)$$

$$8a \longrightarrow X\cdot + 9$$

$$R\cdot + X_2 \longrightarrow RX + X\cdot$$

Termination

$$R\cdot + 8a \longrightarrow RX + 9$$

Scheme 1

[11] D. I. Davies and C. Waring, *Chem. Comm.*, **1965**, 263.
[12] J. K. Kochi, *J. Am. Chem. Soc.*, **87**, 2500 (1965); *J. Org. Chem.*, **30**, 3265 (1965).

It has been found that radical reactions of lead tetraacetate may be initiated by photolysis as an alternative to pyrolysis.[13]

Walling and his colleagues have examined the effect of pressure on the decomposition of peroxides.[14] The small rate enhancement in the decomposition of benzoyl peroxide in chloroform at high pressure is attributed to increased induced decomposition. A product study is consistent with stabilization of PhCOO· (with respect to Ph· + CO_2) at high pressure, and the induced decomposition is held to involve these (PhCOO·) radicals.

The concerted nature of the decomposition of the perester $PhCD_2CO·$ OOBu-t in benzene, is indicated by the deuterium isotope effect (k_H/k_D = 1.17 per D atom) on the first-order decomposition rate.[15] [$^2H_6$]Acetyl peroxide shows no similar effect, and it is concluded that there is little or no methyl-radical character in the transition state for the decomposition of this compound. The isotope effect arises from rehybridization in formation of the carbon radical.

A single-electron transfer mechanism has been preferred for the bimolecular reaction between benzoyl peroxide and dialkylarylamines.[16] Similarly one-electron transfer is written (2) for the decomposition of benzoyl peroxide by iodide in preference to hypoiodite formation.[17a] On the other hand, Kochi

$$(PhCOO)_2 + I^- \longrightarrow PhCOO^- + PhCOO· + I· \tag{2}$$

considers that the decomposition of benzoyl peroxide in acetic acid by lithium chloride does involve formation of benzoyl hypochlorite.[17b] This, as well as molecular chlorine formed from it, can participate in nuclear halogenation of an aromatic substrate.

The decomposition of peroxybenzoyl aryl carbonates (e.g. **10**) is a first-order decomposition which gives benzoyloxy and aryloxy radicals.[18] The latter telomerize, whilst the former appear largely as benzoic acid. The decomposition of aliphatic diacyl peroxides (isobutyryl and cyclohexane-carbonyl) in the presence of 1,1,3,3-bis(biphenylene)-2-phenylallyl as a radical scavenger has been examined.[19] The scavenger did not affect the rate of decomposition but drastically altered the yields of some of the products. A reaction of unknown mechanism was proposed in which short-lived acyloxy radicals were scavenged, and subsequently converted into carboxylic acids

[13] K. Heusler, H. Labhart, and H. Loeliger, *Tetrahedron Letters*, **1965**, 2847.
[14] C. Walling, H. N. Moulden, J. H. Waters, and R. C. Neuman, *J. Am. Chem. Soc.*, **87**, 518 (1965).
[15] T. W. Koenig and W. D. Brewer, *Tetrahedron Letters*, **1965**, 2773.
[16] F. Hrabák and M. Vacek, *Collection Czech. Chem. Commun.*, **30**, 573 (1965).
[17a] G. Tsuchihashi, S. Miyajima, T. Otsu, and O. Simamura, *Tetrahedron*, **21**, 1039 (1965).
[17b] J. K. Kochi, B. M. Graybill, and M. Kurz, *J. Am. Chem. Soc.*, **86**, 5257 (1964); J. K. Kochi and R. V. Subramanian, *ibid.*, **87**, 1508 (1965).
[18] V. A. Dodonov and W. A. Waters, *J. Chem. Soc.*, **1965**, 2459.
[19] R. C. Lamb, J. G. Pacifici, and P. W. Ayers, *J. Am. Chem. Soc.*, **87**, 3928 (1965).

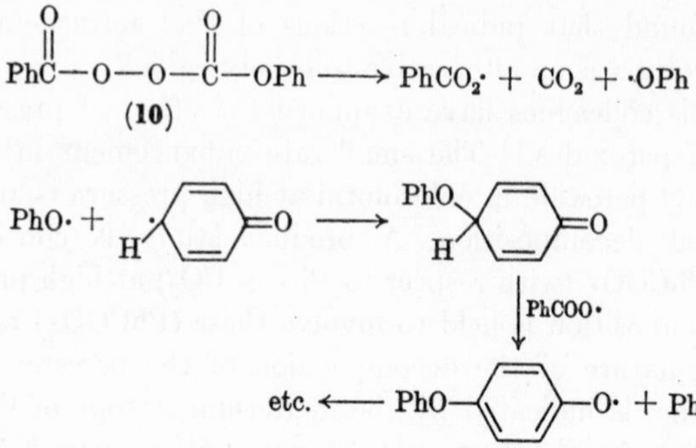

$$PhC(=O)-O-O-C(=O)-OPh \longrightarrow PhCO_2\cdot + CO_2 + \cdot OPh$$
(10)

$$PhO\cdot + \quad \overset{\cdot}{\underset{H}{\bigcirc}}=O \longrightarrow \overset{PhO}{\underset{H}{\bigcirc}}=O$$

$$\downarrow PhCOO\cdot$$

$$etc. \longleftarrow PhO-\bigcirc-O\cdot + PhCO_2H$$

(which were obtained in good yield). In competition with the radical decomposition is a non-radical carboxyl-inversion reaction (to acyl alkyl carbonate). Evidence for this, now familiar, process was found in the negative activation volume for the decomposition. Solvent effects were also studied, and these indicated that the transition states of both processes, particularly the rearrangement, have appreciable polar character. These solvent effects were similar to, but smaller than, those observed for an ionic reaction between pyridine and t-butyl peroxyformate.[20] The similarity is considered to imply a contribution from structure (11) to the transition state for decomposition.

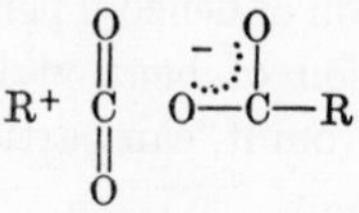

$$R^+ \quad \overset{O}{\underset{O}{\overset{||}{C}}} \quad O\overset{-}{\cdots}\overset{O}{\overset{||}{C}}-R$$

(11)

In an extensive study of the initiation of polymerization of styrene by t-butyl hydroperoxide, Walling and Heaton[21] have unravelled some of the complexities of this system. The peroxide disappears by spontaneous and induced radical decomposition, but to a much greater extent (ca. 90%) by a non-radical reaction which gives styrene oxide. The decomposition is further complicated by association of hydroperoxide molecules by hydrogen-bonding.

Observations on the formation of alkoxy radicals from organic nitrites include pyrolyses of the esters (12) and (13), which both give the corresponding alcohols as major product,[22] and gas-phase pyrolysis of isopropyl nitrite.[23] Decomposition of the resulting isopropoxy radical to methyl radical and acetaldehyde is pressure dependent.[24] Photolysis of steroid 17-nitrites gives

[20] R. E. Pincock, J. Am. Chem. Soc., 87, 1274 (1965).
[21] C. Walling and L. Heaton, J. Am. Chem. Soc., 87, 38 (1965).
[22] P. Gray and M. J. Pearson, J. Chem. Soc., 1964, 5725, 5734.
[23] J. M. Ferguson and L. Phillips, J. Chem. Soc., 1965, 4416.
[24] R. A. Livermore and L. Phillips, Chem. Comm., 1965, 253.

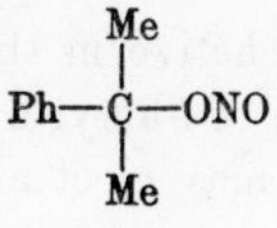

$$\underset{\textbf{(12)}}{\overset{\displaystyle \text{Me}}{\underset{\displaystyle \text{Me}}{\text{Ph}-\text{C}-\text{ONO}}}} \qquad\qquad \underset{\textbf{(13)}}{\overset{\displaystyle \text{Me}}{\underset{\displaystyle \text{Me}}{\text{PhCH}_2-\text{C}-\text{ONO}}}}$$

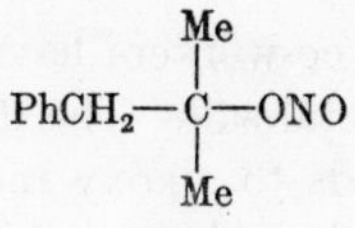

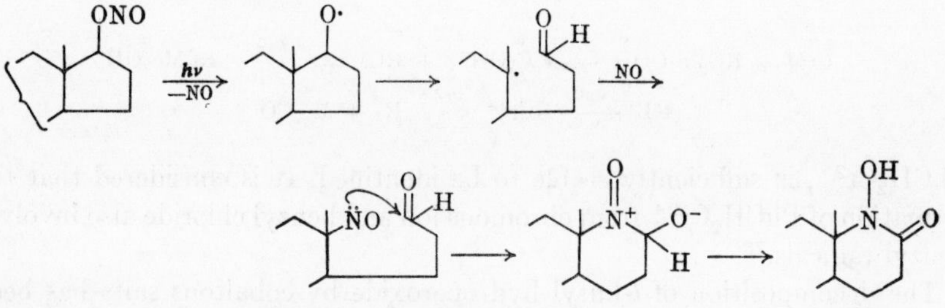

hydroxamic acids (15). The initially formed alkoxy radical cleaves to give the more stable tertiary carbon radical (14).[25]

The cleavage, in solution, of alkoxy radicals [reaction (3)] (formed in the chain decomposition of the corresponding hypochlorites) as a function of the nature of carbon substituents, and of solvent, has been examined by Bacha and Kochi.[26] Hyperconjugative effects appear to be important. For example,

$$\text{RR'R''CO} \cdot \longrightarrow \text{R} \cdot + \text{R'COR''} \qquad\qquad (3)$$

the ethyl group is ejected more readily than is any other group where a 2-hydrogen atom of ethyl is replaced by an alkyl group. Some relative rates of cleavage of primary alkyl groups are given in Table 1. The yields in

Table 1. Relative rates of cleavage of primary alkyl groups from t-alkoxy radicals at 25°.

$CH_3 \cdot$	< 0.005
$CH_3CH_2 \cdot$	1.00
$CH_3CH_2CH_2 \cdot$	0.63—0.74
$CH_3CH_2CH_2CH_2 \cdot$	0.59—0.72
$(CH_3)_2CHCH_2 \cdot$	0.47—0.57
$(CH_3)_3CCH_2 \cdot$	0.42—0.45

cleavage reactions increase with solvent polarity; in hydrogen-bonding solvents there is an unusually pronounced selectivity for which R group is ejected, probably due to hydrogen-bond stabilization of the alkoxy radical.

[25] C. H. Robinson, O. Gnoj, A. Mitchell, E. P. Oliveto, and D. H. R. Barton, *Tetrahedron.* **21**, 743 (1965).
[26] J. D. Bacha and J. K. Kochi, *J. Org. Chem.*, **30**, 3272 (1965).

Kochi and his co-workers have again been active in the field of radical reactions with metal salts. The decomposition of t-alkyl hydroperoxides by chromous ion leads to alkoxy radicals which may eject an alkyl radical;[27] alkyl radicals and unchanged alkoxy radicals are reduced to alkane and alcohol [see reactions (4)]. If $R =$ benzyl, the intermediate species,

$$Cr^{2+} + RCMe_2OOH \longrightarrow CrOH^{2+} + RCMe_2O \cdot \xrightarrow[H_2O]{Cr^{2+}} RCMe_2OH$$

$$RH \xleftarrow{H_2O} RCr^{2+} \xleftarrow{Cr^{2+}} R \cdot + Me_2CO \tag{4}$$

$PhCH_2Cr^{2+}$, is sufficiently stable to be identified. It is considered that the formation of $PhCH_2Cr^{2+}$ from chromous ion and benzyl chloride also involves benzyl radicals.[28]

The decomposition of t-butyl hydroperoxide by cobaltous salts has been studied by Richardson.[29a] When the cobaltous ion was coordinated with EDTA, the rate of peroxide disappearance was both of the first and the second order in the cobaltous ion. To account for this a peroxide–cobaltous ion complex is considered to decompose both spontaneously and by reaction with a second cobaltous species. Both processes give t-butoxy radicals. For cobalt[II] carboxylates a complex mechanistic scheme is developed which incorporates the observed dimeric nature of the carboxylates.

The rate of decomposition of cumene hydroperoxide catalysed by iron salts depends on the nature of the anion. In particular, chelation leads to low rates of decomposition.[29b]

Details have now appeared of a study of the copper salt-catalysed decompositions of peresters induced photochemically.[29c] This procedure allows the examination of radical reactions in low-boiling solvents such as diethyl ether.

Cupric acetate is without effect on the decomposition of azobisisobutyronitrile in acetic acid, which indicates that oxidation of the 2-cyano-2-propyl radical by electron transfer does not occur in this system.[30] Ligand transfer of chlorine from cupric chloride demands much less development of positive charge on the carbon α to the cyano group, and indeed oxidation of the radical by this reagent gives a good yield of 2-cyano-2-propyl chloride. The electron-transfer reagent (cupric acetate) does have some effect when employed in acetonitrile, possibly owing to stabilization of Cu^+ in this solvent, resulting in a larger Cu^+–Cu^{2+} oxidation potential.

[27] J. K. Kochi and P. E. Mocadlo, *J. Org. Chem.*, **30**, 1134 (1965).
[28] J. K. Kochi and D. D. Davis, *J. Am. Chem. Soc.*, **86**, 5264 (1964).
[29a] W. H. Richardson, *J. Am. Chem. Soc.*, **87**, 247, 1096 (1965); *J. Org. Chem.*, **30**, 2804 (1965).
[29b] T. J. Wallace, R. M. Skomoroski, and P. J. Lucchesi, *Chem. Ind. (London)*, **1965**, 1764.
[29c] G. Sosnovsky, *Tetrahedron*, **21**, 871 (1965).
[30] J. K. Kochi and D. M. Mog, *J. Am. Chem. Soc.*, **87**, 522 (1965).

Alkyl radicals from diacyl peroxides have been allowed to react with hydrogen donors (by abstraction to form alkane), in competition with cupric copper (by which they are oxidized to alkenes).[31] The resulting relative rate constants were combined with values of the absolute rate constants for the hydrogen abstraction (extrapolated from gas-phase values) to provide a measure of the rate constants for the electron-transfer oxidation. Typical values for the rate constants for bimolecular oxidation of alkyl radicals by monomeric cupric acetate (the diamagnetic dimer is not effective) were of the order of 10^8 l mole^{-1} sec^{-1} at 57°, approaching the diffusion-controlled limit.

The use of organometallic compound–peroxide systems has been advocated as radical initiators at ambient temperature.[32]

Reactions of Free Radicals

Oxygen Radicals. Several publications have appeared on the nature and reactions of peroxy radicals and autoxidation. In the decomposition of *t*-butyl hydroperoxide induced by di-*t*-butyl peroxyoxalate (**16**), the only radicals involved are *t*-butoxy and *t*-butylperoxy. This system may thus be used to investigate chain-termination that may occur between these radicals

$$t\text{-BuOOCOCOOOBu-}t \longrightarrow 2t\text{-BuO} \cdot + 2CO_2$$
$$\textbf{(16)}$$

$$t\text{-BuO} \cdot + t\text{-BuOOH} \longrightarrow t\text{-BuOH} + t\text{-BuOO} \cdot$$

in autoxidation processes.[33] It was concluded that the only important termination reaction (5) involved formation of di-*t*-butyl peroxide. An

$$2t\text{-BuOO} \cdot \longrightarrow [t\text{-BuO} \cdots O_2 \cdots OBu\text{-}t] \overset{a}{\underset{b}{\longrightarrow}} \begin{cases} 2t\text{-BuO} \cdot + O_2 \\ t\text{-BuOOBu-}t + O_2 \end{cases} \tag{5}$$

independent ESR study[34] showed that, in benzene, the cage recombination (5*b*) occurs at only one-fifth of the rate of (5*a*), and that tetroxide formation is probably rapid and reversible. It was pointed out that the O_2 produced should initially be in the singlet state. Di-*t*-butyl peroxide is also formed in the cage decomposition of perester (**16**), and the extent of this reaction

$$\textbf{(16)} \longrightarrow [t\text{-BuO} \cdot 2CO_2 \cdot OBu\text{-}t] \longrightarrow BuOOBu\text{-}t + 2CO_2$$

path is appreciably greater in solutions of high viscosity.[35] In cumene

[31] J. K. Kochi and R. V. Subramanian, *J. Am. Chem. Soc.*, **87**, 4855 (1965).
[32] C.-H. Wang, R. McNair, and P. Levins, *J. Org. Chem.*, **30**, 3817 (1965).
[33] A. Factor, C. A. Russell, and T. G. Traylor, *J. Am. Chem. Soc.*, **87**, 3692 (1965).
[34] J. R. Thomas, *J. Am. Chem. Soc.*, **87**, 3935 (1965).
[35] R. Hiatt and T. G. Traylor, *J. Am. Chem. Soc.*, **87**, 3766 (1965).

7*

autoxidation, reactions analogous to (5*b*) account for a relatively minor portion of the termination. Much more important are reactions (8–10):[36]

$$Me_2PhCO\cdot \longrightarrow Me\cdot + PhCOMe \qquad (8)$$

$$Me\cdot + O_2 \longrightarrow MeOO\cdot \qquad (9)$$

$$MeOO\cdot + ROO\cdot \longrightarrow CH_2O + O_2 + ROH \qquad (10)$$

The bond-dissociation energies of R–OOR and R–OO· have been measured for a series of groups R.[37]

Radical-chain oxidation of alkanes in the gas-phase has been postulated to involve HOO·. In view of the low strength of the H–OOH bond, this would be expected to show considerable discrimination in favour of reaction at tertiary hydrogen atoms. This has been verified[38] for the initial reaction ($<1\%$) but, as the oxidation proceeds, selectivity falls off, which is consistent with the view that the more reactive HO· has become the chain carrier.

Aromatic amines may inhibit oxidation.[39,40] Observation of deuterium isotope effects in inhibition by 2,4,6-tri-*t*-butyl-phenol, -aniline, and -thio-phenol shows that each reaction is at the functional group.[41]

Further evidence has been obtained for the relatively low reactivity of phenyl radicals towards oxygen,[42] and it has also been calculated that radical adducts (17) of anthracene show a low reactivity towards oxygen.[43]

(17)

Radical Abstraction and Displacement Processes. From a study of the kinetics of hydrogen abstraction from alkanes by the difluoroamino radical it has been calculated that the strength of the F_2N–H bond is ca. 72.5 kcal mole^{-1}.[44] and extension of the work to the reaction with acetone[45] gave a value of 92.1 ± 1 kcal mole^{-1} for the strength of the H–CH$_2$COCH$_3$ bond.

[36] T. G. Traylor and C. A. Russell, *J. Am. Chem. Soc.*, **87**, 3698 (1965).
[37] S. W. Benson, *J. Am. Chem. Soc.*, **87**, 972 (1965).
[38] J. H. Knox and J. M. C. Turner, *J. Chem. Soc.*, **1965**, 3491.
[39] W. R. Yates and J. L. Ihrig, *J. Am. Chem. Soc.*, **87**, 710 (1965).
[40] A. MacLachlan, *J. Am. Chem. Soc.*, **87**, 960 (1965).
[41] D. V. Gardner, J. A. Howard, and K. U. Ingold, *Can. J. Chem.* **24**, 2847 (1964).
[42] K. Tokumaru, H. Horie, and O. Simamura, *Tetrahedron*, **21**, 867 (1965).
[43] L. R. Mahoney, *J. Am. Chem. Soc.*, **87**, 1089 (1965).
[44] J. Grzechowiak, J. A. Kerr, and A. F. Trotman-Dickenson, *Chem. Comm.*, **1965**, 109.
[45] J. Grzechowiak, J. A. Kerr, and A. F. Trotman-Dickenson, *J. Chem. Soc.*, **1965**, 5080.

Examination of the iodine-catalysed isomerization of butenes has allowed the activation energy for hydrogen abstraction by iodine from position 3 of 1-butene to be calculated. Comparison with hydrogen abstraction from $C_{(2)}$ of propane gave ca. 12.6 kcal mole^{-1} as the resonance energy of an allylic radical.[46]

To explain the results of competitive hydrogen abstraction from toluene and cyclohexane in chain chlorination by *t*-butyl hypochlorite, Wagner and Walling[47] were forced to invoke a "phantom intermediate". At first sight, the results appear to require hydrogen abstraction from cyclohexane by resonance-stabilized benzyl radicals. To avoid this, the "phantom intermediate" is proposed as being formed irreversibly from *t*-butoxy radicals and toluene, and reacting either with cyclohexane (as in 11*a*) or by internal hydrogen transfer (as in 11*b*).

$$\text{"Complex"} \underset{\longrightarrow}{\overset{C_6H_{12}}{\longrightarrow}} \begin{array}{l} C_6H_{11}\cdot + PhMe + t\text{-BuOH} \qquad (11a) \\ \\ BuOH + PhCH_2\cdot \qquad\qquad\qquad (11b) \end{array}$$

In a new study of substituent effects in homolytic hydrogen abstraction from substituted toluenes, it was found better to correlate abstraction by *t*-butoxy radicals with σ-substituent constants ($\rho = -0.75$) than with σ^+-constants.[48] There is, however, appreciable deviation from the σ-correlation for a *p*-phenyl substituent, and this is much more pronounced when the abstraction is by a bromine atom. The important contribution of structures such as (18) as well as (19) to the transition state is indicated by these results.

$$[\text{Ar}\overset{\bullet}{\text{C}}\text{H}_2\ \text{H}\!-\!\text{Br}] \qquad [\text{Ar}\overset{+}{\text{C}}\text{H}_2\cdots\text{H}\cdots\overset{-}{\text{Br}}]$$
$$\quad\text{(18)} \qquad\qquad\qquad \text{(19)}$$

A small polar effect, in the same direction as for substituted toluenes, has been observed for hydrogen abstraction by Br· from benzyl methyl ethers.[49] The direction of this effect precludes significant contribution from a species (20) to the incipient radical. On the other hand, Huang and his colleagues[50]

$$\text{Ar}\overset{-}{\text{C}}\text{H}_2\!-\!\overset{+}{\underset{\bullet}{\text{O}}}\!-\!\text{Me}$$
$$\text{(20)}$$

find no appreciable polar effect in hydrogen abstraction from unsymmetrical dibenzyl ethers by Br·. They also examined some reactions of methoxy- and dimethoxy-benzyl radicals.[51]

[46] D. M. Golden, K. W. Egger, and S. W. Benson, *J. Am. Chem. Soc.*, **86**, 5416, 5420 (1964).
[47] P. Wagner and C. Walling, *J. Am. Chem. Soc.*, **87**, 5179 (1965).
[48] R. D. Gilliom and B. F. Ward, *J. Am. Chem. Soc.*, **87**, 3944 (1965).
[49] R. E. Lovins, L. J. Andrews, and R. M. Keefer, *J. Org. Chem.*, **30**, 1577 (1965).
[50] R. L. Huang, H. H. Lee, and M. S. Malhotra, *J. Chem. Soc.*, **1964**, 5947, 5951.
[51] R. L. Huang and K. H. Lee, *J. Chem. Soc.*, **1964**, 5957, 5963.

A Hammett σ-correlation has similarly been found for homolytic conversion of substituted benzaldehydes into the corresponding benzoyl chlorides by sulphuryl chloride.[52] The transition state of the abstraction process is considered to possess a contribution from the polar structure (21).

$$\text{Ar}{-}\overset{+}{\underset{\underset{O}{\|}}{\text{C}}}\cdots\overset{\cdot}{\text{H}}\cdots\bar{\text{S}}\text{O}_2\text{Cl}$$

(21)

In the gas-phase, hydrogen abstraction from nitrogen of dimethylamine by methyl radicals is twenty times faster than from carbon (atom for atom),[53] and with O-methylhydroxylamine abstraction is again predominantly from nitrogen.[54] The ease of hydrogen abstraction by methyl radicals from fluorinated methanes does not increase with fluorine substitution;[55] thus fluorine is without the stabilizing effect that chlorine has on the product radical.

In the competition between abstraction of chlorine from t-alkyl hypochlorite and of bromine from bromotrichloromethane by alkyl radicals, the activation energies favour bromine abstraction, but the pre-exponential factors favour chlorine abstraction.[56] In the same work it was shown that for $\text{R}\cdot = t$-Bu, the reaction $\text{R}\cdot + \text{ClOCMe}_2\text{R} \rightarrow \text{RCl} + \text{Me}_2\text{CO} + \text{R}\cdot$ appears to be concerted, on the grounds of a large relative entropy of activation and failure to isolate 1,1,2,2-tetramethylpropan-1-ol. Solvent effects on the cleavage of alkoxy radicals were also observed.[26] The relative reactivities of different halogen atoms in halogenomethanes towards abstraction by various aryl radicals have been reported.[57]

For reactions where products of aliphatic substitution by radical mechanisms have been studied, one of the most interesting reports is of retention of configuration at a radical centre in a reaction where bridging by a β-halogen substituent is not possible.[58] Photobromination of $(+)$-3-methylvaleronitrile gives the $(+)$-3-bromo nitrile; it is considered that bridging by the cyano group is unlikely, and the result may be explained by hydrogen abstraction by $\text{Br}_3\cdot$ ($\rightarrow \text{HBr} + \text{Br}_2$), followed by a reaction between the resulting carbon radical and bromine which is so rapid that the radical centre has insufficient time to become planar.

Detailed studies have been reported of free-radical chlorination of n-hexane, n-octane, and monochloro-n-hexanes,[59] and of successive side-chain

[52] M. Arai, *Bull. Chem. Soc. Japan.*, **38**, 252 (1965).
[53] P. Gray, A. Jones, and J. C. J. Thynne, *Trans. Farad. Soc.*, **61**, 474 (1965).
[54] J. C. J. Thynne, *Trans. Farad. Soc.*, **60**, 2207 (1964).
[55] G. O. Pritchard, J. T. Bryant, and R. L. Thommarson, *J. Phys. Chem.*, **69**, 664 (1965).
[56] A. A. Zavitsas and S. Ehrenson, *J. Am. Chem. Soc.*, **87**, 2841 (1965).
[57] J. I. G. Cadogan, D. H. Hey, and P. G. Hibbert, *J. Chem. Soc.*, **1965**, 3950.
[58] W. O. Haag and E. I. Heiba, *Tetrahedron Letters*, **1965**, 3679.
[59] N. Colebourne and E. S. Stern, *J. Chem. Soc.*, **1965**, 3599.

chlorinations of *p*-xylene[60] and brominations of 1- and 2-methylnaphthalene.[61] In the xylene reaction, the presence of chlorine slows further reaction both on the same carbon and at the *p*-methyl group. Substitution by bromine similarly slows further substitution of the methylnaphthalenes. Introduction of the first bromine atom is faster for 1-methylnaphthalene than for the 2-isomer by a factor of 3.

In the gas-phase chlorination or bromination of carboxylic acid derivatives, an inductive effect is felt at α- and, to a small extent, β-carbon atoms.[62] The rate of abstraction of α-hydrogen by electrophilic chlorine atoms is reduced to a much greater extent than is that for abstraction by bromine, this being attributed to greater importance of radical stabilization (22) than of the

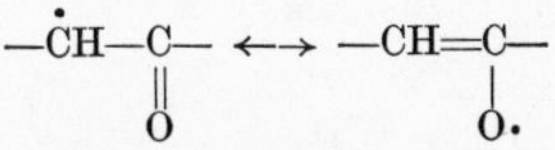

$$-\overset{\cdot}{C}H-\underset{\underset{O}{\|}}{C}- \longleftrightarrow -CH=\underset{\underset{O\cdot}{|}}{C}-$$

(22)

inductive effect for the bromine reaction. In chlorination in acetonitrile solution,[63] the inductive effect is operative further from the carbonyl group than the β-carbon, and it is suggested that this is because solvation of the transition state results in greater charge separation. The opposite result might have been expected because of the reduced inductive effect in a polar medium. Some results are indicated in Table 2.

Table 2. Relative rates of halogenation of carboxylic acid derivatives

Reaction conditions	Relative reactivities
Gas-phase chlorination, 60°	$FCO-CH_2-CH_2-CH_2-CH_3$ 0.02 0.37 1.0 0.24
Gas-phase bromination, 160°	$FCO-CH_2-CH_2-CH_2-CH_3$ 0.42 0.33 1.0 0.01
Gas-phase chlorination, 60°	$FCO-CH_2-CH_2-CH_2-CH_2-CH_3$ 0.05 0.4 1.0 1.0 1.1 0.25
Chlorination in CH_3CN, 52°	$ClCO-CH_2-CH_2-CH_2-CH_2-CH_2-CH_3$ 0.06 0.5 1.0 1.3 1.5 1.1

[60] P. Beltrame and S. Carrà, *Tetrahedron Letters*, **1965**, 3909.
[61] P. R. Taussig, G. B. Miller, and P. W. Storms, *J. Org. Chem.*, **30**, 3122 (1965).
[62] H. Singh and J. M. Tedder, *J. Chem. Soc.*, **1964**, 4737.
[63] H. Singh and J. M. Tedder, *Chem. Comm.*, **1965**, 5.

In the photoinitiated chlorination of *t*-butyl bromide by *t*-butyl hypochlorite at $-78°$, some of the unrearranged product (**23**) accompanies (**24**).[64]

$$
\begin{array}{ccc}
\overset{\displaystyle Me}{\underset{\displaystyle CH_2Cl}{Me\!-\!\!\!\!-\!\!Br}} & & \overset{\displaystyle Me}{\underset{\displaystyle CH_2Br}{Me\!-\!\!\!\!-\!\!Cl}} \\[2pt]
(\mathbf{23}) & & (\mathbf{24})
\end{array}
$$

It was also possible to show that the apparent rearrangement of the intermediate radical proceeds largely, if not entirely, by a dissociation–readdition mechanism (12). The reaction appears to involve free bromine atoms, which may be scavenged by allene; the possibility of direct bimolecular bromine

$$
\underset{(\mathbf{25})}{\overset{\displaystyle Me}{\underset{\displaystyle \dot{C}H_2}{Me\!-\!\!\!\!-\!\!Br}}} \longrightarrow \underset{\displaystyle CH_2}{\overset{\displaystyle Me\quad Me}{\diagdown\!/}} + Br\cdot \longrightarrow \underset{(\mathbf{26})}{\overset{\displaystyle Me}{\underset{\displaystyle CH_2Br}{Me\!-\!\!|\cdot}}} \tag{12}
$$

atom transfer from (**25**) or (**26**) to allene was satisfactorily excluded. Furthermore, the yield of unrearranged product (**23**) increases with the concentration of hypochlorite, and this indicates a finite life-time for the classical radical (**25**). In view of these results, the question of non-classical radicals with bridging bromine atoms in this and related rearrangements must be re-examined.

Evidence has been presented[65] that photochemical bromination of cyclohexane by CCl_3SO_2Br does not involve hydrogen abstraction by $CCl_3SO_2\cdot$, as in the case of CCl_3SO_2Cl, but rather by $\cdot CCl_3$. In the absence of hydrocarbon the sulphonyl chloride is unchanged on photolysis, but the bromo derivative gives $BrCCl_3$ and SO_2. Photochemical chlorination of 2,3-dimethylbutane with $PhICl_2$ gives exclusive substitution at the tertiary carbon atom. This high selectivity is used as evidence that the $Ph\dot{I}Cl$ radical, and not a chlorine atom, is the chain carrier which abstracts hydrogen.[66]

Pyrolysis of gaseous nitric acid at $300°$ in the presence of methane gives nitromethane in a chain reaction (13–15).[67]

$$
HNO_3 \rightleftharpoons HO\cdot + NO_2 \tag{13}
$$

$$
HO\cdot + CH_4 \longrightarrow H_2O + CH_3\cdot \tag{14}
$$

$$
CH_3\cdot + HNO_3 \longrightarrow CH_3NO_2 + HO\cdot \tag{15a}
$$

$$
CH_3\cdot + HNO_3 \longrightarrow CH_3OH + NO_2 \tag{15b}
$$

[64] W. O. Haag and E. I. Heiba, *Tetrahedron Letters*, **1965**, 3683.
[65] R. P. Pinnell, E. S. Huyser, and J. Kleinberg, *J. Org. Chem.*, **30**, 38 (1965).
[66] D. F. Banks, E. S. Huyser and J. Kleinberg, *J. Org. Chem.*, **29**, 3692 (1964).
[67] T. S. Godfrey, E. D. Hughes, and C. K. Ingold, *J. Chem. Soc.*, 1063 (1965).

Publications dealing with intramolecular radical abstraction and substitution include a full account[68] of work on the interception of alkyl-radical intermediates in the Barton reaction, as depicted in (16). Barton and his co-workers also observed that the intramolecular hydrogen transfer exhibited

$$(16)$$

in these reactions was apparently not reproduced during pyrolysis of nitrites. This, they suggested, may be due to the formation of an excited alkoxy radical in the photochemical reaction. Acott and Beckwith have now generated similar alkoxy radicals by ferrous ion-catalysed decomposition of tertiary hydroperoxides and have shown that moderate yields of products of intramolecular hydrogen transfer may be observed [reaction (17)] but that the yields decrease with increasing reaction temperature.[69] It is therefore suggested that failure to obtain similar products on pyrolysis of nitrites is

$$(17)$$

best attributed to the greater contributions of competing side reactions at the high temperatures employed.

Beckwith and Goodrich have found also that photolysis of *N*-chloro amides provides an alternative to that of *N*-iodo amides as a route to lactones.[70] It is interesting that in this reaction significant amounts of δ-lactone accompany the γ-lactone; this necessitates a seven-membered ring in the transition state (27) for the hydrogen transfer, a feature not observed to an appreciable extent in intramolecular reactions of alkoxy or amino radicals. This is felt to reflect the conformational requirement of the amide radical in achieving the required colinear relationship[71] of carbon, hydrogen, and nitrogen in the transition state (27).

Transannular hydrogen transfer is also observed in the photoinitiated

[68] M. Akhtar, D. H. R. Barton, and P. G. Sammes, *J. Am. Chem. Soc.*, 87, 4601 (1965).
[69] B. Acott and A. L. J. Beckwith, *Australian J. Chem.*, 17, 1342 (1964).
[70] A. L. J. Beckwith and J. E. Goodrich, *Australian J. Chem.*, 18, 747 (1965).
[71] E. J. Corey and W. R. Hertler, *J. Am. Chem. Soc.*, 82, 1657 (1960); A. Padwa and C. Walling, *ibid.*, 85, 1597 (1963).

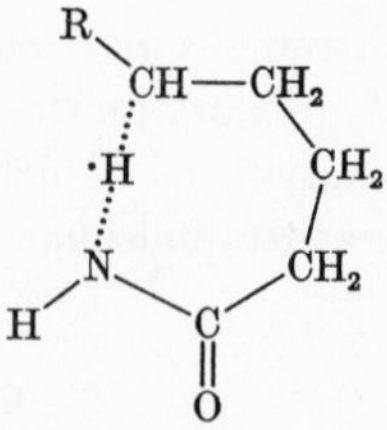

(27)

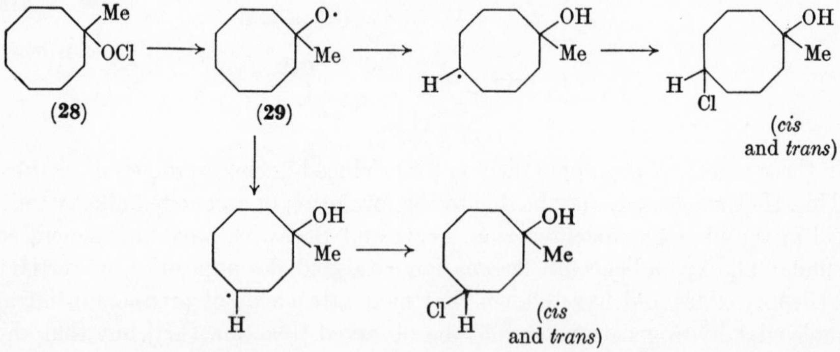

(28)   (29)   (*cis*
                                                                         and *trans*)

                                                                         (*cis*
                                                                         and *trans*)

chain decomposition of hypochlorite (28).[72] The same intermediate radical (29) is formed in the oxidation of 1-methylcyclooctanol by lead tetraacetate, the products being 1,5- and 1,4-epoxy-1-methylcyclooctane in the proportions 3:1. Similar oxidation of cyclooctanol gives only 1,4-epoxycyclooctane, probably reflecting conformational differences between the two alcohols (see also p. 306).

Intramolecular reactions of derivatives of 1,3,3-trimethylcyclohexanol to give the cyclic ether (30) have been reported[73] as proceeding by a diversity

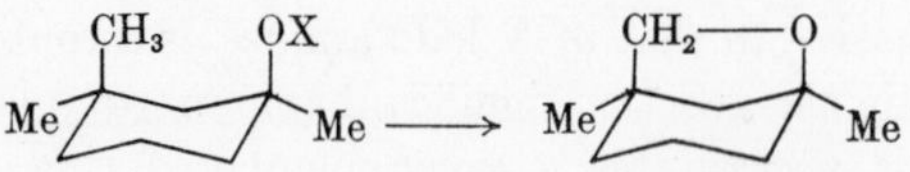

(30)

	X	Reagent	Mechanism
(i)	OH	Cu^+	radical
(ii)	$OSO_2C_6H_4Me$-p (persulphonate)	—	ion pair
(iii)	Br	$h\nu$	radical pair (cage)
(iv)	Br	Ag^+	cationic

[72] A. C. Cope, R. S. Bly, M. M. Martin, and R. C. Petterson, *J. Am. Chem. Soc.*, **87**, 3111 (1965); A. C. Cope, M. Gordon, S. Moon, and C. H. Park, *ibid.*, **87**, 3119 (1965).
[73] R. A. Sneen and N. P. Matheny, *J. Am. Chem. Soc.*, **86**, 5503 (1964).

of mechanisms, though the evidence for each is suggestive rather than conclusive. The silver ion-catalysed hypobromite decomposition now appears to be a radical process, by analogy with other, similar, reactions which give typically free-radical products,[74] and for which ionic pathways may be ruled out.

Cyclization of the tertiary alcohol (31) to the ether (32) by bromine and silver oxide in the dark[75] is an analogous radical process, as shown by loss of optical activity in the intermediate carbon radical.

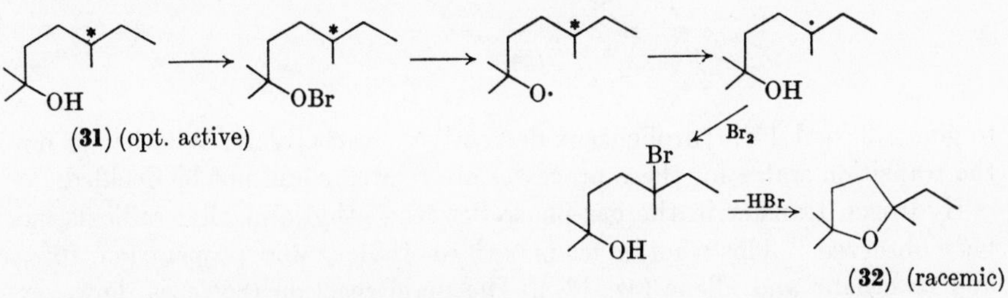

(31) (opt. active)

(32) (racemic)

Radical Coupling and Disproportionation. The exact nature of the transition state for reactions between two radicals is not very well understood. For example, large pre-exponential factors found for disproportionation of alkyl radicals have led to the suggestion that the transition state for disproportionation is a loose one with freedom of motion comparable to that in the transition state for combination.[76] Direct evidence on this problem comes from measurement of isotope effects in the disproportionation of $CH_2TCH_2\cdot$ and $CH_3\dot{C}HCH_2T$.[77a] For both species $k_H/k_T \approx 2$ (at 63°K in solid ethylene). At this temperature a value of $k_H/k_T \approx 10^6$ might have been anticipated for a "head to tail" mechanism (20). The small kinetic isotope

$$CH_3CH_2\cdot \cdots H—CH_2CH_2\cdot \longrightarrow C_2H_6 + C_2H_4 \qquad (20)$$

effect indicates that the C–H bond length in the transition state is essentially the same as in the undisturbed ethyl radical. This, in turn, indicates that the activation energy for the process is very close to zero.

Conflicting evidence comes from a study of hydrogen transfer between, and combination of, ethyl and heptafluoropropyl radicals. These two processes shown an activation-energy difference of greater than 2 kcal mole^{-1}.

[74] M. Akhtar, P. Hunt, and P. B. Dewhurst, *J. Am. Chem. Soc.*, **87**, 1807 (1965).

[75] G. Smolinsky and B. I. Feuer, *J. Org. Chem.*, **30**, 3216 (1965).

[76] See, for example, J. N. Bradley and B. S. Rabinovitch, *J. Chem. Phys.*, **36**, 3498 (1962).

[77a] H. B. Yun and H. C. Moser, *J. Phys. Chem.*, **69**, 1059 (1965); K. W. Watkins and H. C. Moser, *ibid.*, **69**, 1040 (1965).

This is evidently inconsistent with a transition state which is virtually the same for the two processes.[77b]

The same question has been considered for the gas-phase reactions of alkyl radicals with the cyclohexadienyl radical (33).[77c] In this case there is the added complication that two paths are possible for combination reactions,

(33)

to give 1,2- or 1,4-dihydrobenzene derivatives, respectively. Whether or not the transition states for these processes are related could not be decided.

Hydrogen transfer in the gas-phase between ethyl and allyl radicals has been observed.[78] This reaction leads both to ethylene and propene (ca. 10%) and to ethane and allene (*ca.* 4%). The main reaction (85%) is, however, combination to give 1-pentene.

Combination of fluorinated methyl radicals with concomitant elimination of hydrogen fluoride is important only for radicals with a low degree of fluorine substitution.[79] Thus reaction (21) is much more important than reaction (22).

$$CH_2F \cdot + CH_2F \cdot \longrightarrow CH_2 = CHF + HF \qquad (21)$$

$$CF_3 \cdot + CHF_2 \cdot \xrightarrow{\ \ X\ \ } CF_2 = CF_2 + HF \qquad (22)$$

The major product of the gas-phase reaction of toluene with methyl radicals is ethylbenzene. Very little xylene is formed.[80] However, the analogous reaction with phenol gives *o*- and *p*-cresol rather than anisole.[81] The kinetics of these reactions were studied by using a stirred flow reactor, and it was shown that the initially formed keto forms of the cresols had an appreciable life in the gas phase and could participate as such in further reactions with methyl radicals.

Some light has been shed on the rather confused literature concerning the abstraction of oxygen from nitrous oxide by alkyl radicals.[82] In solution, the

[77b] G. O. Pritchard and R. L. Thommarson, *J. Phys. Chem.*, **69**, 1001 (1965).
[77c] D. G. L. James and R. D. Suart, *J. Am. Chem. Soc.*, **86**, 5424 (1964); *J. Phys. Chem.*, **69**, 2362 (1965).
[78] D. G. L. James and G. E. Troughton, *Chem. Comm.*, **1965**, 94.
[79] G. O. Pritchard and J. T. Bryant, *J. Phys. Chem.*, **69**, 1085 (1965).
[80] M. F. R. Mulcahy, D. J. Williams, and J. R. Wilmshurst, *Australian J. Chem.*, **17**, 1329 (1964).
[81] M. F. R. Mulcahy and D. J. Williams, *Australian J. Chem.*, **18**, 20 (1965)
[82] T. N. Bell and K. O. Kutschke, *Can. J. Chem.*, **42**, 2713 (1964).

scavenging of aryl radicals by nitrogen dioxide has been investigated as a possible route to aromatic nitro compounds. Yields were generally low.[83]

Kinetic equations applicable to radical-scavenging experiments have been developed and discussed in terms of a peroxide decomposition.[84]

Miscellaneous Data on Free Radicals. Bartlett and his school have examined the stereochemistry of the 9-decalyl free radical.[85] Both *cis*- and *trans*-9-carbo-*t*-butylperoxydecalin give 9-decalyl radicals by concerted decomposition [reaction (23)] and show identical decomposition rates in cumene. The products formed in dimethoxyethane in the presence of oxygen

$$\text{(23)}$$

at one atmosphere included, in each case, a mixture of *trans*- ca. 90% and *cis*-decalyl hydroperoxide ca. 10%. However, when the oxygen pressure was raised to several hundred atmospheres, the *cis*-peroxyester gave a greatly increased relative yield of *cis*-hydroperoxide (70% at 545 atm.), though the relative yields from the *trans*-peroxyester were unaffected. These results were interpreted in terms of initial formation of tetrahedral decalyl radicals, both of which may, by rehybridization, change to a planar structure (34). At low oxygen pressures, both initial radicals are converted into (34) before reacting, and the proportions of hydroperoxide are governed by steric requirements of approach of oxygen to this species. It is reasonable that conversion of the *trans*-radical into (34) should be faster than conversion of the *cis*-radical because only the latter process requires chair-to-chair inversion of one ring; at high oxygen pressures it appears possible to intercept the tetrahedral *cis*-9-decalyl radical before this inversion can occur. The results lead to an estimated lifetime of 10^{-8} to 10^{-9} sec for the *cis*-radical. These results are consistent with the general view that the preferred conformation of a carbon radical is planar. However, during the year other results have been explained as involving reactions of tetrahedral radicals (see p. 115).

In decarbonylations in carbon tetrachloride (24 and 25), it has been found that the ratios $k_{24}:k_{25}$ can be well correlated with the radical stability of the group R, where this is known.[86] Assuming the generality of this result, the series of experiments has been extrapolated to those in which bridgehead

[83] G. B. Gill and G. H. Williams, *J. Chem. Soc.*, **1965**, 5756.

[84] R. C. Lamb, J. G. Pacifici, and L. P. Spadafino, *J. Org. Chem.*, **30**, 3102 (1965).

[85] P. D. Bartlett, R. E. Pincock, J. H. Rolston, W. G. Schindel, and L. A. Singer, *J. Am. Chem. Soc.*, **87**, 2590 (1965).

[86] D. E. Applequist and L. Kaplan, *J. Am. Chem. Soc.*, **87**, 2194 (1965).

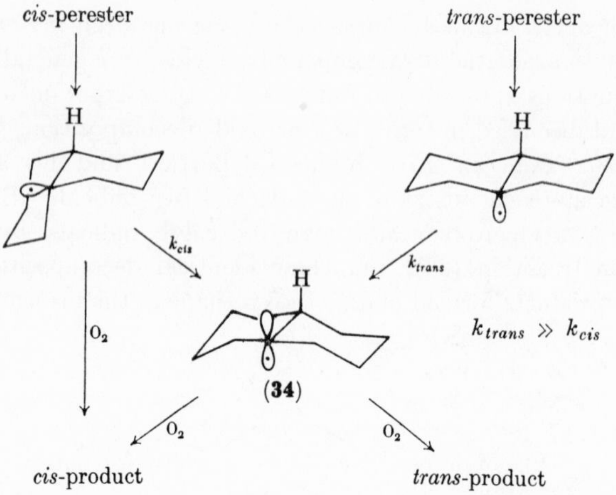

cis-perester   *trans*-perester

$k_{trans} \gg k_{cis}$

(34)

cis-product   *trans*-product

(and therefore non-planar) radicals are formed. From these experiments, it was concluded that the adamantyl radical is, in fact, more stable than the

$$RCHO \xrightarrow{-H\cdot} RCO\cdot -\begin{cases} \xrightarrow{k24} R\cdot \longrightarrow RCl + RH & (24) \\ \xrightarrow[k25]{CCl_4} RCOCl & (25) \end{cases}$$

t-butyl radical (by ca. 1 kcal mole^{-1}), and the bicyclo[2.2.2]octan-1-yl radical has a stability comparable with that of *t*-butyl. The more strained bicyclo-[2.2.1]heptan-1-yl radical, on the other hand, is very much less stable. These results are discussed in terms of a much smaller out-of-plane bending force constant for radicals than for carbonium ions.

Configurational stability of carbanions α to a sulphonyl group is well known (see p. 83). However, similar stability was not exhibited by the α-sulphonyl free radical formed from compound (35).[87]

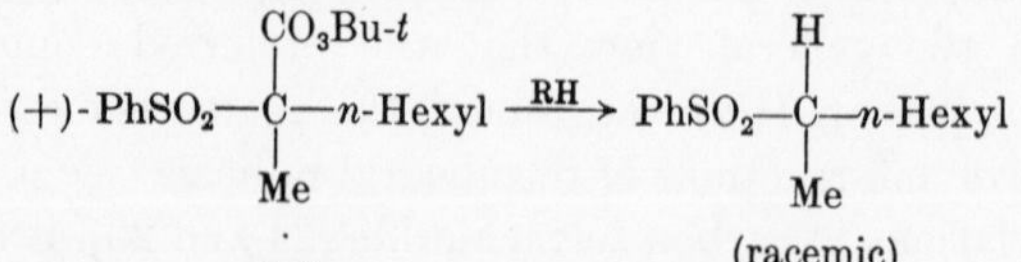

$$(+)\text{-}PhSO_2 - \overset{\overset{\displaystyle CO_3Bu\text{-}t}{|}}{\underset{\underset{\displaystyle Me}{|}}{C}} - n\text{-Hexyl} \xrightarrow{RH} PhSO_2 - \overset{\overset{\displaystyle H}{|}}{\underset{\underset{\displaystyle Me}{|}}{C}} - n\text{-Hexyl}$$

(35)   (racemic)

In a reinvestigation of the anodic acetoxylation of naphthalene, it has been found that electron transfer from naphthalene occurs at a lower potential than from acetate.[88] Thus the acetoxylation probably occurs, not by free-radical substitution (which is in any case unlikely in view of the

[87] E. T. Kaiser and D. F. Mayers, *Tetrahedron Letters*, **1965**, 2767.
[88] M. Leung, J. Herz, and H. W. Salzberg, *J. Org. Chem.*, **30**, 310 (1965).

instability of acetoxy radicals), but by a reaction of the naphthalene radical cation with acetic acid.

In alternating-current electrolyses of solutions containing certain aromatic hydrocarbons the chemiluminescent reaction between aromatic radical anions and radical cations has been observed.[89]

Radical reactions at the cathode have been observed in Kolbe syntheses,[90] and electrolytic generation of the cyano radical has been reported.[91]

A reaction path available (at 300°) for the n-heptyl radical, which involves cleavage of a methyl radical and formation of cyclohexane, has been confirmed.[92] High-temperature (> 300°) reactions of the cyclopentyl radical have also been studied, and the predominant decomposition path has been shown to lead to ethylene and alkyl radicals.[93,94] Loss of a hydrogen atom to give cyclopentene occurs,[93,94] and also observed[94] was a third path involving formation of molecular hydrogen and cyclopentenyl radical.

Reactions of organotin radicals[95,96] and exchange of alkyl radicals at boron have been studied.[97]

Laidler and Eusuf have re-examined the pyrolysis (520–560°) of propionaldehyde.[98] The reaction, though more complex than that of acetaldehyde, is slower and this is probably partly due to a second-order in place of a third-order chain termination, the former being possible with the larger radicals. The effect of nitric oxide was also examined.

In solution, tritium-labelling experiments have permitted interpretation of protonation of naphthalene radical anion in terms of reactions (26)–(28).[99a]

$$ArH\cdot^- + H_2O \longrightarrow ArH_2\cdot + OH^- \tag{26}$$

$$ArH_2\cdot + ArH\cdot^- \longrightarrow ArH_2^- + ArH \tag{27}$$

$$ArH_2^- + H_2O \longrightarrow ArH_3 + OH^- \tag{28}$$

The mechanisms of initiation and termination in the photochemical Hofmann–Loeffler N-chloroamine rearrangement have been examined: initiation involves photolysis of a dichloramine formed by disproportionation.[99b]

[89] E. A. Chandross, J. W. Longworte, and R. E. Visco, *J. Am. Chem. Soc.*, **87**, 3259 (1965); R. E. Visco, and E. A. Chandross, *ibid.*, **86**, 5350 (1964).

[90] L. Rand and A. F. Mohar, *J. Org. Chem.*, **30**, 3885 (1965).

[91] K. Koyama, T. Susuki, and S. Tsutsumi, *Tetrahedron Letters*, **1965**, 627.

[92] N. J. Friswell and B. G. Gowenlock, *Chem. Comm.*, **1965**, 277.

[93] T. F. Palmer and F. P. Lossing, *Can. J. Chem.*, **43**, 565 (1965).

[94] A. S. Gordon, *Can. J. Chem.*, **43**, 570 (1965).

[95] E. J. Kupchik and R. J. Keisel, *J. Org. Chem.*, **29**, 3690 (1964).

[96] W. P. Neumann, R. Sommer, and H. Lind, *Ann. Chem.*, **688**, 14 (1965).

[97] J. Grotewold and E. A. Lissi, *Chem. Comm.*, **1965**, 21.

[98] K. J. Laidler and M. Eusuf, *Can. J. Chem.*, **43**, 268, 278 (1965).

[99a] S. Bank and W. D. Closson, *Tetrahedron Letters*, **1965**, 1349.

[99b] R. S. Neale and M. R. Walsh, *J. Am. Chem. Soc.*, **87**, 1255 (1965).

Stable Radicals and Electron Spin Resonance Studies

Reviews of observations on stable radicals[100] and the use of ESR in organic chemistry[101] have appeared.

Oxygen Radicals. Russell and his co-workers have recorded the ESR spectra of many semidiones of partial structure (36).[102] These species have

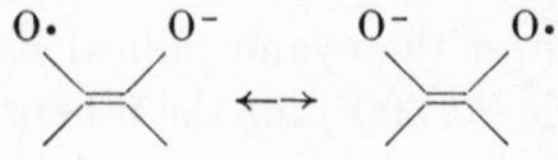

(36)

been obtained by oxidation of ketones with oxygen, in dimethyl sulphoxide containing *t*-butoxide, and of α-bromo ketones by dimethyl sulphoxide and *t*-butoxide. Where two products may be obtained from the first of these reactions, the second leads specifically to the radical ion formed by replacement of bromine. The coupling constants found by examining the spectra of semidiones formed from rigid cyclic molecules such as A-ring steroidal ketones are useful in making stereochemical assignments. Appreciable long-range coupling constants have been assigned in the spectra of bridged semidiones; e.g. in (37) $a_{H_7} = 3.07$ gauss.[103] The semidiones of sterically hindered

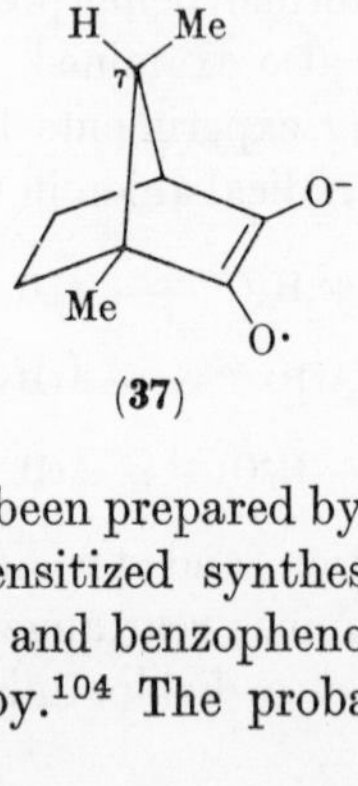

(37)

diketones such as pivalil have been prepared by irradiation of the corresponding acyloin in base. Photosensitized synthesis from benzophenone gives simultaneously the semidione and benzophenone ketyl, which are observed together by ESR spectroscopy.[104] The probable mechanism is shown by

$$Ph_2C{=}O \xrightarrow{h\nu} Ph_2C{=}O^* \longrightarrow Ph_2C{=}O^* \text{ (triplet)} \tag{29}$$

$$Ph_2C{=}O^*\text{(triplet)} + R{-}\overset{\text{-O}}{\underset{|}{C}}{-}\overset{\text{O}}{\underset{\|}{C}}{-}R \longrightarrow Ph_2\overset{\cdot}{C}{-}O^- + R{-}\overset{\cdot\text{O}}{\underset{\|}{C}}{=}\overset{\text{O}^-}{\underset{|}{C}}{-}R \tag{30}$$

[100] A. L. Buchachenko, "Stable Radicals", Consultants Bureau Translation, New York, 1965.
[101] F. Schneider, K. Möbius, and M. Plato, *Angew. Chem. Intern. Ed. Engl.*, **4**, 856 (1965).
[102] G. A. Russell and E. R. Talaty, *J. Am. Chem. Soc.*, **86**, 5345 (1964); E. R. Talaty and G. A. Russell, *ibid.*, **87**, 4867 (1965); G. A. Russell, R. D. Stephens, and E. R. Talaty, *Tetrahedron Letters*, **1965**, 1139.
[103] G. A. Russell and K. Y. Chang, *J. Am. Chem. Soc.*, **87**, 4381 (1965); G. A. Russell, K. Y. Chang, and C. W. Jefford, *ibid.*, **87**, 4383 (1965).
[104] H. C. Heller, *J. Am. Chem. Soc.*, **86**, 5346 (1964).

(29) and (30). Observations of *ortho*-semiquinones,[105] as well as some of their reactions[106] with OH^-, have also been observed by ESR spectroscopy.

The equilibria (31) have been studied in a flow system at high pH.[107] The high equilibrium constants found when electron-withdrawing groups were

$$\text{Quinone} + \text{Dianion} \rightleftarrows 2 \,\text{Semiquinone} \tag{31}$$

present appeared to indicate that such substituents stabilized two semiquinone molecules more than one dianion.

At lower pH, ESR spectroscopy has been used to afford pK values for the protonation of semiquinones.[108] For the proton dissociation of *p*-benzosemiquinone the pK is 4.25. Also, solvent effects on the ESR spectra of semiquinones have been compared with theory.[109]

The free-spin distribution in stable phenoxy radicals has been determined,[110,111] and ESR spectra of stable phenoxy radicals related to lignin have been observed.[112]

The ESR spectra of *t*-butylperoxy radicals have been observed in radical

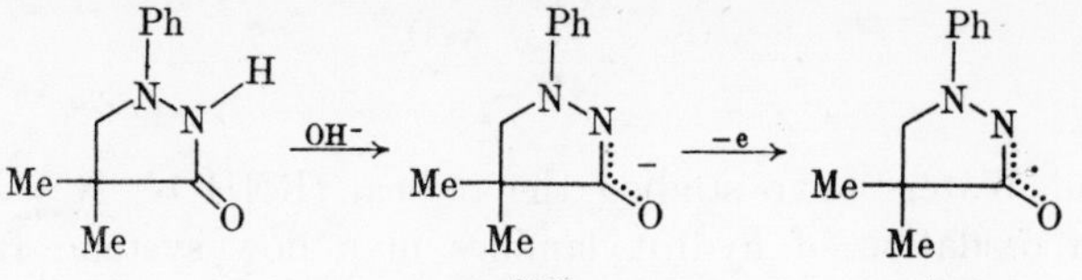

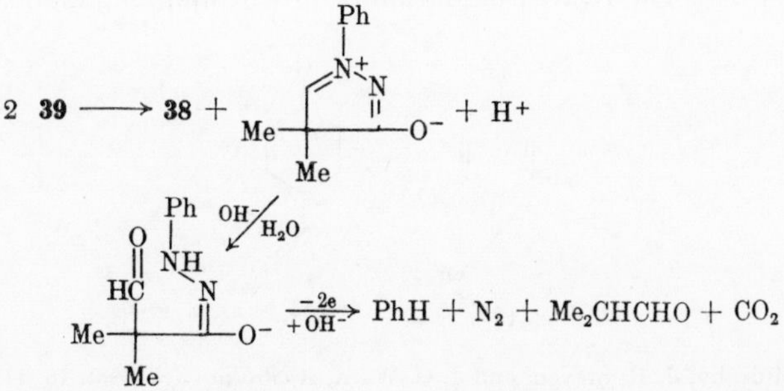

[105] E. Müller, F. Günter, K. Scheffler, P. Ziemek, and A. Rieker, *Ann. Chem.*, **688**, 134 (1965).

[106] T. J. Stone and W. A. Waters, *J. Chem. Soc.*, **1965**, 1488.

[107] C. A. Bishop and L. K. J. Tong, *J. Am. Chem. Soc.*, 87, 501 (1965).

[108] I. Yamazaki and L. H. Piette, *J. Am. Chem. Soc.*, 87, 986 (1965).

[109] E. W. Stone and A. H. Maki, *J. Am. Chem. Soc.*, 87, 454 (1965).

[110] E. Müller, H. Eggensperger, A. Rieker, K. Scheffler, H.-D. Spanagel, H. B. Stegmann, and B. Teissier, *Tetrahedron*, **21**, 227 (1965).

[111] A. Rieker and K. Scheffler, *Tetrahedron Letters*, **1965**, 1337.

[112] C. Steelink, *J. Am. Chem. Soc.*, 87, 2056 (1965).

reactions of t-butyl hydroperoxide,[113] and ESR evidence for hydrogen-bonded hydroxyl radicals in trioxane has been obtained.[114]

The intermediate radical (39) has been detected by ESR spectroscopy in the alkaline ferricyanide-oxidation of 4,4-dimethyl-1-phenylpyrazolidinone.[115]

Nitroxides. Numerous stable nitroxide radicals have been prepared and studied by ESR spectroscopy.[116-121] These include bistrifluoromethylnitrogen oxide, which is a purple gas,[116] the stable diradicals (40a)[117a] and (40b),[117b] and the blue radical anion (41).[118]

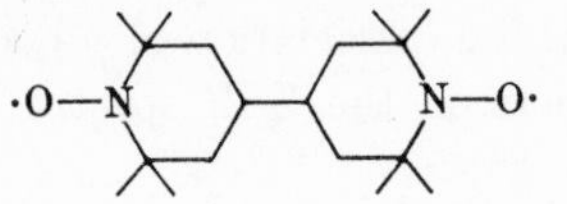

(40a) 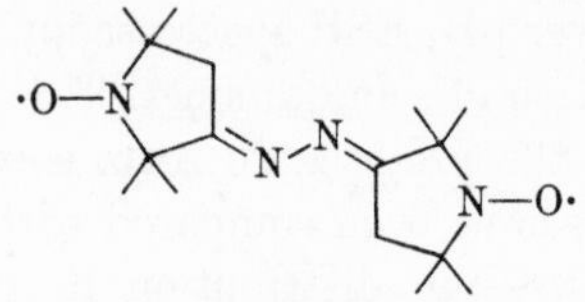   (40b)

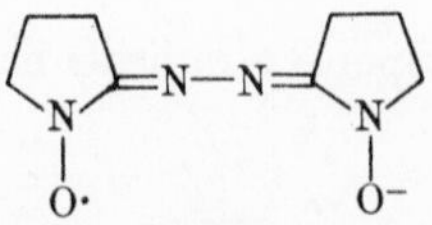

(41)

Gutch and Waters have studied the radical (RNHO·; R = H or Me) obtained on oxidation of hydroxylamines in a flow system. In alkaline carbonate the corresponding radical anions (RNO·⁻) are converted into ·O–NR–CO$_2$⁻.[122] The related oxidation of p-benzoquinone monoxime gives

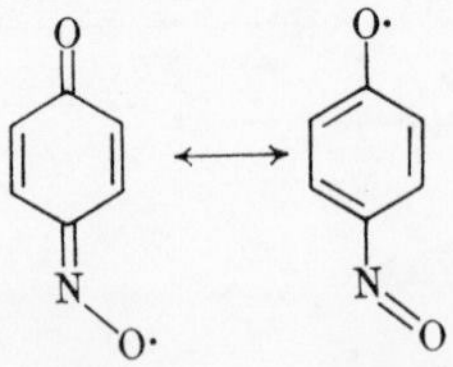

(42)   (43)

[113] M. F. R. Mulcahy, J. R. Steven, and J. C. Ward, *Australian J. Chem.*, 18, 1177 (1965).

[114] H. Yoshida and B. Ranby, *Acta Chem. Scand.*, 19, 1495 (1965).

[115] W. E. Lee, *J. Org. Chem.*, 30, 2571 (1965).

[116] W. D. Blackley and R. R. Reinhard, *J. Am. Chem. Soc.*, 87, 802 (1965).

[117a] E. G. Rosantsev, V. A. Golubev, and M. B. Neiman, *Izv. Akad. Nauk SSSR, Otd. Khim. Nauk.*, 1965, 391, 393; *Chem. Abs.*, 62, 14621 (1965).

[117b] R. M. Dupeyre, H. Lemaire, and A. Rassat, *J. Am. Chem. Soc.*, 87, 3771 (1965).

[118] A. R. Forrester and R. H. Thomson, *J. Chem. Soc.*, 1965, 1224.

[119] F. Tudos, J. Heidt, and J. Ero, *Acta Chim. Acad. Sci. Hung.*, 45, 245 (1965).

[120] H. G. Aurich and F. Baer, *Tetrahedron Letters*, 1965, 2517.

[121] E. G. Rozantzev and L. A. Krinitzkaya, *Tetrahedron*, 1965, 491.

[122] C. J. W. Gutch and W. A. Waters, *J. Chem. Soc.*, 1965, 751.

a radical (42) with $a_N \approx 30$ gauss.[123] This large coupling indicates a relatively minor contribution from (43). The same radical is obtained by alkaline oxidation of NN-dimethyl-p-nitrosoaniline.

Nitrogen Radicals. Diphenylpicrylhydrazyl (DPPH) is a common reference radical in ESR work. A convenient volumetric procedure for the estimation of this radical involves titration with thiosalicylic acid.[124] It has also been shown that DPPH is not a suitable radical scavenger for studying the radical decomposition of peroxybenzoic acid.[125]

Attempts to prepare a diradical related to DPPH gave a diamagnetic product which may be written as (44).[126] However, the "tris-verdazyl" (45) is a stable triradical.[127]

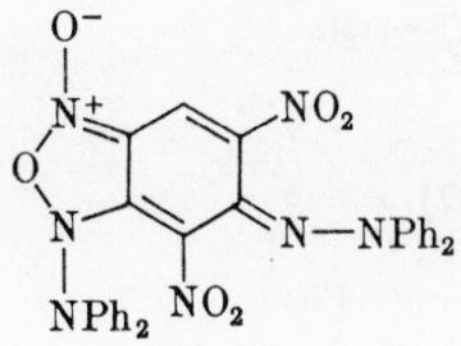

(44)

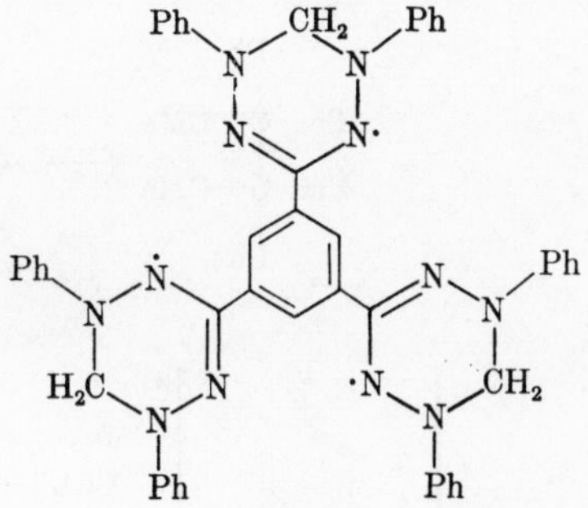

(45)

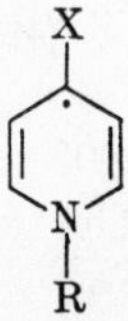

(46)   X = CO_2R', $CONH_2$, CN
        R,R' = alkyl

A range of stable pyridyl radicals (46) have been studied.[128,129] The 3,5-dimethyl-substituted radicals are not stable except where X = CN, indicating the coplanar requirement of other X groups for stabilization of the radical.[129]

[123] W. M. Fox and W. A. Waters, *J. Chem. Soc.*, **1965**, 4628.

[124] J. A. Weil and J. K. Anderson, *J. Chem. Soc.*, **1965**, 5567.

[125] K. Tokumaru, T. Kaziwara, and O. Simamura, *Tetrahedron Letters*, **1965**, 1675.

[126] J. Heidberg and J. A. Weil, *J. Am. Chem. Soc.*, **86**, 5173 (1964).

[127] R. Kuhn, F. A. Neugebauer, and H. Trischmann, *Angew. Chem. Intern. Ed. Engl.*, **4**, 72. (1965).

[128] E. M. Kosower and E. J. Poziomek, *J. Am. Chem. Soc.*, **85**, 2035 (1963); **86**, 5515 (1964); E. M. Kosower and J. L. Cotter, *ibid.*, **86**, 5524 (1964); E. M. Kosower and I. Schwager, *ibid.*, **86**, 5528 (1964).

[129] M. Itoh and S. Nagakura, *Tetrahedron Letters*, **1965**, 417.

Several diarylnitrogen radicals[130] and tetraarylpyrryls[131] have been obtained by reversible dissociation of their dimers, and their ESR spectra have been recorded.

Carbon Radicals. The dissociation of 1,1,2,2-tetraarylethanes having *ortho*-alkyl substituents in the aryl groups has been examined.[132] When each of the four aryl groups is 2,6-diethylphenyl, a solution of the ethane contains an equilibrium concentration of diarylmethyl radicals even at room temperature. The methoxydiphenylmethyl radical (47) has also been observed.[133] In carbon tetrachloride it gives α-methoxybenzhydryl chloride.

Reduction of perchlorate (48) by metal gives the stable red radical (49).[134]

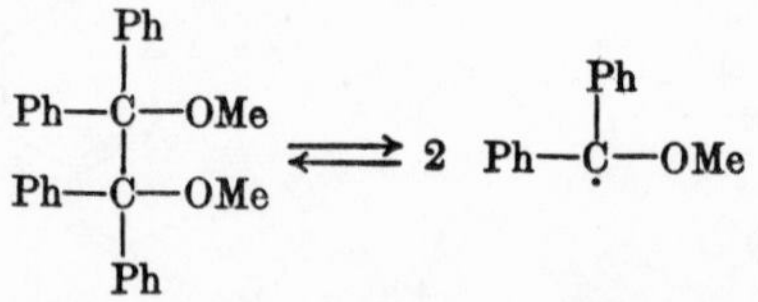

(47)

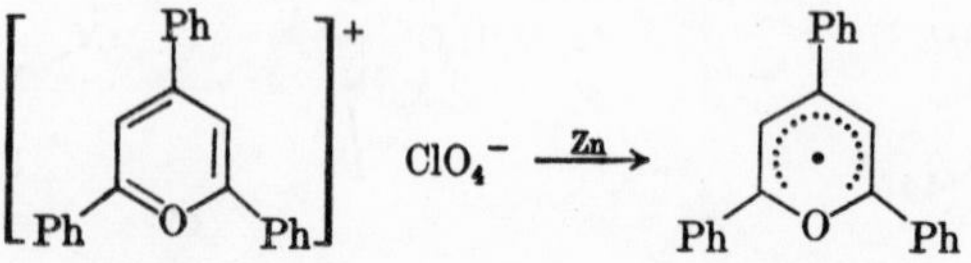

(48)   (49)

Bitropenyl has been found to dissociate above 80° or on photolysis, to give tropenyl (cycloheptatrienyl) free radicals.[135] The variation of the ESR spectrum was studied as a function of temperature.

The ESR spectra of benzyl and hydroxybenzyl radicals formed by reaction of HO· with phenylacetic acid and benzyl alcohol, respectively, have been recorded.[136]

Satisfactory ESR spectra of the phenyl radical have now been obtained,[137] and both these and the electronic spectrum[138] point to a $\pi^6 n$ structure, with the unpaired electron remaining in an $sp^2\sigma$-type orbital.

[130] F. A. Neugebauer and P. H. H. Fischer, *Chem. Ber.*, **98**, 844 (1965).

[131] K. Schilffarth and H. Zimmermann, *Chem. Ber.*, **98**, 3124 (1965).

[132] K. H. Fleurke and W. T. Nauta, *Rec. Trav. Chim.*, **84**, 1059 (1965).

[133] G. E. Hartzell, C. J. Bredeweg, and B. Loy, *J. Org. Chem.*, **30**, 3119 (1965).

[134] V. A. Palchkov, Y. A. Zhdanov, and G. N. Dorofeenko, *Zh. Organ. Khim.*, **1**, 1171 (1965).

[135] G. Vincow, M. L. Morrell, W. V. Volland, H. J. Dauben, and F. R. Hunter, *J. Am. Chem. Soc.*, **87**, 3527 (1965).

[136] H. Fischer, *Naturforsch.*, **20a**, 488 (1965).

[137] J. E. Bennett, B. Mile, and A. Thomas, *Chem. Comm.*, **1965**, 265.

[138] G. Porter and B. Ward, *Proc. Roy. Soc.*, **A287**, 457 (1965).

The 4π-electron system of the pentachlorocyclopentadienyl cation has been shown to have a triplet ground state.[139]

For studies of homogeneous gas kinetics, the walls of the reaction vessel have often been coated with a carbonaceous film in order to minimize surface catalysis. It has been shown that if this film is obtained by pyrolysis of allyl bromide then it gives an appreciable ESR signal, and this demonstrated free radical content should serve as a caution against assuming that the film is indeed inert.[140]

Radical Anions. Russell and Geels have demonstrated the intermediacy of PhNO$\cdot$⁻ in the condensation of nitrosobenzene with phenylhydroxylamine under basic conditions.[141] By showing that there are two different *ortho*-coupling constants in the ESR spectrum of this species and of [PhN=NPh]$\cdot$⁻, it was concluded also that there is restricted rotation about the carbon–nitrogen bonds of these radicals.[142]

In an ESR study of electron-transfer reactions between organic species, high reaction rates were invariably observed, contrary to experience with some inorganic compounds.[143]

The temperature-dependence of the equilibrium (32) has been followed by ESR spectroscopy,[144] and a kinetic study of electron transfer from

$$\text{Na} + \text{Hydrocarbon} \rightleftharpoons [\text{Na}^+ + (\text{Hydrocarbon})\cdot^-] \qquad (32)$$

(anthracene)$\cdot$⁻ and (anthracene)$^{2-}$ to diphenylethylene has been made.[145] The coupling of two $(\text{Ph}_2\text{C–CH}_2)\cdot^-$ radicals is slower than the diffusion-controlled limit by several powers of ten; possibly this is due to dipole repulsion between two ion pairs $[\text{Metal}^+ (\text{Ph}_2\text{C}{=}\text{CH}_2)\cdot^-]$.[146]

The ESR spectrum of the anion radical of toluene is interesting for the low coupling constants of the methyl-hydrogens.[147] These are inconsistent with a hyperconjugative model, but may be interpreted as indicating a small inductive effect of the methyl group.

The ESR spectra of radical anions and radical cations of alternant hydrocarbons are similar, in accord with theoretical predictions. However, the reverse is predicted for the spectra of radical cations and anions of non-alternant hydrocarbons. That this is indeed the case has now been demonstrated for the ions from hydrocarbons (50) and (51).[148]

[139] R. Breslow, R. Hill, and E. Wasserman, *J. Am. Chem. Soc.*, 86, 5349 (1964).
[140] K. A. Holbrook, *Proc. Chem. Soc.*, 1964, 418.
[141] G. A. Russell and E. J. Geels, *J. Am. Chem. Soc.*, 87, 122 (1965).
[142] E. J. Geels, R. Konaka, and G. A. Russell, *Chem. Comm.*, 1965, 13.
[143] J. M. Fritsch, T. P. Layloff, and R. N. Adams, *J. Am. Chem. Soc.*, 87, 1724 (1965).
[144] A. Rembaum, A. Eisenberg, and R. Haack, *J. Am. Chem. Soc.*, 87, 2291 (1965).
[145] J. Jagur-Grodzinski and M. Szwarc, *Proc. Roy. Soc.*, A, 288, 224 (1965).
[146] M. Matsuda, J. Jagur-Grodzinski, and M. Szwarc, *Proc. Roy. Soc.*, A, 288, 212 (1965).
[147] D. Lazdins and M. Karplus, *J. Am. Chem. Soc.*, 87, 920 (1965).
[148] F. Gerson and J. Heinzer, *Chem. Comm.*, 1965, 488.

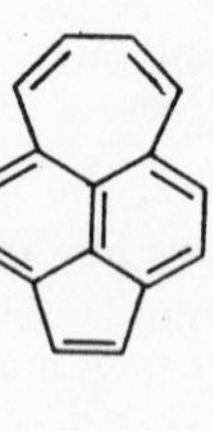

(50) 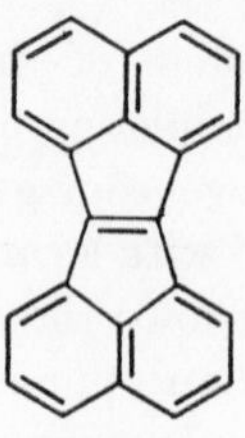   (51)

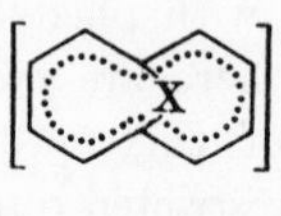

(52) X = CH₂, O 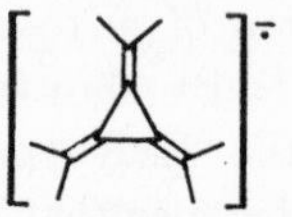   (53) 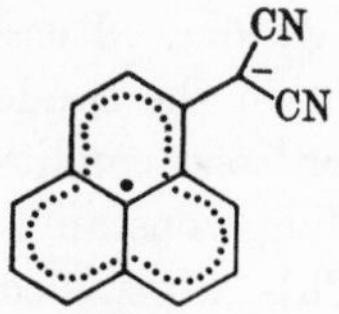   (54)

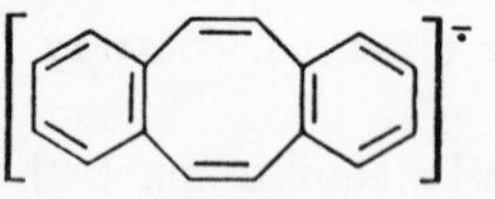

(55) 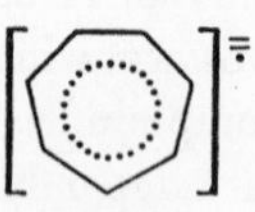   (56)

ESR spectra of the radical anions (52),[149] (53),[150], (54),[151] and (55),[152] and of the dianion radical (56)[153] have been recorded.

The ESR spectra of the radical anions of the benzodiazoles (57) with X = 0, S, and Se show that sulphur and selenium are better able to accept the unpaired electron than is oxygen.[154] Similar results were obtained from a study of the radical anions of dibenzo-furan, -thiophene, and -selenophene. The data are, however, consistent with the absence of significant *d*-orbital interactions.[155] On the other hand, a *d*-orbital model is used successfully to account for the ESR spectra of the radical anions of (58) and (59) where M = Si, Ge, or Sn.[156]

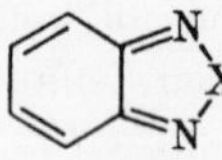

(57) 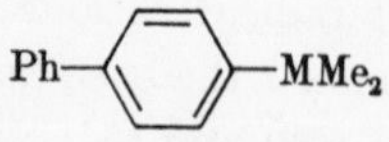   (58) 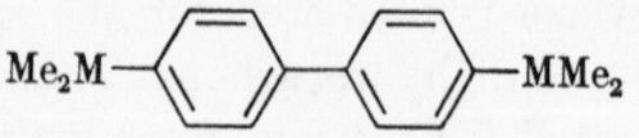   (59)

[149] F. Gerson, E. Heilbronner, W. A. Böll, and E. Vögel, *Helv. Chim. Acta*, 48, 1494 (1965).
[150] F. Gerson, E. Heilbronner, and G. Köbrich, *Helv. Chim. Acta*, 48, 1525 (1965).
[151] T. K. Mukherjee and A. Golubovic, *J. Org. Chem.*, 30, 3166 (1965).
[152] T. J. Katz, M. Yoshida, and L. C. Siew, *J. Am. Chem. Soc.*, 87, 4516 (1965).
[153] N. L. Bauld and M. S. Brown, *J. Am. Chem. Soc.*, 87, 4390 (1965).
[154] E. T. Strom and G. A. Russell, *J. Am. Chem. Soc.*, 87, 3326 (1965).
[155] R. Gerdil and E. A. C. Lucken, *J. Am. Chem. Soc.*, 87, 213 (1965).
[156] M. B. Curtis and A. L. Allred, *J. Am. Chem. Soc.*, 87, 2554 (1965).

The oxygen-17 coupling constant of the nitrobenzene anion radical has been measured.[157]

Janzen has demonstrated spontaneous radical formation when a wide variety of aromatic nitro-compounds are heated above their melting points,[158] the effect being especially pronounced for polynitroaromatic compounds.

Radical Cations. ESR spectra of radical cations of aromatic amines[159] and benzidine derivatives[160] have been recorded. The cation radical of triethylenediamine has of necessity the nitrogen atoms in a pyramidal configuration. This is an unusual situation for a nitrogen cation radical and is probably reflected in the large nitrogen coupling constant ($a_N = 17$ gauss).[161]

The hexamethylbenzene radical cation is observed when hexamethylbenzene is dissolved in oleum or subjected to ultraviolet irradiation in concentrated sulphuric acid.[162] However, if the solution in concentrated sulphuric acid is set aside without irradiation, a different radical slowly appears. This is considered to be (60),[163] and the implications of this result on the mechanism of rearrangements of alkylbenzenes in acids are considered.

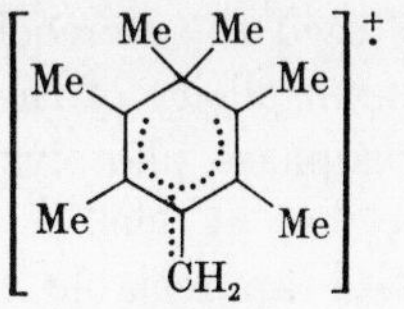

(60)

[157] W. M. Gulick and D. H. Geske, *J. Am. Chem. Soc.*, **87**, 4049 (1965).
[158] E. G. Janzen, *J. Am. Chem. Soc.*, **87**, 3531 (1965).
[159] W. M. Fox and W. A. Waters, *J. Chem. Soc.*, **1964**, 6010.
[160] P. Smejtek, J. Honzl, and V. Metalova, *Collection Czech. Chem. Commun.*, **30**, 3875 (1965).
[161] T. M. McKinney and D. H. Geske, *J. Am. Chem. Soc.*, **87**, 3013 (1965).
[162] R. Hulme and M. C. R. Symons, *Nature*, **206**, 293 (1965); R. Hulme and M. C. R. Symons, *J. Chem. Soc.*, **1965**, 1120.
[163] L. S. Singer and I. C. Lewis, *J. Am. Chem. Soc.*, **87**, 4695 (1965).

CHAPTER 10

Carbenes and Nitrenes

Reviews of intermediates in α-elimination reactions[1] and of carbene additions to alkenes[2] have appeared this year.[3a]

The topics which have received most attention in research papers have been the questions of spin-multiplicity of methylene and other carbenes, and of the nature of the intermediates in α-eliminations catalysed by lithium alkyls, where the formation of free carbenes has been questioned (see, for example, ref. 2a).

In the photolysis of diazomethane or ketene, much evidence has now been accumulated which indicates that triplet methylene may be formed directly, together with the singlet species, when light of relatively long wavelength is used. Much of this work has used the stereochemistry of addition to cis-2-butene as a criterion of the multiplicity of the methylene. Thus Noyes and his co-workers[3b] find that gas-phase photolysis of ketene at 2700 Å gives exclusively singlet methylene, but at 3650 Å non-stereospecific addition to the butene is consistent with appreciable triplet formation. The long-wavelength results are attributed to intersystem crossing of some of the excited ketene molecules to the triplet state, which occurs before decomposition to triplet methylene. The conversion of excited singlet ketene into the triplet state is promoted by collision. Similarly, Rabinovitch, Watkins, and Ring,[4] find that photolysis of diazomethane at 4300 Å gives some triplet methylene. The yields of products derived from triplet methylene increase with pressure, but above about one atmosphere they fall away again. The results at high pressure are attributed to collisional deactivation at a rate greater than that of spin-inversion to triplet diazomethane, and are in accord with the observation that wholly stereospecific (singlet) additions are observed in the liquid phase. By similar, and additional, arguments Frey[5] reaches the same conclusions, and he considers that in photolyses of dia-

[1] W. Kirmse, *Angew. Chem. Intern. Ed. Engl.*, **4**, 1 (1965).
[2] (a) J. I. G. Cadogan and M. J. Perkins, and (b) R. Huisgen, R. Grashey, and J. Sauer, in "The Chemistry of Alkenes", ed. S. Patai, Interscience Publishers, London, New York, and Sydney, 1964.
[3a] A brief survey of carbene chemistry has also appeared: J. O. Schreck, *J. Chem. Educ.* **42**, 260 (1965).
[3b] S. Ho, I. Unger, and W. A. Noyes, *J. Am. Chem. Soc.*, **87**, 2297 (1965).
[4] B. S. Rabinovitch, K. W. Watkins, and D. F. Ring, *J. Am. Chem. Soc.*, **87**, 4960 (1965).
[5] H. M. Frey, *Chem. Comm.*, **1965**, 260.

zomethane at 4360 Å and ketene at 3130 Å some 20–30% of the methylene is initially formed in the triplet state.

Collisional deactivation of singlet methylene to ground-state triplet has been substantiated by Bader and Generosa.[6] These workers also consider the nature of the primary addition product of triplet methylene to an olefin, and suggest that, rather than a diradical (1), "triplet cyclopropane" (2) is a

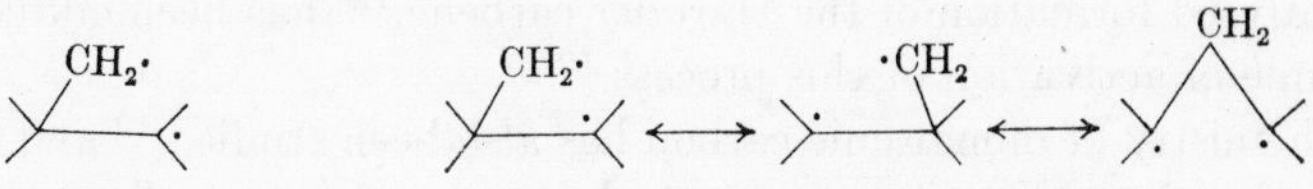

(1)   (2)

better description of the initial adduct. Support for such a concept comes from the photosensitized geometrical isomerization of cyclopropanes studied by Bell,[7] in which proportions of minor by-products are closely similar to those from triplet-methylene additions.

The spin multiplicity of certain substituted carbenes has also received attention. Thus, base-catalysed decomposition of benzaldehyde toluene-*p*-sulphonylhydrazone in the presence of anthracene gives a small yield of compound (3). This is the first example of 1,4-addition of a carbene to anthracene and presumably involves the triplet (cf. A).[8]

The rates of thermal decomposition of diphenyldiazomethane in various solvents, together with the results of a product study, have led to the suggestion of an equilibrium between triplet and singlet diphenylcarbene.[9]

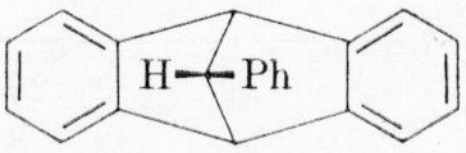

$$Ph_2\dot{C} \rightleftharpoons Ph_2C:$$

(3)   (A)

In the related photolysis of diazofluorene,[10] it is also considered that reactions of two spin-states are observed. The fluorenylidene adds to *cis*-2-butene to give the *cis*- and *trans*-cyclopropane. The proportion of the former may be increased by the presence of oxygen or butadiene, each of which scavenges the triplet state; or it may be lowered by the addition of hexafluorobenzene as an inert solvent which allows more conversion of the initially formed singlet into the triplet fluorenylidene.

[6] R. F. W. Bader and J. I. Generosa, *Can. J. Chem.*, **43**, 1631 (1965).
[7] J. A. Bell, *J. Am. Chem. Soc.*, **87**, 4966 (1965).
[8] H. Nozaki, M. Yamabe, and R. Noyori, *Tetrahedron*, **21**, 1657 (1965).
[9] D. Bethell, D. Whittaker and J. D. Callister, *J. Chem. Soc.*, **1965**, 2466.
[10] M. Jones and K. R. Rettig, *J. Am. Chem. Soc.*, **87**, 4013, 4015 (1965).

In a field almost by itself, is the work of Skell's school on the constituents of carbon vapour. A full report of work on the singlet dicarbene, C_3, has appeared,[11] and it is of interest that this species, in both stages of double addition to an olefin, is much less selective than Hartzler's allenic carbene, $(CH_3)_2C=C=C:$.[12a] It is therefore suggested that the latter is complexed with the Lewis bases present in the medium in which it is formed. However, evidence against such complex-formation, at least in the transition state for base-catalysed formation of the Hartzler carbene,[12b] has been adduced from the volume of activation of this process.[12c]

The chemistry of monatomic carbon has also been studied,[13] and chemical evidence for three distinct spin states has been obtained. Two metastable singlets with half-lives of ca. 2 and ca. 15 seconds (at $-196°$ in a paraffin matrix) are assigned spectroscopic states ^1S and ^1D, respectively. Both added to olefins stereospecifically, to form spiropentanes. Insertion reactions of the ^1S state have also been observed. The more stable ground state of C_1, already identified spectroscopically as ^3P, reacts with an initial molecule of olefin in a stereospecific fashion, and subsequently with a second molecule non-stereospecifically. This double addition has been subjected to a competition study with a series of olefins. Some results are collected in Table 1. The high reactivity of butadiene in the second (triplet) addition may be noted, as well as of the more nucleophilic olefins in the initial singlet reaction.

Table 1. Relative rates of additions of $C_1(^3P)$ to a series of olefins.

Olefin	Initial addition of ⇅C↑↑ (as singlet)	Subsequent addition of △↑↑ (triplet)
=⟋=	1	20
=⟋	5	8
⟍⟋	15	10
=⟨	30	6
⟍=⟍	32	1

Irradiation of both 1- and 2-naphthyldiazomethane in solid solution at $77°K$ gives two triplet species, as shown by their ESR spectra.[14a] These are

[11] P. S. Skell, L. D. Wescott, J. P. Golstein, and R. R. Engel, *J. Am. Chem. Soc.*, **87**, 2829 (1965).

[12a] H. D. Hartzler, *J. Am. Chem. Soc.*, **83**, 4997 (1961).

[12b] W. J. le Noble, *J. Am. Chem. Soc.*, **87**, 2434 (1965).

[12c] For the reaction of this intermediate with some carbanions see A. F. Bramwell, L. Crombie, and M. H. Knight, *Chem. Ind.* (*London*), **1965**, 1265.

[13] P. S. Skell and R. R. Engel, *J. Am. Chem. Soc.*, **87**, (1965), (a) 1135, (b) 2493, (c) 4663.

[14a] A. M. Trozzolo, E. Wasserman, and W. A. Yager, *J. Am. Chem. Soc.*, **87**, 129 (1965).

thought to differ stereochemically, as shown for 1-naphthylmethylene (**4a** and **b**).

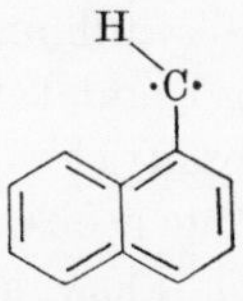

(**4a**) 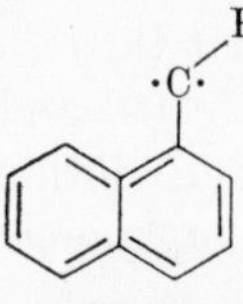   (**4b**)

The question of free carbene versus "carbenoid" intermediates has received considerable attention. In essence, this problem involves the distinction between α-elimination reactions which proceed with formation of free bivalent carbon species (conceivably associated with solvent molecules), and carbenoid processes in which the α-elimination is concerted with product formation, and free carbenes are not involved. In particular, it has been difficult to decide into which category to place the many cyclopropane-forming additions to olefins, where α-elimination is catalysed by a lithium alkyl and where the initial step is assumed to be metallation [e.g. reactions (1)]. Only indirect evidence has previously been brought to bear on the

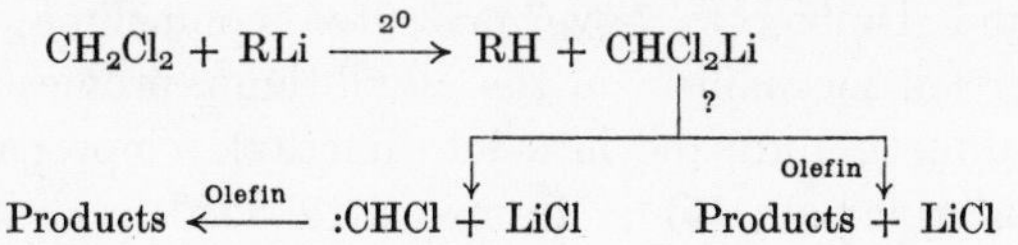

$$CH_2Cl_2 + RLi \xrightarrow{\ 2^0\ } RH + CHCl_2Li$$

$$\text{Products} \xleftarrow{\text{Olefin}} :CHCl + LiCl \xrightarrow{\text{Olefin}} \text{Products} + LiCl \tag{1}$$

product-forming steps. For example, Closs and Coyle showed that mono-chlorocarbene from pyrolysis or photolysis of chlorodiazomethane is a much less discriminating species than that present in the methylene chloride-alkyllithium reaction. The results of this study have been reported in full[14b] and are exemplified by the figures in Table 2.

Table 2. Relative rates of formation of chlorocyclopropanes
(*syn:anti* ratio of products).

Olefin	Reactants	
	CH_2Cl_2/RLi ($-35°$)	$CHClN_2$ (Thermal decomp. $-30°$)
2-Methyl-2-butene	1.76 (1.6)	1.18 (1.0)
Isobutene	1.00	1.00
1-Butene	0.23 (3.4)	0.74 (1.0)

[14b] G. L. Closs and J. J. Coyle, *J. Am. Chem. Soc.*, **87**, 4270 (1965).

8+

Because of the thermal instability of α-halogenoalkyllithium compounds, it has been difficult to make the obvious investigation of kinetic involvement of olefin in the product-forming step [see (1)]. Recently, two groups of workers have found that the α-halogenoalkyllithium reagents are stabilized in tetrahydrofuran at −100° and can, for example, be trapped by carbonation; thus trichloromethyllithium gives trichloroacetic acid. Vigorous decomposition of these organometal reagents is observed, however, when the temperature is raised to ca. −60°. Köbrich, Flory, and Merkle[15] have examined this decomposition of dichloromethyl- and trichloromethyl-lithium in the presence and absence of cyclohexene and ethoxyethylene and have found no evidence for rate enhancement by the olefin. While this result favours free carbenes, different conclusions are reached by Hoeg, Lusk, and Crumbliss[16] who find that at −78° the rates of decomposition of α,α-dichlorobenzyl-lithium and of trichloromethyllithium *are* influenced by cyclohexene. This is apparently not an effect of solvent, as no rate change is produced by addition of cyclohexane. Less nucleophilic olefins such as 1,1-diphenylethylene are also without effect, though cyclopropanes are formed from these olefins during the decomposition at −60°. The results are discussed in terms of a duality of mechanism depending on the reaction conditions and nature of the olefin.

Goldstein and Dolbier[17a] have presented compelling evidence for a carbenoid insertion mechanism in the alkyllithium-promoted conversion of 1,1-diiodo-2,2-dimethylpropane into 1,1-dimethylcyclopropane. The α-halogenolithium intermediate (5) is formed at −116°, and its reactions are compared with those of the hexadeuterated species (6; X = I). The kinetic isotope effect k_H/k_D was calculated from the rates of reaction of the two

$$\underset{\text{(5)}}{\begin{array}{c} H_3C \\ H_3C{-}C{-}CHILi \\ H_3C \end{array}} \xrightarrow{-LiI} \underset{}{\begin{array}{c} H_3C \\ H_3C \end{array}\!\!\triangleright\!\!\!\triangleleft} \qquad \underset{\text{(6)}}{\begin{array}{c} H_3C \\ D_3C{-}C{-}CHXLi \\ D_3C \end{array}}$$

compounds and, separately, for the cyclopropane-forming step by examining the distribution of deuterium in the cyclopropane from (6; X = I). The two values were the same, strongly suggesting that the rate-determining step was the same for the overall reaction as for actual cyclopropane formation. This result allows a concerted one-step mechanism or a rapid pre-equilibrium formation of some intermediate which slowly collapses to cyclopropane. That such an intermediate, if it exists, is not a free carbene was clearly

[15] G. Köbrich, K. Flory, and M. R. Merkle, *Tetrahedron Letters*, 1965, 973.
[16] D. F. Hoeg, D. I. Lusk, and A. L. Crumbliss, *J. Am. Chem. Soc.*, 87, 4147 (1965).
[17a] M. J Goldstein and W. R. Dolbier, *J. Am. Chem. Soc.*, 87, 2293 (1965).

demonstrated by the observation that the deuterium distribution in the dimethylcyclopropane from (6) was markedly dependent on the nature of the halogen X.

The reactivities of a series of alkylethylenes ($RCH=CH_2$; R = Me, Et, i-Pr, or t-Bu) towards "p-tolylcarbenoid" have been studied.[17b] Preferred formation of the sterically least favoured (*cis*) product is observed except where R = t-Bu. This preference is attributed to London and electrostatic interactions in the transition state, as shown in the annexed sketch.

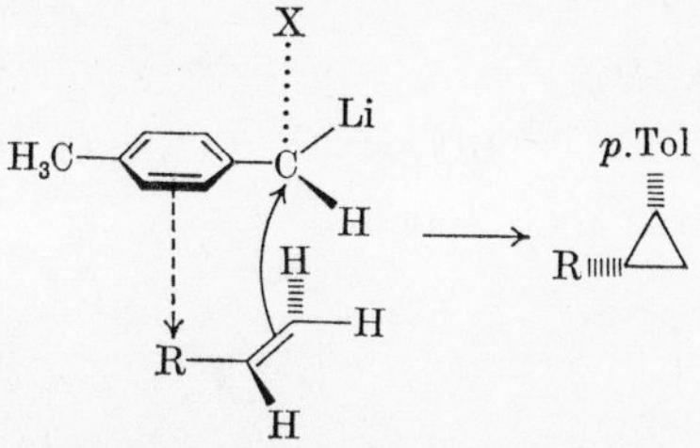

Seyferth's organomercurial reagents for converting olefins to cyclopropanes may react by a carbenoid mechanism, though there is evidence against this.[18,19] If a free carbene is involved it does not appear to come from fragmentations of a halogenomethide anion, as such carbanions could not be trapped by acrylonitrile.[20] Instead, cyanocyclopropanes were formed in the usual way. On the other hand, the iodide-catalysed[21] decomposition of phenyltrihalogenomethylmercury does appear to involve carbanion formation [reactions (2)]. In this reaction, CX_3^- may be trapped by addition

$$I^- + PhHgCX_3 \longrightarrow PhHgI + CX_3^-$$
$$CX_3^- \longrightarrow X^- + :CX_2$$

to acrylonitrile, or by a proton source to give halogenoform. A useful extension of the catalysed reaction is the iodide-catalysed decomposition of Me_3SnCF_3; this provides a convenient source of $:CF_2$ at 80°.[21]

Numerous other new routes to carbene intermediates have appeared during the year. Potentially interesting sources of halogenocarbenes are the halogenodiazirines synthesized by Graham.[22]

[17b] R. A. Moss, *J. Org. Chem.*, **30**, 3261 (1965).

[18] D. Seyferth and J. M. Burlitch, *J. Am. Chem. Soc.*, **86**, 2730 (1964).

[19] O. M. Nefedov and R. N. Shafran, *Izv. Akad. Nauk SSSR, Ser. Khim.*, **1965**, 538; *Chem. Abs.*, **63**, 475 (1965).

[20] D. Seyferth, J. M. Burlitch, R. S. Minasz, J. Y.-P. Mui, H. D. Simmons, A. J. H. Treiber, and S. R. Dowd, *J. Am. Chem. Soc.*, **87**, 4259 (1965).

[21] D. Seyferth, J. Y.-P. Mui, M. E. Gordon, and J. M. Burlitch, *J. Am. Chem. Soc.*, **87**, 681 (1965).

[22] W. H. Graham, *J. Am. Chem. Soc.*, **87**, 4396 (1965).

Gas-phase pyrolysis of chloroform or carbon tetrachloride produces dichlorocarbene which may be trapped by stereospecific addition to an olefin.[23] An interesting route to dichlorocarbene adducts of olefins involves reaction of chloride ion with ethylene oxide in the presence of chloroform and olefin (3–6).[24]

$$\text{Cl}^- + \begin{array}{c} \text{H}_2\text{C} \\ | \\ \text{H}_2\text{C} \end{array}\!\!\!\!\!\!\diagdown\!\!\text{O} \rightleftharpoons \begin{array}{c} \text{CH}_2\text{—O}^- \\ | \\ \text{Cl—CH}_2 \end{array} \tag{3}$$

$$\begin{array}{c} \text{CH}_2\text{—O}^- \\ | \\ \text{Cl—CH}_2 \end{array} + \text{CHCl}_3 \rightleftharpoons \ {}^-\text{CCl}_3 + \begin{array}{c} \text{CH}_2\text{—OH} \\ | \\ \text{Cl—CH}_2 \end{array} \tag{4}$$

$${}^-\text{CCl}_3 \rightleftharpoons \ :\!\text{CCl}_2 + \text{Cl}^- \tag{5}$$

$$:\!\text{CCl}_2 + \bigcirc \!\!\!\parallel \longrightarrow \ \bigcirc\!\!\!\triangleright\!\text{Cl}_2 \tag{6}$$

When phenyllithium is employed as the base catalyst in the formation of chlorocyclopropanes from methylene chloride, some phenylcyclopropane is obtained. It is considered that this involves nucleophilic displacement of chloride from the bivalent intermediate by Ph^- to give phenylcarbene.[25] Phenylcarbene is also proposed as an intermediate in the photolysis of *trans*-stilbene oxide: in the presence of 2-methyl-2-butene geometrical isomers of the cyclopropane (7) are obtained [reactions (7–8)].[26]

$$\begin{array}{c} \text{H} \ \ \text{O} \ \ \text{Ph} \\ \diagdown\!\diagup\diagdown\!\diagup \\ \text{Ph} \ \ \ \ \ \text{H} \end{array} \xrightarrow{h\nu} \begin{array}{c} \text{H} \ \ \text{O} \ \ \text{Ph} \\ \diagdown\!\diagup \cdot \ \cdot \diagdown\!\diagup \\ \text{Ph} \ \ \ \ \ \text{H} \end{array} \longrightarrow \text{PhCHO} + \text{Ph}\ddot{\text{C}}\text{H} \tag{7}$$

$$\text{Ph}\ddot{\text{C}}\text{H} + \ \rangle\!\!=\!\!\langle \ \longrightarrow \ \bigvee\!\!\!\diagup^{\text{CHPh}} \tag{8}$$

$$(90\%)$$

$$(7)$$

Methylene itself may be transferred to an olefin by photolysis of methylene iodide.[27] The stereospecificity of the reaction, and, in particular, the low

---

[23] J. W. Engelsma, *Rec. Trav. Chim.*, **84**, 187 (1965); L. D. Wescott and P. S. Skell, *J. Am. Chem. Soc.*, **87**, 1721 (1965).

[24] F. Nerdel and J. Buddrus, *Tetrahedron Letters*, **1965**, 3585.

[25] O. M. Nefedov, V. I. Shiryaev, and A. S. Khachaturov, *Zh. Obshchei Khim.*, **35**, 509 (1965); *Chem. Abs.*, **63**, 517 (1965).

[26] H. Kristinsson and G. W. Griffin, *Angew. Chem. Intern. Ed. Engl.*, **4**, 868 (1965).

[27] D. C. Blomstrom, K. Herbig, and H. E. Simmons, *J. Org. Chem.*, **30**, 959 (1965).

yields of insertion products, argue against free methylene as an intermediate. On the other hand, on photolysis of both 9,10-dihydro-9,10-methanophenanthrene (8) and of phenylcyclopropane, intermediates are formed which show reactivity patterns characteristic of free methylene.[28] No evidence of

(8)   (9)

methylene formation could be obtained when cycloheptatriene was photolysed. The precursor (8) was prepared from phenanthrene by a simplification of the Simmons–Smith procedure in which zinc replaced zinc–copper couple, dimethoxyethane being the solvent.

Reactions of $PhSO_2\ddot{C}H$[29] and $Me_3Si\ddot{C}H$[30] have been reported, and a convenient synthesis of *gem*-bromophenylcyclopropanes has been described.[31] Dicyanocarbene is formed in the reaction of bromomalononitrile with triethylamine;[32] formed from dicyanodiazomethane,[33] it gives dicyanonorcaradiene (9) with benzene, and, interestingly, shows evidence of reacting at the 2,3- and the 1,9-bond of naphthalene, as well as at the 1,2-bond. A further route to difluorocarbene involves photolysis or pyrolysis of difluorodiazirine.[34] The $:CF_2$ formed reacts stereospecifically with olefins and appears to be rather more selective than $:CCl_2$. In reaction with a given olefin, the proportion of cyclopropane relative to tetrafluoroethylene in the products increases with temperature, in keeping with a finite activation energy for the carbene addition.

α-Elimination from an olefin often results in rearrangement to an acetylene [reaction (9)]. Treatment of bromomethylenecycloalkanes with potassium *t*-butoxide in toluene is not exceptional,[35] in that cycloalkynes can be

$$\underset{R}{\overset{R}{\diagdown}} C = C \underset{X}{\overset{H}{\diagup}} \xrightarrow{-HX} RC \equiv CR \qquad (9)$$

[28] D. B. Richardson, L. R. Durrett, J. M. Martin, W. E. Putnam, S. E. Slaymaker, and I. Dvoretzky, *J. Am. Chem. Soc.*, **87**, 2763 (1965).
[29] R. A. Abramovitch and J. Roy, *Chem. Comm.*, **1965**, 542.
[30] I. A. D'yakonov, I. B. Repinskaya, and G. V. Golodnikov, *Zh. Obshchei Khim.*, **35**, 199 (1965).
[31] R. A. Moss and R. Gerstl, *Tetrahedron Letters*, **1965**, 3445.
[32] J. S. Swenson and D. J. Renaud, *J. Am. Chem. Soc.*, **87**, 1394 (1965).
[33] E. Ciganek, *J. Am. Chem. Soc.*, **87**, 652 (1965).
[34] R. A. Mitsch, *J. Am. Chem. Soc.*, **87**, 758 (1965).
[35] K. L. Erickson and J. Wolinsky, *J. Am. Chem. Soc.*, **87**, 1142 (1965).

isolated if the ring is sufficiently large. With smaller rings the strained cyclic acetylenes may be trapped by diphenylisobenzofuran. Evidence that the initially formed intermediate is a cycloalkylidenecarbene (e.g. **10**) comes

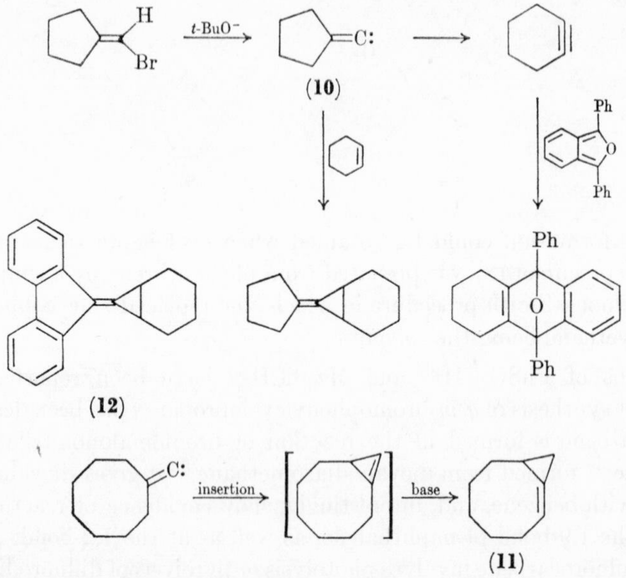

from observation of insertion products (**11**) and of direct trapping with cyclohexene. Fluorenylidenecarbene is a plausible intermediate in the formation of compound (**12**) by diazotization of aminomethylenefluorene with pentyl nitrite in the presence of cyclohexene.[36]

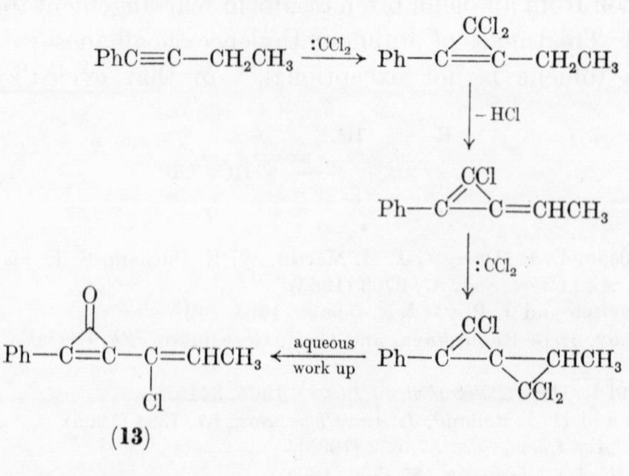

[36] D. Y. Curtin, J. A. Kampmeier, and B. R. O'Connor, *J. Am. Chem. Soc.*, **87**, 863 (1965).

An unusual reaction of dichlorocarbene with an acetylene has been noted by Dehmlow.[37] Rationalization of the apparent insertion product (13) is outlined in the formulae. Dichlorocarbene has also been found to insert into the α-C–H bonds of ethers[38] and to cleave a C–S bond of di-t-butyl disulphide.[39] The Reimer–Tiemann reaction is well known to involve attack on phenoxide ion in the 2-position by $:CCl_2$; thiophenoxide ion reacts with difluorocarbene at sulphur, with formation of a sulphide.[40]

Hydride transfer from a benzylate anion to dichlorocarbene, has been postulated to account for the products formed when chloroform is added to a solution of sodium benzylate in benzyl alcohol.[41] Thus methylene chloride and, after hydrolysis, benzoic acid are reaction products [see (10)]. In addition,

$$PhCH_2O^- + :CCl_2 \longrightarrow PhCHO + {}^-CHCl_2 \qquad (10)$$

$$\downarrow \text{Base} \qquad \qquad \downarrow \text{PhCH}_2\text{OH}$$

$$PhCH_2OCOPh \qquad CH_2Cl_2$$

the formation of benzyl phenylacetate is explained in terms of the rearrangement indicated in (11–13).

$$PhCH_2O^- + :CCl_2 \longrightarrow PhCH_2O\overset{..}{C}Cl + Cl^- \qquad (11)$$

$$PhCH_2O\overset{..}{C}Cl \longrightarrow PhCH_2\overset{+}{C}{=}O + Cl^- \qquad (12)$$

$$PhCH_2\overset{+}{C}{=}O \xrightarrow{PhCH_2OH} PhCH_2\overset{O}{\overset{\|}{C}}{-}OCH_2Ph \qquad (13)$$

A 1,2-hydride shift is the expected mode of reaction of carbene (14),[42] but it is accompanied by the unusual rearrangement shown.

The aprotic base-catalysed decomposition of toluene-p-sulphonylhydrazones (15) and (16) (Bamford–Stevens reaction) occurs under unusually mild conditions, which is regarded as indicative of double-bond participation in carbene formation.[43] Ylid formation by co-ordination of carbenes with the lone-pair electrons on oxygen and sulphur are held to be reasonable first steps in the decomposition of diazoacetic ester in styrene oxide[44] and in the reaction of dichlorocarbene with allyl phenyl sulphide.[45]

Contrary to an earlier report by Frey and Stevens,[46] Friedman and his

[37] E. V. Dehmlow, *Tetrahedron Letters*, **1965**, 4003.
[38] J. C. Anderson, D. G. Lindsay, and C. B. Reese, *J. Chem. Soc.*, **1964**, 4874.
[39] S. Searles and R. E. Wann, *Tetrahedron Letters*, **1965**, 2899.
[40] A. De Cat, R. Van Poucke, R. Pollet, and P. Schots, *Bull. Soc. Chim. Belg.*, **74**, 270 (1965).
[41] J. A. Landgrebe, *Tetrahedron Letters*, **1965**, 105.
[42] P. K. Freeman and D. G. Kuper, *J. Org. Chem.*, **30**, 1047 (1965).
[43] M. Schwarz, A. Besold, and E. R. Nelson, *J. Org. Chem.*, **30**, 2425 (1965).
[44] H. Nozaki, H. Takaya, and R. Noyori, *Tetrahedron Letters*, **1965**, 2563.
[45] W. E. Parham and S. H. Groen, *J. Org. Chem.*, **33**, 728 (1965).
[46] H. M. Frey and I. D. R. Stevens, *Proc. Chem. Soc.*, **1964**, 144.

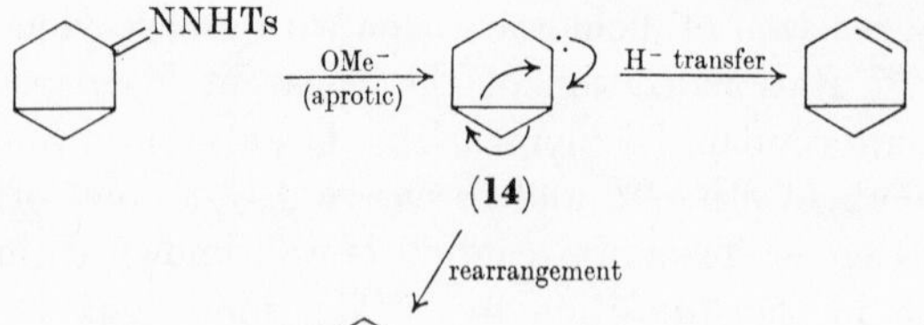
NNHTs
OMe⁻
(aprotic)
H⁻ transfer
rearrangement

(**14**)

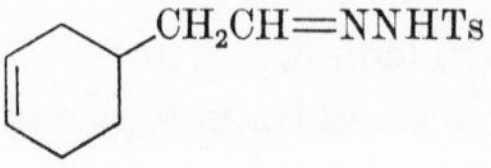
CH₂CH=NNHTs

(**15**)

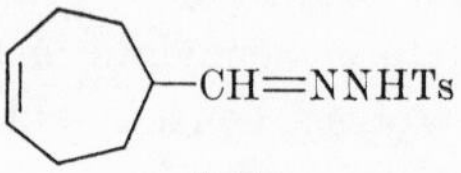
CH=NNHTs

(**16**)

co-workers[47] have shown that bicyclobutane is not a significant product of the aprotic Bamford–Stevens reaction of the toluene-*p*-sulphonylhydrazone of cyclopropanecarboxaldehyde. However, traces of protic solvents (water, ethylene glycol) drastically change the nature of the C_4 decomposition products, the major one then being bicyclobutane. It is suggested that this is formed from an energetic cyclopropylmethyl carbonium ion. Under the aprotic (carbene) conditions, deuterium labelling is consistent with the formation of the major products (cyclobutene and butadiene) directly from the expected carbene (see formulae).

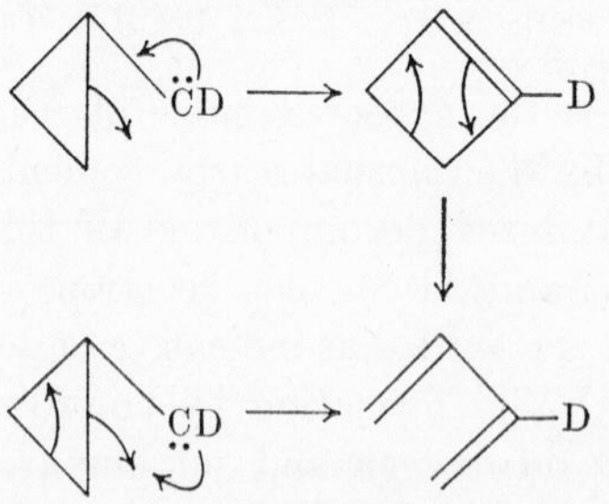
C̈D
D
CD
D

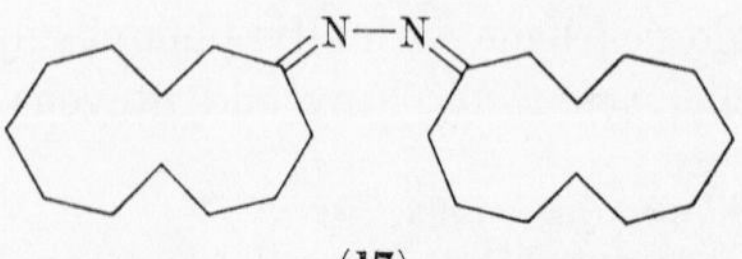
N—N

(**17**)

A halogenocarbene adduct to the 2,3-bond of an indole has been trapped before ring-expansion to the quinoline occurs,[48] and further work on carbene

[47] J. A. Smith, H. Schechter, J. Bayless, and L. Friedman, *J. Am. Chem. Soc.*, **87**, 659 (1965); J. Bayless, L. Friedman, J. A. Smith, F. B. Cook, and H. Schechter, *ibid.*, 661.
[48] H. E. Dobbs, *Chem. Comm.*, **1965**, 56.

addition to pyrroles has also been reported.[49] Dichlorocarbene reacts at the least strained double bond of bicyclopentadiene,[50] and exclusively at a *trans*-double bond of *cis,trans,trans*-cyclododecatriene.[51] With cyclooctene and cyclooctatrienes, normal cyclopropane adducts are formed,[52] and not the derivatives of bicyclo[4,2,1]nonane reported by Sanne and Schlichting.[53]

A conformational effect appears to operate in stabilizing the carbene formed by base-catalysed decomposition of the toluene-*p*-sulphonylhydrazone of cyclododecanone. An appreciable proportion of the intermediate appears to be sufficiently long-lived to participate in a bimolecular reaction leading to the azine (17).[54]

Wanzlick's concept[55] of "nucleophilic carbenes" has been found wanting,[56,57] in that many of the products of reactions of these postulated species can be ascribed to reactions of tetraaminoethylenes, e.g. (18), which are

(18)   (19)   (20)

dimers of the carbenes. However, evidence *has* been presented for possible reaction of the species (19)[58] and (20)[59] in monomeric form. Wanzlick and his colleagues have also reported further results with the tetraaminoethylene system.[60]

Among rearrangements thought to involve carbenes, the Wolff rearrangement of PhCOCH: (21) from photolysis of PhCOCHN$_2$ apparently involves singlet (21). When the diazo compound was subjected to photosensitized decomposition, much acetophenone was formed, presumably by hydrogen abstraction by triplet PhCOĊH.[61] The migratory aptitudes of aryl groups

[49] R. Nicoletti and M. L. Forcellese, *Gazz. Chim. Ital.*, **95**, 83 (1965); *Chem. Abs.*, **63**, 6946 (1965); *Tetrahedron Letters*, **1965**, 3033.
[50] L. Ghosez, P. Laroche, and L. Bastens, *Tetrahedron Letters*, **1964**, 3745.
[51] J. M. Locke and E. W. Duck, *Chem. Ind. (London)*, **1965**, 1727.
[52] D. I. Schuster and F.-T. Lee, *Tetrahedron Letters*, **1965**, 4119.
[53] W. Sanne and O. Schlichting, *Angew. Chem.*, **75**, 156 (1963).
[54] A. P. Krapcho and J. Diamanti, *Chem. Ind. (London)*, **1965**, 847.
[55] H. W. Wanzlick, *Angew. Chem. Int. Ed. Engl.*, **1**, 75 (1962).
[56] D. M. Lemal, R. A. Lovald, and K. I. Kawano, *J. Am. Chem. Soc.*, **86**, 2518 (1964).
[57] H. E. Winberg, J. E. Carnahan, D. D. Coffman, and M. Brown, *J. Am. Chem. Soc.*, **87**, 2055 (1965).
[58] H. Quast and E. Frankenfeld, *Angew. Chem. Intern. Ed. Engl.*, **4**, 691 (1965).
[59] H. Quast and S. Hünig, *Angew. Chem. Intern. Ed. Engl.*, **3**, 800 (1964).
[60] H.-W. Wanzlick, B. Lachmann, and E. Schikora, *Chem. Ber.*, **98**, 3170 (1965).
[61] A. Padwa and R. Layton, *Tetrahedron Letters*, **1965**, 2167.

8*

(p-MeOC$_6$H$_4$ > C$_6$H$_5$ > p-O$_2$NC$_6$H$_4$) in the rearrangement depicted in (14) are consistent with the carbene's being a singlet (electrophilic) species.[62]

$$Ar_3C-\overset{..}{C}H \longrightarrow Ar_2C{=}CHAr \tag{14}$$

However, aprotic base-catalysed decomposition of the toluene-p-sulphonyl-hydrazones (22) causes predominantly ring expansion rather than phenyl migration, even when the product contains a seven-membered ring.[63]

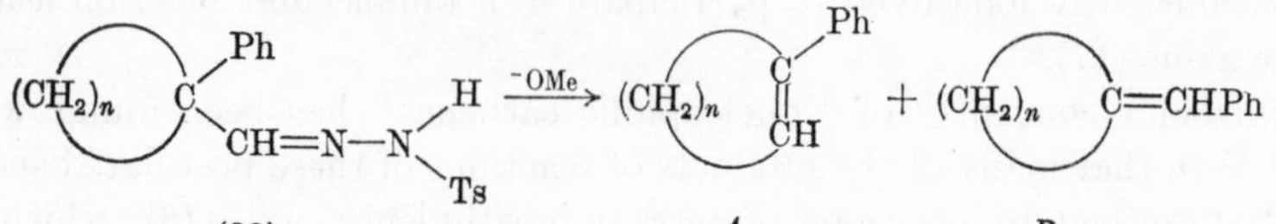

	A	B
$n = 2$	100%	0%
$n = 5$	60%	40%

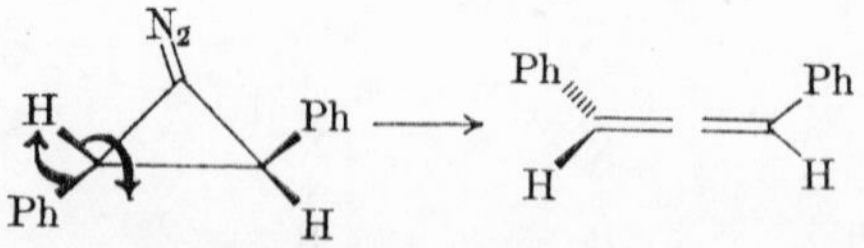

(15)

In the carbenic collapse of a diazocyclopropane to an allene, the configuration of the product appears to be governed by a greater mutual steric interaction between substituents on carbon than between these substituents and the leaving nitrogen molecule [reaction (15)].[64] This type of reaction has now been applied to syntheses of cumulenes.[65]

Rearrangements possibly involving nitrenes have also received attention. Photolysis of pivaloyl azide in cyclohexene gives addition and insertion products of the intermediate nitrene in addition to the rearrangement product, t-butyl isocyanate.[66] Thermal decomposition, however, gives only the isocyanate. The thermal Curtius rearrangement, therefore, may not involve free nitrene intermediates. Photolysis of benzenesulphonyl azide in methanol gives methyl sulphanilate in a novel reaction analogous to the Curtius rearrangement.[67] The possible sulphonylnitrene intermediate may also participate in the decomposition of this azide in norbornadiene, to give a product (23).[68]

[62] P. B. Sargeant and H. Schechter, *Tetrahedron Letters*, **1964**, 3957.
[63] J. W. Wilt, J. M. Kosturik, and R. C. Orlowski, *J. Org. Chem.*, **30**, 1052 (1965).
[64] W. M. Jones and J. W. Wilson, *Tetrahedron Letters*, **1965**, 1587.
[65] F. T. Bond and D. E. Bradway, *J. Am. Chem. Soc.*, **87**, 4977 (1965); L. Skattebøl, *Tetrahedron Letters*, **1965**, 2175; G. Maier, *ibid.*, **1965**, 3603.
[66] W. Lwowski and G. T. Tisue, *J. Am. Chem. Soc.*, **87**, 4022 (1965).
[67] W. Lwowski and E. Scheiffele, *J. Am. Chem. Soc.*, **87**, 4359 (1965).
[68] J. E. Franz and C. Osuch, *Chem. Ind. (London)*, **1964**, 2058.

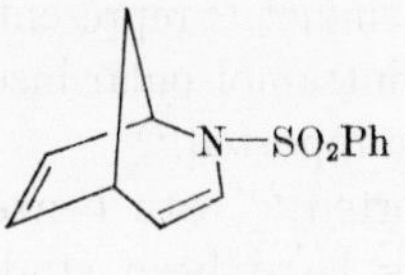

(23) 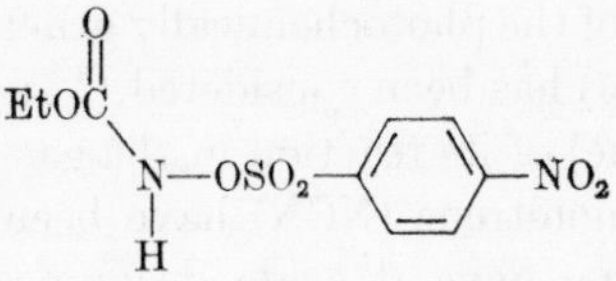   (24) 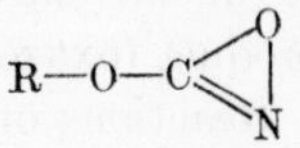   (25)

Acylnitrenes may be involved in a reaction analogous to the Hofmann rearrangement, which is observed when primary amides are oxidized by lead tetraacetate.[69]

Alkoxycarbonylnitrenes continue to receive attention. Direct photolysis of ethyl azidoformate appears to give the singlet nitrene,[70,71] though if the decomposition is photosensitized by acetophenone[71] in cyclohexene the products include ethyl carbamate and bicyclohexenyl, presumably arising from radical reactions of the triplet nitrene (or triplet azide). The reactions of ethoxycarbonylnitrene from base-catalysed decomposition of the ester (24)

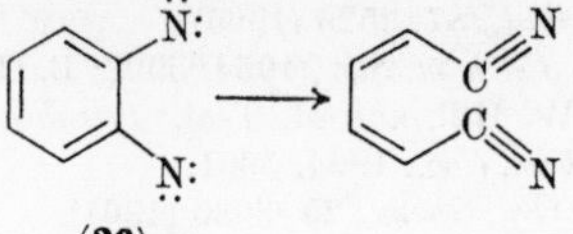

(26)

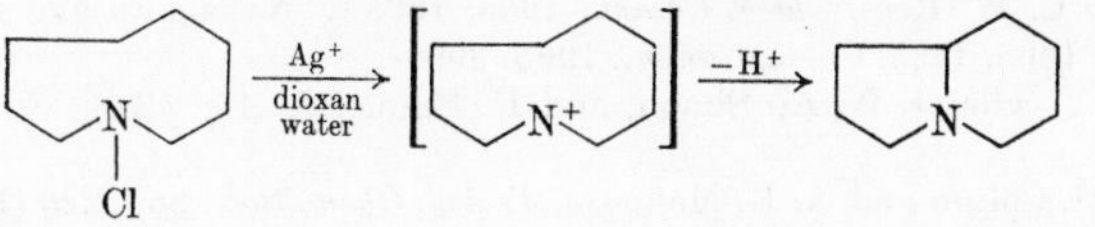

(16)

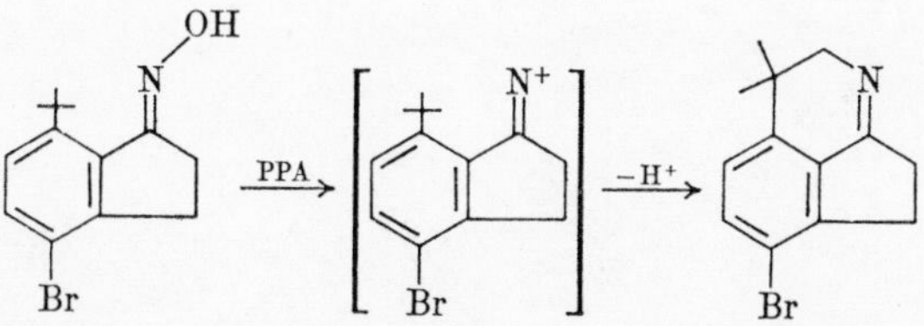

(27)

(17)

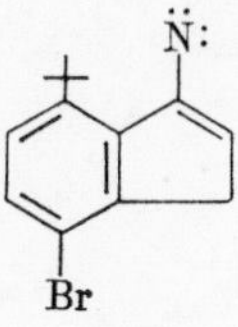

(28)

[69] H. E. Baumgarten and A. Staklis, *J. Am. Chem. Soc.*, **87**, 1141 (1965).
[70] K. Hafner, W. Kaiser, and R. Puttner, *Tetrahedron Letters*, **1964**, 3953.
[71] W. Lwowski and T. W. Mattingly, *J. Am. Chem. Soc.*, **87**, 1947 (1965).

are similar to those of the photochemically generated singlet;[72] representation
of this nitrene as (25) has been considered,[73] and an intramolecular insertion
product (oxazolidone) of its reaction in the gas-phase reported.[74]

Reactions of cyanonitrene (NCN) have been reported;[75] with benzene it
gives 1-cyanoazepine. Several azide decompositions have been studied,[76]
and these include a synthesis of 2-substituted benzimidazoles from *o*-
azidoanils.[77] *o*-Diazidobenzene on pyrolysis gives *cis,cis*-mucononitrile.[78]This
and two other syntheses of the same compound[79] are thought to involve the
dinitrene intermediate (26).

Bivalent nitrenium ions appear to be involved in the insertion reactions
(16)[80] and (17)[81]. Absence of deuterium incorporation from deuterated
polyphosphoric acid into the product (27) appears to exclude the formation of
a nitrene (28).

[72] W. Lwowski and T. J. Maricich, *J. Am. Chem. Soc.*, **87**, 3630 (1965).
[73] D. W. Cornell, R. S. Berry, and W. Lwowski, *J. Am. Chem. Soc.*, **87**, 3626 (1965).
[74] R. Kreher and D. Kühling, *Angew. Chem. Intern. Ed. Engl.*, **4**, 69 (1965).
[75] A. G. Anastassiou, H. E. Simmons, and F. D. Marsh, *J. Am. Chem. Soc.*, **87**, 2296 (1965);
     F. D. Marsh and H. E. Simmons, *ibid.*, **87**, 3529 (1965).
[76] R. K. Smalley and H. Suschitzky, *J. Chem. Soc.*, **1964**, 5922; B. Coffin and R. F. Robbins,
     ibid., **1965,**, 1252; J. H. Hall, J. W. Hill, and H. Tsai, *Tetrahedron Letters*, **1965**, 2211;
     B. Coffin and R. F. Robbins, *J. Chem. Soc.*, **1964**, 5901.
[77] L. Krbechek and H. Takimoto, *J. Org. Chem.*, **29**, 3630 (1964).
[78] J. H. Hall, *J. Am. Chem. Soc.*, **87**, 1147 (1965).
[79] C. D. Campbell and C. W. Rees, *Chem. Comm.*, **1965**, 192; K. Nakagawa and H. Onoue,
     Tetrahedron Letters, **1965**, 1433; *Chem. Comm.*, **1965**, 396.
[80] O. E. Edwards, O. Vocelle, J. W. Ap Simon, and F. Haque, *J. Am. Chem. Soc.*, **87**, 678
     (1965).
[81] P. J. Lansbury, J. G. Colson and N. R. Mancuso, *J. Am. Chem. Soc.*, **86**, 5225 (1964).

CHAPTER 11

Reactions of Aldehydes and Ketones and their Derivatives[1]

Formation and Reactions of Acetals and Ketals

A definitive paper on substituent effects in the hydrolysis of benzaldehyde acetals has appeared.[2] Both diethyl acetals (1) and 2-phenyl-1,3-dioxolanes (2) were studied. The former reacted 30–35 times the faster owing to more favourable entropies of activation. The entropies of activation for the

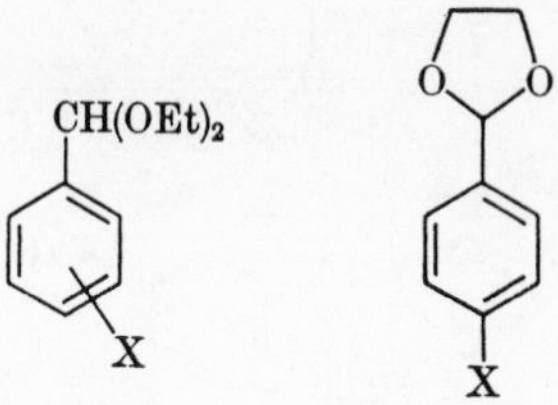

(1)   (2)

2-aryl-1,3-dioxolanes were in fact all negative ($\Delta S^{\ddagger} = -7.3$ to -9.6 e.u.) but the mechanism was still considered to be A-1 and reasons for the negative values were discussed. Curved plots of log k against σ were obtained for both series of compounds with *para*-substituents, but a straight line was obtained with the *meta*-substituted diethyl acetals, ρ being -3.35. In particular, the point for the p-methoxy compound fell above this line but when σ^+ constants were used the point for this compound fell below the line. It seemed, therefore, as if the σ^+ constants were overcompensating for the interaction of the substituent with the carbonium ion. In the Reviewers' opinion, the reason for this may be that the measured rate constant is a composite of the equilibrium constant for protonation (correlated by σ) and the rate constant (correlated by σ^+).

The effect of ring size on the rates of hydrolysis of cyclic benzaldehyde acetals has also been determined, by studying the hydrolyses of compounds (3)–(6);[3] the rate varies with ring size in the order $7 > 5 > 6$, as shown.

[1] For a review on certain aspects see W. P. Jencks, *Progr. Phys. Org. Chem.*, **2**, 63 (1964).
[2] T. H. Fife and L. K. Jao, *J. Org. Chem.*, **30**, 1492 (1965).
[3] J. Kovář, J. Šteffkova, and J. Jarý, *Collection Czech. Chem. Commun.*, **30**, 2793 (1965).

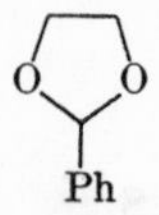

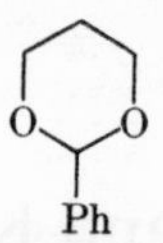

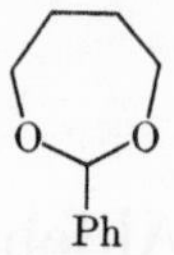

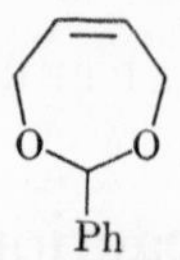

	Ph	Ph	Ph	Ph
$10^5 k$ (sec^{-1}) at pH 1·95 at 25°	238	39·2	1000	87·3
	(3)	(4)	(5)	(6)

Introduction of methyl groups into the glycol residue of the five- and six-membered-ring compounds was shown to cause a rate decrease; *gem*-dimethyl groups were especially effective in this respect. Other extensive kinetic investigations on the hydrolyses of 1,3-dioxolanes[4] and 1,3-dioxanes[5] have been reported.

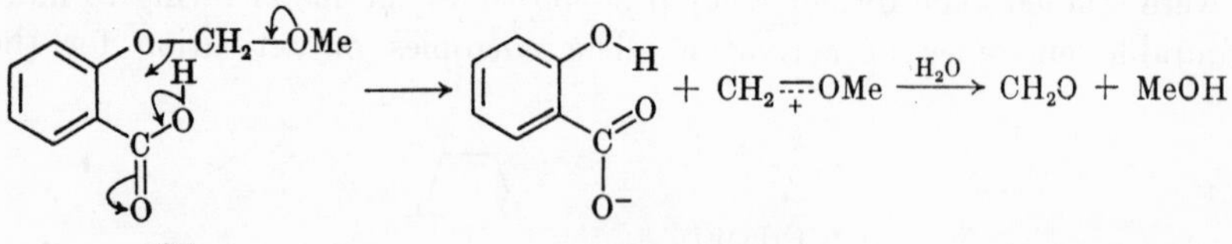

(7)

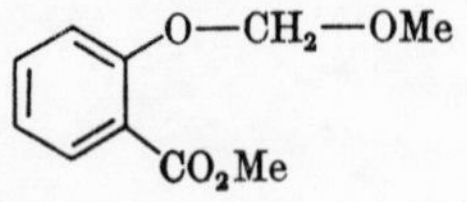

(8) 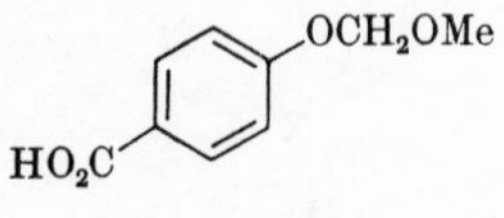   (9) 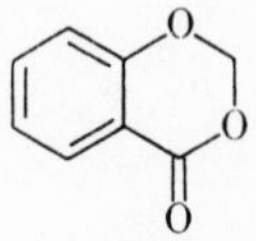   (10)

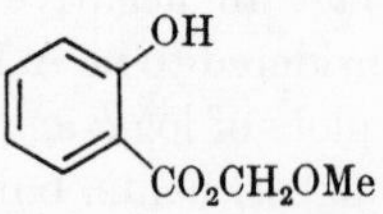

(11)

Intramolecular acid catalysis has been reported for the hydrolysis of *o*-methoxymethoxybenzoic acid (7).[6] In the pH range 3.1–5.5 the reaction follows a rate law, Rate = k[Un-ionized form]; at pH 4.11 the rate is 650 times greater than that for methyl *o*-methoxymethoxybenzoate (8), and at pH 4.08 it is 300 times faster than that for *p*-methoxymethoxybenzoic acid

[4] F. Aftalion, M. Hellin, and F. Coussemant, *Bull. Soc. Chim. France*, **1965**, 1497; M. F. Shostskovskii, A. S. Atavin, B. A. Trofimov, and V. I. Lavrov, *Izv. Sibirsk. Otd. Akad. Nauk SSR, Ser. Khim. Nauk*, **1965**, 93; *Chem. Abs.*, **63**, 11,280 (1965); J. Leslie and D. Hamer, *J. Chem. Soc.*, **1965**, 5769.

[5] M. Garnier, F. Aftalion, D. Lumbroso, M. Hellin, and F. Coussemant, *Bull. Soc. Chim. France*, **1965**, 1512; F. Aftalion, D. Lumbroso, M. Hellin, and F. Coussemant, *ibid.*, **1965**, 1950, 1958.

[6] B. Capon and M. C. Smith, *Chem. Comm.*, **1965**, 523.

(9). The possibility that the reaction involved nucleophilic participation by the ionized carboxyl group and specific hydronium ion-catalysis was excluded by studying the hydrolysis of the compounds (10) and (11) which would be intermediates.

Nucleophilic participation in the acid-catalysed rupture of the acetal bonds of the acyclic dimethyl acetals of glucose (12) and galactose has been reported.[7] In dilute aqueous acid these compounds, besides undergoing hydrolysis, also yield methyl furanosides (e.g. 13) by a ring-closure reaction. It was estimated that the rates of these ring closures were faster than the expected unassisted rates of hydrolysis, and it was therefore concluded that they were synchronous processes, the hydroxyl group providing anchimeric assistance. It has also been shown[8] that hydrolysis of the diethyl acetal of

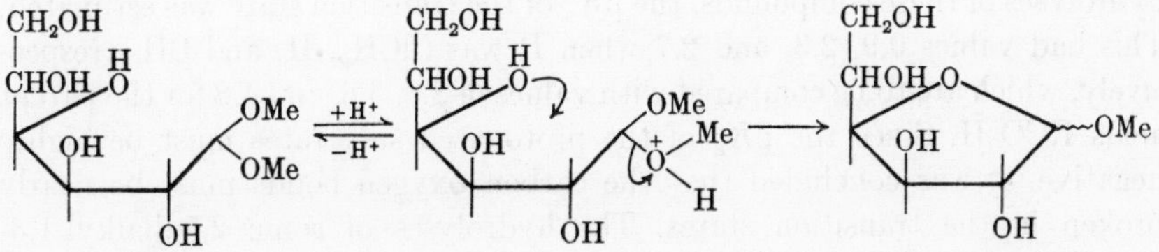

methylthioacetaldehyde proceeds at a somewhat enhanced rate, which was attributed to participation by the methylthio group, as illustrated.

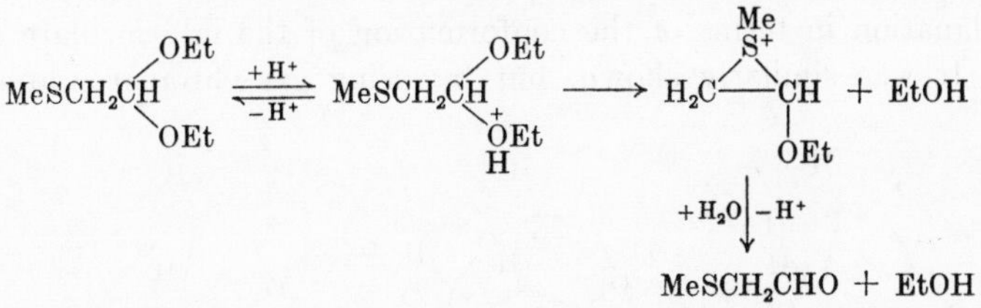

There have been several investigations of the hydrolysis of acylals, R.CO.OCHR'–OR". The pH-rate profile for the hydrolysis of 4-ethoxy-4-butyrolactone (14) has been obtained by Fife.[9] Below about pH 3 the rate is proportional to the concentration of hydrogen ions, and above pH 10 proportional to the concentration of hydroxyl ion, but from pH 5 to 9 the rate profile shows a plateau, with the rate independent of pH. It was thought that this plateau did not correspond to catalysis by water since neither imidazole nor acetate ion had a catalytic effect. Instead, a spontaneous

[7] B. Capon and D. Thacker, *J. Am. Chem. Soc.*, **87**, 4199 (1965).
[8] J. C. Speck, D. J. Rynbrandt, and I. H. Kochevar, *J. Am. Chem. Soc.*, **87**, 4979 (1965).
[9] T. H. Fife, *J. Am. Chem. Soc.*, **87**, 271 (1965).

decomposition in which the carboxylate ion acts as a leaving group was preferred as depicted in the annexed formulae. The entropy of activation for

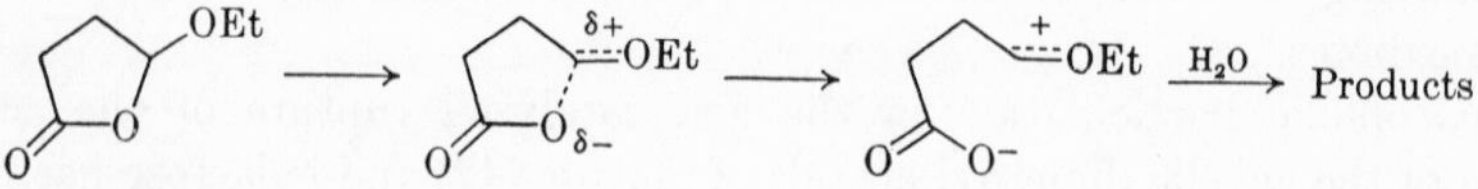

(14)

this reaction is -18.5 e.u., indicating a high degree of solvation for the transition state, and hence bond breaking is presumably well advanced. A similar conclusion has been reached by Salomaa in an investigation of the hydrolyses of a series of methoxymethyl esters, $MeOCH_2OCOR$.[10] From a comparison of the rate constants for the spontaneous and acid-catalysed hydrolyses of these compounds, the pK_a of the transition state was estimated. This had values 0.9, 2.3, and 2.7 when R was $ClCH_2$, H, and CH_3, respectively, which are to be compared with values of 2.8, 3.8, and 4.8 for the parent acids RCO_2H. Since the pK_a of the protonated substrates must be highly negative, it was concluded that the carbon–oxygen bonds must be nearly broken in the transition states. The hydrolyses of some 2,5-dialkyl-1,3-dioxolones have also been investigated.[11]

It has been shown by monitoring the reaction solution by NMR spectroscopy that treatment of 1,4-anhydroerythritol (15) with benzaldehyde in nitromethane containing toluene-p-sulphonic acid yields preferentially the acetal (16) (Ph *endo*), which then equilibrates with its isomer (17) (Ph *exo*).[12] An explanation in terms of the conformation of the intermediate ion was offered. It was similarly shown, but by using gas chromatography, that

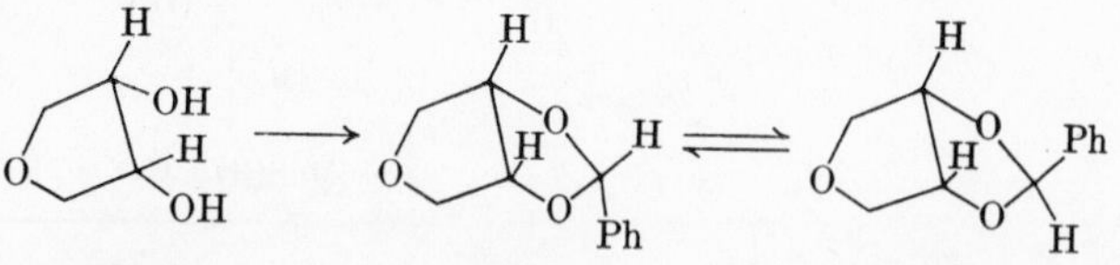

(15)　　　　　　　　(16)　　　　　　　　(17)

glycerol and acetaldehyde yield the *cis*- and *trans*-dioxolanes (18) first and the dioxanes (19) later.[13] The equilibria between these isomers were also studied. Further, the reactions of several *O*-isopropylidene derivatives of alditols have been investigated.[14]

[10] P. Salomaa, *Acta Chem. Scand.*, **19**, 1263 (1965).

[11] P. Salomaa and K. Sallinen, *Acta Chem. Scand.*, **19**, 1054 (1965).

[12] F. S. Al-Jeboury, N. Baggett, A. B. Foster, and J. M. Webber, *Chem. Comm.*, **1965**, 222.

[13] G. Aksnes, P. Albriktsen, and P. Juvvik, *Acta Chem. Scand.*, **19**, 920 (1965).

[14] N. Baggett, K. W. Buck, A. B. Foster, R. Jefferis, B. H. Rees, and J. M. Webber, *J. Chem. Soc.*, **1965**, 3382.

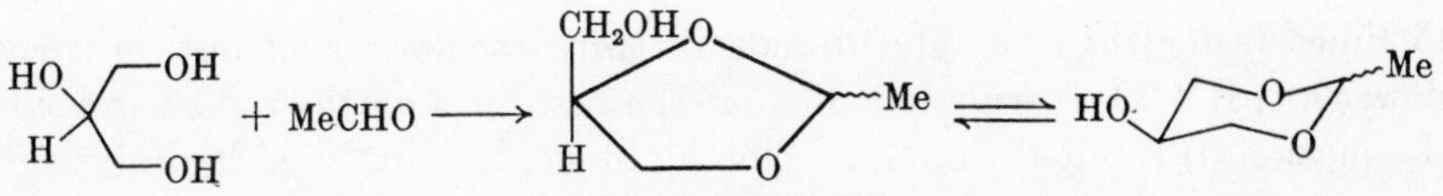

(18)   (19)

Hydrolysis of the 5,6-isopropylidene group of 1,2:5,6-di-O-isopropylidene-glucofuranose (20) occurs about 80 times faster than that of 1,2-O-isopropylideneglucofuranose (21).[15] When the latter compound is allowed to react with methanolic acid, glucose, and not methyl furanosides, is formed and hence it was considered that the hydrolysis involved fission of the bond between the oxygen and the isopropylidene residue.

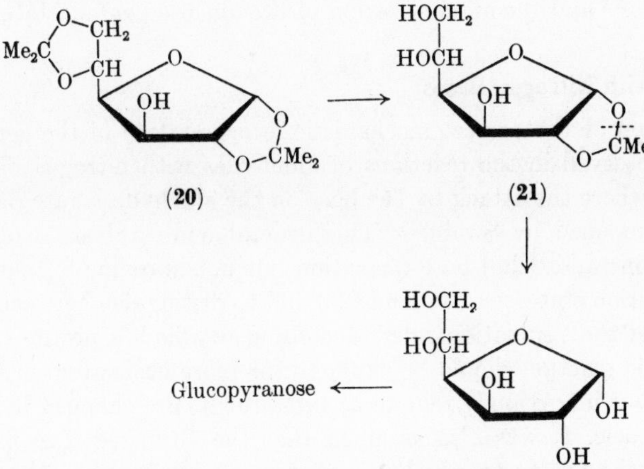

(20)   (21)

Glucopyranose ⟵

An extensive investigation of the kinetics of the hydrolysis of glucuronides in 0.5M-sulphuric acid has been reported.[16] The pH-dependence of the rate of hydrolysis of 2-naphthyl β-D-glucuronide (22) at 90° has been determined.[17]

(22)   (23)

The reaction follows a rate law: (i) Rate $= k_1$[Un-ionized glucuronide] $+ k_2$[Un-ionized glucuronide]h_0, while the corresponding glucoside (23) shows a rate law (ii) Rate $= k$[Glucoside]h_0. This results in the glucoside's reacting

[15] P. M. Collins, *Tetrahedron*, **21**, 1809 (1965).
[16] T. E. Timell, W. Enterman, F. Spencer, and E. J. Soltes, *Can. J. Chem.*, **43**, 2296 (1965).
[17] B. Capon and B. C. Ghosh, *Chem. Comm.*, **1965**, 586.

45 times faster than the glucuronide in M-hydrochloric acid but 35 times slower at pH 4.79. It was thought that the first term on the right-hand side of equation (i) resulted from a specific hydrogen ion-catalysed hydrolysis of the ionized form of the glucuronide which proceeds about 1500 times faster than that of the un-ionized form, the reaction being particularly sensitive to the polar effects of substituents at $C_{(5)}$.

Structure and reactivity in the acid-catalysed hydrolysis of glycosides has been discussed.[18]

Other reactions which have been investigated include the alkaline hydrolysis of aryl 2-deoxyglucosides[19] and dhurrin,[20] the reaction of acetone with sorbose,[21] the alcoholysis of acetals,[22] the hydrolysis of the acetals of α-ketoglutaric acid and phenylpyruvic acid,[23] the reaction between chloral and alcohols,[24] and the anomerization of the glucose pentaacetates.[25]

Reactions with Nitrogen Bases

The question of the correct mechanistic interpretation of the general-acid-catalysis observed in the reactions of aldehydes with nitrogen bases under conditions where the attack by the base on the aldehyde is rate-determining has been reopened by Swain.[26a] This general-acid-catalysis is observed in semicarbazone and Schiff-base formation but not in oxime formation.[1] Two likely transition states are (24) and (25) and to distinguish between them use was made of the "solvation rule," according to which a proton transferred in an organic reaction should be closer to the more basic atom in the transition state, but increasingly remote as substituents are changed to make this atom less basic. It was also assumed that the Brønsted α-coefficient is a measure of the extent to which the proton is transferred in the transition state, e.g. $\alpha = 0.5$ means that the proton is half-transferred. Two limiting examples of transition state (24) were considered, namely, (26) and (27). If the transition state resembles reactants (as 26) the proton will be located on the anion A^-, with $\alpha < 0.5$. If the transition state resembles products, the proton will be on the alkoxide oxygen, with $\alpha > 0.5$. The electron-donating power of R, as measured by the pK_a of RNH_2, increases in the order $R = NH_2 \cdot CO \cdot NH < Ph < OH$, so that according to the "reacting bond"

[18] M. S. Feather and J. F. Harris, *J. Org. Chem.*, **30**, 153 (1965).
[19] R. J. Ferrier, W. G. Overend, and A. E. Ryan, *J. Chem. Soc.*, **1965**, 3484.
[20] C. H. Mao and L. Anderson, *J. Org. Chem.*, **30**, 603 (1965).
[21] T. Maeda, M. Kiyokawa, and K. Tokuyama, *Bull. Chem. Soc. Japan*, **38**, 332 (1965).
[22] I. Jansson, *Suomen Kemistilehti*, **38**, 117 (1965).
[23] H. Plieninger, L. Arnold, and W. Hoffmann, *Chem. Ber.*, **98**, 1761 (1965).
[24] R. B. Jensen and E. C. Munksgaard, *Acta Chem. Scand.*, **18**, 1896 (1964).
[25] E. P. Painter, *Chem. Ind. (London)*, **1965**, 1380.
[26a] C. G. Swain and J. C. Worosz, *Tetrahedron Letters*, **1965**, 3199; C. G. Swain, D. A. Kuhn, and R. L. Schowen, *J. Am. Chem. Soc.*, **87**, 1553 (1965).

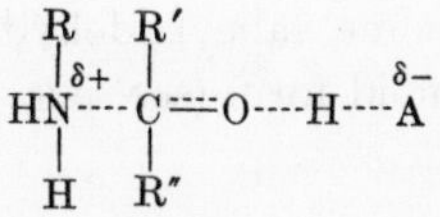

R   R'
|δ+ | δ−
HN----C===O----H----A

H   R"
(24)

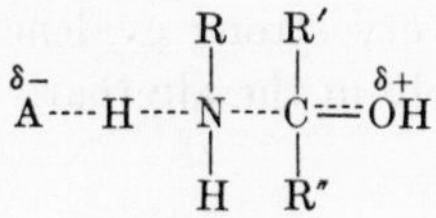

R   R'
δ− | | δ+
A----H----N----C===OH

H   R"
(25)

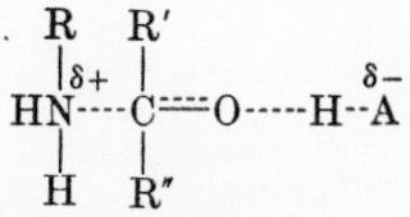

. R   R'
|δ+ | δ−
HN----C===O----H--A

H   R"
(26)

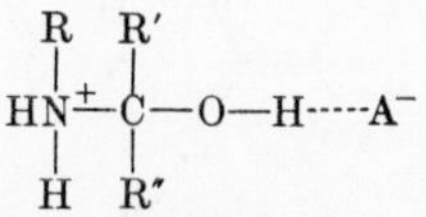

R   R'
| |
HN—±—C—O—H----A−

H   R"
(27)

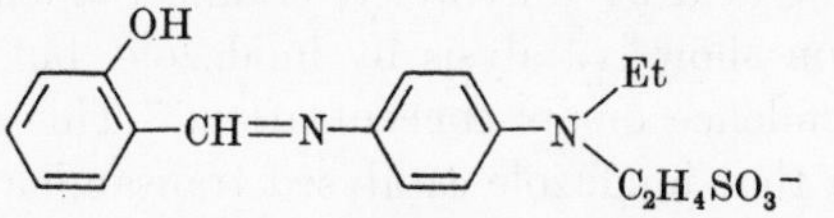

(28)

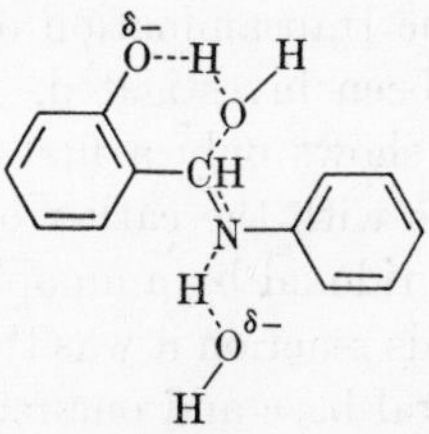

(29)

rule[26b] the nearest reacting bond (N–C) should become longer and the more remote reacting bond (C–O) shorter along this series. The transition will, therefore, become more like reactants and the carbonyl-oxygen less basic. By the solvation rule, the proton should then be more remote from this oxygen but closer to A−, and hence α should fall. This was, in fact, what had been found, since with R = $NH_2 \cdot CO \cdot NH$, Ph, and OH, α = 0.25, 0.20, and < 0.15 (i.e. general-acid-catalysis was not detected), respectively. With transition state (25), however, making R more electron-donating should bring the proton closer to the nitrogen, and α should rise. Transition state (24) was, therefore, preferred. To the objection that this would mean that the reaction of the protonated aldehyde with semicarbazide would be a diffusion-controlled process it was argued that protonated aldehyde might not lie on the reaction path and that in (24) it "is stabilized by the proximity of the amine and the anion, and therefore may be stabler than the protonated aldehyde or a transition state solvated by water alone."

The rate of hydrolysis of the Schiff base (28) increases about 360 times from pH's where the phenolic group is un-ionized to those at which it is ionized.[27] Similar but smaller rate differences were observed with other Schiff bases of salicylaldehyde. These results were interpreted as indicating intramolecular catalysis by the phenolic group, and (29) and (30) were considered as possible transition states. It should, however, be noted that

26b C. G. Swain and E. R. Thornton, *J. Am. Chem. Soc.*, **84**, 817 (1962).
27 R. L. Reeves, *J. Org. Chem.*, **30**, 3129 (1965).

there is now very strong evidence that some salicylaldehyde Schiff bases exist appreciably in the alternative quinonoid form (see, e.g., ref. 28).

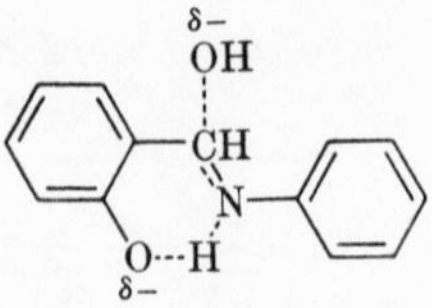

(30)

The transamination of L-glutamic acid by 3-hydroxypyridine-4-aldehyde has been investigated. The reaction shows catalysis by imidazole, but the rate shows only a first-order dependence on its concentration.[29] This contrasts with the earlier observation that imidazole-catalysed transamination of pyridoxal by aminophenylacetic acid is of the second order in imidazole. In this reaction it was thought that one molecule of imidazole was acting as a general base and one molecule of imidazolinium as a general acid, but in the reaction now studied it was thought that the role of acid catalyst was taken over by the internal hydroxyl group (see 31).

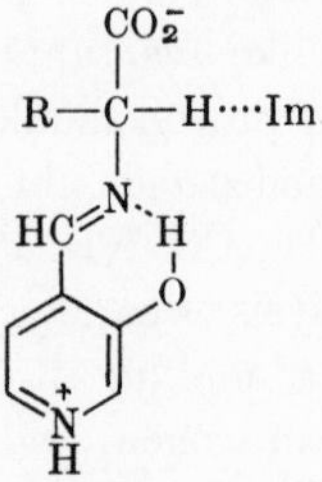

(31)

The effect of micelle-forming detergents on the rate of hydrolysis of N-(p-chlorobenzylidene)-1,1-dimethylethylamine have been investigated, and the rate constants for the individual steps in the micellar phases evaluated.[30] Three detergents giving anionic, cationic, and neutral micelles were used, but no rate enhancements due to incorporation into the micellar phases were observed. The most striking effects were observed with the anionic detergent, for which rate constants for the attack of hydroxide ion on the protonated Schiff base and for the decomposition of the carbinolamine intermediate in the micellar phase were markedly reduced (see also p. 80).

[28] G. O. Dudek and E. P. Dudek, *Chem. Comm.*, **1965**, 464.
[29] J. W. Thanassi, A. R. Butler, and T. C. Bruice, *Biochemistry*, **4**, 1463 (1965).
[30] M. T. A. Behme and E. H. Cordes, *J. Am. Chem. Soc.*, **87**, 260 (1965).

The reaction of aromatic aldehydes with ^{18}O-labelled phenylhydroxylamine to yield nitrones, $ArCH=N(O)Ph$, showed that all the oxygen of the nitrone came from the original phenylhydroxylamine.[31] This is quite different to the condensation of phenylhydroxylamine with nitrosobenzene, which gave azoxybenzene in which only half of the oxygen came from the hydroxylamine.[32] (See p. 219.)

Hydrolyses of the N-arylglucosylamines (32) and (34) are preceded by their rapid interconversion via an intermediate Schiff base (33).[33,34] The hydrolyses are general-acid-catalysed and the plots of rates against acidity

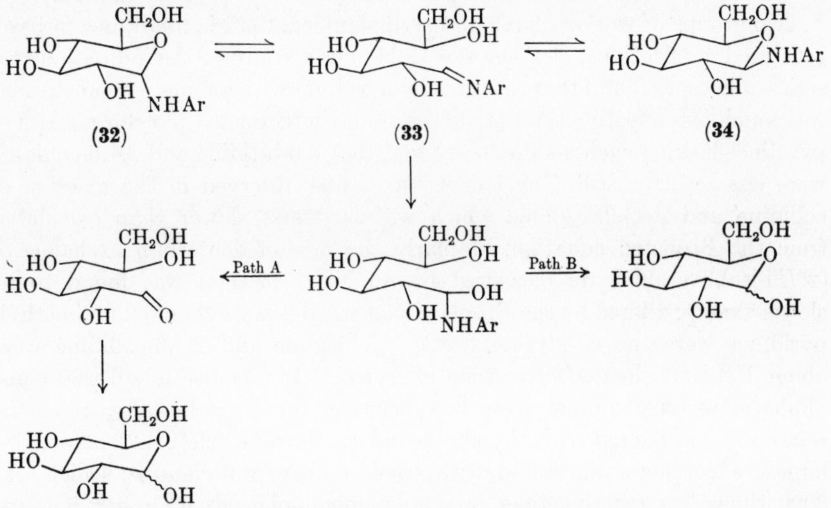

show maxima. These results were interpreted as indicating that the hydrolyses also proceed through the Schiff base. Consistently with this, the anomeric composition of the glucose formed in the hydrolysis of p-hydroxyphenyl- and p-tolyl-glucosylamines was found to be $62 \pm 5\%$ α and $38 \pm 5\%$ β, which is the composition expected from the ring closure of *aldehydo*-glucose.[33] There is also the possibility that the final step is an intramolecular displacement of the arylamino group (Path B),[34] but in the Reviewers' opinion this is more likely to involve formation of a five- rather than a six-membered ring. Other investigations of the anomerization of glycosylamines[35] and of their tetramethyl ethers[36] have been reported.

[31] L. A. Neiman, V. I. Maimind, and M. M. Shemyakin, *Tetrahedron Letters*, **1965**, 3157.
[32] See also G. A. Russell and E. J. Geels, *J. Am. Chem. Soc.*, **87**, 122 (1965).
[33] B. Capon and B. E. Connett, *J. Chem. Soc.*, **1965**, 4497.
[34] H. Simon and D. Palm, *Chem. Ber.*, **98**, 433 (1965).
[35] T. Jasinski, K. Smiataczowa, and J. Sokolowski, *Roczniki Chem.*, **39**, 827 (1965).
[36] R. Bognár, P. Nánási, and A. Lippták, *Acta Chim. Acad. Sci. Hung.*, **45**, 47 (1965).

Other reactions which have been investigated include the condensation of acetaldehyde with urea[37] and of formaldehyde with urea and with semicarbazide,[38] osazone formation,[39] the hydrolysis of some 1-silylpropyl-2-imidazolines,[40] the hydrolysis of β-keto imines and their Cu[II] derivatives,[41] and the *cis–trans* isomerization of *para*-substituted *N*-benzylideneanilines.[42]

Enolization and Related Reactions

Extensive investigations continue to be made on the detailed mechanism of enolization; reaction rates are generally determined as rates of halogenation under conditions where prior enolization is rate-determining.

Two groups of workers have observed significant steric hindrance in base-catalysed enolizations. Feather and Gold[43] in a study of the iodination of a series of ketones found that catalysis by pyridines with a 2-methyl substituent was much less effective than predicted from the Brønsted plot for a series of pyridines lacking such a substituent and that 2,6-lutidine and 2,4,6-collidine were less reactive still. The largest effect was observed in the reaction of collidine and cyclohexanone which was 50 times slower than calculated from the Brønsted equation. Similarly the rate of deuterium exchange of [2²*H*]-isobutyraldehyde measured by an NMR method was found to be slower than predicted by the Brønsted plot when 2-methyl- and 2,6-dimethyl-pyridines were the catalysing bases; 2,6-lutidine and 2,4,6-collidine were about 100 times less effective than expected.[44] It was also found that some aliphatic tertiary amines were less reactive, for their basicity, than the relatively unhindered trimethylamine and 1,4-diazabicyclo[2.2.2]octane. The largest effect here was found with tris-2-hydroxypropylamine, which was 3000 times less reactive than predicted; intramolecular hydrogen bonding may also contribute to this. Different views were expressed by the two groups as to the geometry of the transition state. Feather and Gold[43] suggest that the breaking C–H bond and the carbonyl group are in a planar *trans*-conformation, as (35), "with some kind of conjugative interaction between the carbonyl group and the heteroaromatic ring which is formally indicated by one of the dotted lines." Hine and his co-workers[44] prefer the more orthodox structure (36) in which the carbon–hydrogen bond being broken is held almost parallel to the π-orbitals of the carbonyl group.

[37] Y. Ogata, A. Kawasaki, and N. Okumura, *J. Org. Chem.*, **30**, 1636 (1965).
[38] B. Glutz and H. Zollinger, *Angew. Chem. Intern. Ed. Engl.*, **4**, 440 (1965).
[39] M. M. Shemyakin, V. I. Maimind, K. M. Ermolaev, and E. M. Bamdas, *Tetrahedron*, **21**, 2771 (1965); I. Dijong and F. Micheel, *Ann. Chem.*, **684**, 216 (1965).
[40] J. C. Saam and H. M. Bank, *J. Org. Chem.*, **30**, 3350 (1965).
[41] D. F. Martin and F. F. Cantwell, *J. Inorg. Nucl. Chem.*, **26**, 2219 (1964).
[42] G. Wettermark, J. Weinstein, J. Sousa, and L. Dogliotti, *J. Phys. Chem.*, **69**, 1584 (1965).
[43] J. A. Feather and V. Gold, *J. Chem. Soc.*, **1965**, 1752.
[44] J. Hine, J. G. Houston, J. H. Jensen, and J. Mulders, *J. Am. Chem. Soc.*, **87**, 5050 (1965).

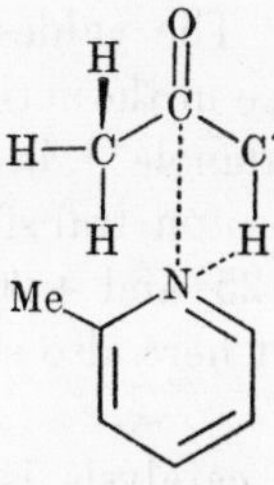

(35) 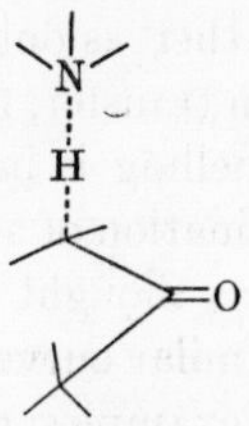   (36)

An interesting effect, presumably also steric in origin, has been observed in the base-catalysed deuterium exchange of isofenchone and camphor which undergo exclusive monodeuteration to yield the *exo*-deuterated species (37) and (38).[45] Norcamphor is, however, reported to undergo appreciable dideuteration.

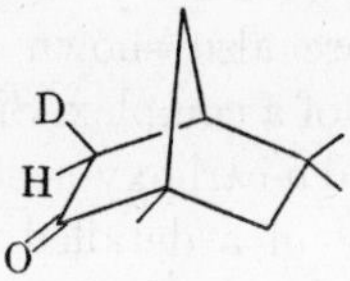

(37) 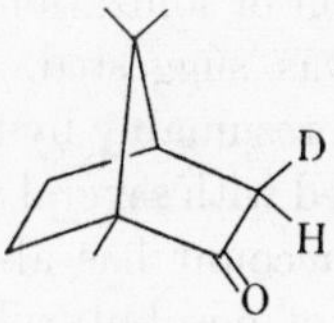   (38)

Deuterium isotope effects on the rates of ionization of a series of β-dicarbonyl compounds as measured by their rates of bromination have been determined.[46] The most extensively discussed reaction was that of ethyl α-methylacetoacetate. It gave a good Brønsted plot for catalysis by five carboxylate ions, and the values of k_H/k_D varied smoothly with their basic strength from 3.84 to 6.25. Since all the anions were weaker bases than the anion of α-methylacetoacetate an increase in the basic strength of the catalyst makes the transition state more symmetrical and hence this variation in k_H/k_D is in accord with Westheimer's analysis.[47] It was, however, considered that other factors besides the zero-point energy of the symmetrical vibration of the transition state should be taken into account in a theoretical analysis of these effects.

The Arrhenius plot for the bromination of diisopropyl ketone in alkaline solutions between 0° and 50° shows marked curvature, which was attributed to quantum-mechanical tunnelling through the energy barrier.[48] This conclusion has, however, been shown to be incorrect by measurements of the

[45] A. F. Thomas and B. Willhalm, *Tetrahedron Letters*, 1965, 1309.
[46] R. P. Bell and J. E. Crooks, *Proc. Roy. Soc.*, A, **286**, 285 (1965).
[47] F. H. Westheimer, *Chem. Rev.*, **61**, 265 (1961).
[48] J. R. Hulett, *J. Chem. Soc.*, 1965, 430.

rate of detritiation of the tritiated ketone.[49] The values of k_H/k_T are low (1.65 at 50°) and there is only a small difference in the activation energies for proton and trition transfer, $E_T - E_H = 500$ cal mole^{-1}, in disagreement with a view that tunnelling is important in the proton transfer. The Arrhenius plot for the bromination of acetone between $-25°$ and $+20°$ was also curved but tunnelling was thought to be unimportant here also since hexadeuterio-acetone gave a similar curved plot.[50]

Several more examples of intramolecular catalysis is enolization have been reported. The iodination of pyruvic acid, HP, is zero order in iodine and, in contrast to the iodination of acetone and other ketones, zero order in hydrogen ions;[51a,b] there is also a small contribution to the total rate from the reaction of the pyruvate anion so that $r = 4.4 \times 10^{-7}[\text{HP}] + 9.4 \times 10^{-8}[\text{P}^-]$ mole l^{-1} sec^{-1} at 25°. These rate constants are considerably larger than for the enolization of biacetyl, $k = 5.5 \times 10^{-9}$ sec^{-1}, as determined[51c] by the rate of deuterium exchange, and intramolecular catalysis of the enolization or ionization of the methyl group by both the CO_2H and CO_2^- groups was suggested. Zinc ions were also shown to catalyse the enolization,[51b] presumably by the formation of a complex. Similar behaviour has been observed with several o-hydroxy- and o-carboxy-acetophenones.[51d,e] A preliminary account has also appeared[51f] of a detailed investigation of the enolization of o-isobutyrylbenzoic acid. In the pH region 2.5—10 the predominant reaction involves intramolecular base-catalysis by the ionized o-carboxylate ion.

The factors which influence the direction of bromination of 2-alkylcyclo-alkanones and their ketals have been discussed.[52]

The rate constants for the reactions of the enols of cyclohexanone and acetone with bromine have been determined by a "direct" method.[53a] This involves carrying out the reaction in neutral solution where the enolization is slow, and measuring the rate of reaction of bromine with the enol initially present by a coulamperometric method. The values obtained were: for cyclohexanone, $2.3 \pm 0.2 \times 10^7$ l mole^{-1} min^{-1}; and for acetone, between 4×10^8 and 7×10^8 l mole^{-1} min^{-1}; in reasonably good agreement with earlier measurements by a different method. The rate constants for the reaction

[49] J. R. Jones, *Trans. Farad. Soc.*, **61**, 2456 (1965).
[50] J. R. Hullett, *J. Chem. Soc.*, **1965**, 1166; J. R. Jones *Trans. Farad. Soc.*, **61**, 95 (1965).
[51a] W. J. Albery, R. P. Bell, and A. L. Powell, *Trans. Farad. Soc.*, **61**, 1194 (1965).
[51b] A. Schellenberger and G. Hübner, *Chem. Ber.*, **98**, 1938 (1965).
[51c] W. D. Walters, *J. Am. Chem. Soc.*, **63**, 2850 (1941).
[51d] A. Schellenberger, G. Oehme, and G. Hübner, *Chem. Ber.*, **98**, 3578 (1965).
[51e] G. Hübner, *Angew. Chem. Intern. Ed. Engl.*, **4**, 881 (1965).
[51f] M. L. Bender and E. T. Harper, Abstracts of Papers, 150th Meeting A.C.S., September, 1965, p. 15S; see also *J. Am. Chem. Soc.*, **87**, 5625 (1965).
[52] E. W. Garbisch, *J. Org. Chem.*, **30**, 2109 (1965).
[53a] J. E. Dubois and G. Barbier, *Bull. Soc. Chim. France*, **1965**, 682.

of the enol of ethyl acetoacetate and its α-bromo derivative with bromine have similarly been determined to be 8.10×10^6 and 6.90×10^4 l mole^{-1} min^{-1}.[53b]

The kinetics of bromination of a number of enols and their anions have also been determined by working at very low bromine concentrations.[54] Surprisingly, the rate constants for the bromination of the enols of dimedone, 2-bromodimedone, 3-methyltetronic acid, and 3-bromotetronic acid all have rate constants similar to those for their anions. Since the rate constants all

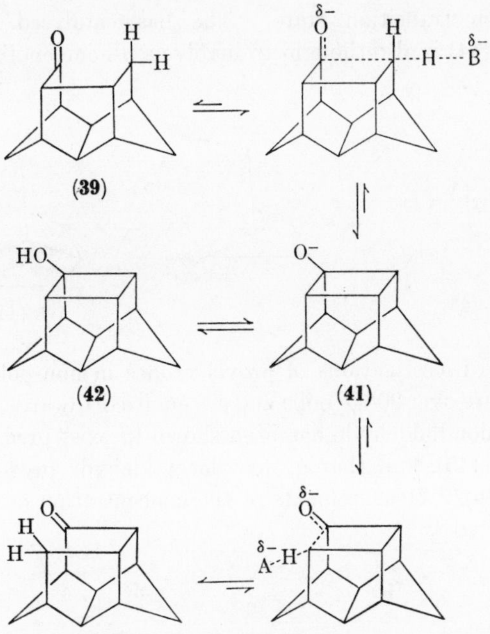

lie in the range 5×10^6 to 22×10^6 l mole^{-1} sec^{-1} it seems unlikely that these reactions are diffusion-controlled. Instead it was thought that the reactivities of these enolate anions really are low, owing perhaps to the same factors which make them, for enolate ions, such weak bases.

The rate constants for the ionization of some enols and for the reverse reactions have been determined by relaxation methods.[55]

Two more examples of homoenolization[56] have been reported. Treatment of the bird-cage ketone (39) with potassium *t*-butoxide in *t*-butyl alcohol at

[53b] J.-E. Dubois and P. Alcais, *Compt. Rendus*, **260**, 887 (1965).
[54] R. P. Bell and G. G. Davis, *J. Chem. Soc.*, **1965**, 353.
[55] M. Eigen, G. I. Ilgenfritz, and W. Kruse, *Chem. Ber.*, **98**, 1623 (1965).
[56] A. Nickon and J. L. Lambert, *J. Am. Chem. Soc.*, **84**, 4604 (1962).

ca. 200° yields the isomeric ketone (40) in a first-order process.[57,58] The final
reaction mixture is reported by one group[58] to contain more than 99% of
(40) at 175—200° and by the other[57] to contain approximately 4% of (39)
at 250°. The reaction is thought to proceed via the anion (41), and generation
of this ion from the bird-cage alcohol (42) is reported to yield either > 99%
of (40)[58] at 100° or a mixture of (39) and (40) in the ratio of 4:96 at 250°.[57]
It was considered[58] that "From the general rate level observed for the
disappearance of (39), ... that the carbonyl group aids the proton removal
from (39) by delocalizing the developing ionic charge in proceeding to the
homoenolization transition state." The base-catalysed isomerization of
ketone (43) to (44) is also thought to involve a β-homoenolization process, as
shown.[59]

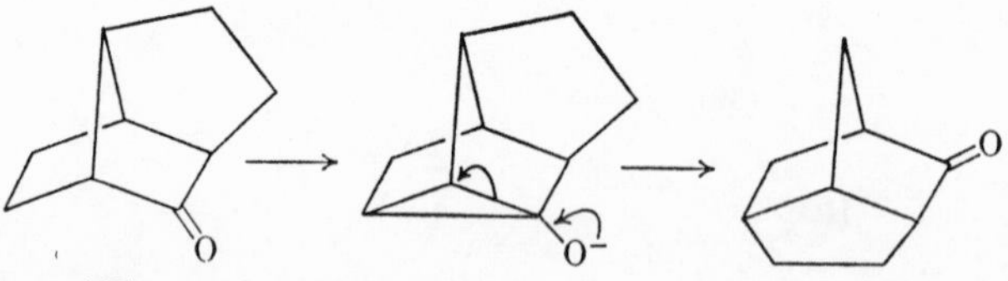

(43)   (44)

Spectroscopic investigations of aroylacetones in non-polar solvents indicate that they are over 90% enolic and are enolized towards the aryl group.[60]
The enol of malondialdehyde has been shown to exist predominantly in the
sym-trans-form (45), and that of acetylacetaldehyde predominantly in the
sym-cis-form (46).[61] Measurements of the concentration of other enols have
also been reported.[62]

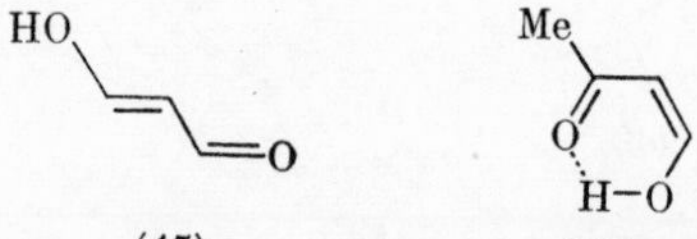

(45)   (46)

The factors which influence the direction of enolization of cyclic α-formyl-
ketones have been discussed by Garbisch.[63]

[57] T. Fukunaga, *J. Am. Chem. Soc.*, **87**, 916 (1965).

[58] R. Howe and S. Winstein, *J. Am. Chem. Soc.*, **87**, 915 (1965).

[59] A. Nickon, H. Kwasnik, T. Swartz, R. O. Williams, and J. B. DiGiorgio, *J. Am. Chem. Soc.*, **87**, 1615 (1965).

[60] J. U. Lowe and L. N. Ferguson, *J. Org. Chem.*, **30**, 3000 (1965).

[61] A. A. Bothner-By and R. K. Harris, *J. Org. Chem.*, **30**, 254 (1965).

[62] S. J. Rhoads and C. Pryde, *J. Org. Chem.*, **30**, 3212 (1965); M. T. Rogers and J. L. Burdett, *Can. J. Chem.*, **43**, 1516 (1965).

[63] E. W. Garbisch, *J. Am. Chem. Soc.*, **87**, 505 (1965).

The rearrangement of 9,10-anthraquinols to 10-hydroxy-9-anthrones,[64] and the photoenolization of 2-methylbenzophenone[65] and o-methylaceto-phonone,[66] have also been investigated.

In a detailed mechanistic investigation[67] of the acid-catalysed isomerization of Δ^5-3-keto-steroids (47) into Δ^4-3-keto-steroids (48), two mechanisms were considered, one involving the formation of an enol (path A) and one involving addition of a proton to $C_{(6)}$ and removal of a proton from $C_{(5)}$ of the resulting

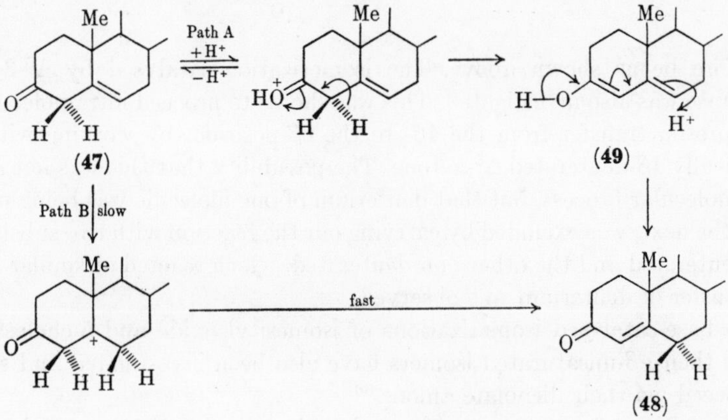

carbonium ion (path B). The kinetic isotope effect on dideuteration at position 4 was $k_H/k_D = 4.1$, and the solvent isotope effect was $k_{D_2O}/k_{H_2O} = 1.56$. These results were taken to indicate a slow removal of the proton from $C_{(4)}$ and a pre-equilibrium proton transfer, i.e. they support path A but not path B. Evidence was also provided which indicated that the enol (49) would undergo protonation at $C_{(6)}$ and not at $C_{(4)}$. First, the corresponding enol ether was used as a model for the enol and treatment of this with DCl or DOAc gave a Δ^4-3-ketone containing one atom of deuterium at the axial 6β-position. Also the enolization of the Δ^4-ketone was studied in the presence of DCl and this proceeded with incorporation of one atom of deuterium into the 6β-position with less than 0.1 atom at $C_{(4)}$. This was essentially the same as in a Δ^4-ketone obtained on isomerization of Δ^5-ketone in the presence of DCl. By working with Δ^5-ketone specifically β-deuterated at $C_{(4)}$ it was shown that there was a slight preference for loss of the α-proton at this position; $k_{4\beta H}/k_{4\alpha H} = 0.86$. The preferred stereochemistry of the isomerization may

[64] A. D. Broadbent and E. F. Sommerman, *Tetrahedron Letters*, **1965**, 2649; A. D. Broadbent, *Chem. Comm.*, **1965**, 107.

[65] E. F. Ullman and K. R. Huffman, *Tetrahedron Letters*, **1965**, 1863.

[66] G. Wettermark, *Photochem. Photobiol.*, **4**, 621 (1965).

[67] S. K. Malhotra and H. J. Ringold, *J. Am. Chem. Soc.*, **87**, 3228 (1965).

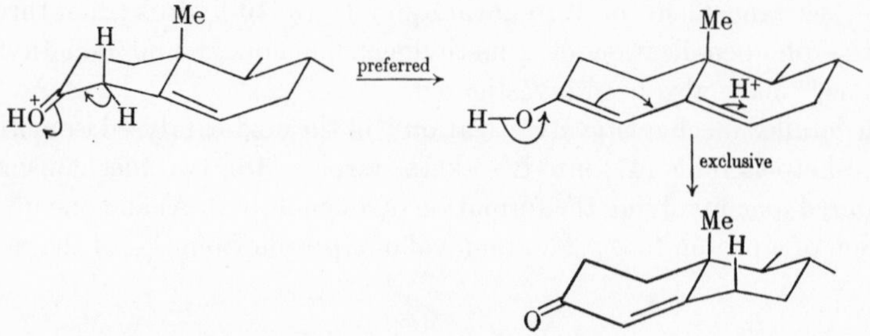

therefore be as shown above. The isomerization catalysed by Δ^5-3-keto-isomerase was also investigated. This was shown to proceed intramolecularly, with proton transfer from the 4β- to the 6β-position, by working with the specifically 4β-deuterated Δ^5-ketone. The possibility that this was not a true intramolecular process, but that deuterium of one molecule was being passed on to the next, was excluded by carrying out the reaction with two substrates, one deuterated and the other non-deuterated, which reacted at similar rates; no transfer of deuterium was observed.

The base-catalysed isomerizations of isomesityl oxide and 5-cholesten-3-one to their $\alpha\beta$-unsaturated isomers have also been investigated and shown to proceed via their dienolate anions.[68]

The position of protonation of a series of cyclic dienol ethers (Scheme 1)

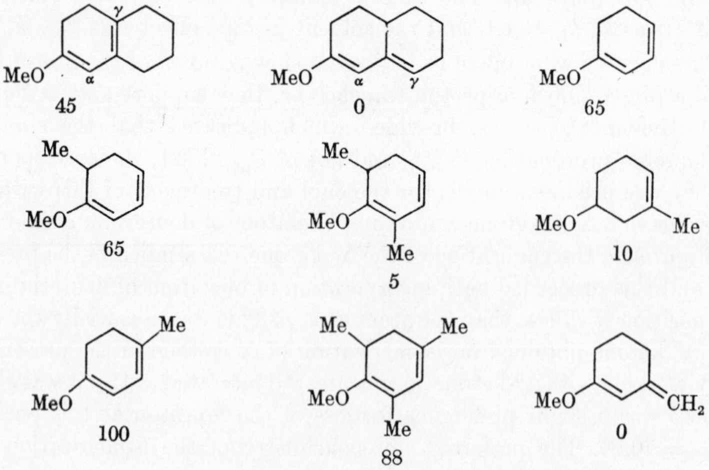

Figures are percentage α-protonation

Scheme 1

[68] H. C. Volger and W. Brackman, *Rec. Trav. Chim.*, **84**, 1017 (1965).

has been determined.[69] Strikingly, the transoid ethers undergo exclusive γ-protonation. This was attributed to a considerable degree of bonding in the transition state, conjugation of the product being an important factor in determining its energy. However, this must be less important in the protonation of the homoannular dienol ethers which undergo appreciable, and even in one instance exclusive, α-protonation. The suggestion thus arose that their transitions states are more "reactant like" and it was shown that the relative amounts of α- and γ-protonation were correlated by the differences in the charge densities at these positions as determined by molecular-orbital calculations.

The mechanism of the hydrolysis of the vinyl ether, 2-ethoxy-1-cyclopentenecarboxylic acid (50), has been studied.[70] The reaction gives 2-oxocyclopentanecarboxylic acid (51) which then undergoes decarboxylation to

cyclopentanone. The hydrolysis of the ether shows general-acid-catalysis, gives a value of $k_{D_2O}/k_{H_2O} = 0.20$—0.25 for catalysis by general acids and 0.35 for catalysis by hydronium ion; and the entropy of activation for catalysis by hydronium ion is -14.5 e.u. These results were taken to indicate a rate-determining proton transfer as shown. The 2-ethoxy-1-cyclopentenecarboxylate ion reacts 209 times faster than the unionized form and it was suggested that this was the result of participation by the carboxylate ion to give the β-lactone illustrated. Other kinetic investigations of the hydrolysis of enol ethers[71] and enol esters[72] have been reported.

The mechanism of the hydrolysis of enol ethers is very similar to that of the structurally related enamines which has been investigated by Stamhuis and Maas.[73] The compounds studied were 4-(2-methylpropenyl)morpholine and

[69] N. A. J. Rogers and A. Sattar, *Tetrahedron Letters*, **1965**, 1471.

[70] T. H. Fife, *J. Am. Chem. Soc.*, **87**, 1084 (1965).

[71] D. M. Jones and N. F. Wood, *J. Chem. Soc.*, **1964**, 5400; M. F. Shostakovsky, A. S. Atavin, B. V. Prokop'ev, B. A. Trofimov, V. I. Lavrov, and N. M. Deriglazov, *Izv. Akad. Nauk, SSSR, Otd. Khim. Nauk*, **1965**, 1485; I. N. Azerbaev, A. S. Atavin, B. A. Trafimov, and R. D. Yakubov, *Izv. Akad. Nauk, Kaz. SSR, Otd. Khim. Nauk*, **14**, 80 (1964).

[72] J. A. Langrebe, *J. Org. Chem.*, **30**, 2997 (1965).

[73] E. J. Stamhuis and W. Maas, *J. Org. Chem.*, **30**, 2156 (1965).

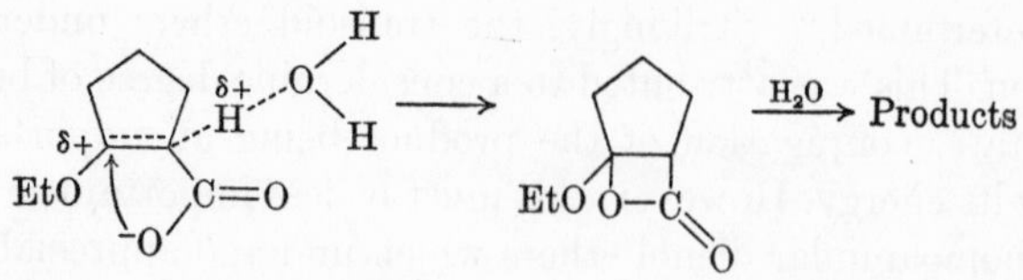

1-(2-methylpropenyl)-piperidine and -pyrrolidine. These undergo protonation on nitrogen, with pK_a 5.47, 8.35, and 8.84, respectively,[74] but these protonations are parasitic equilibria and do not lie on the reaction coordinate for hydrolysis. The reactive species are the unprotonated forms (52) which

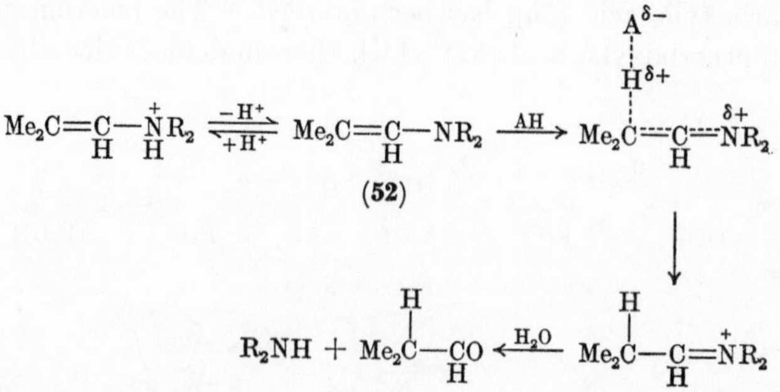

Scheme 2

undergo a general-acid-catalysed hydrolysis with $k_{D_2O}/k_{H_2O} = 0.11$ for catalysis by acetic acid and 0.40 for catalysis by hydronium ion. The mechanism shown in Scheme 2, involving a rate-determining proton transfer to the β-carbon was proposed. The entropies of activation for catalysis by H_2O, AcOH, and H_3BO_3 are all negative, which is consistent with this mechanism, but, while that for catalysis by hydronium ions of the hydrolysis of the morpholino-compound is slightly negative (-4 ± 5 e.u.), those for the piperidino- and the pyrrolidino-compound are $+29 \pm 7$ and $+26 \pm 6$ e.u., respectively; the reason for this is not clear.

The hydrolysis of enamines in D_2O is a convenient method for the preparation of 2-deuterated ketones.[75] Under certain conditions more than one deuterium atom is introduced, which suggests that the β-protonation (see Scheme 2) is reversible.

The enamine, methyl 2-aminovinyl ketone, exists preferentially in the *cis*- and *trans*-enamine forms rather than in the alternative imine form.[76] Evidence for the transient existence of the latter was, however, provided

[74] E. J. Stamhuis, W. Maas, and H. Wynberg, *J. Org. Chem.*, **30**, 2160 (1965).
[75] J. P. Schaefer and D. S. Weinberg, *Tetrahedron Letters*, **1965**, 1801.
[76] J. Dabrowski and J. Terpiński, *Tetrahedron Letters*, **1965**, 1363.

by the observation that the enamine undergoes a specific α-deuteration when it is dissolved in D_2O and left for 3 days; the N,N-dimethyl derivative was unaffected by this treatment. The deuterium exchange was, however, slower than the *cis–trans* isomerization.

The conversion of 2-methyl-2-(3-oxybutyl)cyclohexane-1,3-dione (53) into the ketol (59) is catalysed by primary and secondary, but not by tertiary,

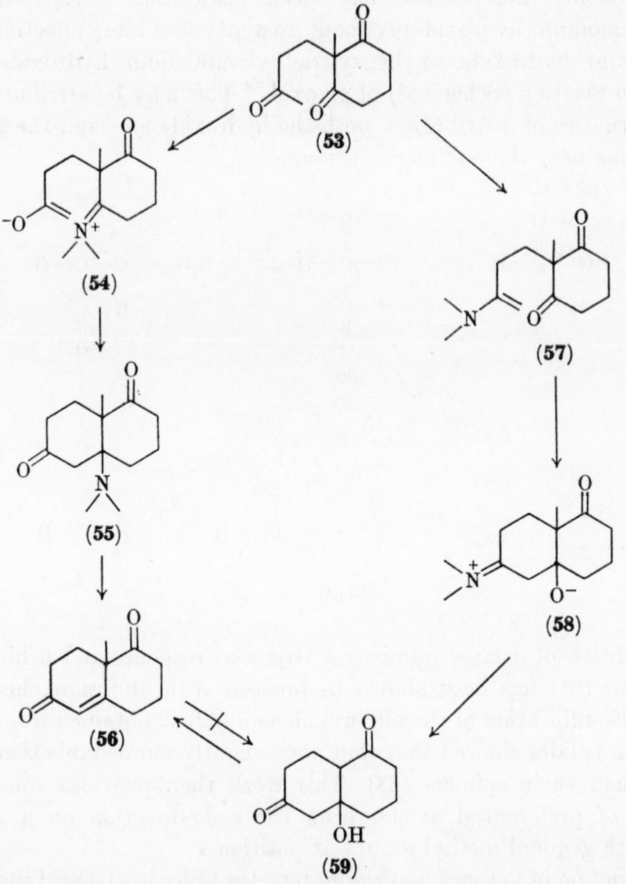

amines; pyrrolidine is particularly effective.[77] Two possible pathways involving condensation with the amine via an immonium salt (54) or an enamine (57) were considered. It was thought that the immonium salt would probably react via the intermediate tertiary amine (55) and its enedione (56), but the latter was shown not to give the ketol under the reaction conditions. The

[77] T. A. Spencer, H. S. Neel, T. W. Fletchtner, and R. A. Zayle, *Tetrahedron Letters*, 1965, 3889.

pathway involving the enamine was therefore preferred. In fact, the ketol (59) underwent slow dehydration to the dienone, and again pyrrolidine was particularly effective and tertiary amines ineffective as catalysts, and so it was suggested that this reaction also involved an enamine.

Other Reactions

The interesting observation has been made that poly[triethyl(vinylbenzyl)ammonium hydroxide] is about a twenty-fold more effective catalyst than sodium hydroxide or benzyltriethylammonium hydroxide for the Cannizarro reaction (Scheme 3) of glyoxal.[78] This may be attributed to two non-uniformities of distribution, both the hydroxide ions and the anion (60) congregating near the polymeric cation.

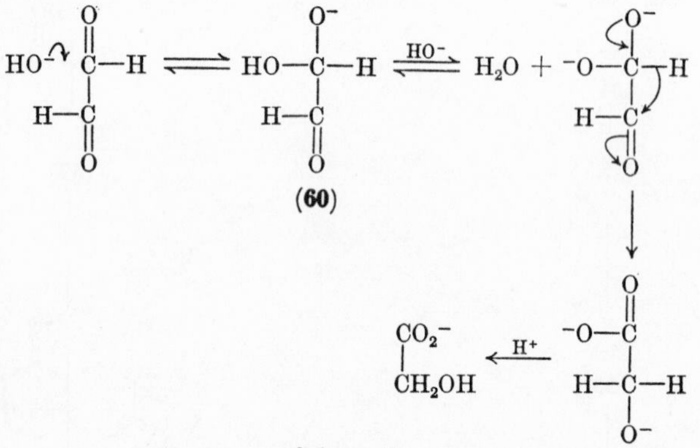

Scheme 3

The addition of a large number of Grignard reagents and lithium alkyls to camphor (61) has been shown to proceed with almost exclusive *endo*-attack.[79] Equilibration of the alkynyl alcohol (62) so obtained from alkynyl-magnesium halides showed that they were slightly more stable thermodynamically than their epimers (63). This work then provides some further examples of preferential attack from the *endo*-direction on a norbornyl system with geminal methyl groups at position 7.

The reduction of ketones by lithium tetrakis-(dihydro-1-pyridyl)aluminate (LDPA) (64) in pyridine has been further investigated. In contrast to reduction by borohydride in hydroxylic solvents, diaryl ketones are reduced more rapidly by LDPA than are either dialkyl or arylalkyl ketones.[80] This difference was thought to be associated with the fact that the carbonyl

[78] C. L. Arcus and B. A. Jackson, *Chem. Ind. (London)*, **1964**, 2022.
[79] M. L. Capmau, W. Chodkiewicz, and P. Cadiot, *Tetrahedron Letters*, **1965**, 1619.
[80] P. T. Lansbury and R. E. MacLeay, *J. Am. Chem. Soc.*, **87**, 831 (1965).

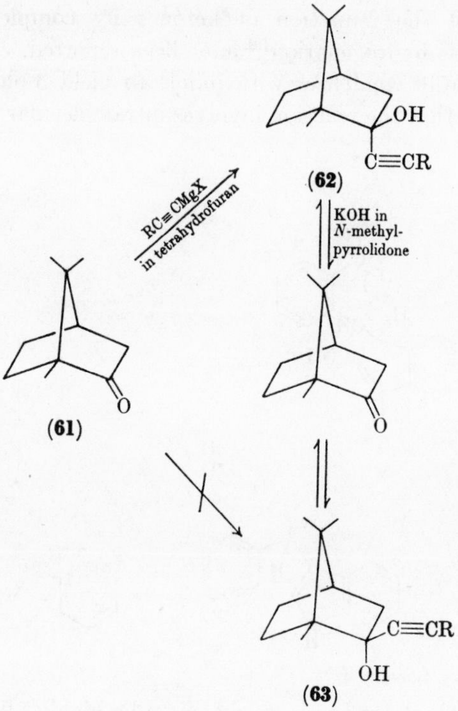

% at equilibrium

58·2

(62)

KOH in
N-methyl-
pyrrolidone

$RC \equiv CMgX$
in tetrahydrofuran

(61)

3·5

37·9

(63)

group is hydrogen-bonded in hydroxylated solvents, so that with aryl ketones resonance of the type **(67)** ↔ **(68)** is important; and this reduces their reactivity. In reductions by LDPA in pyridine, however, it is thought

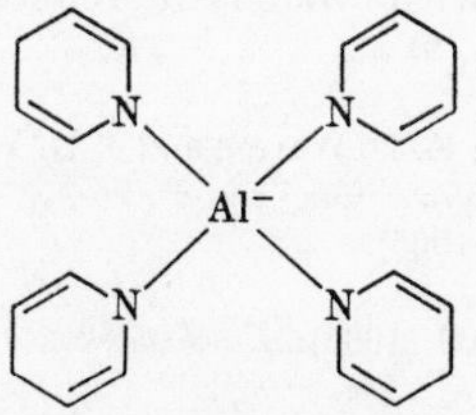

Al^- Li^+

(64)

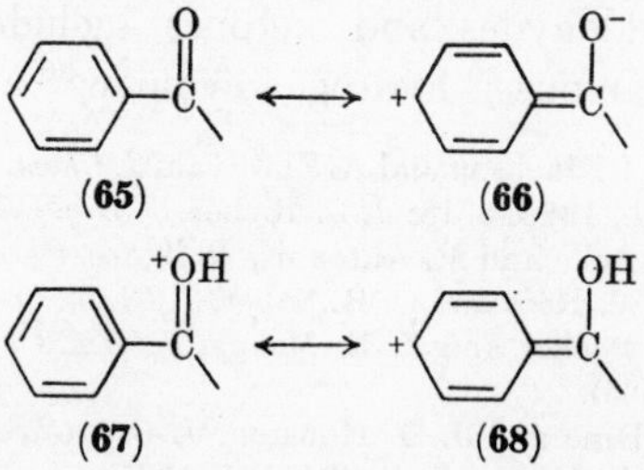

(65)   **(66)**

(67)   **(68)**

that the free ketone is being reduced, and that resonance of the type **(65)** ↔ **(66)** is much less important, so that the reactivity is now controlled by the inductive effect of the aryl group. In support of this view it was found that diaryl ketones are reduced more rapidly than dialkyl ketones by sodium and lithium borohydride in pyridine.[81]

[81] P. T. Lansbury, R. E. Macleay, and J. O. Peterson, *Tetrahedron Letters*, **1964**, 311.

9+

Other investigations of the reduction of ketones by complex metal hydrides[82] and by catalytic hydrogenation[83] have been reported.

Phthaldehydic acid readily condenses with indole to yield 3-phthalidyl-indole.[84] It was suggested that the reaction involves intramolecular catalysis as shown in Scheme 4.

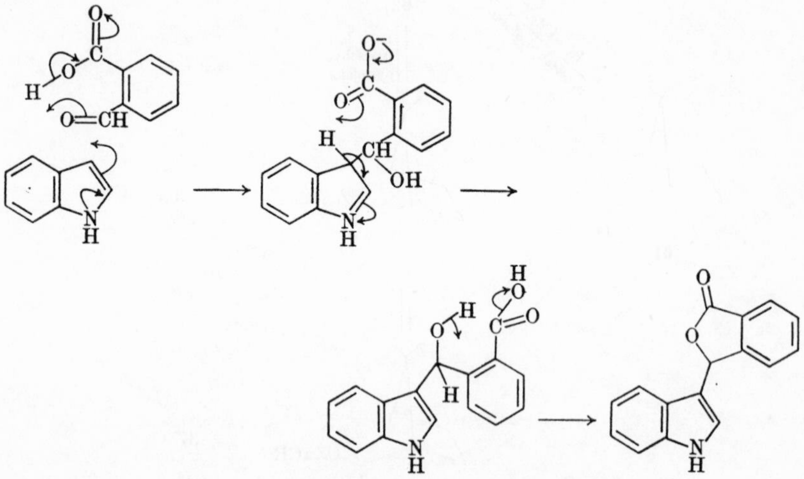

Scheme 4

Erythrocyte carbonic anhydrase has been shown to be highly effective in catalysing the hydration of acetaldehyde.[85] The hydration and aldonization of isobutyraldehyde[86] and the hydration of pyruvic acid[87] have been investigated by NMR spectroscopy.

Other investigations on additions across the carbon–oxygen double bond of aldehydes and ketones include reactions with Grignard reagents,[88] carbanions,[89] hydrogen cyanide,[90] and acetate ion.[91]

[82] W. R. Jackson and A. Zurqiyah, *J. Chem. Soc.*, **1965**, 5280; K. D. Warren and J. R. Yandle, *ibid.*, **1965**, 5518; J.-C. Richer, *J. Org. Chem.*, **30**, 324 (1965).

[83] M. Acke and M. Anteunis, *Bull. Soc. Chim. Belg.*, **74**, 41 (1965).

[84] C. W. Rees and C. R. Sabet, *J. Chem. Soc.*, **1965**, 680.

[85] Y. Pocker and J. E. Meany, *J. Am. Chem. Soc.*, **87**, 1809 (1965); *Biochemistry*, **4**, 2535 (1965).

[86] J. Hine and J. G. Houston, *J. Org. Chem.*, **30**, 1328 (1965); J. Hine, J. G. Houston, and J. H. Jensen, *ibid.*, **30**, 1184 (1965).

[87] V. Gold, G. Socrates, and M. R. Crampton, *J. Chem. Soc.*, **1964**, 5888.

[88] R. Kimbrough and R. D. Hancock, *Chem. Ind.* (*London*), **1965**, 1180; H. O. House and W. L. Respess, *J. Org. Chem.*, **30**, 301 (1965); R. D'Hollander and M. Anteunis, *Bull. Soc. Chim. Belg.*, **74**, 71 (1965).

[89] E. M. Kaiser and C. R. Hauser, *Chem. Ind.* (*London*), **1965**, 1299; M. Laloi, M. Rubinstein, and P. Rumpf, *Bull. Soc. Chim. France*, **1964**, 3235.

[90] N. H. P. Smith, *J. Chem. Soc.*, **1965**, 4499; K. L. Servis, L. K. Oliver, and J. D. Roberts, *Tetrahedron*, **21**, 1827 (1965); O. H. Wheeler, *J. Org. Chem.*, **29**, 3634 (1964).

[91] L. G. Gruen and P. T. McTigue, *Australian J. Chem.*, **18**, 1299 (1965).

The kinetics of the reactions of a series of substituted benzaldehyde *p*-nitrophenylhydrazones (69) with bromine to yield the hydrazidic bromides (71) have been measured.[92] The rate constants were correlated to the σ

$$\text{ArCH}=\text{N·C}_6\text{H}_4\text{·NO}_2 \qquad \underset{\underset{+}{\overset{|}{\text{Br}}}}{\text{ArCH}-\text{N·C}_6\text{H}_4\text{·NO}_2} \qquad \text{ArCBr}=\text{N·C}_6\text{H}_4\text{·NO}_2$$

$$(69) \qquad\qquad\qquad (70) \qquad\qquad\qquad (71)$$

constants, except those for the *p*-methoxy and *p*-hydroxy compounds which fell off the line when σ or σ^+ constants were used; the ρ-value was -0.62 and the entropies of activation were in the range -24.0 to -33.1 e.u. It was suggested that the azabromonium ion (70) was an intermediate. The solvolyses of the hydrazidic halides have also been investigated.[93]

Other reactions which have been investigated include: acetylation reactions with isopropenyl acetate;[94] dissociation of 2,2-dinitropropanol to formaldehyde and the anion of 1,1-dinitroethane;[95] deuteration of some 1-methyl-4-pyridones;[96] equilibration of isomeric enolate anions;[97] the Mannich reaction;[98] decarbonylation of cinnamaldehyde and *trans*-α-substituted cinnamaldehydes;[99] base-catalysed ethynylation of aldehydes and ketones;[100] borate- and carbonate-catalysed peroxidic cleavage of benzil;[101] acid-catalysed cleavage of cyclopropyl ketones related to lumisantonin;[102] thermal retroaldo condensation of β-hydroxy ketones;[103] alkaline degradation of 9-formylmethyl-6,7-dimethylisoalloxazine;[104] and reactions of carbonyl compounds with arsonium ylids.[105]

[92] A. F. Hegarty and F. L. Scott, *Tetrahedron Letters*, **1965**, 3801.
[93] F. L. Scott and J. B. Aylward, *Tetrahedron Letters*, **1965**, 841.
[94] W. B. Smith and T.-K. Chem., *J. Org. Chem.*, **30**, 3095 (1965).
[95] T. N. Hall, *J. Org. Chem.*, **30**, 3157 (1965).
[96] P. Beak and J. Bonham, *J. Am. Chem. Soc.*, **87**, 3365 (1965).
[97] H. O. House and B. M. Trost, *J. Org. Chem.*, **30**, 1341, 2502 (1965).
[98] J. E. Fernandez, J. S. Fowler, and S. J. Glaros, *J. Org. Chem.*, **30**, 2787 (1965).
[99] N. E. Hoffman and T. Puthenpurackal, *J. Org. Chem.*, **30**, 420 (1965).
[100] R. J. Tedeschi, *J. Org. Chem.*, **30**, 3045 (1965).
[101] H. Kwart and N. J. Wegemer, *J. Am. Chem. Soc.*, **87**, 511 (1965).
[102] P. J. Kropp, *J. Am. Chem. Soc.*, **87**, 3914 (1965).
[103] G. G. Smith and B. L. Yates, *J. Org. Chem.*, **30**, 2067 (1965).
[104] P.-S. Song, E. C. Smith, and D. E. Metzler, *J. Am. Chem. Soc.*, **87**, 4181 (1965).
[105] A. W. Johnson and J. O. Martin, *Chem. Ind. (London)*, **1965**, 1726.

CHAPTER 12

Reactions of Acids and their Derivatives

Carboxylic Acids

The stimulus given to the study of nucleophilic reactions of carboxylic acids and their derivatives by Bender's classic review[1] continues to be felt.

Several kinetic demonstrations of the intervention of tetrahedral intermediates have been reported. For instance, the hydrolysis of ethyl trifluorothiolacetate in the pH range 0—6 follows the rate law:[2]

$$r/[\text{Ester}] = k_{\text{obs.}} = 1/(2.31 + 6.6a_{\text{H}})$$

the rate decreasing with increasing hydrogen-ion activity between pH 2 and 0. It was also shown that acetate and formate ions act as general-base catalysts. These results were interpreted as indicating a mechanism (1) which involves the general-base-catalysed nucleophilic attack of water at the ester bond and the unsymmetrical partitioning of a tetrahedral intermediate which

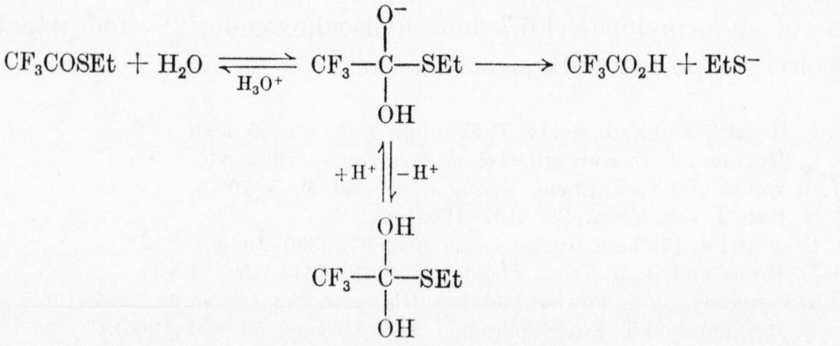

$$\text{CF}_3\text{COSEt} + \text{H}_2\text{O} \underset{\text{H}_3\text{O}^+}{\rightleftharpoons} \text{CF}_3-\overset{\overset{\displaystyle \text{O}^-}{|}}{\underset{\underset{\displaystyle \text{OH}}{|}}{\text{C}}}-\text{SEt} \longrightarrow \text{CF}_3\text{CO}_2\text{H} + \text{EtS}^- \qquad (1)$$

$$+\text{H}^+ \updownarrow -\text{H}^+$$

$$\text{CF}_3-\overset{\overset{\displaystyle \text{OH}}{|}}{\underset{\underset{\displaystyle \text{OH}}{|}}{\text{C}}}-\text{SEt}$$

collapses spontaneously to products and is converted back to ester via general-acid catalysis. Attempts to demonstrate inhibition by acetic acid and formic acid were, however, inconclusive.

The reaction of formamide with hydroxylamine is higher than first-order in hydroxylamine, but the plots of the apparent second-order rate constants against concentration of hydroxylamine at different constant pH's are curves, showing that the order with respect to hydroxylamine decreases with increas-

---

[1] M. L. Bender, *Chem. Rev.*, **60**, 53 (1960).

[2] L. R. Fedor and T. C. Bruice, *J. Am. Chem. Soc.*, **86**, 5697 (1964); **87**, 4138 (1965).

ing concentration, i.e. at low concentrations the reaction is of the second order in hydroxylamine but at high concentrations tends to become of first order.[3] It was similarly shown that catalysis by imidazole was small at high concentrations of hydroxylamine but large at low concentrations. These results were interpreted as indicating a change in rate-determining step from one that is sensitive to base catalysis at low concentrations to one that is relatively insensitive at high concentrations.

The pH–rate profiles for the cleavage of diethyl acetylmalonate and diethyl acetylethylmalonate to acetic acid and diethyl malonate or diethyl ethylmalonate (reaction 2) have been interpreted as indicating that at high pH's the rate-determining step is hydration and at low pH's is cleavage of the carbon–carbon bond.[4a]

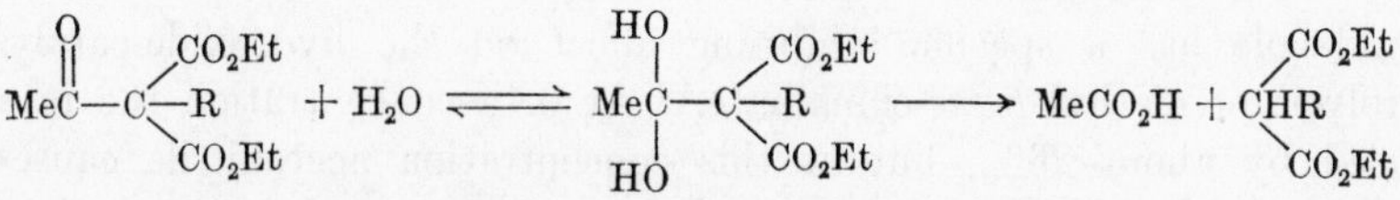

$$
\underset{\substack{\text{MeC}}}{\overset{\substack{O\\\parallel}}{}}\!\!-\!\underset{\substack{|\\\text{CO}_2\text{Et}}}{\overset{\substack{\text{CO}_2\text{Et}\\|}}{\text{C}}}\!\!-\!\text{R} \ + \ \text{H}_2\text{O} \ \rightleftharpoons \ \underset{\substack{|\\\text{HO}}}{\overset{\substack{\text{HO}\\|}}{\text{MeC}}}\!\!-\!\underset{\substack{|\\\text{CO}_2\text{Et}}}{\overset{\substack{\text{CO}_2\text{Et}\\|}}{\text{C}}}\!\!-\!\text{R} \ \longrightarrow \ \text{MeCO}_2\text{H} + \underset{\substack{\\\text{CO}_2\text{Et}}}{\overset{\substack{\text{CO}_2\text{Et}\\}}{\text{CHR}}} \qquad (2)
$$

Another example of a rate-limiting decomposition of a tetrahedral intermediate is found in the hydrolysis of 2-thiazolines.[4b] A wide range of these compounds has been studied by Schmir,[5] and their hydrolyses all show bell-shaped pH–rate profiles. The mechanism shown in Scheme 1 was proposed, with attack of water on the protonated thiazoline rate limiting on the alkaline side of the pH–rate profile and decomposition of the tetrahedral intermediate rate limiting on the acid side.

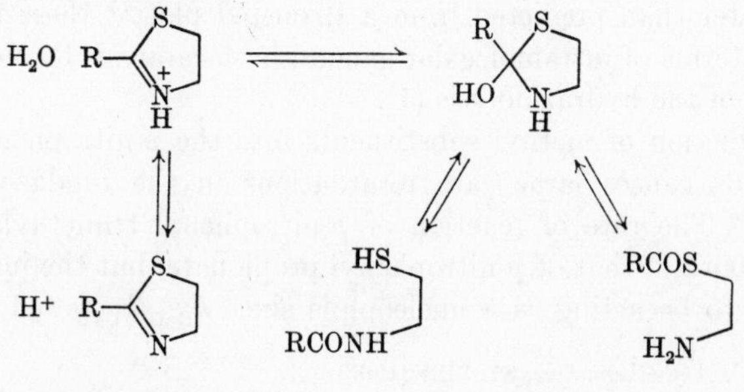

Scheme 1

[3] W. P. Jencks and M. Gilchrist, *J. Am. Chem. Soc.*, **86**, 5616 (1964).
[4a] G. E. Lienhard and W. P. Jencks, *J. Am. Chem. Soc.*, **87**, 3855 (1965).
[4b] R. B. Martin, R. I. Hedrick, and A. Parcell, *J. Org. Chem.*, **29**, 3197 (1964), and references cited therein.
[5] G. L. Schmir, *J. Am. Chem. Soc.*, **87**, 2743 (1965).

The pH–rate profile for the hydrolysis of trifluoroacetanilide has been determined[6] for a wide range of alkaline pH's. The levelling off of rate at high pH is not due to formation of the unreactive anion, as previously suggested,[7] since it occurs well above the pK_a of the anilide. In fact, the anion is quite reactive and hydrolyses in a reaction which is first-order in anion and first-order in hydroxide ion. This behaviour was also attributed[6] to a change in rate-determining step from decomposition of the tetrahedral intermediate at low pH's to its formation at high pH's. The hydrolysis of N-methyltrifluoroacetanilide has also been studied.[8]

The kinetics of the reaction of methyl chloroformate with silver nitrate have been interpreted in terms of a mechanism involving attack of nitrate ion to form a tetrahedral intermediate which then loses a chloride ion with and without electrophilic assistance from Ag^+.[9]

Imidazole has a specific inhibiting effect on the hydroxide-catalysed hydrolysis of methyl *trans*-cinnamate.[10] At 0.4M-concentration the rate is reduced by about 26%, but at this concentration acetonitrile causes a decrease of only 3%. It was suggested that the ester and imidazole form a complex that is relatively unreactive towards hydroxide ion.

According to Bruice and Willis[11] the rate law for the aminolysis of phenyl acetate by n-butylamine does not show a term which is second-order in amine, in disagreement with the report by Jencks and Carriuolo.[12]

The rate law for the reaction of phenyl acetate with hydrazine shows second-order terms that were interpreted as indicating general-acid and general-base catalysis by a second molecule of hydrazine.[13a] The reactions with 2-dimethylaminoethylhydrazine and 3-dimethylaminopropylhydrazine are, however, first-order in these reactants, and the rate constants are about 10^3 fold greater than predicted from a Brønsted plot;[11] these results were explained in terms of intramolecular general-base catalysis by the dimethyl-amino group of the hydrazinolysis (1).

The introduction of methyl substituents into the p-nitrophenyl esters of aliphatic acids causes large rate retardations in the imidazole-catalysed hydrolyses.[13b] The rate of reaction of p-nitrophenyl trimethylacetate was only about 100th of that of p-nitrophenyl propionate, but the imidazole was still thought to be acting as a nucleophile since $k_{H_2O}/k_{D_2O} = 1.15$. A quite

[6] P. M. Mader, *J. Am. Chem. Soc.*, **87**, 3191 (1965).
[7] S. S. Biechler and R. W. Taft, *J. Am. Chem. Soc.*, **79**, 4927 (1957).
[8] R. L. Schowen and G. W. Zuorick, *Tetrahedron Letters*, **1965**, 3839.
[9] D. N. Kevill and G. H. Johnson, *J. Am. Chem. Soc.*, **87**, 928 (1965).
[10] K. A. Connors and J. A. Mollica, *J. Am. Chem. Soc.*, **87**, 123 (1965).
[11] T. C. Bruice and R. G. Willis, *J. Am. Chem. Soc.*, **87**, 531 (1965).
[12] W. P. Jencks and J. Carriuolo, *J. Am. Chem. Soc.*, **82**, 675 (1960).
[13a] T. C. Bruice and S. J. Benkovic, *J. Am. Chem. Soc.*, **86**, 418 (1964).
[13b] T. H. Fife, *J. Am. Chem. Soc.*, **87**, 4597 (1965).

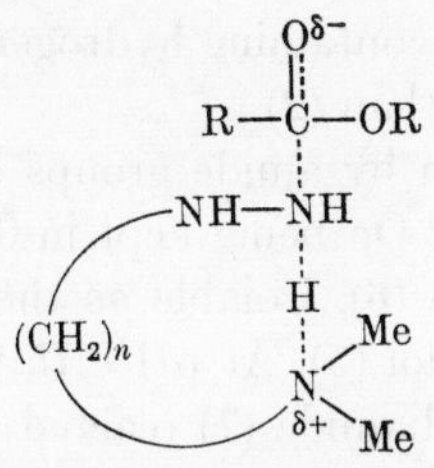

(1)

different pattern of reactivity was, however, found for the imidazole-catalysed hydrolysis of a series of *N*-acylimidazoles: these show catalysis by imidazole and the imidazolinium ion, and α-methyl substituents have only a very small effect on the rate with, e.g., k_{Im} for trimethylacetylimidazole about 2.4 times that for propionylimidazole. β-Methyl substituents, however, cause rate decreases, and k_{Im} for the hydrolysis of *N*-propionyl imidazole is about 800 times that for triethylacetylimidazole. The rate-enhancing effect of the α-methyl substituents was attributed to steric inhibition of resonance in the initial state caused by the carbonyl groups being forced out of the plane of the imidazole ring. In addition, it was thought that the alkyl groups might cause the water around the molecule to be more highly ordered and that this would favour a mechanism involving proton-abstraction from a remote water molecule with concerted transfers through a chain.[13b]

Intramolecular nucleophilic catalysis by the amide group has been shown to occur in the acid-catalysed hydrolysis of *o*-benzamido-*N*,*N*-dicyclohexyl-benzamide (2), which is 10^4 times faster than that of *N*,*N*-dicyclohexyl-benzamide.[14] The postulated intermediate (3) was formed when the amide (2)

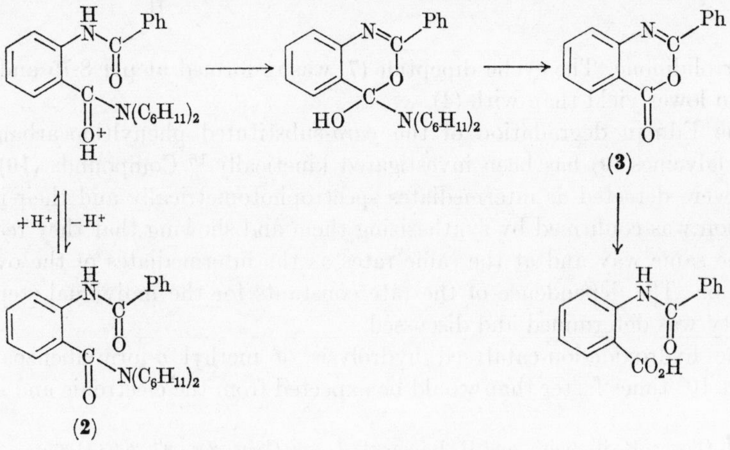

(3)

(2)

[14] T. Cohen and J. Lipowitz, *J. Am. Chem. Soc.*, **86**, 5611 (1964).

was treated with dry dioxan containing hydrogen chloride and was shown to be hydrolysed more rapidly than (2).

Transannular participation by amide groups has been observed with the cyclic diamides (4) and (7).[15] On being kept in 0.1M-HCl at 20° for 10 days, diamide (4) yields the imide (6), isolable as the hydrochloride and formed presumably through the cyclol (5). At pH 8.75 the imide rapidly re-formed the cyclic diamide (4). The diamide (7) reacted much less readily, but with 0.1M-acid at 90° formed 1-(4-aminobutyryl)-2-pyrrolidinone (8) and some

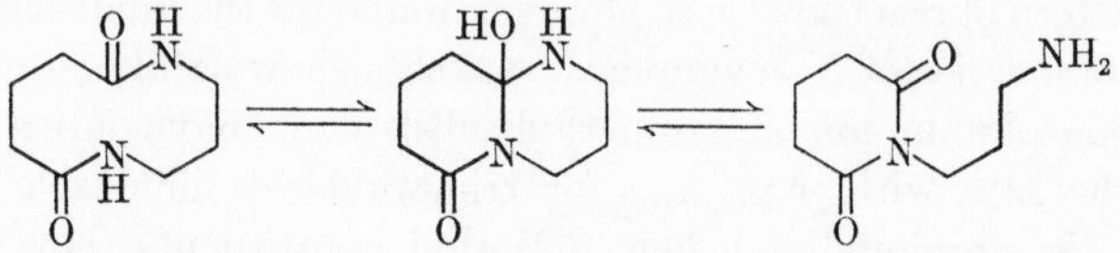

2-pyrrolidinone. The cyclic dipeptide (7) was re-formed at pH 8.75 and 20°, but in lower yield than with (4).

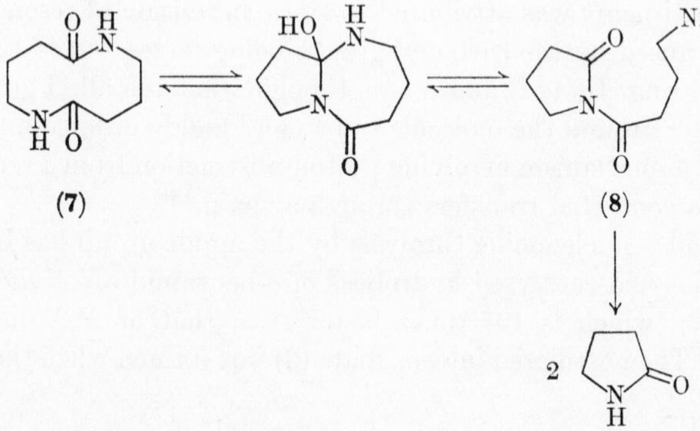

The Edman degradation of the *para*-substituted phenylthiocarbamoyl-leucylglycines (9) has been investigated kinetically.[16] Compounds (10) and (11) were detected as intermediates spectrophotometrically and their intervention was confirmed by synthesizing them and showing that they reacted in the same way and at the same rates as the intermediates of the overall reaction. The dependence of the rate constants for the individual steps on acidity was determined and discussed.

The hydroxide-ion-catalysed hydrolysis of methyl *o*-formylbenzoate is about 10^5 times faster than would be expected from the electronic and steric

[15] G. I. Glover, R. B. Smith, and H. Rapoport, *J. Am. Chem. Soc.*, **87**, 2003 (1965).
[16] D. Bethell, G. E. Metcalfe, and R. C. Sheppard, *Chem. Comm.*, **1965**, 189.

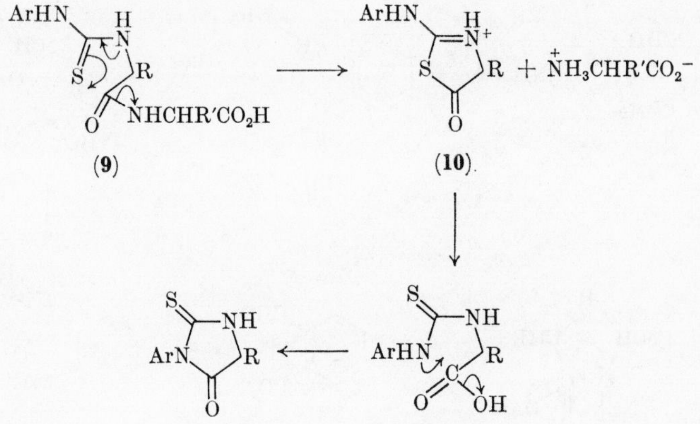

(9) → (10)

(11)

effects of the *o*-formyl group.[17] The reaction is thought to involve attack of hydroxide ion on the formyl group to yield the adduct (12) which then

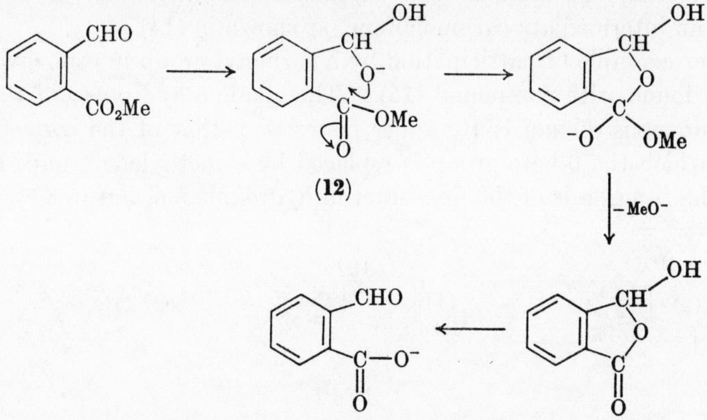

(12)

undergoes an intramolecular reaction. It was estimated that the rate constant for the hydrolysis ($k_2 = 2000$ l mole^{-1} sec^{-1}) was of the same order of magnitude as that expected for the rate constant for the hydration of the aldehyde by $^-$OH, and so it is possible that this is the rate-determining step. Morpholine is also a very effective catalyst and 3-morpholinophthalide (13) was shown to be an intermediate. Catalysis involving an analogous interaction in which the catalyst is an aldehyde and the substrate the ester of an amino acid has also been reported.[18] It was observed that benzaldehyde was six

[17] M. L. Bender, J. A. Reinstein, M. S. Silver, and R. Mikulak, *J. Am. Chem. Soc.*, **87**, 4545 (1965).
[18] B. Capon and R. Capon, *Chem. Comm.*, **1965**, 502.

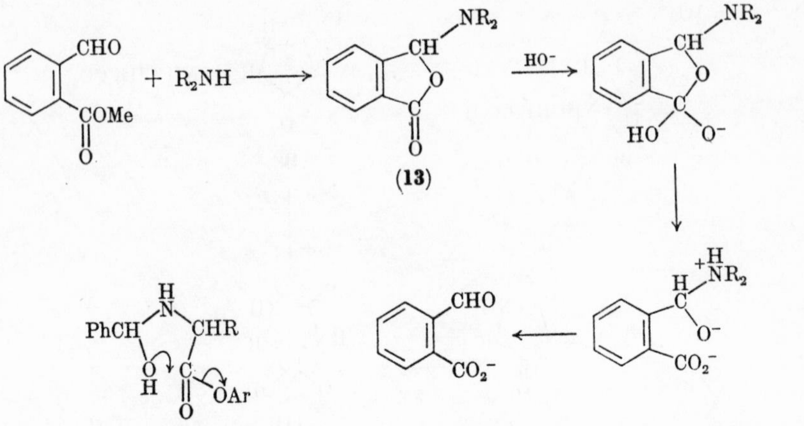

(13)

(14)

times more effective as catalyst than imidazole for the hydrolysis of the
p-nitrophenyl ester of leucine but had no catalytic effect on the hydrolysis
of p-nitrophenyl acetate. It was suggested that the reaction proceeded
through an intermediate carbinolamine, as shown in **(14)**.

Another example of participation by a carbonyl group in ester hydrolysis
has been found with compound **(15)**.[19] The alkaline hydrolysis of this ester
in 50% aqueous dioxan is 10^4 times faster than that of the corresponding
ester in which the 6-keto group is replaced by a methylene group. In 50%
dioxan the reaction is of the first order in hydroxide ion, but in 80% dioxan

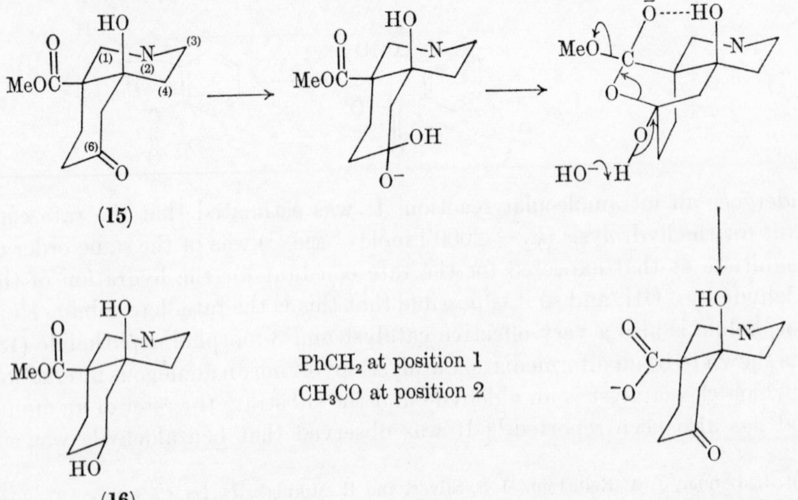

(15)

(16)

PhCH$_2$ at position 1
CH$_3$CO at position 2

[19] H. G. O. Becker, J. Schneider, and H.-D. Steinleitner, *Tetrahedron Letters*, **1965**, 3761.

of the second order. In the Reviewers' opinion this could result from the rate-determining step changing from the first to the last. There is also the possibility that the hydroxyl group provides intramolecular acid-catalysis. Neighbouring-group participation has also been observed in the hydrolysis of the ester (16) with an equatorial hydroxyl group at position 6 which undergoes hydrolysis about 10^4 times faster than its epimer.

Participation by a carbonyl group subsequent to its interaction with a nucleophile has also been shown to occur in the Hinsberg thiophene synthesis.[20] Condensation of benzil with diethyl thiodiacetate yielded the thiophene half-ester (17), not the diester; and when ^{18}O-labelled benzil was used, the resulting half-ester, on decarboxylation, yielded CO_2 containing 50% of the ^{18}O originally present in the benzil. The annexed mechanism was therefore suggested.

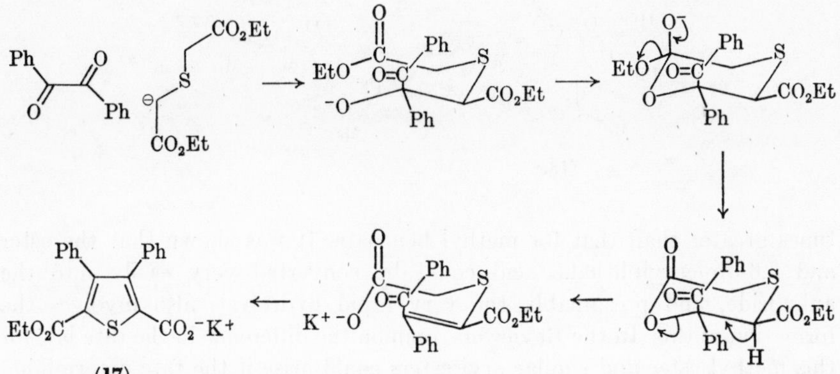

(17)

The pathway for esterification of a series of methyl-substituted *o*-benzoylbenzoic acids by methanolic hydrogen chloride has been investigated.[21] It was found, inter alia, that *o*-benzoylbenzoic acid itself reacts mainly via the pathway (18a) → (18b), to yield the pseudo-ester which is then converted into the normal ester. 2-Benzoyl-3,6-dimethylbenzoic acid, however, which exists mainly in the hydroxy-lactone from (18c) and whose pseudo-ester is the more stable, yields normal ester initially.

A detailed kinetic investigation of the hydrolysis of methyl hydrogen 3,6-dimethylphthalate (19) has been reported.[22] In the pH range 3—5 the reaction follows a rate law, $r = k$[Un-ionized form]. This contrasts with the behaviour of the half aryl esters of dicarboxylic acids which normally show a rate law: $r = k$[Ionized form].[22] It was estimated that at pH 5 the rate was about 10^5

[20] H. Wynberg and H. J. Kooreman, *J. Am. Chem. Soc.*, **87**, 1739 (1965).
[21] M. S. Newman and C. Courduvelis, *J. Org. Chem.*, **30**, 1795 (1965).
[22] L. Eberson, *Acta Chem. Scand.*, **18**, 2015 (1964).

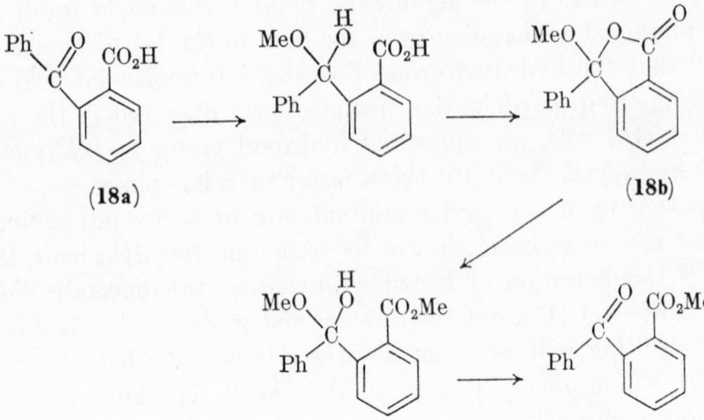

(18a)   (18b)

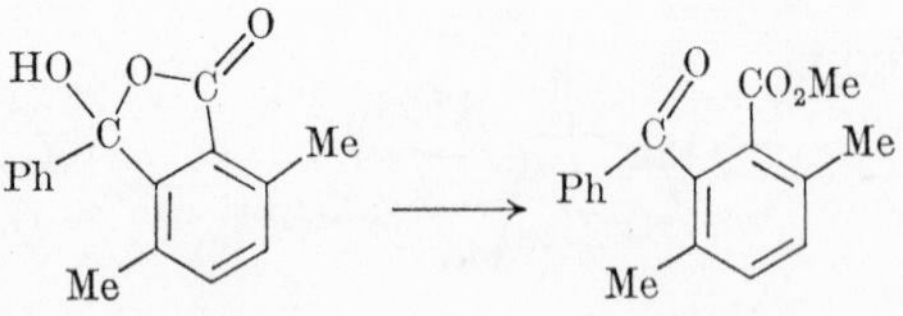

(18c)

times greater than that for methyl benzoate. It was shown that the ester and 3,6-dimethylphthalic acid could be converted very easily into the anhydride, and presumably the very rapid hydrolysis also involves the formation of this. In the Reviewers' opinion the difference in the rate law for this methyl ester and similar aryl esters could arise if the rate-determining

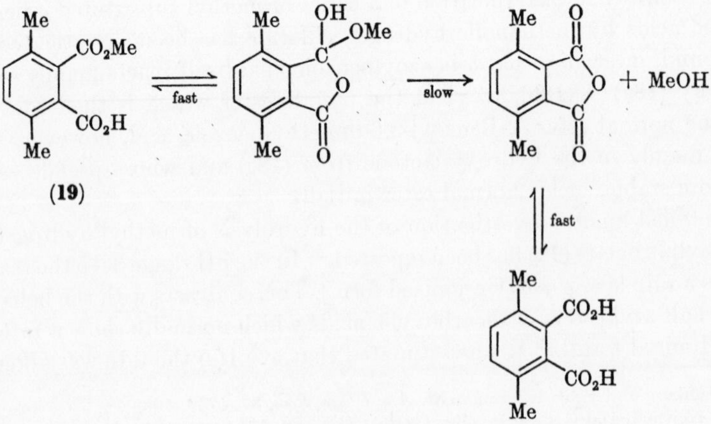

(19)

Scheme 2

step for the latter were formation of the tetrahedral intermediate and for the former the decomposition of this to products, as shown in Scheme 2.

The rate of hydrolysis of ester (19) is very much greater than that of the unsubstituted methyl hydrogen phthalate,[23] presumably, because the methyl groups in the former constrain the carboxylic acid and ester groups to a conformation favourable for ring closure. A similar effect has been observed[24] in the lactonization of 2-(hydroxymethyl)benzoic acids (20). Here all sub-

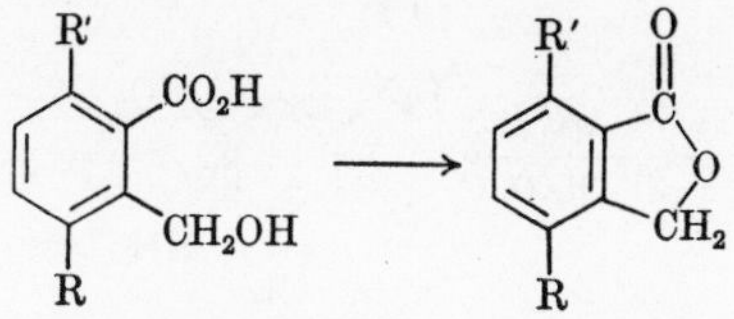

(20)

stituents (except fluorine) in positions 3 and 6 enhance the rate, irrespective of their electronic effect. The largest effect was observed on introduction of 3,6-methyl substituents, which resulted in a 315-fold rate enhancement.

The introduction of substituents at position 3 of *p*-bromophenyl hydrogen

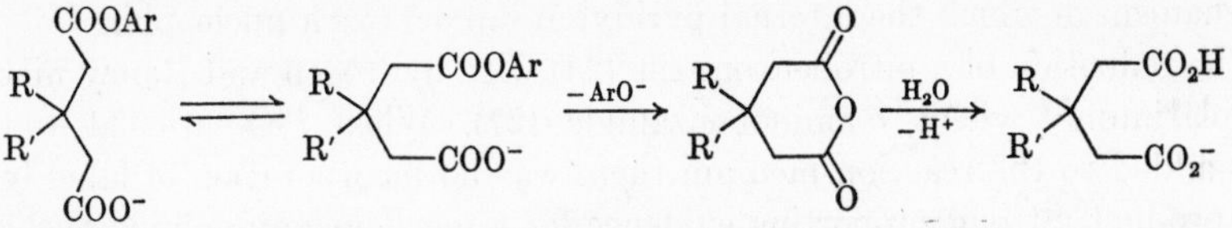

(21)

Scheme 3

glutarate enhances the rate of the intramolecularly catalysed hydrolyses (Scheme 3).[25] It was suggested that this effect was caused by a decrease in the population of the kinetically unprofitable, extended rotamer conformation (21). The phenyl group exhibits a much smaller steric effect than

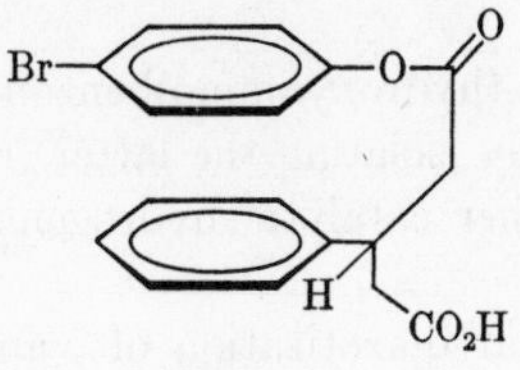

(22)

[23] M. L. Bender, F. Chloupek, and M. C. Neveu, *J. Am. Chem. Soc.*, **80**, 5384 (1958).
[24] J. F. Bunnett and C. F. Hauser, *J. Am. Chem. Soc.*, **87**, 2214 (1965).
[25] T. C. Bruice and W. C. Bradbury, *J. Am. Chem. Soc.*, **87**, 4838, 4846, 4851 (1965).

expected from its size, and it was suggested that this is the result of hydrophobic bonding between it and the *p*-bromophenoxyl group, as shown in (**22**), which results in an increase in the population of the extended conformations. The kinetics of the hydrolyses of the 3,3-dialkylglutaric anhydrides were also investigated.

Neighbouring-group participation by the pyridyl group has been observed in the hydrolysis of [*N*-(3,5-dinitro-2-pyridyl)alanyl]glycine (**23**). In 6M-HCl its rate of hydrolysis is about 1000 times greater than that of the 3,5-

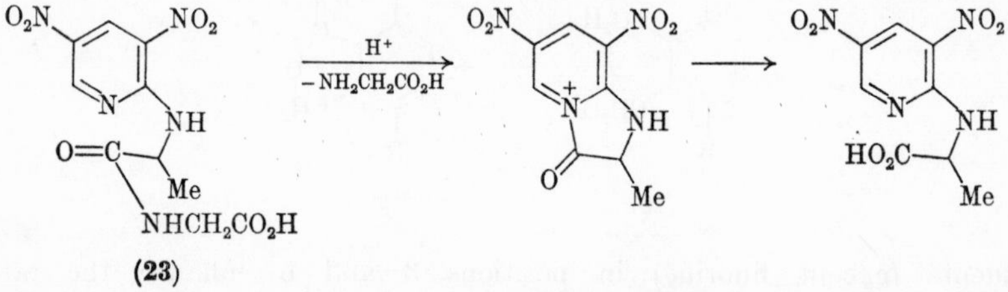

O_2N
NO_2
N
NH
O=C
Me
$NHCH_2CO_2H$
H⁺
$-NH_2CH_2CO_2H$
O_2N
NO_2
N
NH
O
Me
O_2N
NO_2
N
NH
HO_2C
Me

(**23**)

dinitro-4-pyridyl analogue and about 500 times greater than that of 2,4-dinitrophenylalanylglycine.[26] The plot of log k against H_0 is a straight line of slope 1 and the Bunnett w-value is 0.0. This contrasts with the w-values of 1—2.5 normally observed in amide hydrolysis but is consistent with a mechanism in which the internal pyridyl group acts as a nucleophile.

The reduction of *o*-nitrobenzonitrile (**24**) with hydrogen and Raney nickel or platinum[27] yields *o*-aminobenzamide (**27**). When ^{18}O-enriched water was added to the reaction medium there was no incorporation of label into the product. Strong supporting evidence for a mechanism involving cycliza-

CN
NO_2
CN
OH
NH
NH
C
O
NH
NH_2
C
O
NH_2

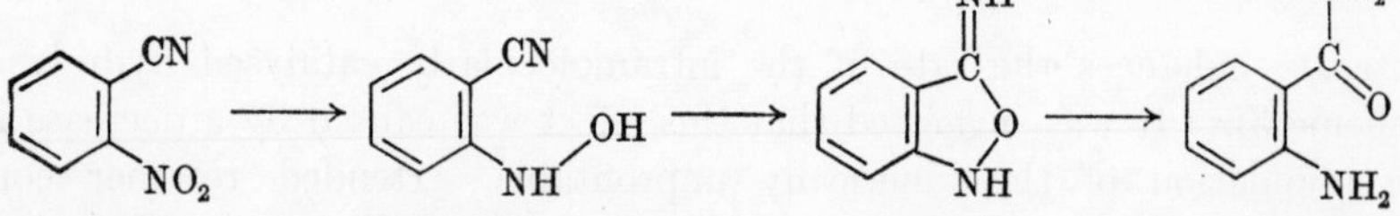

(**24**) (**25**) (**26**) (**27**)

tion of the intermediate *o*-(hydroxyamino)benzonitrile (**25**) to 3-aminoanthranil (**26**) was obtained by isolating the latter from the reaction medium and showing that on further catalytic hydrogenation it yielded *o*-aminobenzamide.

It has been observed that diazotization of *o*-aminophenyl 2,6-dimethylbenzoate results in rapid hydrolysis of the ester group;[28] this reaction

[26] A. Signor and E. Bordignon, *J. Org. Chem.*, **30**, 3447 (1965).
[27] H. Musso and H. Schröder, *Chem. Ber.*, **98**, 1562, 1577 (1965).
[28] D. J. Triggle and S. Vickers, *Chem. Comm.*, **1965**, 544.

presumably proceeds with acyl–oxygen fission since, when tne diazotization is carried out in ethanol with isopentyl nitrite, ethyl and isopentyl 2,6-dimethylbenzoate are obtained. The authors preferred a mechanism written

as shown. Consistent with this, *p*-aminophenyl 2,6-dimethylbenzoate may be diazotized without hydrolysis of the ester group.

Other reactions involving neighbouring-group participation which have been studied include the hydrolysis of salicylamide,[29] aspirin,[30] methyl tetrahydro-4-hydroxyfuran-2-carboxylates,[31] and of *N*-methylhippuric acid,[32] cyclization of cyanophenylacetonitrile,[33] and several acyl migrations.[34]

A striking example of selective catalysis brought about by the preferential binding of substrate to catalyst has been reported by Letsinger and Klaus.[35] It was shown that poly-(1-vinylimidazole) is a good catalyst for the hydrolysis of the ester groups of the negatively charged co-polymer (28) of acrylic acid and 2,4-dinitrophenyl *p*-vinylbenzoate, with a catalytic constant about 20 times greater than that for the hydrolysis of 2,4-dinitrophenyl *p*-isopropyl-benzoate. The catalytic constant for the 1-methylimidazole-catalysed hydrolysis of the latter was, however, slightly larger than that for the ester group of the co-polymer. Protonated and unprotonated imidazole groups are necessary for the catalytic action of the poly-(1-vinylimidazole) on the co-polymer (28) since the catalytic effect is a maximum when the former is partly protonated. The rate did not increase indefinitely with increasing concentration of substrate or catalyst when one of these was varied and the other held constant, but instead the rate reached a steady value. To account for these results it was suggested that a complex is formed involving binding

[29] T. C. Bruice and D. W. Tanner, *J. Org. Chem.*, **30**, 1668 (1965).
[30] E. Mario and J. Gerraughty, *J. Pharm. Sci.*, **54**, 321 (1965).
[31] T. Yamanaka, A. Ichihara, K. Tanabe, and T. Matsumoto, *Tetrahedron*, **21**, 1031 (1965).
[32] E. Schätzle and M. Rottenberg, *Experientia*, **21**, 373 (1965).
[33] J. Zabicky, *Chem. Ind.* (*London*), **1965**, 1651.
[34] K. Ponsold and B. Häfner, *Chem. Ber.*, **98**, 1487 (1965); G. Adam and K. Schreiber, *Angew. Chem. Intern. Ed. Engl.*, **4**, 74 (1965); S. J. Angyal and G. J. H. Melrose, *J. Chem. Soc.*, **1965**, 6494, 6501; C. B. Reese and D. R. Trentham, *Tetrahedron Letters*, **1965**, 2459, 2467; P. J. Garegg, *Arkiv Kemi*, **23**, 255 (1965).
[35] R. L. Letsinger and I. S. Klaus, *J. Am. Chem. Soc.*, **87**, 3380 (1965).

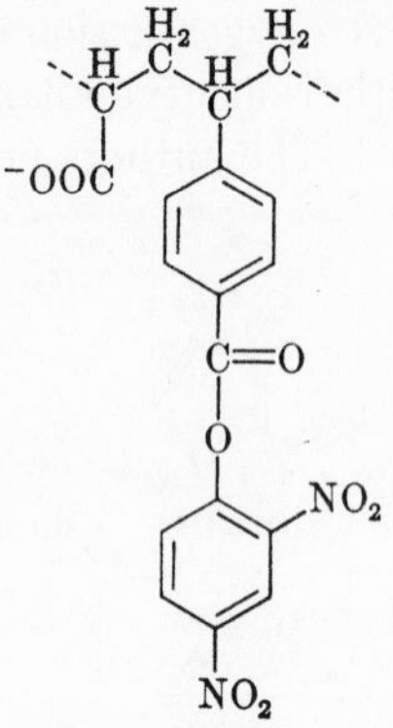

$$\begin{array}{c} \underset{\text{H}}{\overset{\text{H}_2}{\text{C}}} \underset{\text{H}}{\overset{\text{H}_2}{\text{C}}} \\ -\text{OOC} \end{array}$$

(28)

between the cationic sites on the catalyst and the anionic sites on the substrate, and this view is supported by the observation that the reaction is
inhibited by poly(acrylic acid) which competes with the co-polymer for
binding sites on the poly-(1-vinylimidazole). Similar but smaller effects have
been observed in the hydrolysis of 3-acetoxy-N-trimethylanilinium iodide
catalysed by a co-polymer of 4-vinylimidazole and acrylic acid,[36] the
hydrolysis of sodium 4-acetoxy-3-nitrobenzenesulphonate catalysed by poly-
(5-vinylbenzimidazole),[37] and the dehydration of carbonic acid catalysed by
poly-(1-vinylimidazole).[38]

Association of substrate and catalyst has also been achieved by making
them both ligands to the same metal ion.[39] Zinc pyridinecarbaldoxime anion
was shown to catalyse the hydrolysis of 8-acetoxyquinoline-5-sulphonate
(29) by a mechanism (Scheme 4) involving acyl transfer within a catalyst-
substrate complex (30). Under the conditions used the hydrolysis of (31) was
about ten times slower than acetyl transfer from (29), but nevertheless the
overall rate of hydrolysis was increased by a factor of ten.

The hydrolysis of methyl orthobenzoate is subject to catalysis by sodium
dodecyl sulphate as a result of incorporation of the substrate into the
micelles.[40] Below the critical micelle concentration the rate increases
approximately as the fourth power of the sodium dodecyl sulphate, suggesting
the formation of substrate-induced micelles. Above the critical micelle
concentration the rate increases less rapidly and approaches a maximum, a
result which suggests saturation of the substrate with catalyst. Since the

[36] C. G. Overberger, R. Sitaramaiah, T. St. Pierre, and S. Yaroslavsky, *J. Am. Chem. Soc.*,
    87, 3270 (1965).
[37] C. G. Overberger, T. St. Pierre, and S. Yaroslavsky, *J. Am. Chem. Soc.*, 87, 4310 (1965).
[38] H. P. Gregor and K.-J. Liu, *J. Am. Chem. Soc.*, 87, 1678 (1965).
[39] R. Breslow and D. Chipman, *J. Am. Chem. Soc.*, 87, 4195 (1965).
[40] M. T. A. Behme, J. G. Fullington, R. Noel, and E. H. Cordes, *J. Am. Chem. Soc.*, 87, 266
    (1965).

solvent deuterium isotope effect is unchanged it appears that the mechanism of the reaction in the presence of detergent is similar to that in its absence. It was suggested that the observed catalysis could be the result of either electrostatic stabilization of the developing carboxonium ion in the transition state or of a locally increased concentration of hydrogen ions close to the substrate–detergent complex.

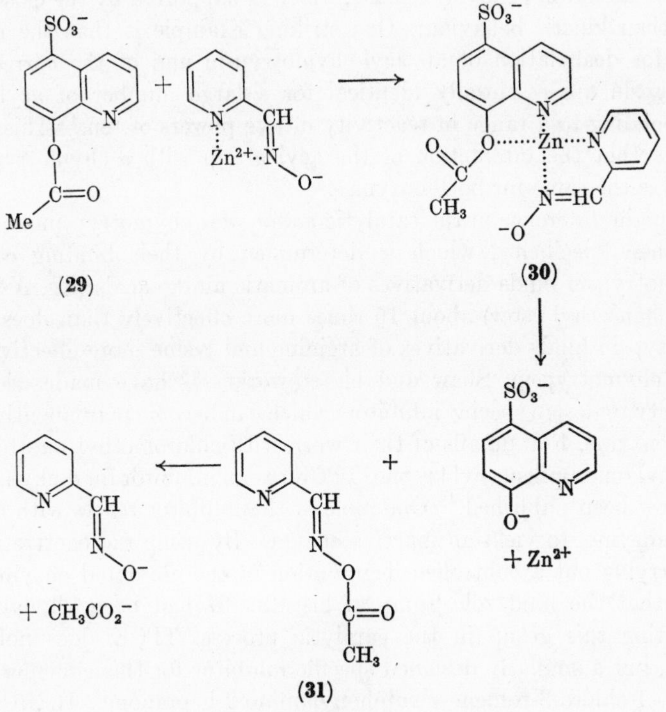

Scheme 4

There have been several kinetic investigations of reactions in frozen systems.[41]

The mechanism of action of the proteolytic enzymes has been reviewed.[42]

Similarities in the amino acid sequences of chymotrypsin and trypsin have been discussed.[43] In the sequences of the precursors trypsinogen and chymotrypsinogen A, 100 residues are identical, and 127 are similar; this

---

[41] H. E. Alburn and N. H. Grant, *J. Am. Chem. Soc.*, **87**, 4174 (1965); N. H. Grant and H. E. Alburn, *Biochemistry*, **4**, 1913 (1965); R. E. Pincock and T. E. Kiovsky, *J. Am. Chem. Soc.*, **87**, 2072 (1965).

[42] M. L. Bender and F. J. Kézdy, *Ann. Rev. Biochem.*, **27**, 49 (1965).

[43] F. Šorm, V. Holeyšovský, O. Mikeš, and V. Tomášek, *Collection Czech. Chem. Commun.*, **30**, 2103 (1965); B. S. Hartley, J. R. Brown, D. L. Kauffman, and L. B. Smillie, *Nature (London)*, **207**, 1157 (1965).

corresponds to 40% and 51%, respectively, of the sequences. In addition, four of the five disulphide bridges of chymotrypsinogen are exactly homologous in position with four bridges of trypsinogen, two of the three histidines of trypsin parallel the two histidines of α-chymotrypsin, and the sequential distances between the histidines and the active serine are identical. These similarities between the two enzymes suggest that the mechanism of their catalytic action is probably similar, which is supported by the close similarity of their kinetic behaviour. One striking example is that the rate constants for deacylation of an acyl-chymotrypsin and of the corresponding acyl-trypsin are essentially identical for a large number of acyl groups, corresponding to a range in reactivity of five powers of ten.[44] This strongly suggests that the interaction of the acyl group with α-chymotrypsin and trypsin is the same for both enzymes.

The main difference in the catalytic action of α-chymotrypsin and trypsin is in their specificity, which is determined by their binding constants. α-Chymotrypsin binds derivatives of aromatic amino acids (e.g. N-acetyl-L-tryptophan ethyl ester) about 15 times more effectively than does trypsin, while trypsin binds derivatives of arginine and lysine more effectively than does α-chymotrypsin. Shaw and his co-workers[45] have made use of this specificity to design specific inhibitors which can become permanently bonded to the enzyme. Full details of their work with chloromethyl L-α-(toluene-p-sulphonylamido)phenethyl ketone (TPCK) as an inhibitor for α-chymotrypsin have now been published.[45] One mole of the inhibitor reacts with one mole of the enzyme, to yield an inactive enzyme. By using radioactive inhibitor and carrying out a controlled degradation of the alkylated enzyme it was shown that the imidazole group of histidine-57 had been alkylated, thus implicating this group in the catalytic process. TPCK does not inhibit trypsin, but a similarly designed specific inhibitor for this enzyme, namely, 7-amino-1-chloro-3-toluene-p-sulphonylamido-2-heptanone, $H_2N(CH_2)_4CH$-$(NHTos)COCH_2Cl(TLCK)$, was shown[46] to react with the imidazole group of histidine 46; significantly the position of this histidine in the trypsin sequence corresponds exactly to the position of Histidine-57 in the chymotrypsin sequence.

By imposing a number of constraints on the amino acid sequence of α-chymotrypsin, Bender has shown that the molecule can take up a conformation in which a hydrophobic cavity is formed by the cycle between Cys-42 and Cys-58 which are linked by a disulphide bridge.[47] It was suggested that

[44] M. L. Bender, J. V. Killheffer, and F. J. Kézdy, *J. Am. Chem. Soc.*, **86**, 5330 (1964).
[45] E. B. Ong, E. Shaw, and G. Schoellmann, *J. Biol. Chem.*, **240**, 694 (1965).
[46] E. Shaw, M. Mares-Guia, and W. Cohen, *Biochemistry*, **4**, 2219 (1965); V. Tomášek, E. S. Severin, and F. Šorm, *Biochem. Biophys. Res. Commun.*, **20**, 545 (1965).
[47] M. L. Bender, J. V. Killheffer, and F. J. Kézdy, *J. Am. Chem. Soc.*, **86**, 5331 (1964).

this forms the major binding site, with the sequence from Cys-191 to Cys-201 serving as a subsidiary binding site.

The ultraviolet spectra of cinnamoyl-, furylacryloyl-, and indolylacryloyl-chymotrypsin and -subtilisin have been held to be consistent only with structures in which an imidazole group is acylated.[48] This conclusion is, however, unlikely to be correct, since at low pH's the hydrolysis of cinnamoyl-chymotrypsin is slower than that of cinnamoylimidazole itself.

Degradation studies of the monofurylacryloyl derivatives of two subtilisin enzymes have shown the furylacryloyl group to be present as *O*-furylacryloylserine in the sequence $Asp(NH_2)$–Gly–Thr–Ser–Met.[49]

There have been many other investigations of the mechanism of action of chymotrypsin[50] and trypsin.[51]

A partial amino acid sequence for papain has been reported.[52] It was concluded from an investigation of their ultraviolet spectra that the acyl enzymes from papain and ficin with methyl thionhippurate,[53] and from papain with 1-*trans*-cinnamoylimidazole,[54] are formed by acylation of a thiol group. By alkylating papain with the specific inhibitor,

[48] H. F. Noller and S. A. Bernard, *Biochemistry*, **4**, 1118 (1965).

[49] S. A. Bernhard, S. J. Lau, and H. Noller, *Biochemistry*, **4**, 1108 (1965).

[50] F. J. Kézdy and M. L. Bender, *Biochemistry*, **4**, 104 (1965); P. W. Inward and W. P. Jencks, *J. Biol. Chem.*, **240**, 1986 (1965); H. Weiner and D. E. Koshland, *ibid.*, **240**, PC.2764 (1965); R. Biltonen, R. Lumry, V. Madison, and H. Parker, *Proc. Nat. Acad. Sci. U.S.*, **54**, 1412 (1965); A. N. Glazer, *ibid.*, **54**, 171 (1965); A. Yapel and R. Lumry, *J. Am. Chem. Soc.*, **86**, 4499 (1964); R. L. Biltonen and R. Lumry, *ibid.*, **87**, 4208 (1965); T. Inagami, S. S. York, and A. Patchornik, *ibid.*, **87**, 126 (1965); W. B. Lawson and H.-J. Schramm, *Biochemistry*, **4**, 377 (1965); T. Inagami and J. M. Sturtevant, *ibid.*, **4**, 1330 (1965); J. B. Jones, C. Niemann, and G. E. Hein, *ibid.*, **4**, 1735 (1965); H. I. Arbash and C. Niemann, *ibid.*, **4**, 99 (1965); A. M. Gold, *ibid.*, **4**, 897 (1965); S. G. Cohen and S. Y. Weinstein, *J. Am. Chem. Soc.*, **86**, 5326 (1964); S. A. Bernhard and Z. H. Tashjian, *ibid.*, **87**, 1806 (1965); R. M. Epand and I. B. Wilson, *J. Biol. Chem.*, **240**, 1104 (1965); A. Y. Moon, J. Mercouroff, and G. P. Hess, *ibid.*, **240**, 717 (1965); J. B. Jones, T. Kunitake, C. Niemann, and G. E. Hein, *J. Am. Chem. Soc.*, **87**, 1777 (1965).

[51] M. L. Bender, F. J. Kézdy, and J. Feder, *J. Am. Chem. Soc.*, **87**, 4953 (1965); M. L. Bender and F. J. Kézdy, *ibid.*, **87**, 4954 (1965); M. L. Bender, F. J. Kézdy, and J. Feder, *ibid.*, **87**, 4955 (1965); T. E. Barman and H. Gutfreund, *Proc. Nat. Acad. Sci. U.S.*, **53**, 1243 (1965); M. L. Bender, J. V. Killheffer, and R. W. Roeske, *Biochem. Biophys. Res. Commun.*, **19**, 161 (1965); S. A. Bernhard and H. Gutfreund, *Proc. Nat. Acad. Sci. U.S.*, **53**, 1238 (1965); J. A. Stewart and J. E. Dolson, *Biochemistry*, **4**, 1086 (1965); T. F. Spande and B. Witkop, *Biochem. Biophys. Res. Commun.*, **21**, 131 (1965); T. Inagami, *J. Biol. Chem.*, **240**, PC. 3453 (1965); W. R. Finkenstadt and M. Laskowski, *ibid.*, **240**, PC.962 (1965); M. Mares-Guia and E. Shaw, *ibid.*, **240**, 1579 (1965); M. Gemperli, W. Hofmann, and M. Rottenberg, *Helv. Chim. Acta*, **48**, 939 (1965).

[52] R. Frater, A. Light, and E. L. Smith, *J. Biol. Chem.*, **240**, 253 (1965); A. Light and J. Greenberg, *ibid.*, **240**, 258 (1965); J. R. Kimmel, H. J. Rogers, and E. L. Smith, *ibid.*, **240**, 266 (1965); M. A. McDowall and E. L. Smith, *ibid.*, **240**, 281 (1965); A. Light, R. Frater, J. R. Kimmel, and E. L. Smith, *Proc. Nat. Acad. Sci. U.S.*, **52**, 1276 (1964).

[53] G. Lowe and A. Williams, *Biochem. J.*, **96**, 189, 194, 199 (1965); *Proc. Chem. Soc.*, **1964**, 140.

[54] M. L. Bender and L. J. Brubacher, *J. Am. Chem. Soc.*, **86**, 5333 (1964).

$TosNHCH_2COCH_2Cl$, this thiol group was identified[55] as being that of cysteine-25. Kinetic studies of the papain-catalysed hydrolyses of α-N-benzoyl-L-arginine ethyl ester and amide are also consistent with a mechanism involving the formation of an acyl enzyme.[56] The pH–rate profile for acylation is bell-shaped, showing that an acid–base pair with pK_a's 4.24—4.29 and 8.49—8.35 intervenes in the catalytic process. The pH–rate profile for deacylation is, however, sigmoid, indicating the intervention of a group of pK_a 3.91. It was suggested that the group of $pK \approx 4$ was a carboxyl group and the group of $pK \approx 8$ a thiol group.[56]

It has been shown that the papain-catalysed hydrolysis of ethyl [*carbonyl*-^{18}O]hippurate proceeds without ^{18}O-exchange, a result which is also consistent with a reaction involving an acyl enzyme.[57]

Investigations which add to our knowledge of the mechanism of the action of acetylcholinesterase[58] and pepsin[59] have also been reported.

There have been several important discussions of the mechanism of the hydrolysis of orthoesters.[60–62] These reactions show general-acid catalysis with high Brønsted α-values and solvent-deuterium isotope effects k_{H_2O}/k_{D_2O} = 0.5. Substituent effects are small and sometimes opposite to that expected; thus, ethyl orthobenzoate is hydrolysed more slowly than ethyl orthoformate and orthoacetate. The general-acid catalysis has been attributed[60,61] to an S_E2 mechanism involving a rate-determining proton transfer, possibly concerted with carbon–oxygen bond-breaking (Scheme 5). It was pointed

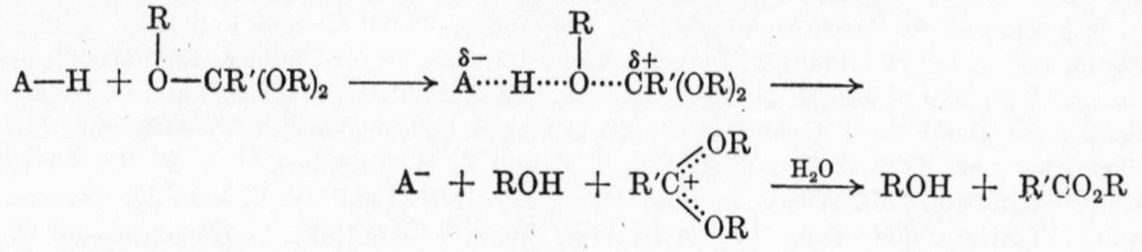

$$A\!-\!H + \overset{\overset{\displaystyle R}{\displaystyle |}}{O}\!-\!CR'(OR)_2 \longrightarrow \overset{\delta-}{A}\cdots H\cdots \overset{\overset{\displaystyle R}{\displaystyle |}}{\overset{\delta+}{O}}\cdots CR'(OR)_2 \longrightarrow$$

$$A^- + ROH + R'\overset{\displaystyle OR}{\underset{\displaystyle OR}{C^+}} \xrightarrow{H_2O} ROH + R'CO_2R$$

Scheme 5

out that it is not unreasonable to expect proton transfers between oxygen atoms to be rate-determining steps when these occur between bases of widely

[55] S. S. Husain and G. Lowe, *Chem. Comm.*, **1965**, 345.

[56] J. R. Whitaker and M. L. Bender, *J. Am. Chem. Soc.*, 87, 2728 (1965); see also K. Brocklehurst, *Chem. Comm.*, **1965**, 234.

[57] J. F. Kirsch and E. Katchalski, *Biochemistry*, 4, 884 (1965).

[58] M. L. Bender and J. K. Stoops, *J. Am. Chem. Soc.*, 87, 1622 (1965); R. M. Krupka, *Biochem. Biophys. Res. Commun.*, 19, 531 (1965); R. M. Krupka, *Biochemistry*, 4, 429 (1965).

[59] M. S. Silver, *J. Am. Chem. Soc.*, 87, 1627 (1965); M. S. Silver, J. L. Denburg, and J. J. Steffens, *ibid.*, 87, 886 (1965).

[60] C. A. Bunton and R. H. De Wolfe, *J. Org. Chem.*, 30, 1371 (1965).

[61] A. J. Kresge and R. J. Preto, *J. Am. Chem. Soc.*, 87, 4593 (1965).

[62] A. M. Wenthe and E. H. Cordes, *J. Am. Chem. Soc.*, 87, 3173 (1965).

different strength since the free energy of activation in one direction must be at least equal to the free-energy difference between the two conjugate acids involved in the transfer. On these grounds it was estimated that the value of the second-order rate constant for the hydrolysis of ethyl ortho-carbonate, $3 \times 10^2 \, l \, mole^{-1} \, sec^{-1}$, would be approximately the same as that expected for proton transfer. The value of the solvent isotope effect in these reactions, $k_{H_2O}/k_{D_2O} = 0.5$, appears to afford an argument against an S_E2 mechanism and to favour an A–1 mechanism. It has, however, been suggested that, if proton transfer is almost complete in the transition state, the primary isotope effect will be close to unity but that the inverse secondary isotope effect[63a] will be close to its maximum value and that the overall isotope effect could, therefore, be less than unity. The effect of substituents was taken to indicate that the transition states have little carbonium ion character. It has been reported that in the gas phase the stabilization of the cations does not vary in the series $MeOCMe_2{}^+$, $(MeO)_2CH^+$, $(MeO)_2CMe^+$, and $(MeO)_3C^+$.[63b] If then these results can be extrapolated to solution, the transition states for the orthoester hydrolyses could still have carbonium-ion character and be consistent with the observed substituent effects.[62]

From the observations that spontaneous and acid-catalysed hydrolyses of trimethylacetic anhydride are slower than the corresponding reactions of acetic anhydride, and that these reactions have strongly negative entropies of activation, it has been concluded that they all proceed by bimolecular mechanisms.[64] The decrease in the rate of hydrolysis of cyclic anhydrides with increasing acid concentration[65] has been investigated further,[66] this behaviour has now been observed also with succinic, tetramethylsuccinic, maleic, and phthalic anhydride but is not found with glutaric, homophthalic, camphoric, or *cis*-3-cyclohexene-1,2-dicarboxylic anhydride. The spontaneous and acid-catalysed hydrolyses are both very sensitive to salt effects, and to account for the observed decreases in rate with increasing acid concentration it is presumed that rate-enhancement of the acid catalysis is less than the rate decrease from the negative salt effect.

The spontaneous hydrolysis of diacetyl sulphide proceeds at about one-half of the rate for acetic anhydride.[67] The former hydrolysis is much less susceptible to acid-catalysis than the latter, the rate being only about 10%

[63a] A. J. Kresge and D. P. Onwood, *J. Am. Chem. Soc.*, **86**, 5014 (1964).

[63b] Unpublished observations by R. W. Taft reported in ref. 62.

[64] C. A. Bunton and J. H. Fendler, *J. Org. Chem.*, **30**, 1365 (1965).

[65] J. Koskikallio and A. Ervasti, *Suomen Kemistilehti*, **35B**, 213 (1962); C. A. Bunton, J. H. Fendler, N. A. Fuller, S. Perry, and J. Rocek, *J. Chem. Soc.*, **1963**, 5361.

[66] C. A. Bunton, J. H. Fendler, N. A. Fuller, S. Perry, and J. Rocek, *J. Chem. Soc.*, **1965**, 6174.

[67] J. Hipkin and D. P. N. Satchell, *J. Chem. Soc.*, **1965**, 1057.

greater in 3M-perchloric acid than in water; this results from cancellation
of a small catalytic effect by a negative salt effect. Hydrolysis of substituted
benzoic acids in dioxan–water mixtures,[68] and of diketene,[69] additions
to ketene,[70] and acetyl exchange between anhydrides,[71] have also been
investigated.

Other reactions which have been investigated include: those between
carboxylic acids and diphenyldiazomethane;[72] oxygen exchange of oxalic
acid[73] and glycine;[74] hydrolysis of thiol carboxylic acids;[75] conversion of
peroxybenzoic acid into acetyl benzoyl peroxide in acetic acid;[76] esterification of acetylenic and fluoro acids;[77] reaction of acetyl chloride and
chloroacetyl chloride with phenols in acetonitrile;[78] reaction of acetyl
chloride and L-menthol in liquid sulphur dioxide;[79] solvolysis of the ethyl
chloroformate;[80] hydrolysis of bornyl and isobornyl acetate;[81] ester saponifi-
cation in water–dimethyl sulphoxide mixtures;[82] reaction of esters with
sodium hydroxide in anhydrous dimethyl sulphoxide;[83] alkaline hydrolysis
of methyl furan- and thiophene-carboxylates,[84] of ethyl azulenecarboxyl-
ates,[85] of glycine ethyl ester,[86] and of ethyl acetate with sulphur-containing
substituents;[87] neutral hydrolysis of chloromethyl and methyl dichloro-
acetate;[88] cupric ion-catalysed hydrolysis of ethyl picolinate;[89] ester exchange

[68] J. Koskikallio, *Acta Chem. Scand.*, **18**, 2248 (1964).

[69] J. M. Briody and D. P. N. Satchell, *J. Chem. Soc.*, **1965**, 3778.

[70] J. M. Briody and D. P. N. Satchell, *Chem. Ind.* (*London*), **1965**, 1427.

[71] M. Sheinblatt and S. Alexander, *J. Am. Chem. Soc.*, **87**, 3905 (1965); T. G. Bonner, E. G.
Gabb, P. McNamara, and B. Smethhurst, *Tetrahedron*, **21**, 463 (1965); L. Ötvös, F. Dutka,
and H. Tüdós, *Acta Chim. Acad. Sci. Hung.*, **43**, 53 (1965).

[72] R. A. More O'Ferrall, W. K. Kwok, and S. I. Miller, *J. Am. Chem. Soc.*, **86**, 5553 (1964);
C. K. Hancock and E. Foldvary, *J. Org. Chem.*, **30**, 1180 (1965); A. Buckley, N. B. Chapman,
and J. Shorter, *J. Chem. Soc.*, **1965**, 6310.

[73] C. O'Connor and D. R. Llewellyn, *J. Chem. Soc.*, **1965**, 2197.

[74] C. O'Connor and D. R. Llewellyn, *J. Chem. Soc.*, **1965**, 2669.

[75] J. Hipkin and D. P. N. Satchell, *Tetrahedron*, **21**, 835 (1965).

[76] Y. Ogata, Y. Furuya and K. Aoki, *Bull. Chem. Soc. Japan*, **38**, 838 (1965).

[77] J. Radell, B. W. Brodman, A. Hirshfeld, and E. D. Bergmann, *J. Phys. Chem.*, **69**, 928 (1965).

[78] J. M. Briody and D. P. N. Satchell, *J. Chem. Soc.*, **1965**, 168.

[79] N. Tokura and F. Akiyama, *Bull. Chem. Soc. Japan*, **37**, 1723 (1964).

[80] A. Kivinen, *Acta Chem. Scand.*, **19**, 845 (1965).

[81] C. A. Bunton, K. Khaleeluddin, and D. Whittaker, *J. Chem. Soc.*, **1965**, 3290.

[82] E. Tommila, *Suomen Kemistilehti*, **37B**, 117 (1964); D. D. Roberts, *J. Org. Chem.*, **30**, 3516
(1965).

[83] W. Roberts and M. C. Whiting, *J. Chem. Soc.*, **1965**, 1290.

[84] S. Oae, N. Furukawa, T. Watanabe, Y. Otsuji, and M. Hamada, *Bull. Chem. Soc. Japan*,
38, 1247 (1965).

[85] P. Leermakers and W. A. Bowman, *J. Org. Chem.*, **29**, 3708 (1964).

[86] M. Robson, *Nature*, **208**, 265 (1965).

[87] R. P. Bell and B. A. W. Coller, *Trans. Farad. Soc.*, **61**, 1445 (1965).

[88] N. J. Cleve and E. K. Euranto, *Suomen Kemistilehti*, **37B**, 126 (1964); E. K. Euranto, *ibid.*,
38A, 25 (1965).

[89] A. Agren, *Acta Pharm. Suecica*, **2**, 87 (1965); *Chem. Abs.*, **63**, 1675 (1965).

of co-ordinated esters;[90] metal ion-catalysed hydrolysis of potassium ethyl oxalate;[91] hydrolysis of ethyl acetate in concentrated sulphuric acid;[92] acid-catalysed hydrolysis of benzyl acetate,[93] *t*-butyl acetate, and 1-methoxy-1-methylethyl acetate;[94] methanolysis of 2,4-dinitrophenyl acetate;[95] cleavage of benzyloxycarbonyl and cycloalkyloxycarbonyl derivatives of amino acids and of amino esters;[96] aminolysis of acyl derivatives of oximes and amidoximes;[97] acid-catalysed hydrolysis of acetamide,[98] glycylglycylglycine,[99] lactams,[100] cyanamide,[101] ethyl benzimidates,[102] and hydantoins;[103] alkaline and acid hydrolysis of amides with structures analogous to that of vitamin B_{12};[104] alkaline hydrolysis of urea,[105] and of *N*-acyl derivatives of 5-ethyl-1-methyl-5-phenylbarbituric acid;[106] hydrolysis of monothiosuccinimide to succinimide,[107] and of *N*-alkyl-*N*-nitrosoureas;[108a] nucleophilic reactions of 1-dimethylcarbamoylpyridinium chloride;[108b] reactions of CO_2 and COS with amines;[109] reaction between acetamide and alkyl nitrites in dimethylformamide;[110] hydrolysis of diaryl carbonates[111] and diethyl dicarbonate;[112] racemization of acyl-peptides;[113] hydrolysis of 1,2,4-triazoles;[114] imidazole-catalysed acetylation of serine and threonine

[90] R. P. Houghton and D. J. Pointer, *J. Chem. Soc.*, **1965**, 4214.

[91] R. W. Hay and N. J. Walker, *Nature*, **204**, 1189 (1965).

[92] D. Jaques, *J. Chem. Soc.*, **1965**, 3874.

[93] H. Sadek and F. Y. Khalil, *Suomen Kemistilehti*, **38B**, 55 (1965); *Chem. Abs.*, **63**, 2864 (1965).

[94] T. Yrjana, *Suomen Kemistilehti*, **37B**, 108 (1964).

[95] W. R. Ali, A. Kirkien-Konasiewicz, and A. Maccoll, *J. Chem. Soc.*, **1965**, 6409.

[96] K. Bláha and J. Rudinger, *Collection Czech. Chem. Commun.*, **30**, 585, 599 (1965). R. B. Homer, R. B. Moodie, and H. N. Rydon, *J. Chem. Soc.*, **1965**, 4403.

[97] R. Buyle, *Helv. Chim. Acta*, **47**, 2444 (1964).

[98] J. Koskikallio, *Acta Chem. Scand.*, **18**, 1831 (1964); R. J. L. Martin, *Australian J. Chem.*, **18**, 807 (1965); P. D. Bolton and I. R. Wilson, *ibid.*, **18**, 795 (1965); K. Yates and J. C. Riordan, *Can. J. Chem.*, **43**, 2328 (1965).

[99] D. A. Long and T. G. Truscott, *Trans. Farad. Soc.*, **61**, 531 (1965).

[100] R. S. Muromova, *Zh. Vses. Khim. Obshchestva, im. D. I. Mendeleeva*, **10**, 102 (1965); *Chem. Abs.*, **62**, 16005 (1965).

[101] B. R. Mole, J. P. Murray, and J. G. Tillett, *J. Chem. Soc.*, **1965**, 802.

[102] R. H. DeWolfe and F. B. Augustine, *J. Org. Chem.*, **30**, 699 (1965).

[103] H. B. Milne and W. D. Kilday, *J. Org. Chem.*, **30**, 67 (1965).

[104] R. Bonnett, J. A. Raleigh, and D. G. Redman, *J. Am. Chem. Soc.*, **87**, 1600 (1965).

[105] K. R. Lynn, *J. Phys. Chem.*, **69**, 687 (1965).

[106] J. Bojarski, W. Kahl, and M. Melzacka, *Roczniki Chem.*, **39**, 875 (1965).

[107] D. T. Witiak, T.-F. Chin, and J. L. Lach, *J. Org. Chem.*, **30**, 3721 (1965).

[108a] E. R. Garrett, S. Goto, and J. F. Stubbins, *J. Pharm. Sci.*, **54**, 119 (1965).

[108b] S. L. Johnson and K. A. Rumon, *J. Am. Chem. Soc.*, **87**, 4782 (1965).

[109] M. M. Sharma, *Trans. Farad. Soc.*, **61**, 681 (1965).

[110] Z. Kricsfalussy and A. Bruylants, *Bull. Soc. Chim. Belg.*, **74**, 17 (1965).

[111] G. D. Cooper, H. T. Johnson, and B. Williams, *J. Org. Chem.*, **30**, 3989 (1965).

[112] A. Kivinen, *Suomen Kemistilehti*, **38**, 106 (1965).

[113] I. Antonovics and G. T. Young, *Chem. Comm.*, **1965**, 398.

[114] H. Gehlen and J. Schmidt, *Ann. Chem.*, **682**, 123 (1965).

derivatives by acetic anhydride;[115] reactions of isocyanates[116] and isothio-cyanates;[117] and cyclization of 4-oxo derivatives to 10*H*- and 10-alkyl-pyrimido[5,4-*b*][1,4]benzothiazines.[118]

The significance of proton mobility in acid-catalysed esterification has been discussed.[119]

There have been several investigations of decarboxylation.[120]

Non-carboxylic Acids[121]

Most of the work this year has been on derivatives of phosphoric acid and of alkylphosphonic acids.

It has been shown that five-membered cyclic phosphonate esters[122] as well as phosphate esters[123] undergo hydrolysis very rapidly. The relative rates of hydrolysis of the five-membered and six-membered cyclic phostonates (**32**) and (**33**) and sodium ethyl phosphonate are $5 \times 10^4 : 3 : 1$ in acid and $6 \times 10^5 : 24 : 1$ in alkali. Tracer studies showed that the phosphonates undergo exclusive phosphorus–oxygen fission but that acyclic phosphonates undergo 50% carbon–oxygen fission, so that the relative rates of hydrolysis of the phosphonates at phosphorus are slightly higher than given above.

[115] F. Schneider, *Experientia*, **21**, 316 (1965).

[116] H. van Landeghem and I. de Aguirre, *Bull. Soc. Chim. France*, **1965**, 1328; A. J. Leusink and J. G. Noltes, *Rec. Trav. Chim.*, **84**, 585 (1965); A. M. Kardos, J. Volke, and P. Kristián, *Collection Czech. Chem. Commun.*, **30**, 931 (1965); I. de Aguirre and J.-C. Jungers, *Bull. Soc. Chim. France*, **1965**, 1316; J. N. Greenshields, R. H. Peters, and R. F. T. Stepto, *J. Chem. Soc.*, **1964**, 5101; L. Rand, B. Thir, S. L. Reegen, and K. C. Frisch, *J. Appl. Polymer. Sci.*, **9**, 1787 (1965); A. Farkas and P. F. Strohm, *Ind. Eng. Chem. Fundamentals*, **4**, 32 (1965); O. V. Nesterov and S. G. Entelis, *Kinetika i Kataliz*, **6**, 178 (1965).

[117] L. Drobnica and J. Augustin, *Collection Czech. Chem. Commun.*, **30**, 1221 (1965); E. Dyer and R. B. Pinkerton, *J. Appl. Polymer. Sci.*, **9**, 1713 (1965); L. Drobnica and J. Augustin, *Collection Czech. Chem. Commun.*, **30**, 1618 (1965).

[118] B. Roth and J. F. Bunnett, *J. Am. Chem. Soc.*, **87**, 340 (1965).

[119] H. Zimmermann and J. Rudolph, *Angew. Chem. Intern. Ed. Engl.*, **4**, 40 (1965).

[120] F. Cramer and W. Kampe, *J. Am. Chem. Soc.*, **87**, 1115 (1965); E. M. Kosower and P. C. Huang, *ibid.*, **87**, 4645 (1965); D. S. Noyce, S. K. Brauman, and F. B. Kirby, *ibid.*, **87**, 4335 (1965); D. S. Noyce, L. Gortler, M. J. Jorgenson, F. B. Kirby and E. C. McGoran, *ibid.*, **87**, 4329 (1965); D. S. Noyce and R. A. Heller, *ibid.*, **87**, 4325 (1965); C. G. Over-berger, N. Weinshenker, and J.-P. Anselme, *ibid.*, **87**, 4119 (1965); G. A. Hall and E. S. Hanrahan, *J. Phys. Chem.*, **69**, 2402 (1965); J. Watson and P. Haake, *J. Org. Chem.*, **30**, 1122 (1965); R. W. Hay, *Australian J. Chem.*, **18**, 337 (1965); R. W. Hay and S. J. Harvie, *ibid.*, **18**, 1197 (1965); J. J. Sims, *J. Am. Chem. Soc.*, **87**, 3511 (1965); T. Hanafusa, L. Birladeanu, and S. Winstein, *ibid.*, **87**, 3511 (1965); G. J. Litchfield and G. Shaw, *Chem. Comm.*, **1965**, 563; J. K. Kochi, *J. Org. Chem.*, **30**, 3265 (1965); G. W. Kosicki, S. N. Lipovac, and R. G. Annett, *Can. J. Chem.*, **42**, 2806 (1964); A. T. Blades and M. G. H. Wallbridge, *J. Chem. Soc.*, **1965**, 792; W. H. Starnes, *J. Am. Chem. Soc.*, **86**, 5603 (1964).

[121] For reviews of the mechanisms of reactions of phosphate esters, see: (*a*) J. R. Cox and O. B. Ramsay, *Chem. Rev.*, **64**, 317 (1964); and (*b*) W. P. Jencks, Brookhaven Symposia in Biology, No. 15, p. 134.

[122] A. Eberhard and F. H. Westheimer, *J. Am. Chem. Soc.*, **87**, 253 (1965).

[123] J. Kumamoto, J. R. Cox, and F. H. Westheimer, *J. Am. Chem. Soc.*, **78**, 4858 (1956).

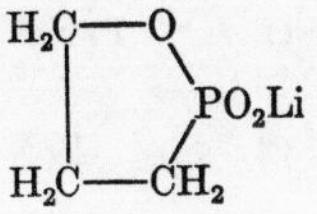

(32) 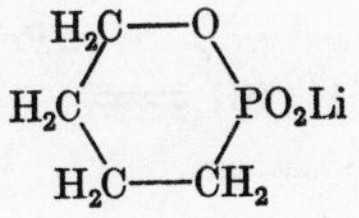 (33)

The high rates of hydrolysis of five-membered cyclic phosphates and phosphonates are thought to be mainly the result of angle strain. It is to be hoped, then, that the X-ray determination of the structure of ethylene methyl phosphate (1-methoxy-1,3,2-dioxophospholidine)[124] will throw light on the nature of this strain. Calculations of the bond angles of ethylene methyl phosphate before the X-ray analysis were found to be in good agreement with it.[125]

A very large rate enhancement has also been observed in the alkaline hydrolysis of the five-membered cyclic sulphate, pyrocatechol cyclic sulphate (34), which is 2×10^7 times faster than that of its open-chain analogue diphenyl sulphate (35).[126] This rate difference was presumed to reflect the difference for attack of hydroxide ion at sulphur since attack on carbon was thought to be very unlikely.

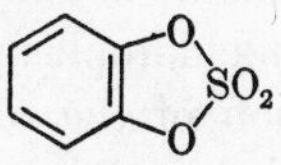

(34) 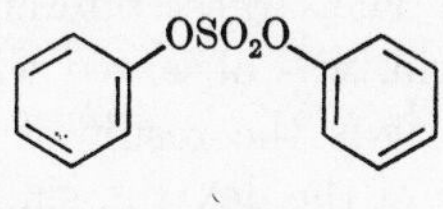 (35)

The hydrolyses of N,N,N',N'-tetramethyl- (36) and N,N'-dipropyl-phosphorodiamidic chloride (37) and their reactions with nucleophiles (e.g. pyridine) proceed at similar rates in neutral and slightly acid solutions, but the hydroxide ion-catalysed hydrolysis of (37) is at least 4×10^6 times faster

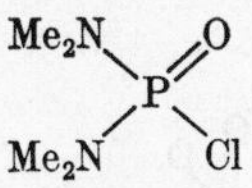

(36) 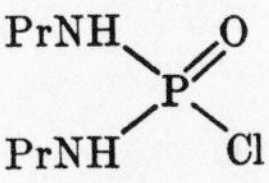 (37)

than that of (36),[127] and it is therefore assumed that compound (37) can react by a mechanism involving proton abstraction as shown in Scheme 6. Attempts to trap the metaphosphate derivative (38) were, however, unsuccessful.[127] The mechanism of Scheme 6 has been criticized[128] since it was observed

[124] T. A. Steitz and W. N. Lipscomb, *J. Am. Chem. Soc.*, **87**, 2488 (1965).
[125] D. A. Usher, E. A. Dennis, and F. H. Westheimer, *J. Am. Chem. Soc.*, **87**, 2320 (1965).
[126] E. T. Kaiser, I. R. Katz, and T. F. Wulfers, *J. Am. Chem. Soc.*, **87**, 3781 (1965).
[127] P. S. Traylor and F. H. Westheimer, *J. Am. Chem. Soc.*, **87**, 553 (1965).
[128] D. B. Coult and M. Green, *J. Chem. Soc.*, **1964**, 5478.

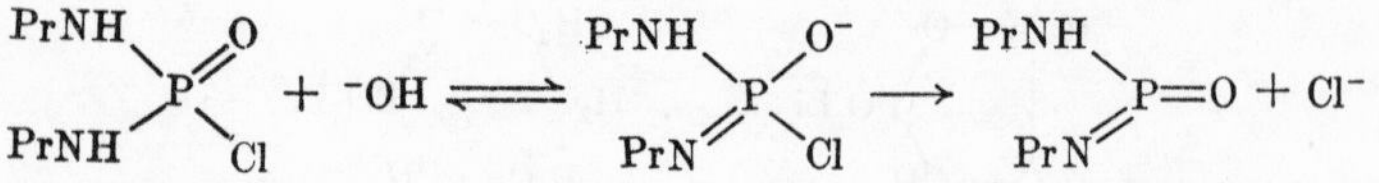

Scheme 6

$$\text{(38)}$$

that addition of hydrogen peroxide to alkaline solutions of p-nitrophenyl NN'-diphenyl- and NN'-dimethylphosphorodiamidate enhanced the rate of formation of p-nitrophenol.[128] Generally HOO^- is a stronger nucleophile, but a weaker base, than HO^-,[129] so that if HO^- were acting as a base, as in Scheme 6, addition of hydrogen peroxide would result in a rate decrease. The observed rate increase was, therefore, taken to indicate that HO^- acts as a nucleophile in the alkaline hydrolysis of these phosphorodiamidates. An elimination-addition mechanism similar to that of Scheme 6 has also been proposed for the alkaline hydrolysis of p-nitrophenyl N-methylcarbamate.[130]

The rate of the calcium-ion-catalysed cleavage of diaryl pyrophosphates is enhanced in the presence of cyclodextrins.[131] The β-dextrin (7 glucose residues, $\phi = 8$ Å) is more effective than the γ-dextrin (8 glucose residues, $\phi = 10$ Å), which is in turn more effective than the α-dextrin (6 glucose residues, $\phi = 6$ Å). The largest rate enhancement, 200-fold in the presence of 1 mol of β-dextrin, was observed with di-p-chlorophenyl pyrophosphate. This suggests that it is the result of inclusion of the phosphate into the hydrophobic cavity of the dextrin, since this is known to occur readily with halogen-substituted organic compounds; in accord with this, transfer of phosphate to a hydroxyl group of the dextrin was observed. Transfer of the inside phosphate group was demonstrated by showing that in the presence of the dextrin the hydrolysis of P^1-p-chlorophenyl P^2-ethyl pyrophosphate yielded phosphorylated dextrin, ethyl phosphate, and p-chlorophenol. The reaction can, therefore, be formulated as shown in Scheme 7.

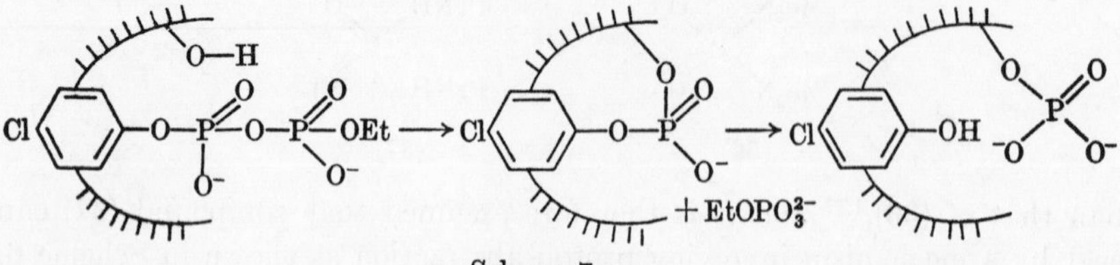

Scheme 7

The hydrolysis of the mixed phosphoric carbonic anhydride (39) proceeds with carbon–oxygen fission and is catalysed by bases and metal ions.[132]

[129] Cf. R. G. Pearson and D. N. Edgington, *J. Am. Chem. Soc.*, **84**, 4607 (1962).
[130] M. L. Bender and R. B. Homer, *J. Org. Chem.*, **30**, 3975 (1965).
[131] N. Hennrich and F. Cramer, *J. Am. Chem. Soc.*, **87**, 1121 (1965).
[132] D. L. Griffith and M. Stiles, *J. Am. Chem. Soc.*, **87**, 3710 (1965).

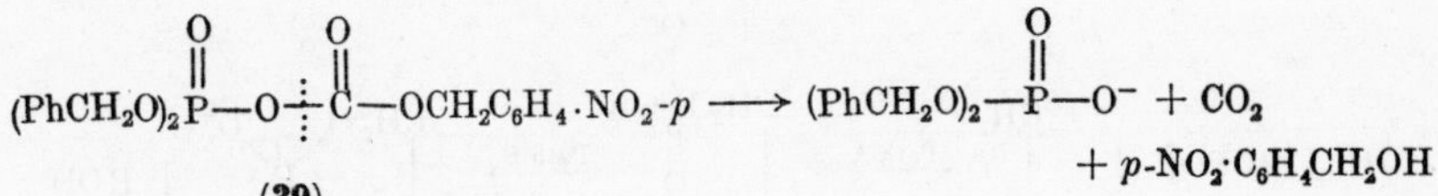

$$(PhCH_2O)_2 \overset{O}{\underset{\|}{P}} - O \overset{\vdots}{-} \overset{O}{\underset{\|}{C}} - OCH_2C_6H_4 \cdot NO_2\text{-}p \longrightarrow (PhCH_2O)_2 - \overset{O}{\underset{\|}{P}} - O^- + CO_2$$
$$+ \ p\text{-}NO_2 \cdot C_6H_4CH_2OH$$

(39)

The suggestion of Ramirez and his co-workers[133] that the extremely rapid hydrolysis of acetoin dimethyl phosphate (40) proceeds by a mechanism involving the dioxaphosphorane (42) has been criticized. It was thought that,

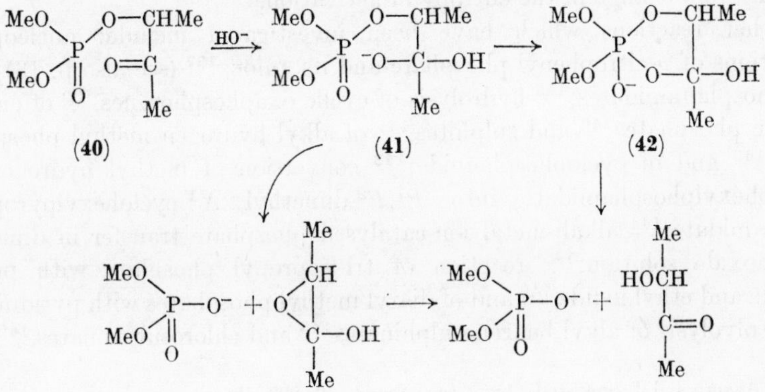

(40) (41) (42)

rather than yielding dimethyl phosphate, the ion (42) would expel MeO^-; therefore the possibility of attack by the negatively charged oxygen of (41) on carbon rather than on phosphorus was considered. Support for a mechanism involving this was obtained by treating compound (40) with sodium methoxide in methanol, when the ketal (43) was formed.[134]

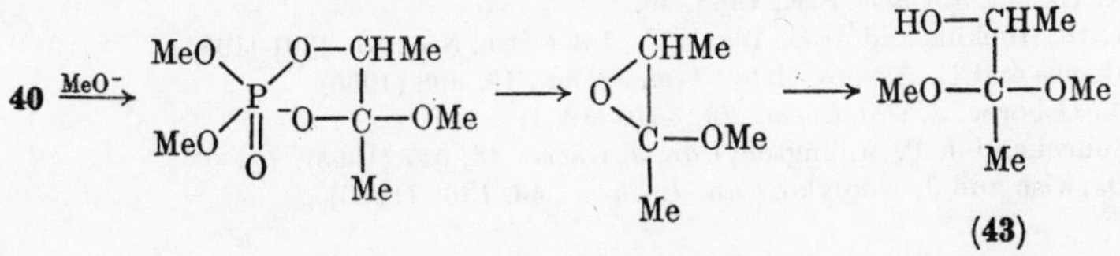

Treatment of cyclohexyl esters of 2-hydroxyalkyl phosphates also frequently results in attack by the ionized hydroxyl group on carbon rather than on phosphorus.[135] With esters of 2-hydroxypropyl phosphate (44), attack on carbon (path A) occurs when R is cyclohexyl, but on phosphorus (path B) when it is phenyl.[135]

[133] F. Ramirez, B. Hansen, and N. B. Desai, *J. Am. Chem. Soc.*, **84**, 4588 (1962).
[134] H. Witzel, A. Botta, and K. Dimroth, *Chem. Ber.*, **98**, 1465 (1965); see also ref. 121a.
[135] D. M. Brown and D. A. Usher, *J. Chem. Soc.*, **1965**, 6547, 6558.

$$H^+ + ROPO_3^{2-} + \underset{\underset{MeCH}{|}}{\overset{CH_2}{\diagdown}O} \xleftarrow{\text{Path A}} \underset{MeCH\cdot OH}{\overset{CH_2OPOR}{\underset{\|}{\overset{\|}{O}}}} \xrightarrow{\text{Path B}} \underset{MeCH\cdot O}{\overset{CH_2\ O}{\diagup}} \overset{O^-}{\underset{O}{\diagdown P}} + ROH$$

(44)

The rate of the hydroxide-ion-catalysed hydrolysis of phosphonate esters $RPO(OEt)_2$ increases as R is changed from CH_3 to $ClCH_2$ to Cl_2CH, owing mainly to a change in the entropy of activation.[136]

Other reactions which have been investigated include: nucleophilic reactions of *p*-nitrophenyl phosphate and its anions[137] (see also p. 131) and of phosphoramidates;[138] hydrolysis of cyclic oxaphosphoranes,[139] of enediol cyclic phosphates[140] and sulphites,[141] of alkyl hydrogen methyl phosphonates,[142] and of cyclophosphamide;[143] conversion of methyl hydrogen *N*-cyclohexylphosphamidate into P^1,P^2-dimethyl N^1-cyclohexylpyrophosphoramidate;[144] alkali-metal ion-catalysed phosphate transfer in dimethyl sulphoxide solution;[145] reaction of tri-isopropyl phosphite with propyl iodide and ethyl iodide,[146] and of diaryl methyl phosphates with pyridine;[147] and solvolysis of alkyl benzenesulphinates[148] and chlorosulphonates.[149]

[136] G. Aksnes and J. Songstad, *Acta Chem. Scand.*, **19**, 893 (1965).
[137] A. J. Kirby and W. P. Jencks, *J. Am. Chem. Soc.*, **87**, 3209, 3217 (1965).
[138] W. P. Jencks and M. Gilchrist, *J. Am. Chem. Soc.*, **87**, 3199 (1965).
[139] F. Ramirez and A. V. Patwardhan, N. B. Desai, and S. R. Heller, *J. Am. Chem. Soc.*, **87**, 549 (1965).
[140] F. Ramirez, O. P. Madan and C. P. Smith, *J. Am. Chem. Soc.*, **87**, 670 (1965).
[141] P. A. Bristow, M. Khowaja, and J. G. Tillett, *J. Chem. Soc.*, **1965**, 5779.
[142] L. Keay, *Can. J. Chem.*, **43**, 2637 (1965).
[143] O. M. Friedman, S. Bien, and J. K. Chakrabarti, *J. Am. Chem. Soc.*, **87**, 4978 (1965).
[144] N. K. Hamer, *J. Chem. Soc.*, **1965**, 46.
[145] E. A. H. Hopkins and J. H. Wang, *J. Am. Chem. Soc.*, **87**, 4391 (1965).
[146] G. Aksnes and D. Aksnes, *Acta Chem. Scand.*, **19**, 898 (1965).
[147] D. W. Osborne, *J. Org. Chem.*, **29**, 3570 (1964).
[148] E. Buncel and J. P. Millington, *Can. J. Chem.*, **43**, 547 (1965).
[149] D. Darwish and J. Noreyko, *Can. J. Chem.*, **43**, 1366 (1965).

CHAPTER 13

Photochemistry

The appearance, in 1963, of the first volume of "Advances in Photochemistry"[1] reflects the rapid expansion of work on this subject in recent years; this expansion has been further shown by the I.U.P.A.C.-sponsored symposium[2] in 1964 and a Gordon Conference in 1965. Recent reviews deal with phototropy (photochromism)[3] and the light-induced formation of acids from cyclic ketones.[4] The present chapter is principally concerned with reactions in solution.

The chromophore which has received most attention in photochemical reactions is undoubtedly the carbonyl group. The molecular gymnastics which lead to product formation on irradiation of 2,5-cyclohexadienones are, by now, well known. However, the mechanisms of the processes involved are not so well understood. They have frequently been discussed in terms of carbonium-ion rearrangements involving the grouping ^+C—C=C—O^-. Zimmerman and his co-workers have argued[5] that the $n \to \pi^*$ excited state (in which an electron from a non-bonding orbital on oxygen is promoted to an anti-bonding π-orbital)[6] must render the β-carbon atom electron-rich rather than electron-deficient. A valence-bond structure for the excited state would then be $\dot{C}$—C=C—$\ddot{O}$:. Zimmerman and his co-workers have now disclosed the relatively simple photochemical dienone rearrangement shown in reactions (1).[7] Compound (1B) undergoes acid-catalysed rearrangement with exclusive migration of the phenyl group. However, in the photochemical rearrangement it is predominantly the cyanophenyl group that migrates, in accord with a radical mechanism. That the non-migrating group does not

---

[1] "Advances in Photochemistry", Vol. I—III, Interscience Publishers Inc., New York, London, Sydney, 1963—64.

[2] *Pure Appl. Chem.*, **9**, 461—621 (1964).

[3] R. Exelby and R. Grinter, *Chem. Rev.*, **65**, 247 (1965).

[4] G. Quinkert, *Angew. Chem. Intern. Ed. Engl.*, **4**, 211 (1965).

[5] H. E. Zimmerman in ref. 1, Vol. I.

[6] Another important point is that this separation of electrons in formation of the excited state will lead to a species whose reactivity will probably not differ significantly according to whether the two odd electrons are spin-paired (singlet state) or not (triplet state). This point has been made by Zimmerman (ref. 5, p. 197) and by Hammond (footnote 12 in ref. 26); subsequently, however, Zimmerman reported evidence for triplet intermediates in rearrangements of 2,5-cyclohexadienone (*J. Am. Chem. Soc.*, **86**, 1436 (1964).

[7] H. E. Zimmerman, R. C. Hahn, H. Morrison, and M. C. Wani, *J. Am. Chem. Soc.*, **87**, 1939 (1965).

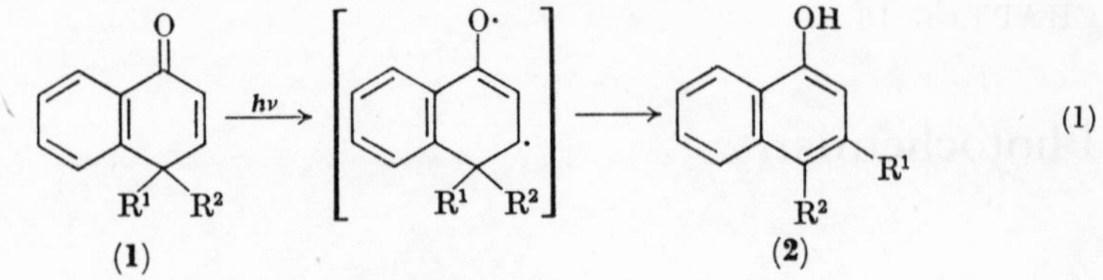

(1)

A: $R^1 = Ph$, $R^2 = Me$
B: $R^1 = p\text{-}CNC_6H_4$, $R^2 = Ph$

control the rearrangement was demonstrated by the exclusive phenyl migration exhibited by dienone (**1A**). More direct evidence for the radical nature of intermediates in the photochemistry of cyclohexadienones has, however, been obtained by Schuster and Patel.[8] In particular, dienone (**3**) on photolysis in hydrocarbon solvents gives *p*-cresol and hexachloroethane.

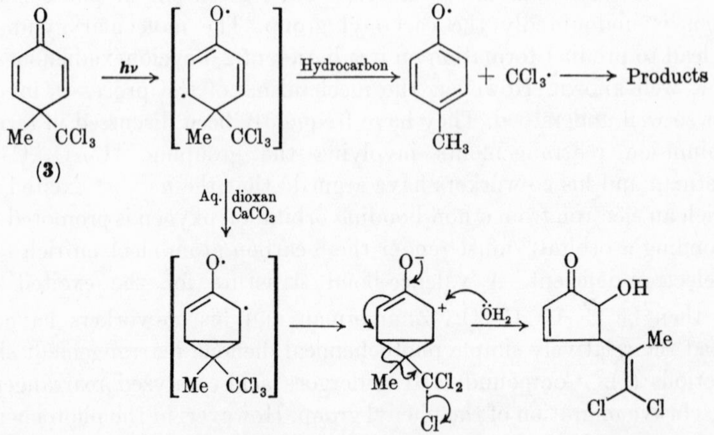

In aqueous dioxan, in the presence of calcium carbonate, a different product is formed, presumably by bond alteration and $\pi^* \rightarrow n$ electron demotion, followed by hydrolysis.[9]

Yet another, hitherto unreported, mode of rearrangement is observed on irradiation of the sterically crowded dienone (**4**),[10] as illustrated.

The photochemical transformations of steroidal dienones of partial structure (**5**) have also been studied.[11]

The photochemistry of benzophenones continues to attract attention, both in its own right and because of the use of these compounds as photo-

[8] D. I. Schuster and D. J. Patel, *J. Am. Chem. Soc.*, **87**, 2515 (1965).
[9] J. King and D. Leaver, *Chem. Comm.*, **1965**, 539.
[10] B. Miller and H. Margulies, *Chem. Comm.*, **1965**, 314.
[11] B. Nann, H. Wehrli, K. Schaffner, and O. Jeger, *Helv. Chim. Acta*, **48**, 1680 (1965).

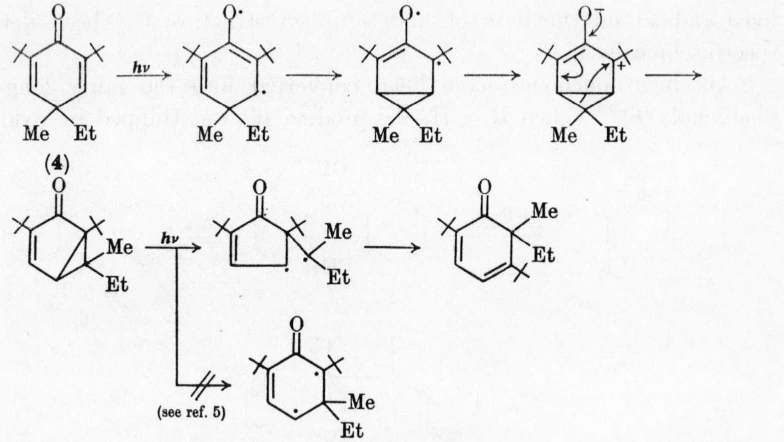

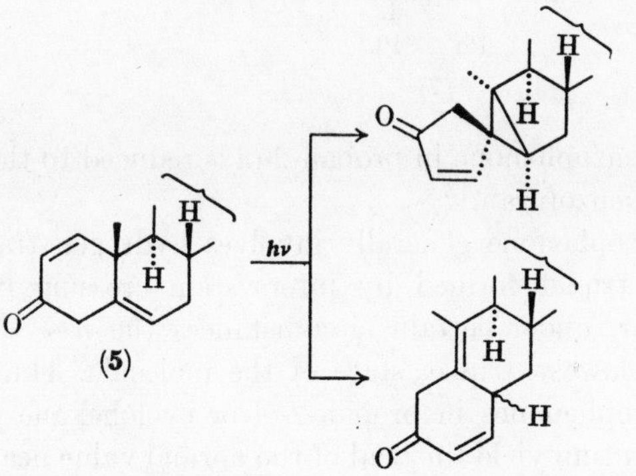

(4)

(see ref. 5)

(5)

sensitizers. The photoreduction of benzophenone in propan-2-ol to give benzpinacol is well known. A similar process has now been observed for primary and for secondary amines (which have an α-hydrogen atom)[12] and has been adapted as an excellent route for conversion of amines into

$$>CH-NH- \xrightarrow{Ph_2CO*} Ph_2\overset{.}{C}OH + >\overset{.}{C}-NH-$$

$$>\overset{.}{C}-NH- \xrightarrow{Ph_2CO} >=N- + Ph_2\overset{.}{C}OH$$

$$2Ph_2\overset{.}{C}OH \longrightarrow HOCPh_2CPh_2OH$$

$$>=N- \xrightarrow{H_2O} >=O$$

Scheme 1

carbonyl compounds (Scheme 1). In the analogous reaction in ether, a long-

[12] S. G. Cohen and R. J. Baumgarten, *J. Am. Chem. Soc.*, **87**, 2996 (1965).

lived radical intermediate of undetermined structure has been detected spectroscopically.[13]

2-Alkylbenzophenones have been converted into the fairly long-lived photoenols (6).[14] When R = H, the product (6) was trapped by oxidation

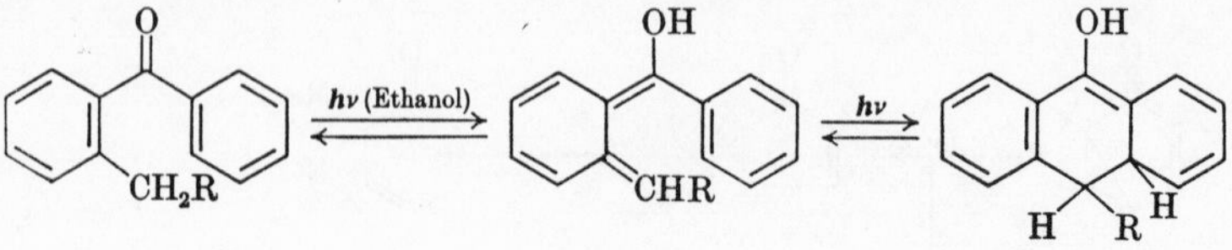

(6)

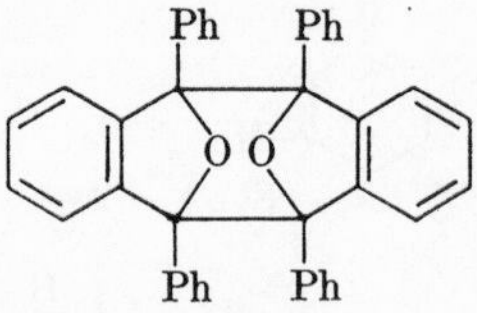

(7)

to anthrone. 2-Benzoylbenzophenone in propan-2-ol is reduced to the photo-dimer (7) of diphenylisobenzofuran.[15]

The reduction of benzophenone generally involves hydrogen transfer to oxygen of the $n \to \pi^*$ triplet formed by intersystem crossing from the $n \to \pi^*$ singlet. However, under certain circumstances the $n \to \pi^*$ triplet appears not to be the lowest triplet state of the molecule. Thus photo-reduction of phenylbenzophenones in propan-2-ol or cyclohexane proceeds with a relatively low quantum yield (instead of the normal value near unity), and this is attributed to excitation of the $\pi \to \pi^*$ triplet of these compounds.[16] With some amino- and alkylamino-benzophenones, the quantum yield in propan-2-ol is zero, which is ascribed to excitation of a low-energy charge-transfer triplet. Because of the large charge separation associated with this excited state, its energy is strongly solvent-dependent; thus in cyclohexane the $n \to \pi^*$ triplet lies below the charge-transfer triplet, and in this solvent the aminobenzophenones are photoreduced with a high quantum efficiency.[16]

Hammond and his colleagues have continued their work on energy transfer in photochemical systems. Their recent work has included a study of the quenching of benzophenone triplets by metal chelates, as a function of the ligand (quenching of singlet excited benzophenone was observed in some instances).[17] Intramolecular energy transfer in molecules containing two

[13] H. Mauser, U. Sproesser, and H. Heitzer, *Chem. Ber.*, **98**, 1639 (1965).
[14] E. F. Ullman and K. R. Huffman, *Tetrahedron Letters*, **1965**, 1863.
[15] P. Courtot and D. H. Sachs, *Bull. Soc., Chim. France*, **1965**, 2259.
[16] G. Porter and P. Suppan, *Trans. Farad. Soc.*, **61**, 1664 (1965).
[17] G. S. Hammond and R. P. Foss, *J. Phys. Chem.*, **68**, 3739 (1964); R. P. Foss, D. O. Cowan, and G. S. Hammond, *ibid.*, **68**, 3747 (1964).

non-conjugated chromophores has also been examined.[18a] For example, it was demonstrated that triplet excitation of the benzophenone moiety of the ketone (8) was cleanly transferred to the naphthalene nucleus. Singlet excitation is transferred in the reverse direction with 75% efficiency. At the concentrations employed, intermolecular transfer was shown to be negligible.

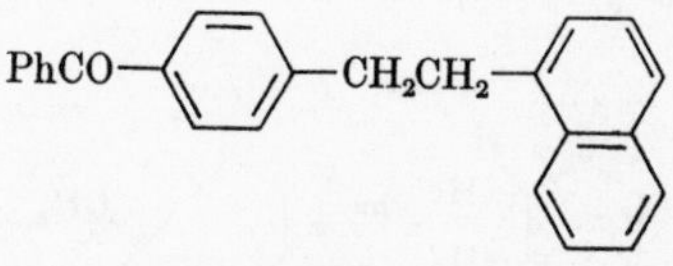

(8)

These conclusions were drawn largely from physical data. Morrison has reported a chemical consequence of intramolecular energy transfer.[18b] Excitation of the $n \rightarrow \pi^*$ state of the non-conjugated *trans*-5-hepten-2-one leads to isomerization to the *cis*-isomer.

New studies of photolysis of cyclic ketones have been reported.[19-22]

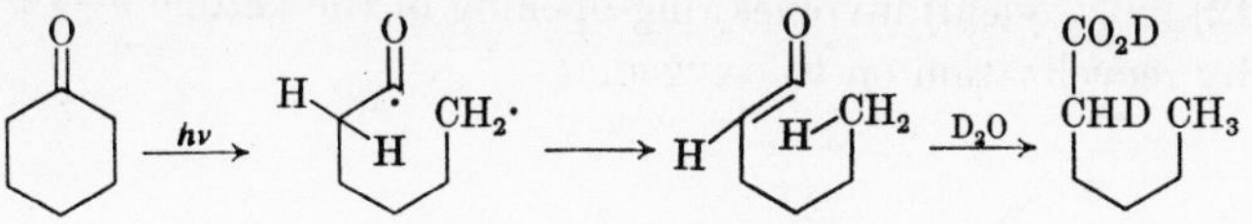

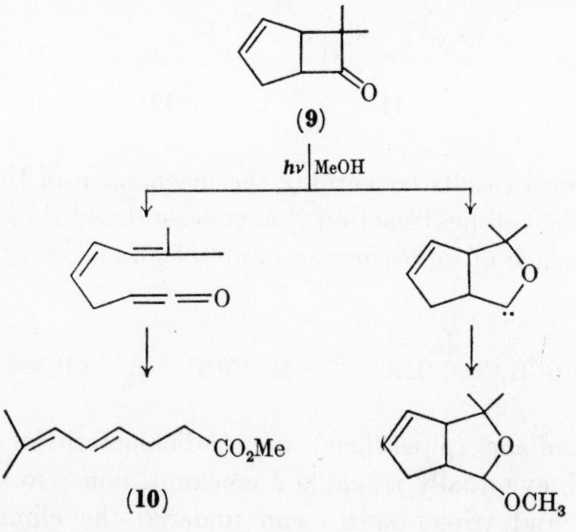

(9)

(10)

[18a] A. A. Lamola, P. A. Leermakers, G. W. Byers, and G. S. Hammond, *J. Am. Chem. Soc.*, **87**, 2322 (1965).

[18b] H. Morrison, *J. Am. Chem. Soc.*, **87**, 932 (1965).

[19] R. F. Klemm, D. N. Morrison, P. Gilderson, and A. T. Blades, *Can. J. Chem.*, **43**, 1934 (1965).

[20] G. O. Schenck and F. Schaller, *Chem. Ber.*, **98**, 2056 (1965).

[21] H. U. Hostettler, *Tetrahedron Letters*, **1965**, 687.

[22] R. Srinivasan and S. E. Cremer, *J. Am. Chem. Soc.*, **87**, 1647 (1965).

10+

Irradiation of cyclohexanone in deuterium oxide gives 2-deuteriohexanoic acid, substantiating the ketene mechanism for its formation as in the annexed formulae.[20] A ketene intermediate is also postulated for the conversion of the ketone (9) into the ester (10);[21] the cyclic acetal may arise by way of a carbene intermediate. A minor product of irradiation of liquid cyclohexanone is 2-methylcyclopentanone.[22] This ring contraction appears

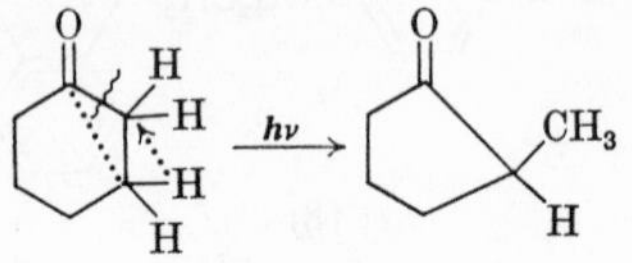

to be concerted (and may involve the excited singlet). (The dotted lines in the formulae indicate only the atom and bond migrations.)

Further results on the photolysis of cyclobutanediones[23] and of other cyclic 1,3-diketones[24,25] have appeared. Cookson and his colleagues[25] have suggested that the efficient conversion of the diketone (11) into the enol lactone (12) (50% yield) involves ring-opening of the ketone $n \to \pi^*$ triplet, followed by recyclization on to oxygen.

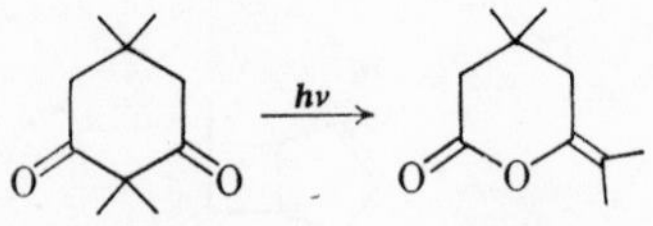

                (11)   (12)

Earlier equivocal results concerning the mechanism of the photochemical cleavage of acyclic ketones (reaction 2) have been resolved by observation[26,27] that in the presence of increasing amounts of an efficient triplet-quenching

$$\text{MeCOCH}_2\text{CH}_2\overset{\text{H}}{\underset{|}{\text{C}}}\text{CRR}' \xrightarrow{h\nu} \text{MeCOCH}_2\text{—H} + \text{CH}_2\text{=CRR}' \qquad (2)$$

agent [1,3-pentadiene (piperylene) or *cis*-dichloroethylene] the yield of acetone falls off, eventually reaching a constant, non-zero value. Thus both excited singlet and triplet states can undergo the elimination reaction; furthermore, the proportion of reaction proceeding through the triplet

[23] I. Haller and R. Srinivasan, *J. Am. Chem. Soc.*, **87**, 1144 (1965); N. J. Turro, P. A. Leermakers, H. R. Wilson, D. C. Neckers, G. W. Byers, and G. F. Vesley, *ibid.*, **87**, 2613 (1965).
[24] H. U. Hostettler, *Tetrahedron Letters*, **1965**, 1941.
[25] R. C. Cookson, A. G. Edwards, J. Hudec, and M. Kingsland, *Chem. Comm.*, **1965**, 98.
[26] P. J. Wagner and G. S. Hammond, *J. Am. Chem. Soc.*, **87**, 4009 (1965).
[27] T. J. Dougherty, *J. Am. Chem. Soc.*, **87**, 4011 (1965).

depends on the nature of the ketone. The photolysis of isopropyl methyl ketone, for which reaction (2) is not available, has also been examined.[28]

Products of photochemical reactions of quinones are formed by intra-[29-31] and inter-molecular[32] hydrogen abstraction by oxygen in the excited quinone, and by 1,4-cycloaddition of the carbonyl bond to dienes.[33] *p*-Benzoquinone with aldehydes gives *C*-acylquinols.[34]

Since van Tamelen and Pappas's report[35] that photoisomerization of 1,2,4-tri-*t*-butylbenzene gives a derivative (13) of bicyclo[2.2.0]hexadiene (Dewar benzene), several other examples of photoisomerizations of benzene derivatives have been disclosed. Some of these merely involve positional isomerizations, as in the conversion of *ortho*- into *meta*- and *para*-di-*t*-butylbenzene,[36,37] and of 1,2,4,5- into the interesting 1,2,3,5-tetra-*t*-butylbenzene.[38] However, these positional isomerizations probably all involve skeletal rearrangements of the benzene ring by way of Dewar, Ladenburg, or benzvalene (see below) intermediates or excited states. Skeletal rearrangement of ring-labelled mesitylene during photoisomerization to 1,2,4-trimethylbenzene has been clearly demonstrated (reaction 3),[39] and a similar mechanism has been proposed for the photorearrangement of 2-phenylthiophene

(3)

(4)

(reaction 4).[40] These later results are all somewhat overshadowed by the isolation by Wilzbach and Kaplan of compounds with Dewar (13), Ladenburg

[28] A. Zahra and W. A. Noyes, *J. Phys. Chem.*, **69**, 943 (1965).
[29] A. T. Shulgin and H. O. Kerlinger, *Tetrahedron Letters*, **1965**, 3355.
[30] C. M. Orlando and A. K. Bose, *J. Am. Chem. Soc.*, **87**, 3782 (1965).
[31] D. W. Cameron and R. G. F. Giles, *Chem. Comm.*, **1965**, 573.
[32] M. B. Rubin and P. Zwitkowits, *Tetrahedron Letters*, **1965**, 2453.
[33] J. A. Barltrop and B. Hesp, *J. Chem. Soc.*, **1965**, 5182.
[34] J. M. Bruce and E. Cutts, *Chem. Comm.*, **1965**, 2.
[35] E. E. van Tamelen and S. P. Pappas, *J. Am. Chem. Soc.*, **84**, 3789 (1962).
[36] E. E. van Tamelen, *Angew Chem. Intern. Ed. Engl.*, **4**, 738 (1965).
[37] A. W. Burgstahler, P.-L. Chien, and M. O. Abdel-Rahman, *J. Am. Chem. Soc.*, **86**, 5281 (1964).
[38] E. M. Arnett and J. M. Bollinger, *Tetrahedron Letters*, **1964**, 3803.
[39] L. Kaplan, K. E. Wilzbach, W. G. Brown, and S. S. Yang, *J. Am. Chem. Soc.*, **87**, 675 (1965).
[40] H. Wynberg and H. van Driel, *J. Am. Chem. Soc.*, **87**, 3998 (1965).

(14), and benzvalene (15) structures after irradiation of both 1,2,4- and 1,3,5-tri-*t*-butylbenzene.[41] It has been pointed out that the existence of Dewar benzenes may be attributed to violation of orbital symmetry requirements in the isomerization to the more stable aromatic systems.[36]

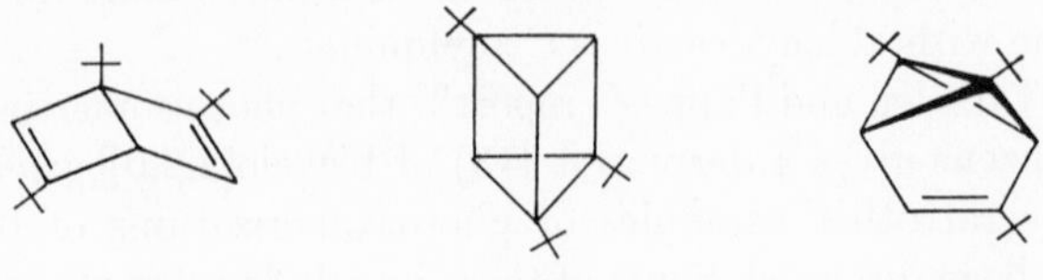

(13)   (14)   (15)

Photoaddition of maleic anhydride to benzene derivatives involves excitation of a charge-transfer complex.[42] The initial photoadduct (16) of benzene with dichloromaleic anhydride (DCMA) can lose hydrogen chloride, leading ultimately to the novel 2:2 adduct (17).[43] Irradiation (at 2537 Å) of a solution of cyclobutene in benzene gives the Dewar benzene adduct (18).[44]

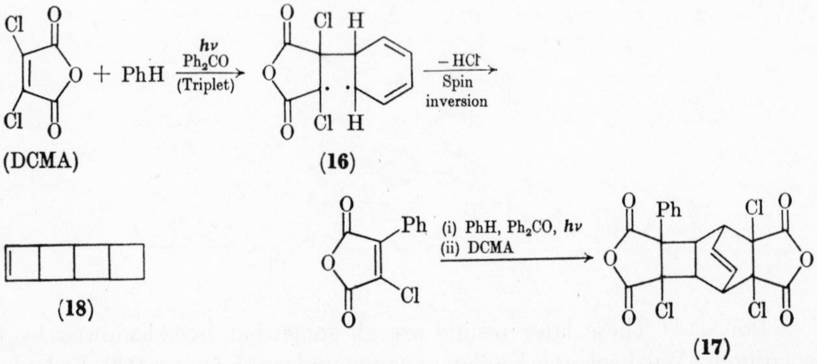

Butadiene shows an absorption maximum at 2100 Å, and it had been assumed that the limit of the associated $\pi \rightarrow \pi^*$ transition was at ca. 2300 Å. It has now been shown that excitation at 2537 Å is adequate to raise the molecule to its first excited singlet state, in which form it reacts to give appreciably different products from those obtained in triplet photosensitized reactions.[45] The range of monomeric products from irradiation of dienes is wide. Examples are shown in the annexed formulae. Some of these reactions

[41] K. E. Wilzbach and L. Kaplan, *J. Am. Chem. Soc.*, **87**, 4005 (1965).

[42] D. Bryce-Smith and A. Gilbert, *J. Chem. Soc.*, **1965**, 918.

[43] G. B. Vermont, P. X. Riccobono, and J. Blake, *J. Am. Chem. Soc.*, **87**, 4024 (1965).

[44] R. Srinivasan and K. A. Hill, *J. Am. Chem. Soc.*, **87**, 4653 (1965).

[45] R. Srinivasan and F. I. Sonntag, *J. Am. Chem. Soc.*, **87**, 3778 (1965).

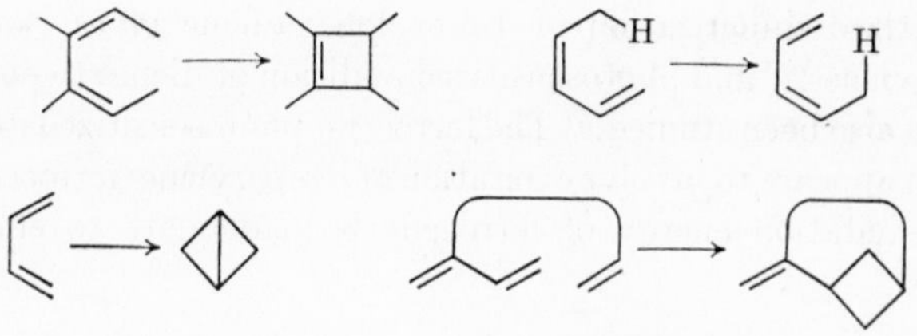

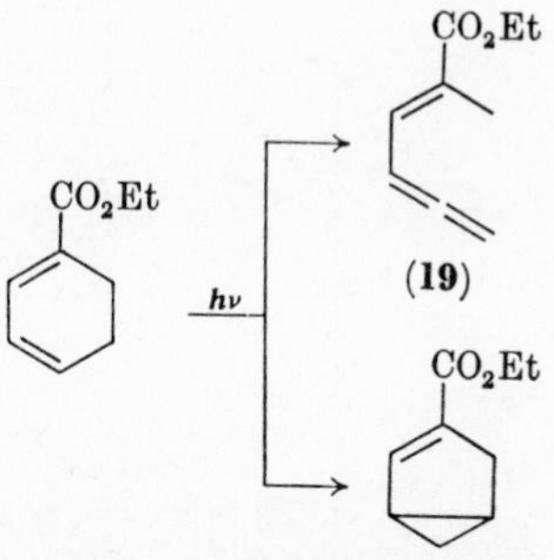

have recently been examined by Crowley.[46] Ethyl 1,3-cyclohexadiene-carboxylate, on irradiation in ether, gives the allene derivative (19) and the bicyclic isomer (20).[47]

Hammond's group has studied the photosensitized dimerization of isoprene and of butadiene, using a wide range of sensitizers.[48a] With sensitizers with triplet excitation energy > 60 kcal mole^{-1}, the dimers are almost exclusively divinylcyclobutanes, but when the sensitizers have excitation energy in the range 50—60 kcal. mole^{-1}, the dimers contain as much as 60% of vinyl-cyclohexene though the quantum efficiency is unaffected. These results are consistent with triplet excitation of *cisoid*-diene, requiring some 10 kcal mole^{-1} less energy than excitation of the *transoid*-diene. The excited diene molecules retain the geometry of the 2,3-bond,[48b] and *cisoid* excited states lead to vinylcyclohexenes, *transoid* to divinylcyclobutanes. It is hard to see how these results with so wide a range of sensitizers could be correlated in terms of triplet energy transfer via bond formation, as suggested by Schenck and Steinmetz.[49] On the other hand, they are satisfied by a direct energy-transfer mechanism. Conceivably Schenck's mechanism could operate in the reaction (which proceeds with very much reduced quantum yield) when sensitizers of less than 50 kcal mole^{-1} triplet energy are employed.

[46] K. J. Crowley, *Tetrahedron*, **21**, 1001 (1965).

[47] H. Prinzbach and E. Druckrey, *Tetrahedron Letters*, **87**, 2959 (1965).

[48a] R. S. H. Liu, N. J. Turro, and G. S. Hammond, *J. Am. Chem. Soc.*, **87**, 3406 (1965).

[48b] In the $\pi \rightarrow \pi^*$ state there is increased double-bond character between $C_{(2)}$ and $C_{(3)}$ relative to the ground state.

[49] G. O. Schenck and R. Steinmetz, *Bull. Soc. Chim. Belges*, **71**, 781 (1962).

Photosensitized dimerization of 1,3-cyclohexadiene gives isomeric octa-hydrobiphenylenes,[50] and photosensitized addition of dienes to polyhalogeno-ethylenes has also been studied.[51] The ferrocene-photosensitized isomerization of piperylene appears to involve excitation of a piperylene–ferrocene complex; the triplet excitation energy of ferrocene is inadequate to effect the iso-merization.[52]

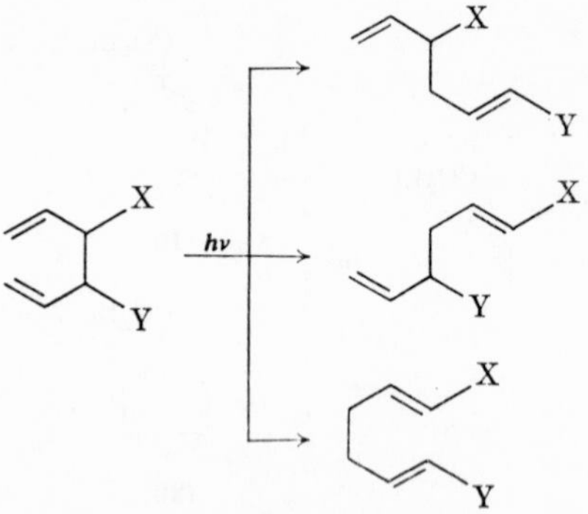

Scheme 2

Scheme 3

Cookson and his co-workers[53] have discussed examples of the generalized photochemical rearrangements shown in Schemes 2 and 3. Two recent observations which are encompassed by these generalizations are shown in reactions (5)[54] and (6).[55]

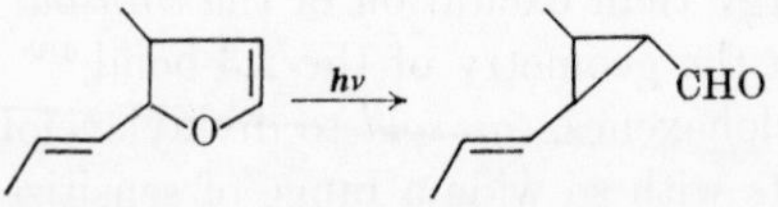

(5)

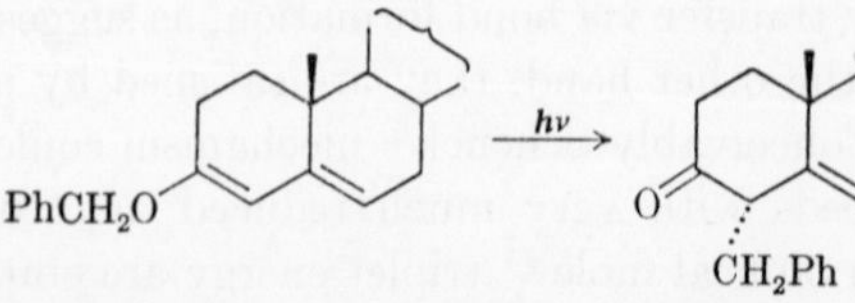

(6)

[50] D. Valentine, N. J. Turro, and G. S. Hammond, *J. Am. Chem. Soc.*, **86**, 5202 (1964).
[51] N. J. Turro and P. D. Bartlett, *J. Org. Chem.*, **30**, 1849 (1965).
[52] J. J. Dannenberg and J. H. Richards, *J. Am. Chem. Soc.*, **87**, 1626 (1965).
[53] R. C. Cookson, V. N. Gogte, J. Hudec, and N. A. Mirza, *Tetrahedron Letters*, **1965**, 3955.
[54] J. Wiemann, N. Thoai, and F. Weisbuch, *Tetrahedron Letters*, **1965**, 2983.
[55] J. T. Pinhey and K. Schaffner, *Chem. Comm.*, **1965**, 579.

The *cis–trans* photoisomerization of azobenzene has been treated theoretic-ally[56] and has also been contrasted with the photosensitized isomerization.[57] Isomerization of *cis*- and *trans*-2-butene in the gas-phase sensitized by triplet benzene has been examined,[58] as has the photoisomerization of 4-methoxy-4'-nitrostilbene.[59] Irradiation of the *trans*-isomer of the latter compound appears to generate an intermediate capable of inducing the *cis* → *trans*

(21)

reaction. The cyano nitrone **(21)** cyclizes to an oxazirane when irradiated, but in a photosensitized reaction only geometrical isomerization was observed.[60] Photochemical isomerization of *N*-(*o*-[and *p*-]hydroxybenzyl-idine)aniline and related compounds gives unstable *cis*-isomers which rapidly revert to the *trans*-compound with a half-life of *ca.* 1 sec at room temperature.[61] This short life of the *cis*-isomer is probably responsible for the reported failure to cyclize anils to phananthridines. Where a *cis*-structure is mandatory, as in **(22)**, the phenanthridine is obtained in good yield (though the cyclization proceeds very much more slowly than the corresponding

(22)

synthesis of 9-phenylphenanthrene from triphenylethylene).[62] Further phenanthrene syntheses,[63] conversion of *o*-terphenyl to triphenylene,[64] and cyclizations of azobenzenes to benzocinnolines[65] have also been detailed.

The photochemistry of three-membered ring compounds has aroused con-siderable interest recently. Photolysis of benzylcyclopropane,[66] and extrusion

[56] D. R. Kearns, *J. Phys. Chem.*, **69**, 1062 (1965).
[57] L. B. Jones and G. S. Hammond, *J. Am. Chem. Soc.*, **87**, 4219 (1965).
[58] M. Tanaka, T. Terao, and S. Sato, *Bull. Chem. Soc., Japan*, **38**, 1645 (1965).
[59] D. Schulte-Frohlinde and H. Güsten, *Z. Physik. Chem. (Frankfurt)*, **45**, 209 (1965).
[60] K. Koyano and I. Tanaka, *J. Phys. Chem.*, **69**, 2545 (1965).
[61] D. G. Anderson and G. Wettermark, *J. Am. Chem. Soc.*, **87**, 1433 (1965).
[62] F. B. Mallory and C. S. Wood, *Tetrahedron Letters*, **1965**, 2643.
[63] M. V. Sargent and C. J. Timmons, *J. Chem. Soc.*, **1964**, 5544.
[64] N. Kharasch, T. G. Alston, H. B. Lewis, and W. Wolf, *Chem. Comm.*, **1965**, 242.
[65] G. M. Badger, N. C. Jamieson, and G. E. Lewis, *Australian J. Chem.*, **18**, 190 (1965); G. M. Badger, C. P. Joshua, and G. E. Lewis, *ibid.*, **18**, 1639 (1965).
[66] P. A. Leermakers and G. F. Vesley, *J. Org. Chem.*, **30**, 539 (1965).

of ethylene from the cyclopropane ring of a 3-cyclopropylacrylic ester[67] have been reported. The interconversion of 1,2-diphenylcyclopropanes and 1,3-diphenylpropenes[68] by irradiation in benzene or cyclohexane requires hydrogen and/or phenyl migration and presumably involves an intermediate

$$Ph \overset{\cdot}{\diagup}\diagdown\overset{\cdot}{\diagdown} Ph$$

$$(23)$$

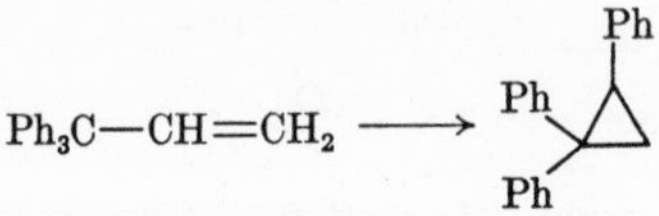

$$Ph_3C-CH=CH_2 \longrightarrow \quad \underset{Ph}{\overset{Ph}{\diagup}}\!\!\!\triangle \qquad\qquad (7)$$

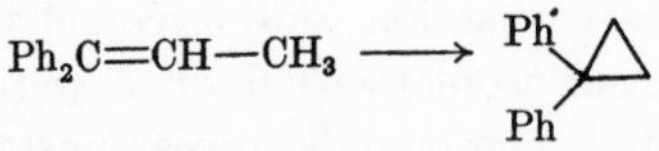

$$Ph_2C=CH-CH_3 \longrightarrow \quad \underset{Ph}{\overset{Ph^{\cdot}}{\diagup}}\!\!\!\triangle \qquad\qquad (8)$$

biradical (23). Further examples[69] of this rearrangement require specifically phenyl (7) and hydrogen (8) migration.

Extrusion of sulphur from dibenzoylstilbene episulphide,[70] and deamination of *trans*-2-benzoyl-1-cyclohexyl-3-phenylaziridine,[71] have been reported by Padwa and his co-workers. (See also ref. 28, p. 229.)

Particularly interesting is Hammond and Cole's observation[72a] of asymmetric induction in the photosensitized equilibration of *cis*- and *trans*-1,2-diphenylcyclopropanes; use of the optically active sensitizer (24) led to an equilibrium mixture containing roughly equal proportions of the two isomers;

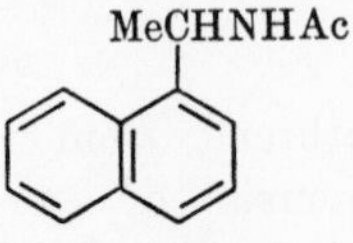

$$\text{MeCHNHAc}$$

$$(24)$$

however the *trans*-isomer showed a specific rotation of ca. 25° (the degree of optical purity could not be determined). This result clearly demonstrates the intimate association between substrate and sensitizer which is required for energy transfer. The reaction selected for study shows an energy-transfer

[67] M. J. Jorgenson and C. H. Heathcock, *J. Am. Chem. Soc.*, **87**, 5264 (1965).

[68] G. W. Griffin, J. Covell, R. C. Petterson, R. M. Dodson, and G. Klose, *J. Am. Chem. Soc.*, **87**, 1410 (1965).

[69] G. W. Griffin, A. F. Marcantonio, H. Kristinsson, R. C. Petterson, and C. S. Irving, *Tetrahedron Letters*, **1965**, 2951.

[70] A. Padwa and D. Crumrine, *Chem. Comm.*, **1965**, 506.

[71] A. Padwa and L. Hamilton, *J. Am. Chem. Soc.*, **87**, 1821 (1965).

[72a] G. S. Hammond and R. S. Cole, *J. Am. Chem. Soc.*, **87**, 3256 (1965).

rate much below the diffusion-controlled limit,[72b] thus permitting the stereochemical selectivity to be observed.

A further interesting isomerization that may be photochemically induced (or photosensitized) is the racemization of sulphones.[73] For the reaction to occur, at least one substituent on sulphur must be an aryl group.

The photosensitized isomerization of azobenzene was mentioned above. Azoalkanes or biacetyl quench the triplet state of acetone, but here the transfer of energy causes decomposition (to alkyl radicals and nitrogen).[74] The photosensitized homolysis of the O–O bond of peroxides has also been observed;[75] only di-*t*-butyl peroxide failed to react, and some peroxides (e.g. *trans*-4-*t*-butylcyclohexanecarbonyl peroxide), whose normal decomposition follows an ionic pathway, suffered homolysis in the photosensitized reaction.

The photolysis of azobisisobutyronitrile is known to take a free-radical course similar to that of pyrolysis. However, photolysis in benzene in the presence of silica gel gave tetramethylsuccinonitrile as the sole product [in the

$$\overset{\displaystyle CN}{\underset{\displaystyle (25)}{Me_2C-N=C=CMe_2}}$$

absence of silica gel a 60% yield of the ketene imine (25) is obtained].[76] The quantum yield was unaffected by the silica gel, whose influence was ascribed to reaction in an adsorbed "super-cage".

Other novel techniques include the use of 2,3-diphenylindenone to detect short-lived triplet states in solution,[77] and the examination of photochemical reactions in KBr pellets.[78] The change in a reaction on passing from solution to the relatively rigid medium of the pellet (and on varying the pressure within the pellet) can give information on the nature of the excited state participating in the photochemical reaction.

The stereochemical requirements of hydroperoxide formation by photosensitized oxygenation of olefins has been studied by Nickon's school.[79]

[72b] One of the electronic transitions involved is a non-vertical one (violating the Franck–Condon principle).

[73] K. Mislow, M. Axelrod, D. R. Rayner, H. Gotthardt, L. M. Coyne, and G. S. Hammond, *J. Am. Chem. Soc.*, **87**, 4958 (1965).

[74] R. E. Rebbert and P. Ausloos, *J. Am. Chem. Soc.*, **87**, 1847 (1965).

[75] C. Walling and M. J. Gibian, *J. Am. Chem. Soc.*, **87**, 3413 (1965).

[76] P. A. Leermakers, L. D. Weis, and H. T. Thomas, *J. Am. Chem. Soc.*, **87**, 4403 (1965).

[77] E. F. Ullman, *J. Am. Chem. Soc.*, **86**, 5357 (1964).

[78] J. K. S. Wan, R. N. McCormick, E. J. Baum, and J. N. Pitts, *J. Am. Chem. Soc.*, **87**, 4409 (1965).

[79] A. Nickon, N. Schwartz, J. B. DiGiorgio, and D. A. Widdowson, *J. Org. Chem.*, **30**, 1711 (1965); A. Nickon and W. L. Mendelson, *ibid.*, **30**, 2087 (1965); *J. Am. Chem. Soc.*, **87**, 3921 (1965).

10*

This process is different from the radical oxygenation, in that the hydroperoxide is always formed with allylic migration of the double bond. In cyclic systems a quasiaxial allylic hydrogen is abstracted. This process may be slowed by 1,3-diaxial interactions. Examination of the rate of the reaction as a function of the substituents on the double bond has shown that the attacking species is strongly electrophilic[80] and is presumably singlet oxygen (e.g. tetramethylethylene is oxidized 5500 times faster than cyclohexene). A reasonable reaction mechanism is therefore that outlined in Scheme 4.

$$S \xrightarrow{h\nu} S*$$

$$S* \longrightarrow S* \text{ (triplet)}$$

$$S* \text{ (triplet)} + O_2 \longrightarrow S + O_2 \text{ (singlet)}$$

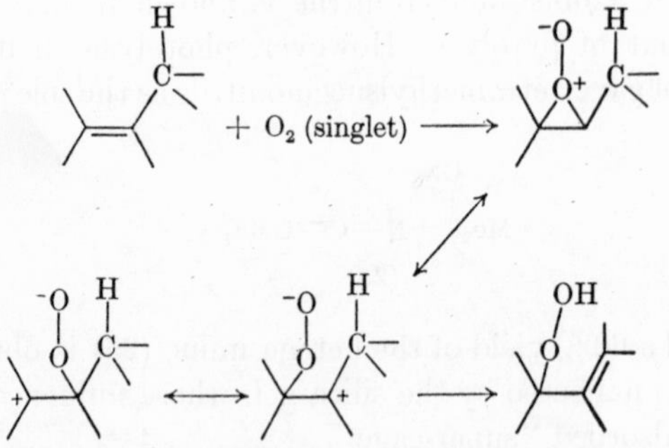

Scheme 4   (S = sensitiser)

The photo-Fries rearrangement of aryl esters has received further attention.[81] Photolysis of aryl formates gives relatively high yields of non-acylated phenol,[82] the existence of radical intermediates being shown by oxidation of the solvent, ethanol, to acetaldehyde. A related photo-anilide rearrangement (to give *o*- and *p*-acylanilines) has also been reported,[83] but its mechanism is not properly understood.

Other miscellaneous photochemical reactions reported include demonstration of intramolecular hydrogen transfer to a nitro-group after $n \rightarrow \pi^*$ excitation of *o*-nitrotoluene,[84] and photochemical interactions in 2'-substituted 2-nitrobiphenyls.[85] Nitrobenzyl anions have been observed as intermediates in the photolytic conversion of nitrophenyl acetates into

[80] K. R. Kopecky and H. J. Reich, *Can. J. Chem.*, **43**, 2265 (1965).
[81] R. A. Finnegan and J. J. Mattice, *Tetrahedron*, **21**, 1015 (1965).
[82] W. M. Horspool and P. L. Pauson, *J. Chem. Soc.*, **1965**, 5162.
[83] D. Elad, D. V. Rao, and V. I. Stenberg, *J. Org. Chem.*, **30**, 3252 (1965).
[84] H. Morrison and B. H. Migdalof, *J. Org. Chem.*, **30**, 3996 (1965).
[85] E. C. Taylor, B. Furth, and M. Pfau, *J. Am. Chem. Soc.*, **87**, 1400 (1965).

nitrotoluenes.[86] The dimethyldihydropyrene (26) is isomerized to the metacyclophane (27) by visible light.[87] A highly stereospecific dimerization

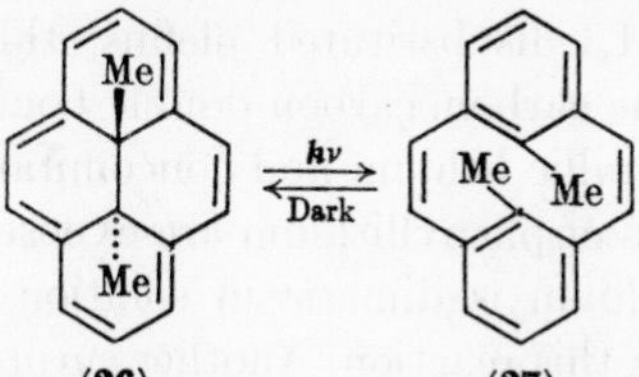

(26) (27)

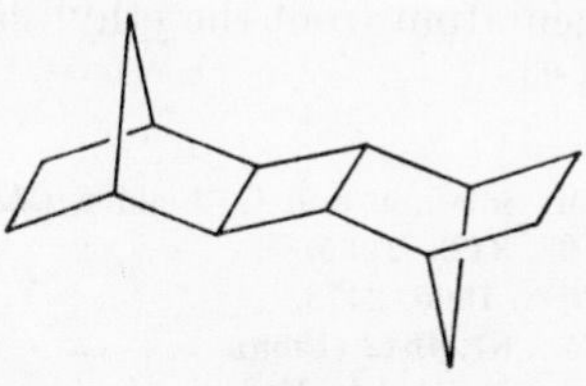

(28)

of norbornene to give the *exo,trans,exo*-product (28) was observed when irradiation was effected in the presence of cuprous halides;[88] photosensitized dimerization gives mainly the *exo,trans,endo*-isomer. Photolysis of nitrosyl chloride in the presence of cycloalkanes produces the cycloalkanone oxime,

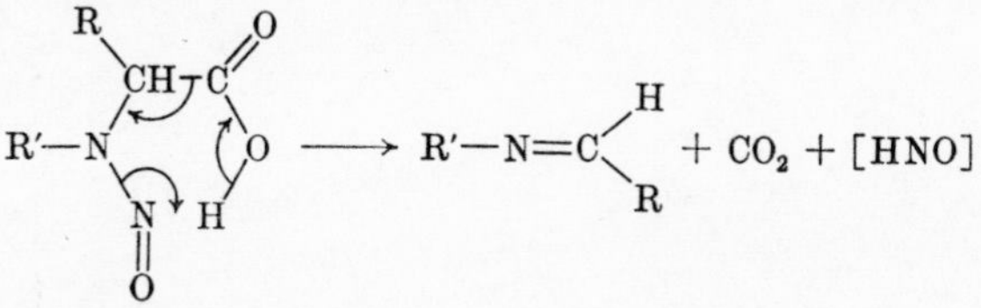

$$R'-N=C\begin{smallmatrix}H\\R\end{smallmatrix} + CO_2 + [HNO] \qquad (9)$$

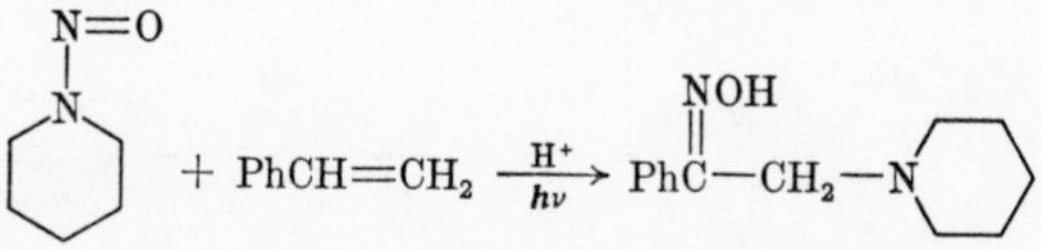

(10)

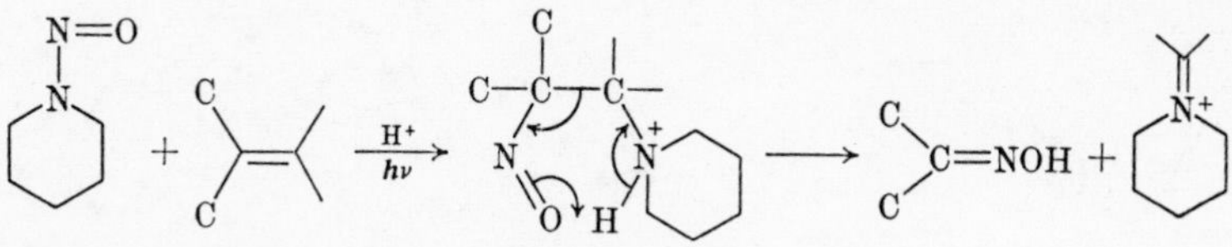

(11)

[86] J. D. Margerum, *J. Am. Chem. Soc.*, **87**, 3772 (1965).

[87] H.-R. Blattman, D. Meuche, E. Heilbronner, R. J. Molyneux, and V. Boekelheide, *J. Am. Chem. Soc.*, **87**, 130 (1965).

[88] D. J. Trecker, J. P. Henry, and J. E. McKeon, *J. Am. Chem. Soc.*, **87**, 3261 (1965).

usually in excellent yield.[89] The photolysis of N-nitroso amines requires acid-catalysis, but α-amino acid derivatives experience intramolecular catalysis (reaction 9).[90] Photolysis of nitrosoamines in olefins is shown in (10) and (11).[91] For 1,1-disubstituted olefins, this provides an excellent method for cleaving the carbon–carbon double bond.

The extrusion of metallic lithium and concomitant formation of biphenyl when ethereal solutions of phenyllithium are exposed to ultraviolet light are intriguing.[92] Phenyllithium is dimeric in solution, which fact is probably of major significance in this reaction. Another surprising result, that appears to be well substantiated by isotopic-labelling experiments, is the reported photolysis of an α-hydrogen atom from the ethyl group of monoethyl phosphate in aqueous solution.[93]

[89] E. Müller, H. G. Padeken, M. Salamon and G. Fielder, *Chem. Ber.*, **98**, 1893 (1965); E. Müller and G. Fiedler, *ibid.*, **98**, 3493 (1965).
[90] Y. L. Chow, *Tetrahedron Letters*, **1965**, 2473.
[91] Y. L. Chow, *J. Am. Chem. Soc.*, **87**, 4642 (1965).
[92] E. E. van Tamelen, J. I. Brauman, and L. E. Ellis, *J. Am. Chem. Soc.*, **87**, 4964 (1965).
[93] M. Halmann and I. Platzner, *J. Chem. Soc.*, **1965**, 5380.

CHAPTER 14

Oxidations[1] and Reductions

Oxidation of Olefins

According to the Criegee mechanism for ozonolysis (Scheme 1) the primary ozonide (1) decomposes to a zwitterion (2) and a carbonyl fragment (3) which then recombine to give the normal ozonide (4). An unsymmetrical olefin could give two zwitterions and two carbonyl fragments and hence in

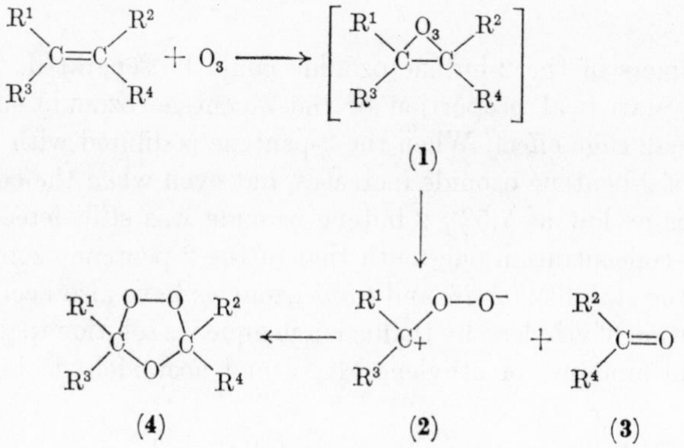

(1)

(4)   (2)   (3)

Scheme 1

principle three ozonides, each capable of existing as a *cis–trans*-pair. This behaviour has, however, only rarely been observed,[2a] and hence it was considered that the zwitterion and carbonyl fragments were generally formed in a cage. Loan, Murray, and Story have now made a detailed investigation of the ozonolysis of pure 2-pentene at $-70°$ and shown that this yields 2-butene ozonide, 2-pentene ozonide, and 3-hexene ozonide in the ratio $1:2.42:0.67$ (see Scheme 2; statistical ratio $= 1:2:1$).[2b] The ozonides were separated by gas chromatography and were identified with NMR spectroscopy as mixtures of the *cis* and *trans* ozonides in almost equal amounts;

---

[1] The mechanism of the oxidation of organic compounds has been reviewed recently: (a) W. A. Waters, "Mechanisms of Oxidation of Organic Compounds", Methuen and Co. Ltd., London, 1964; (b) R. Stewart, "Oxidation Mechanisms: Applications to Organic Chemistry", W. A. Benjamin, Inc., New York, 1964; (c) K. B. Wiberg, *Survey of Progress in Chemistry*, 1, 211 (1963).

[2] (a) Cf. G. Riezebos, J. C. Grimmelikhuysen, and D. A. van Dorp, *Rec. Trav. Chim.*, 82, 1234 (1963); (b) L. D. Loan, R. W. Murray, and P. R. Story, *J. Am. Chem. Soc.*, 87, 737 (1965); see also R. W. Murray, P. R. Story, and L. D. Loan, *ibid.*, 87, 3025 (1965).

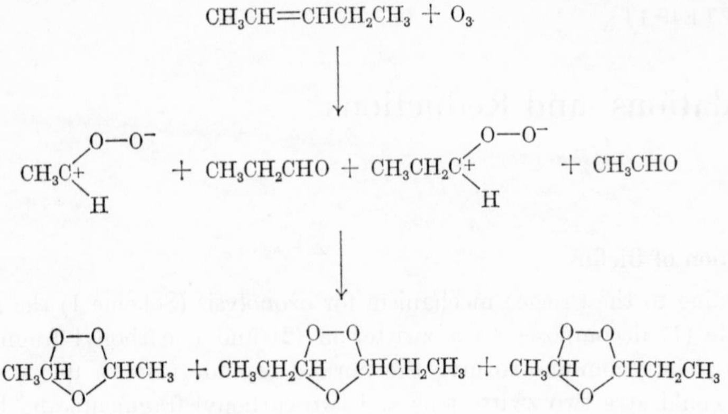

$$CH_3CH\!=\!CHCH_2CH_3 + O_3$$

Scheme 2

only the isomers of the 2-butene ozonide could be separated. The slightly higher than statistical proportion of the 2-pentene ozonide suggests that there is a small cage effect. When the 2-pentene is diluted with pentane the proportion of 2-pentene ozonide increases, but even when the concentration of pentene is as low as 7.5% 2-butene ozonide was still detectable, being formed at a concentration one-tenth that of the 2-pentene ozonide. Factors controlling the stability of *cis*- and *trans*-ozonides have also been discussed.[3]

The oxidation of ethylene by thallic ion in aqueous solution to give approximately equal amounts of ethylene glycol and acetaldehyde is of the first

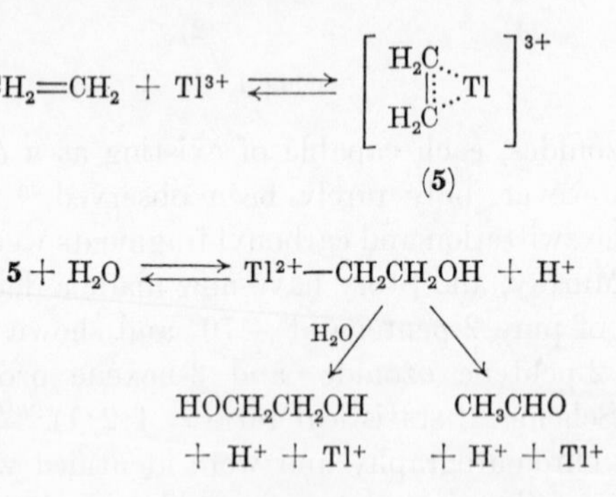

(5)

Scheme 3

order in each reactant.[4] The rate is strongly increased, and the proportion of glycol in the product decreased, with increasing electrolyte concentration. The mechanism was written as shown in Scheme 3. The rate increase with

[3] F. L. Greenwood, *J. Org. Chem.*, **30**, 3108 (1965).
[4] P. M. Henry, *J. Am. Chem. Soc.*, **87**, 990 (1965).

electrolyte concentration was shown to be an inverse function of the activity of water. This was discussed from the alternative viewpoints that hydration of the thallic ion increases its reactivity, and that the activity coefficient of the thallic ion is affected more than that of the transition state by electrolyte concentration. The introduction of methyl substituents into ethylene caused rate enhancements consistent with the mechanism of Scheme 3.[5]

The oxidation of olefins by nitric acid is specifically catalysed by vanadic ion.[6] The formation of 2-nitrosocyclohexyl nitrate dimer (6) and 2-oxocyclohexyl nitrate (7) in solutions of cyclohexene in nitric acid containing nitrogen tetroxide was demonstrated. It was proposed that these are hydrolysed to 2-hydroxycyclohexanone (8) which then undergoes cleavage with vanadic ion.

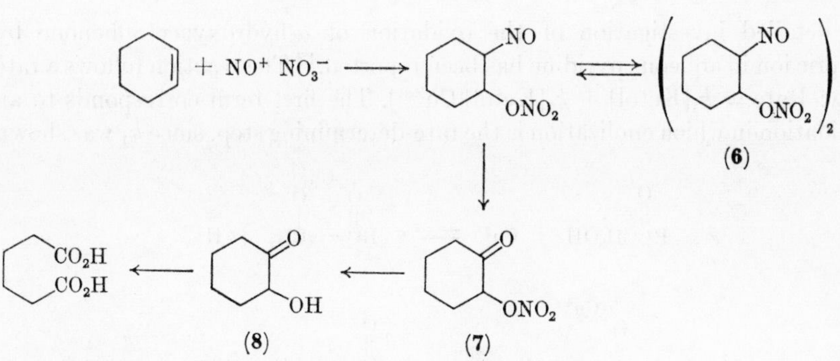

Support for the view that the oxidation of olefins by mercuric acetate proceeds via organomercurials has been provided by the observation that 1- and 2-butene yield almost exclusively the secondary acetate (11).[7] It was

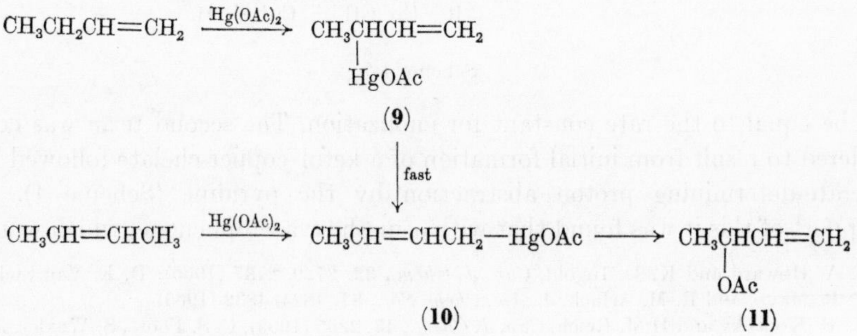

[5] P. M. Henry, *J. Am. Chem. Soc.*, **87**, 4423 (1965).
[6] J. E. Franz, J. F. Herber, and W. S. Knowles, *J. Org. Chem.*, **30**, 1488 (1965).
[7] Z. Rappoport, P. D. Sleezer, S. Winstein, and W. G. Young, *Tetrahedron Letters*, **1965**, 3719.

shown that butenylmercuric acetate (10) undergoes demercuration to give 1-methylallyl acetate and that the secondary butenylmercuric acetate (9) rapidly gives the primary mercuric acetate (10) in the presence of mercuric salts. Hence oxidation of both 1- and 2-butene would lead to the primary mercurial (10), which would then rearrange to give the observed secondary acetate.

Other investigations of the oxidation of olefins include studies of autoxidation,[8] photosensitized olefin oxidations,[9] high-temperature oxidation of [^{14}C]propene,[10] oxidation of olefins by peroxides catalysed by copper salts,[11] and oxidation of perfluoropropene by oxygen atoms generated by mercury-sensitized decomposition of nitrous oxide.[12]

Oxidation of Hydroxyl and Carbonyl Compounds

A detailed investigation of the oxidation of α-hydroxyacetophenone by cupric ion in aqueous pyridine has been reported.[13] The reaction follows a rate law: Rate = k_1[Ketol] + k_2[Ketol][Cu^{2+}]. The first term corresponds to an oxidation in which enolization is the rate-determining step, since k_1 was shown

$$
\underset{\text{RCCH}_2\text{OH}}{\overset{\text{O}}{\|}} + \text{Cu}^{2+} \rightleftharpoons \underset{\text{RC}\underset{\text{}}{\quad}\text{CH}_2}{\overset{\text{O}^{\cdots}\overset{\text{Cu}^+}{\text{O}}}{\|\quad|}} + \text{H}^+
$$

$$
\underset{R-\overset{}{C}\underset{|}{\quad}\overset{}{C}-H}{\overset{\text{Cu}^+}{\overset{\text{O}^{\cdots}\text{O}}{\|\quad|}}} \xrightarrow[\text{pyridine}]{\text{Slow}} \underset{R-\overset{\text{O}}{\overset{\|}{C}}-\overset{\bullet}{C}\text{HOH}}{} + \text{Cu}^+
$$

$$
\downarrow \text{Cu}^{2+}
$$

$$
\underset{R-\overset{\text{O}}{\overset{\|}{C}}-\overset{\text{O}}{\overset{\|}{C}}\text{H}}{} + \text{Cu}^+ + \text{H}^+
$$

Scheme 4

to be equal to the rate constant for enolization. The second term was considered to result from initial formation of a ketol–copper chelate followed by a rate-determining proton abstraction by the pyridine (Scheme 4). In support of this it was found that with α-methoxyacetophenone, which would

[8] J. A. Howard and K. U. Ingold, *Can. J. Chem.*, **32**, 2729, 2737 (1965); D. E. Van Sickle, F. R. Mayo, and R. M. Arluck, *J. Am. Chem. Soc.*, **87**, 4824, 4832 (1965).

[9] K. R. Kopecky and H. J. Reich, *Can. J. Chem.*, **43**, 2265 (1965); C. S. Foote, S. Wexler, and W. Ando, *Tetrahedron Letters*, **1965**, 4111.

[10] E. R. White, H. G. Davis, and E. S. Hammack, *J. Am. Chem. Soc.*, **87**, 1175 (1965).

[11] J. K. Kochi and H. E. Mains, *J. Org. Chem.*, **30**, 1862 (1965).

[12] D. Saunders and J. Heicklen, *J. Am. Chem. Soc.*, **87**, 4062 (1965).

[13] K. B. Wiberg and W. G. Nigh, *J. Am. Chem. Soc.*, **87**, 3849 (1965).

form a complex with cupric ion much less readily, the oxidation proceeded more slowly. The addition of bipyridyl up to a concentration equal to that of the cupric ion caused an increase in the rate of oxidation of the α-hydroxy-acetophenone paralleling the enhancing effect of bipyridyl on the electrode potential of the copper(II)–copper(I) couple. Concentrations of bipyridyl above this cause a rate decrease which was ascribed to the fact that, as the other co-ordination sites of the cupric ion become filled by this ligand, complex-formation with the ketol is restricted.

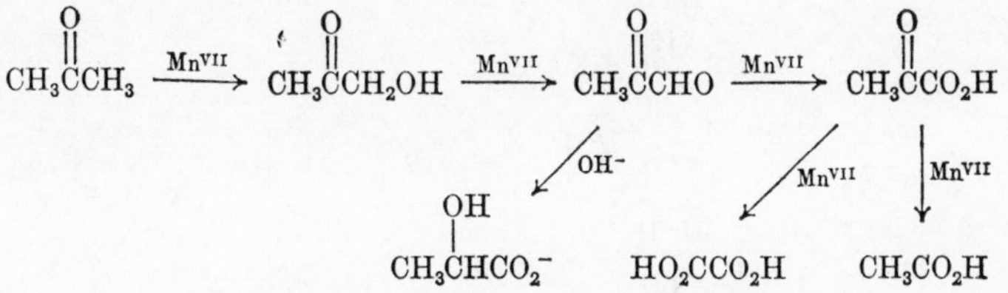

$$CH_3\overset{O}{\overset{\|}{C}}CH_3 \xrightarrow{Mn^{VII}} CH_3\overset{O}{\overset{\|}{C}}CH_2OH \xrightarrow{Mn^{VII}} CH_3\overset{O}{\overset{\|}{C}}CHO \xrightarrow{Mn^{VII}} CH_3\overset{O}{\overset{\|}{C}}CO_2H$$

$$\xrightarrow{OH^-} \qquad \xrightarrow{Mn^{VII}} \qquad \downarrow Mn^{VII}$$

$$CH_3\overset{OH}{\overset{|}{C}}HCO_2^- \qquad HO_2CCO_2H \qquad CH_3CO_2H$$

Scheme 5

As determined by an initial-slope method, the rate law for the oxidation of acetone by permanganate approximates to: $d[MnO_4^-]/dt = -k[\text{Acetone}][HO^-][MnO_4^-]$.[14] This was interpreted as indicating oxidation of the enolate ion, and the most likely reaction sequence was thought to be that of Scheme 5. All the compounds in this Scheme except pyruvaldehyde, which is oxidized rapidly under the reaction conditions, were shown by isotopic dilution analysis to be present in the reaction mixture. Their individual rates of oxidation

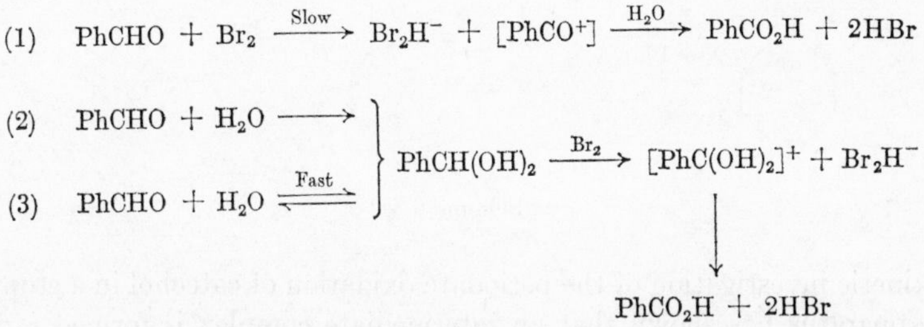

(1) $\quad PhCHO + Br_2 \xrightarrow{\text{Slow}} Br_2H^- + [PhCO^+] \xrightarrow{H_2O} PhCO_2H + 2HBr$

(2) $\quad PhCHO + H_2O \longrightarrow$

(3) $\quad PhCHO + H_2O \underset{\text{Fast}}{\rightleftharpoons}$ $\Big\}$ $PhCH(OH)_2 \xrightarrow{Br_2} [PhC(OH)_2]^+ + Br_2H^-$

$$\downarrow$$

$$PhCO_2H + 2HBr$$

were also studied and shown to be consistent with Scheme 5, except that for acetol. The concentration of this compound calculated from its rate of oxidation and the rate of oxidation of acetone was about four times the observed concentration; it was thought that some acetone is oxidized directly to pyruvaldehyde, by-passing acetol.

The effect of substituents on the rate of oxidation of benzaldehyde by

[14] K. B. Wiberg and R. D. Geer, *J. Am. Chem. Soc.*, **87**, 5202 (1965).

bromine is very small with electron-withdrawing substituents deactivating.[15]
This result excludes the mechanism of McTigue and Sime[16] (1) and suggests
one in which attack by water is on the aldehyde-carbon (2 or 3).

Cleavage of the pinacol (12) by chromic acid is 120 times faster than that of
the cyclopentyl analogue (13).[17] This result was interpreted as indicating that

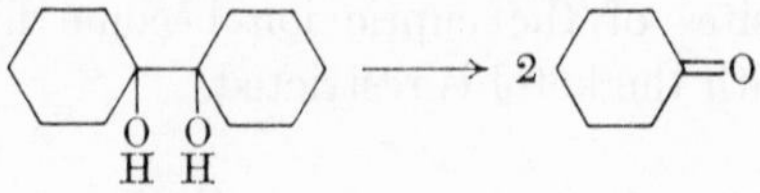

(12)

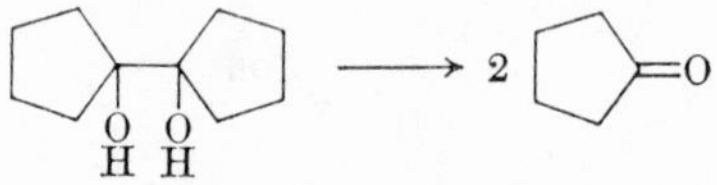

(13)

the transition state cannot be "product like" but that, instead, formation
of the cyclic chromate ester is rate-determining. The slower rate for (13)
was then considered to result because formation of the cyclic ester tends to
arrest the pseudorotation of its cyclopentane rings. The chromic acid oxidation of benzyl alcohol[18] and of 2-propanol in trifluoroacetic acid[19] have
also been investigated.

The oxidation of cyclo-octanol with lead tetra-acetate gave 1,4-epoxy-
cyclooctane, and of cycloheptanol gave 1,4-epoxycycloheptane, thought to
be formed as shown in Scheme 6.[20]

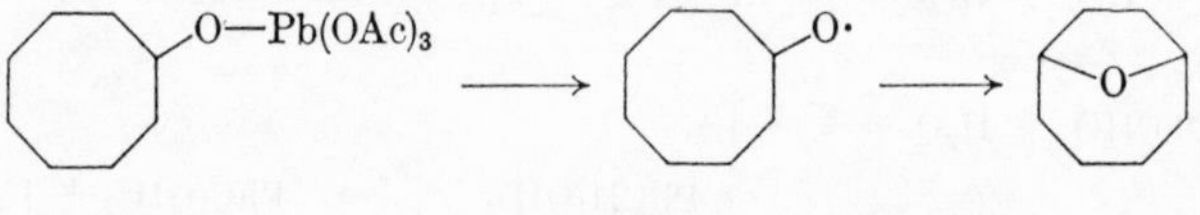

Scheme 6

A kinetic investigation of the periodate oxidation of catechol in a stopped-
flow apparatus has shown that an intermediate complex is formed rapidly
in a second-order reaction ($k_2 = 1.7 \times 10^3$ l mole^{-1} sec^{-1}) and that this
then decomposes to products in a first-order reaction ($k_1 = 0.65$ sec^{-1}) (see

[15] I. R. L. Barker and R. H. Dahm, *Chem. Comm.*, **1965**, 194.
[16] P. T. McTigue and J. M. Sime, *J. Chem. Soc.*, **1963**, 1303.
[17] H. Kwart and D. Bretzger, *Tetrahedron Letters*, **1965**, 3985.
[18] G. V. Bakore, K. K. Banerjee, and R. Shanker, *Z. Physik. Chem.* (**Frankfurt**), **45**, 129 (1965).
[19] D. G. Lee and D. T. Johnson, *Can. J. Chem.*, **43**, 1952 (1965).
[20] A. C. Cope, M. Gordon, S. Moon, and C. H. Park, *J. Am. Chem. Soc.*, **87**, 3119 (1965); R. M.
Moriarty and H. G. Walsh, *Tetrahedron Letters*, **1965**, 465.

$$\underset{\text{H}_4\text{IO}_6}{\overset{\text{IO}_4^-}{\text{or}}} + \underset{\text{OH}}{\overset{\text{OH}}{\bigcirc}} \xrightarrow{k_2} \underset{\text{complex}}{\text{Intermediate}} \xrightarrow{k_1} \underset{\text{O}}{\overset{\text{O}}{\bigcirc}} + \text{IO}_3^-$$

Scheme 7

$$\underset{(14)}{\overset{\text{O}}{\bigcirc}}\text{IO}_3^- \qquad \underset{(15)}{\overset{\text{O}}{\bigcirc}}\text{IO}_4\text{H}_2^-$$

Scheme 7).[21] Structures (14) and (15) were considered for the intermediate complex.

Other reactions which have been investigated include: Oppenauer oxidation;[22] Baeyer–Villiger oxidation of Δ^4-3-keto-steroids;[23] photochemical oxidation of heptanol;[24] lead tetra-acetate oxidation of alcohols;[25] oxidation of cyclohexanol,[26] benzaldehyde, and acetaldehyde[27] by molecular oxygen; cuprous-chloride-catalysed autoxidation of *o*-cresol-,[28] oxidative coupling of 2,6-xylenol;[29] and oxidation of reducing sugars by ferricyanide.[30]

Other Reactions

It has been shown[31] that diethyl 2,6-dimethyl-1,4-dihydropyridine-3,5-dicarboxylate (17) is oxidized to the pyridine by 2-mercaptobenzophenone (16). The reaction is promoted by air, peroxides, and ferrous ion and does not involve direct transfer of a hydride ion. *p*-Mercaptobenzophenone, *o*-hydroxybenzophenone, *o*-aminobenzophenone, *o*-thiomethylbenzophenone and *o*-mercaptoacetophenone do not react. A mechanism (Scheme 8) involving formation of a radical (18) followed by transfer of a hydrogen atom from the neighbouring *o*-thiol group was suggested; the necessity of a thiol group is thus explained, as also is the failure of *o*-mercaptoacetophenone to react since the radical from this, analogous to (18), would be much less stable.

[21] E. T. Kaiser and S. W. Weidman, *Tetrahedron Letters*, **1965**, 497.
[22] L. Ötvös, L. Gruber and J. Meisel-Ágoston, *Acta Chim. Acad. Sci. Hung.*, **43**, 149 (1965); B. J. Yager and C. K. Hancock, *J. Org. Chem.*, **30**, 1174 (1965).
[23] J. T. Pinhey and K. Schaffner, *Tetrahedron Letters*, **1965**, 601.
[24] J. Lemaire and M. Niclause, *Comp. Rend.*, **260**, 2203 (1965).
[25] M. L. Mihailović, Ž. Ceković, Z. Maksimović, D. Jeremić, L. Lorenc, and R. I. Mamuzić, *Tetrahedron*, **21**, 2799 (1965).
[26] C. Parlant, I. Sérée de Roch, and J. C. Balaceanu, *Bull. Soc. Chim. France*, **1964**, 3161.
[27] Z. Csuros, J. Morgos, and B. Losonczi, *Acta Chim. Acad. Sci. Hung.*, **43**, 271, 297 (1965).
[28] Y. Ogata and T. Morimoto, *Tetrahedron*, **21**, 2791 (1965).
[29] G. D. Cooper, H. S. Blanchard, G. F. Endres, and H. Finkbeiner, *J. Am. Chem. Soc.*, **87**, 3996 (1965).
[30] N. Nath and M. P. Singh, *J. Phys. Chem.*, **69**, 2038 (1965).
[31] K. A. Schellenberg and F. H. Westheimer, *J. Org. Chem.*, **30**, 1859 (1965).

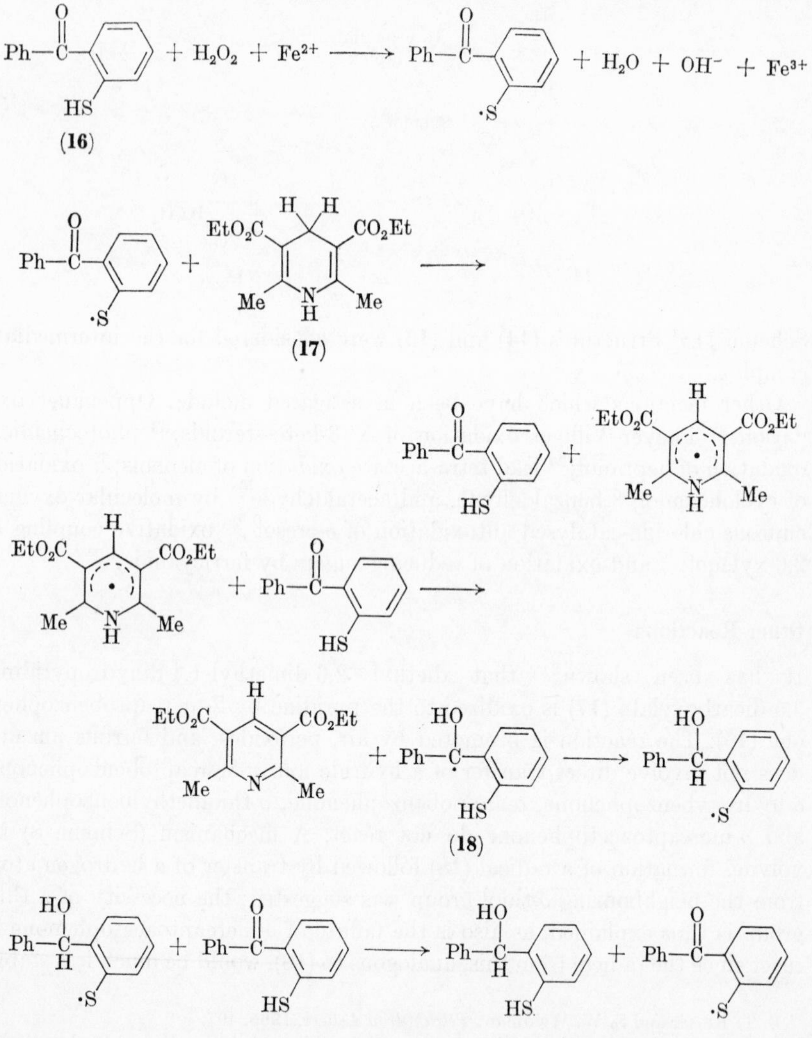

$$Ph-\overset{O}{\underset{HS}{C}}\!\!-\!\!C_6H_4 + H_2O_2 + Fe^{2+} \longrightarrow Ph-\overset{O}{\underset{\cdot S}{C}}\!\!-\!\!C_6H_4 + H_2O + OH^- + Fe^{3+}$$

(16)

$$Ph-\overset{O}{\underset{\cdot S}{C}}\!\!-\!\!C_6H_4 + \underset{Me\;\underset{H}{N}\;Me}{\overset{H\;\;H}{EtO_2C\diagup\diagdown CO_2Et}} \longrightarrow$$

(17)

$$\underset{Me\;\underset{H}{N}\;Me}{\overset{H}{EtO_2C\diagup\diagdown CO_2Et}} + Ph-\overset{O}{\underset{HS}{C}}\!\!-\!\!C_6H_4 \longrightarrow$$

$$Ph-\overset{O}{\underset{HS}{C}}\!\!-\!\!C_6H_4 + \underset{Me\;\underset{H}{N}\;Me}{\overset{H}{EtO_2C\diagup\diagdown CO_2Et}}$$

$$\underset{Me\;\underset{H}{N}\;Me}{\overset{H}{EtO_2C\diagup\diagdown CO_2Et}} + Ph-\overset{O}{\underset{HS}{C}}\!\!-\!\!C_6H_4 \longrightarrow Ph-\overset{HO}{\underset{\cdot S}{\underset{|}{C}}}\!\!-\!\!C_6H_4$$

(18)

$$Ph-\overset{HO}{\underset{\cdot S}{\underset{H}{C}}}\!\!-\!\!C_6H_4 + Ph-\overset{O}{\underset{HS}{C}}\!\!-\!\!C_6H_4 \longrightarrow Ph-\overset{HO}{\underset{HS}{\underset{H}{C}}}\!\!-\!\!C_6H_4 + Ph-\overset{O}{\underset{\cdot S}{C}}\!\!-\!\!C_6H_4$$

Scheme 8

The effect of complex formation on the oxidation of reduced N-heteroaromatic compounds by molecular oxygen has also been studied.[32]

The researches of Norman and his co-workers into hydroxyl radical oxidations have been extended to alicyclic compounds[33] and to a number of benzenoid aromatics.[34] ESR spectra of radicals formed by abstraction of

[32] W. Bartok and H. Pobiner, *J. Org. Chem.*, **30**, 274 (1965).
[33] W. T. Dixon and R. O. C. Norman, *J. Chem. Soc.*, **1964**, 4850.
[34] W. T. Dixon and R. O. C. Norman, *J. Chem. Soc.*, **1965**, 4857.

hydrogen from five- and six-membered alicyclic compounds showed that the spin coupling between the unpaired electron and the nucleus of a hydrogen atom on an adjacent carbon is dependent on the dihedral angle between the C–H bond and the singly occupied p-orbital. The radical centre was assumed to be planar-trigonal. It was estimated that the rate constant for the chair-to-chair interconversion of radicals from six-membered rings is approximately 5×10^8 sec^{-1}. Interestingly this is very close to the value estimated for chair inversion of the *cis*-9-decalyl radical (ref. 85, Chap. 9), though in that case rehybridization at the radical centre was postulated.

Hydroxyl radicals add to benzene and several other simple aromatic compounds, and ESR spectra of hydroxycyclohexadienyl radicals were observed. However, toluene and phenol give benzyl and phenoxy radicals, respectively.[34]

The pH-rate profile for the oxidation of methyl phenyl sulphoxide by perbenzoic acid shows that the rate law is: Rate $= k[\text{PhSOMe}][\text{PhCO}_3{}^-]$.[35] This was interpreted as indicating the mechanism illustrated.

$$\text{PhCO}_3{}^- + \underset{\underset{\text{Me}}{|}}{\overset{\overset{\text{Ph}}{|}}{\text{S}}}{=}\text{O} \longrightarrow \left[\text{PhCO}_2\text{O}-\underset{\underset{\text{Me}}{|}}{\overset{\overset{\text{Ph}}{|}}{\text{S}}}-\text{O} \right]^- \longrightarrow \text{PhCO}_2{}^- + \underset{\text{Me}}{\overset{\text{Ph}}{\diagdown}}\text{SO}_2$$

The oxidation of sulphoxides to sulphones has also been investigated by Eliel and his co-workers.[36]

The oxidation of dihydrolipoic acid (19) by flavine mononucleotide (20; R = ribityl) has been investigated as a model for the enzyme, lipoyl dehydrogenase.[37] The reaction is first-order in each reactant and the rate increases with increasing pH. This was interpreted as indicating a reaction between the monoanion of the dihydrolipoic acid and the flavine as shown in Scheme 9. On the grounds that the effect on the rate of varying R paralleled closely the effect on the rates of several other reactions which are known to involve two-electron transfers, it was concluded that this oxidation also involved a two-electron transfer. The oxidation of polyvinyl mercaptan and related dithiols by molecular oxygen has also been investigated.[38]

There have also been studies of the oxidation of sulphides to sulphoxides,[39]

[35] R. Curci and G. Modena, *Tetrahedron Letters*, **1965**, 863.

[36] E. L. Eliel, E. W. Della, and M. Rogić, *J. Org. Chem.*, **30**, 855 (1965).

[37] I. M. Gascoigne and G. K. Radda, *Chem. Comm.*, **1965**, 211.

[38] C. G. Overberger, K. H. Burg, and W. H. Daly, *J. Am. Chem. Soc.*, **87**, 4125 (1965).

[39] C. R. Johnson and D. McCants, *J. Am. Chem. Soc.*, **87**, 1109 (1965); K. Mislow, M. M. Green, and M. Raban, *ibid.*, **87**, 2761 (1965); T. J. Wallace, H. Pobiner, F. A. Baron, and A. Schriesheim, *J. Org. Chem.*, **30**, 3147 (1965); T. J. Wallace and F. A. Baron, *ibid.*, **30**, 3520 (1965); T. J. Wallace and A. Schriesheim, *Tetrahedron*, **21**, 2271 (1965).

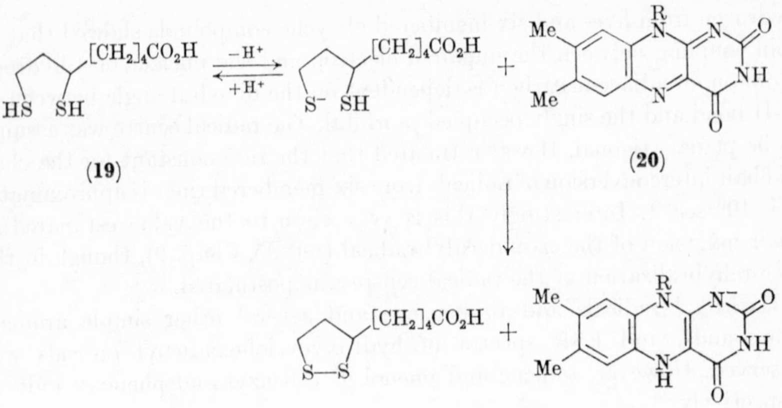

(19)   (20)

Scheme 9

and of the reduction of alkylsulphinyl-carboxylic acids with hydriodic acid,[40] the asymmetric oxidation of diphenyl disulphide to phenyl benzenethiol-sulphinate,[41] the disproportionation of aromatic sulphinic acids,[42] and the reaction of thiols with sulphoxides.[43]

The oxidative cleavage of the cyclopropane ring of bicyclo[2.1.0]pentane by lead tetra-acetate and by thallium triacetate proceeds exclusively with internal cleavage to yield derivatives of cyclopentane.[44] However, cleavage

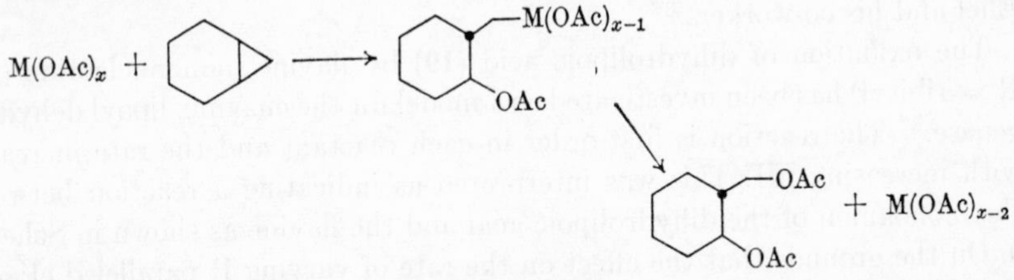

Scheme 10

of bicyclo[3.1.0]hexane and bicyclo[4.1.0]heptane yields appreciable quantities of cyclopentylmethyl and cyclohexylmethyl derivatives. The reaction was discussed in terms of a mechanism involving electrophilic attack of the metal salt to give an organometallic compound which then decomposes solvolytically (see, e.g., Scheme 10).[44]

[40] S. Allenmark, *Acta Chem. Scand.*, **19**, 1 (1965).
[41] J. L. Kice and G. B. Large, *Tetrahedron Letters*, **1965**, 3537.
[42] J. L. Kice, D. C. Hampton, and A. Fitzgerald, *J. Org. Chem.*, **30**, 882 (1965).
[43] T. J. Wallace and J. J. Mahon, *J. Org. Chem.*, **30**, 1502 (1965).
[44] R. J. Ouellette, A. South, and D. L. Shaw, *J. Am. Chem. Soc.*, **87**, 2602 (1965).

Other oxidations which have been investigated include oxidative decarboxylation of pentanoic acid by lead tetra-acetate[45] and of carboxylic acids by cobaltic salts,[46] oxidation of stable carbanions[47] and aromatic proton complexes,[48] and autoxidation of tetralin[49a] and phenylhydroxylamine.[49b]

There have been relatively few mechanistic studies of reduction reactions. Further examples have been recorded of reduction by complex metal hydrides in which the hydroxyl group of an organic compound is first converted into an oxymetal hydride group, which then transfers a hydride ion intramolecularly.[50] Franzus and Snyder[51] report that reduction of norbornadien-7-ol (21) or its acetate with lithium aluminium hydride results only in reduction of the double bond *syn* to the 7-substituent. Reduction of the alcohol (21) with lithium aluminium deuteride followed by reaction with H_2O yields *anti*-norbornen-7-ol with one deuterium atom in the *exo*-position; but when the

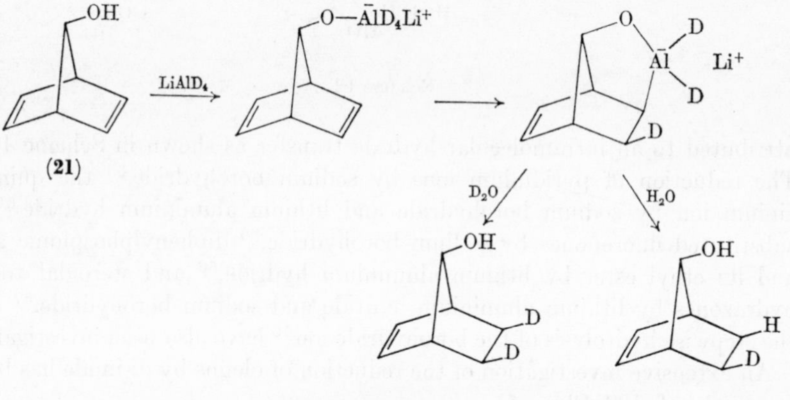

Scheme 11

mixture after reduction is allowed to react with D_2O *anti*-norbornen-7-ol with two deuterium atoms in the *exo*-positions is obtained. The reaction was therefore formulated as shown in Scheme 11. It was also found that, whereas reduction of the double bond of *syn*-norbornen-7-ol and its acetate proceeded

[45] J. K. Kochi, *J. Am. Chem. Soc.*, **87**, 3609 (1965).
[46] A. A. Clifford and W. A. Waters, *J. Chem. Soc.*, **1965**, 2796.
[47] J. G. Pacifici, J. F. Garst, and E. G. Janzen, *J. Am. Chem. Soc.*, **87**, 3014 (1965).
[48] H. M. Buck, H. P. J. M. Dekkers, and L. J. Oosterhoff, *Tetrahedron Letters*, **1965**, 505.
[49a] J. A. Howard and K. U. Ingold, *Can. J. Chem.*, **43**, 2724 (1965).
[49b] Y. Ogata and T. Morimoto, *J. Org. Chem.*, **30**, 597 (1965).
[50] For previous examples see R. B. Woodward, M. P. Cava, W. D. Ollis, A. Hunger, H. U. Daeniker, and K. Schenker, *Tetrahedron*, **19**, 263 (1963); D. M. S. Wheeler and M. M. Wheeler, *J. Org. Chem.*, **27**, 3796 (1962).
[51] B. Franzus and E. I. Snyder, *J. Am. Chem. Soc.*, **87**, 3423 (1965).

rapidly, their *anti*-isomers were quite unreactive. The observation that the ester group as well as the ketone group of the ester of the Windaus acid (partial structure **22**) could be reduced by sodium borohydride was also

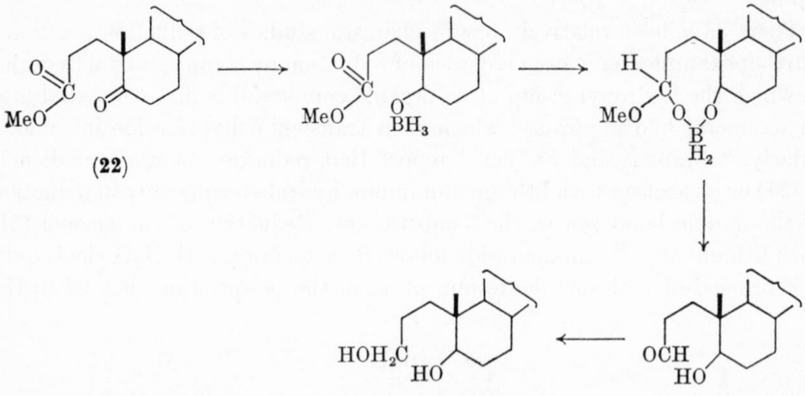

Scheme 12

attributed to an intramolecular hydride transfer as shown in Scheme 12.[52] The reduction of pyridinium ions by sodium borohydride,[53] the quinolizinium ion by sodium borohydride and lithium aluminium hydride,[54] 2-substituted fluorenones by sodium borohydride,[55] diphenylphosphinic acid and its ethyl ester by lithium aluminium hydride,[56] and steroidal tosylhydrazones by lithium aluminium hydride and sodium borohydride,[57] and the stepwise hydrolysis of the borohydride ion[58] have also been investigated.

An extensive investigation of the reduction of olefins by di-imide has been reported (ref. 132, Chap. 5).

Other reactions which have been studied include the reduction of $\Delta^{1(9)}$-2-octalone by dissolving metals,[59] cyclopropyl ketones by lithium in liquid ammonia,[60] adrenochrome by ascorbic acid,[62] and aromatic nitro-compounds by carbon monoxide,[63] the electrolytic reduction of porphins, metal

[52] E. C. Pesterfield and D. M. S. Wheeler, *J. Org. Chem.*, **30**, 1513 (1965).
[53] P. S. Anderson, W. E. Krueger, and R. E. Lyle, *Tetrahedron Letters*, **1965**, 4011.
[54] T. Miyadera and Y. Kishida, *Tetrahedron Letters*, **1965**, 905.
[55] J. A. Parry and K. D. Warren, *J. Chem. Soc.*, **1965**, 4049.
[56] K. B. Mallion and F. G. Mann, *J. Chem. Soc.*, **1964**, 6121.
[57] M. Fischer, Z. Pelah, D. H. Williams, and C. Djerassi, *Chem. Ber.*, **98**, 3236 (1965).
[58] J. A. Gardiner and J. W. Collat, *J. Am. Chem. Soc.*, **87**, 1692 (1965).
[59] M. J. T. Robinson, *Tetrahedron*, **21**, 2475 (1965).
[60] T. Norin, *Acta Chem. Scand.*, **19**, 1289 (1965).
[61] O. Červinka and O. Kříž, *Collection Czech. Chem. Commun.*, **30**, 1700 (1965).
[62] G. L. Mattok, *J. Chem. Soc.*, **1965**, 4728.
[63] J. E. Kmiecik, *J. Org. Chem.*, **30**, 2014 (1965).

porphins,[64] and cyclo-octatetraene,[65] and the reaction of *p*-nitrobenzaldehyde with sodium hydride.[66]

The reduction of organic compounds by hydrazine has been reviewed.[67]

[64] D. W. Clack and N. S. Hush, *J. Am. Chem. Soc.*, **87**, 4238 (1965).
[65] R. D. Allendoerfer and P. H. Rieger, *J. Am. Chem. Soc.*, **87**, 2336 (1965).
[66] G. E. Lewis, *J. Org. Chem.*, **30**, 2433 (1965).
[67] A. Furst, R. C. Berlo, and S. Hooton, *Chem. Rev.*, **65**, 51 (1965).

Author Index

Subject Index